Population Biology of Plants

JOHN L. HARPER

School of Plant Biology
University College of North Wales, Bangor

1977

ACADEMIC PRESS

London · New York · San Francisco

A Subsidiary of Harcourt Brace Jovanovich, Publishers

ACADEMIC PRESS INC. (LONDON) LTD
24–28 Oval Road,
London NW1

U.S. Edition published by
ACADEMIC PRESS INC.
111 Fifth Avenue,
New York, New York 10003

Library of Congress Catalog Card Number: 76-016973

ISBN: 0-12-325852-9

Text set in 11/12 pt. IBM Baskerville, printed by photolithography,
and bound in Great Britain at The Pitman Press, Bath

Population Biology
of Plants

Preface

The first significant paper in population biology was written by a botanist (Nägeli, 1874) and ignored — the subject has developed almost entirely in the hands of zoologists. It is concerned with the numbers of organisms and with the consequences of these numbers. It deals with birth and death rates, immigration and emigration, with the consequences of exponential growth rates, the processes of colonization and the stresses that result from overcrowding. Malthus (1798) made it clear that the essential properties of populations were common to plants and animals. There are problems in studying populations of plants that have scared botanists away from a concern with numbers. In particular, it is frustrating that it is often very hard to decide what, in a population of plants, should be counted: is it the product of a seed, the rooted shoot, the branch or even some finer scale module of plant growth? These, however, are not really uniquely botanical problems; exactly the same problems arise for a student of colonial Hymenoptera (does he count colonies or insects?). Population biology has developed as a largely zoological discipline not because all animals are easy to count, but because species have been deliberately chosen for study that can be counted (e.g. fruit flies, hares, voles, titmice). In many ways plants make better material for the study of populations — plants stand still to be counted and do not have to be trapped, shot, chased or estimated.

There was an extraordinary, very brief period in the history of plant ecology when three of the most distinguished plant ecologists started, apparently independently, to study the population biology of plants in ways that were a breathtaking departure from tradition. In Russia, Sukatschew (1928) grew biotypes of dandelions at different densities and in different combinations and studied survivorship. In the U.S.A., F. E. Clements studied the reaction to density of a variety of species and wrote a classic textbook on "Plant Competition" (Clements *et al.*, 1929). In Britain, Tansley (1917) grew populations of bedstraws (*Galium* spp.) in pure and mixed stands and followed the struggle for existence between them on different soils. Tansley and Adamson (1925) employed a perturbation technique (exclosure of grazing animals) to show the role of the biotic factor in a natural plant community. By the end of the 1920s all these ecologists had left the field of population biology and experimentation. Virtually all subsequent population biology was with animals while plant ecologists concentrated on the description of vegetation, the evolution of ecotypes, the physiology of adaptation, and more recently on productivity.

Population biology became a zoological subject though it was continued with plants by agronomists. It was a revelation to me in 1945, as an undergraduate botanist, to meet Charles Elton and the group at the Bureau of Animal Populations at Oxford and discover that, for zoologists, populations (natural and laboratory) were appropriate subjects for experiment and theory. It was a second revelation to discover, during a few months at the Welsh Plant Breeding Station, that plants could be counted, that their populations could be controlled and manipulated and that, to the agronomist, there was nothing immoral in taking a plant community to pieces to discover how it worked.

This book attempts to bring together some of our present knowledge of plants that might be relevant to understanding their population biology. It is heavily larded with examples from forestry and agriculture, because these are the sources of most of the facts. In the absence of a coherent body of theory and practice from the pure science, the applied plant sciences developed the population biology needed for managing commercial populations of crops and trees.

I had in mind, when writing the book, a reader who was a zoologist, already with some knowledge of the population biology of animals and to whom I wanted to explain the ways in which the population biology of plants was sometimes different from and sometimes the same as that of animals. I have therefore deliberately avoided much of the background of general demographic theory which is beautifully treated in the books of Williamson (1972), Wilson and Bossert (1971) and Pianka (1974).

I hope there is some evidence throughout the book of the influence of Darwin's "Origin of Species". I believe that Chapter 3 of the "Origin" is the best ecological text ever written and I toyed with the idea of pirating it as the first chapter of this book. It provides more than any other work, a spyglass on the world of nature; "nothing in biology has meaning except in the light of evolution" (Dobzhansky, 1973). There is also much in this book that I owe to the writings of Robert MacArthur. There is little in what he wrote, right and wrong, that has not coloured the population biologist's view of the world. If he had thought and written more about plants this would have been a better book.

I am immensely grateful to the many students and overseas visitors who have worked with me at Oxford and Bangor. I can never disentangle which ideas were mine and which were theirs — I suspect that most came from them (G. R. Sagar, I. H. McNaughton, J. N. Clatworthy, D. Gajic, G. Orshan, J. Ogden, P. D. Putwain, M. Ross, M. Cahn, R. Naylor, B. Trenbath, J. White, M. Obeid, P. Cavers, R. Oxley, R. Turkington, A. R. Watkinson, J. Noble, J. T. Williams, J. Sheldon, K. A. Haizel, F. A. Bazzaz and R. Mack). I owe much to Dr G. L. Stebbins and Professor G. E. Blackman, F.R.S. for their stimulation and encouragement to work in an area that was at one time rather unpopular.

The writing of a book requires both stimulus and some periods of peace: both of these were given to me by the University of Massachusetts, in an invitation to give a series of 24 lectures on the population biology of plants, and four months' stay in Amherst in 1969 in which to do nothing else. I am also deeply grateful to the Centre National de la Recherche Scientifique for the opportunity to spend 6 months at the Centre D'Etudes Phytosociologiques et Ecologiques at Montpellier in 1975, when I was again free simply to think and write. The last four chapters formed the major part of the Prather Lectures which I delivered at Harvard University in the winter of 1974.

Sections of the manuscript were read, criticized and much improved by Dr G. L. Stebbins, Dr P. Jacquard, Dr T. B. Reynoldson, Dr G. R. Sagar and Dr J. Noble; errors and omissions are all mine. Typing and the preparation of indexes were major tasks — much of the typing was done from dictation, by Miss K. Coutts and Miss P. Harris. The indexes were prepared by Anthea Duckett and Catherine Sellek. I am very grateful for this help, and particularly to Barbara Field of Academic Press for her catalytic role in the process of publication. Above all, I thank my wife who not only helped with the typing but also tolerated quite out-rageous periods of silence and preoccupation while I was thinking what to write.

February 1977   J. L. H.

Acknowledgements

This book includes material reproduced from original sources in a wide range of journals and other publications. I am deeply grateful to authors, learned societies and publishers who freely granted permission for reproduction. The sources of the figures are as follows:

Acta Forestalia Fennica: 23/15
Agricultural Meteorology: 10/13
Agronomy Journal: 10/10, 10/11, 10/20
Allgemeine Forstzeitschrift: 21/1, 21/2
American Journal of Botany: 2/2a
American Naturalist: 15/8, 15/9, 21/6, 21/7, 21/11, 22/6, 23/4
    Copyright © 1968, 1970, 1973, 1976 by the University of Chicago
Annales d'Amélioration des Plantes: 8/19, 8/20
Annals of Applied Biology: 4/8, 8/8, 10/21, 11/1c, 11/26, 16/4
Annals of Botany: 1/1, 1/2, 3/4, 4/5, 4/6, 10/2
Archiv für Meterologie Geophysik und Bioklimatologie: 10/3
Australian Forestry: 2/2c
Australian Journal of Agricultural Research: 6/1, 6/26, 6/27, 10/6, 10/7, 11/1a, 11/2a, 11/3, 11/4, 11/6, 11/8, 11/9, 13/7
Avdelningen för Skogsekologi: 20/23, 20/24
BioScience: 19/6
British Veterinary Journal: 13/5
Brookhaven Symposia in Biology, U.S. Dept. of Commerce: 17/1, 23/3
Bulletin of the Torrey Botanical Club: 4/4

Canadian Entomologist: 13/8

Commonwealth Scientific and Industrial Research Organization
Wild Life Research: 13/6

Crop Science: 5/12, 5/13, 6/8, 7/2, 10/9

Ecological Monographs: 20/11, 20/12, 20/18, 20/19, 20/20, 20/21,
20/22

Ecology: 7/19, 10/14, 18/10, 18/11, 20/3, 20/9, 20/10, 20/15,
20/16, 11/III, 12/IV, 15/I, 18/I.

Empire Journal of Experimental Agriculture: 14/2, 14/7

Euphytica: 8/17, 8/18

Evolution: 13/3

Forestry Chronicle: 20/7, 20/8

Forest Science: 7/11, 7/13, 7/16, 20/1, 20/17

Heredity: 23/16, 23/17, 24/1, 24/3, 24/4

International Congress of Entomology: 12/4

International Copper Research Association, Inc.: 12/1

Irish Forestry: 7/14, 7/15

Jaarboek I. B. S.: 12/7 (Instituut voor Biologisch en Scheikundig
Onderzoek van Landbouwgewassen).

Journal of Agricultural Research: 7/12

Journal of the Arnold Arboretum: 23/12

Journal of Applied Ecology: 9/19, 11/16, 12/6, 14/8, 15/10, 18/3

Journal of Ecology: 3/1, 3/2, 4/9, 4/10, 5/1, 5/2, 5/3, 5/4, 5/6,
5/7, 5/8, 5/10, 5/14, 6/10, 6/11, 6/12, 6/13, 6/14, 6/15, 6/24,
6/25, 6/28, 6/29, 6/30, 6/31, 7/1, 7/6, 7/10, 8/2, 9/1, 9/2, 9/3,
9/12, 9/13, 9/16, 11/7, 14/1, 15/5, 18/1, 18/7, 19/6, 19/12,
19/13, 19/14, 19/15, 19/16, 19/17, 19/18, 19/19, 19/21, 19/22,
19/24, 19/25, 19/27, 21/3, 21/4, 21/10, 22/3, 23/6, 23/10,
23/11, 23/13, 23/14

Journal of Experimental Botany: 1/3, 1/4, 1/5, 1/6, 11/5

Journal of the Faculty of Science, University of Tokyo, 6/19, 10/5,
10/8

Journal of the Indian Botanical Society, 14/5, 14/6, 14/7

Journal of the Institute of Polytechnics and the Journal of Biology,
Osaka City University: 6/4, 6/5, 6/20, 6/21, 6/22, 6/23

Journal of Theoretical Biology: 14/12

Madroño: 20/14

Nature: 3/5, 12/2

Nature Conservancy: 13/2

New Phytologist: 1/7, 2/3, 2/4, 4/7, 6/18, 9/14, 9/15, 23/14
New Zealand Journal of Agricultural Research: 14/4
Oecologia Plantarum: 8/21
Oikos: 15/7, 19/5, 19/7, 19/8, 19/9, 19/10, 20/5
Oregon State University, Dept. of Forest Management: 5/17, 5/18
Outlook on Agriculture: 13/1
Phytocoenosis, 23/7, 23/8
Proceedings of the British Weed Control Conference: 5/11
Proceedings of the National Academy of Science: 18/VII
Proceedings of the New Zealand Ecological Society: 6/9, 6/16
Science: 11/V (copyright © 1970 by the American Association for
    the Advancement of Science.)
Symposia of the British Ecological Society: 10/17, 23/21
Symposia of the Royal Entomological Society: 23/20
Symposia of the Society for Experimental Biology: 6/3, 8/4, 10/3,
    10/4, 10/15, 24/2
Taxon: 23/5
U.S.D.A. Forest Service: 2/8
University of Texas Press: 13/4
Verslagen van het landbouwkundig onderzoek in Nederland: 8/13,
    8/14, 8/15, 9/4, 9/5, 9/6, 9/8, 9/9, 9/10, 9/11
Weed Research: 3/3, 6/17
Weed Science: 10/12
Welsh Plant Breeding Station, Technical Bulletin: 14/3

I also thank the University of Wales and authors for permission to reproduce material from the theses of J. Brook, 23/II; P. B. Cavers,
9/17, 9/18; J. Foster, 5/15, 5/16, 14/9, 14/10; Khan, 7/5; M. Mortimer,
2/5, 2/7; M. Obeid, 8/7; J. Ogden, 21/5; J. Sarukhán, 19/20; R. Turkington, 24/5; A. R. Watkinson, 18/8, 18/9; J. White, 5/5, and the University of Oxford and authors for material from theses of G. R. Sagar,
19/23, 23/9, and J. N. Clatworthy, 6/2, 8/6.

Quotations are reprinted with permission as follows: from "Stability and
Complexity of Model Ecosystems" by Robert M. May (copyright © by
Princeton University Press.) Reprinted by Princeton University Press and
from "Ecology of Desert Plants" by F. W. Went in Scientific American,
1975. Other quotations are acknowledged in the text.

Chapter Summaries

Chapter 1

Simple experimental models of population growth, using algae or duckweeds, illustrate the concepts of intrinsic rate of natural increase of a population (r) and the carrying capacity of a population (K). The behaviour of a mixed population is not predictable from these measures but depends on specialized features of morphology and "natural history" of the species.

The growth of a colony of duckweed is both the growth of a population *and* the growth of an individual plant that falls to pieces as it grows. This suggests that the population biology of higher plants needs to take account of demography at two levels — the numbers of genets (products of individual zygotes) in a population and the number of modules of structure that compose each genet. The modules may be leaf plus bud, ramets, tillers or branch units. These modules have birth and death rates that represent a level of population behaviour distinct from but interacting with the birth and death rates of genets.

The growth of a population of plants in nature differs from classical models of population growth because it occurs in bursts of establishment and reproduction determined by life cycle events. A diagrammatic model of the phases in development of a plant population needs to take into account phases of dispersal and dormancy, the degree of persistence of seeds after dispersal and the environmental heterogeneity that determines which seeds are recruited into a seedling population. In their subsequent growth plants may interfere with each other, causing death or hindering development of whole genets *and* of the modules that compose each plant. After a period of growth the plant may convert a large part of accumulated resources to reproduction and die or spread the reproductive activity over a series of years. This process releases new seeds for dispersal and the environment may have been altered in the meantime by feedback effects from the growing plants so that vegetational succession and/or microevolutionary change take place in the populations. These elements are incorporated into a simple model.

Chapter 2

The numbers of an organism are determined by the number of "habitable" sites and by the speed with which they are discovered and colonized. Discovery depends on the spatial distribution of the habitable areas and on the dispersibility of the propagules. Thus dispersal can play a critical role in determining population size. Most acts of seed dispersal place seeds close to the parent and seed density falls off steeply with distance: the curve relating seed density to distance is commonly close to an inverse square law but there is wide variation — the dispersal pattern from isolated plants differs strongly from that of plants in a population. The shape of the dispersal/distance curve can determine whether a population colonizes as an advancing front (a horizon) or as a series of outpost populations.

Dispersibility has a cost in expenditure on plumes or pappus, on swollen attractive receptacles or ovaries. Such expenditure may be interpreted as some measure of the fitness advantage that ancestors have gained by placing descendants at a distance from, rather than close to the parent. Animal-dispersed seeds are often deposited in groups in specialized habitats. Animals (particularly birds) include species that disperse seeds and species that digest seeds, and the consequences to a plant population are quite different. The evolution of hard and soft gizzards in birds is presumably a process that has coevolved with attractive or protective fruits.

Seed dispersal is generally favoured by high release: tall inflorescences are an alternative to specialized dispersal structures on the seeds. Dispersal of seeds in space is an alternative to dispersal in time. It may sometimes be a fitter strategy for seeds to remain dormant with long life and no dispersal. This will be the case when the probability of a suitable environment reappearing locally is greater than that of finding a habitable site at a distance.

Chapter 3

Dormant phases occur in the lives of most organisms and in plants occur as interruptions in the growth sequence either in the embryonic phase of the seed or as intermissions in the cycle of growth. Dormancy implies that the organism temporarily opts out of the struggle for existence, ceases temporarily to pre-empt environmental resources and delays leaving progeny. Dormancy may be associated with a dispersal phase or the growth phase, usually both.

The triggering of onset and end of the dormant phase is sometimes innate (characteristic dormancies of genotypes and species), enforced (environmental unsuitability, e.g. cold or drought) or induced. In environments with seasonally predictable hazards the breaking of dormancy is usually stimulated by an external clock, e.g. photoperiod. In an uncertain environment dormancy is usually broken by the return of favourable conditions. Intraspecific variation occurs in the nature of seed dormancy, taking the form of genetic polymorphism or more often somatic polymorphism in which seeds with different dormancies are produced by different parts of the plant or inflorescence.

Chapter 4

Most seeds fall onto and some are incorporated into soil. Seeds in the "seed bank" are usually most abundant near the soil surface and numbers decline rapidly with depth. Small seeds tend to be washed deeper than large seeds and some are buried by the activities of worms and other animals. The seed bank under plant communities is a biased history of the past vegetation of the area — biased by the different dispersibility, reproductive output and seed longevity of the species. Colonizing and pioneer species tend to leave large reserves of seed in the soil, particularly weeds of agricultural and horticultural land. Tree species, especially those of advanced successions and stable environments, are poorly represented in the seed bank: they have short lives in the soil.

Seeds are lost from the seed bank by germination, predation and decay. The rate of loss is exponential in the few cases studied. The role of predators in reducing the seed bank is scarcely known but the activity of seed-rotting pathogens is likely to increase with the time for which a species occupies an area and seed rotting is affected by both the temperature and moisture status of the soil. Thus the physical conditions of the environment can determine the probability of a seed surviving by its effects on the activity of seed decomposers.

Seed longevity is heritable and the optimal fraction of a seed population that germinates or remains dormant is a function of the predictability of the environment.

Chapter 5

The environment immediately surrounding a seed determines whether it receives the conditions, resources and stimuli needed for germination. At the scale of size of a seed, soil environments are very heterogeneous. This heterogeneity in the microenvironment, combined with the extreme subtlety of germination requirements, can itself determine the numbers and variety of seedlings that are recruited from the seed bank into a population of growing seedlings. Changes in the gross physical environment, e.g. temperature, soil type, compaction and exposure, change the frequency of "safe sites" in the soil environment and so alter the probabilities of a seed forming an established seedling as well as the representation of different species. The causal events that determine whether a seed forms a seedling or not are usually untraceable when the plant has developed and the real causes of distribution and abundance will often be missed when mature vegetation is studied.

The growing plant is an important factor in modifying the abundance and nature of "safe sites"; shading, litterfall, chemical change of the soil and the encouragement or discouragement of pathogens and predators cause a feedback of growing vegetation on the nature of the environmental sieve that determines which and how many seeds are recruited into subsequent populations of seedlings.

Chapter 6

Neighbouring plants interfere with each other's activities according to their age, size and distance apart. Such density stress affects the birth rates and death rates of plant parts. As plants in a population develop, the biomass produced becomes limited by the rate of availability of resources so that "yield" per unit area becomes independent of density — the carrying capacity of the environment. The stress of density increases the risk of mortality to whole plants as well as their parts and the rate of death becomes a function of the growth rate of the survivors. Self-thinning in populations of single species regularly follows a 3/2 power equation that relates the mean weight per plant to the density of survivors. Density-stressed populations tend to form a hierarchy of dominant and ± stressed subordinate individuals and the frequency distribution of plant weights becomes log-normal (or bimodal). The death risk is concentrated within the classes of suppressed individuals. A clonal plant such as a perennial grass may react to density by a reduced birth rate of its parts and an increased death rate. The effect is that the number of modules (e.g. tillers) per unit area of land becomes independent of genet density, although the number of genets present continuously declines.

Chapter 7

The effects of density do not fall equally on all parts of a plant. In general the size of parts (e.g. leaves or seeds) is much less plastic than the number of parts (e.g. branches). The stress created by the proximity of neighbours may be absorbed in an increased mortality risk for whole plants or their parts, reduced reproductive output, reduced growth rate, delayed maturity and reproduction. Even classically biennial species may be converted into long-lived individuals by the stress of density. Foresters use the term density in a special sense, to signify the integrated stresses within a community rather than the number of individuals per unit area. The effect of proximity of neighbours in a forest is essentially the same as in a herbaceous community except that the traces of dead parts persist for longer. The form of a crowded tree is determined by the high death rates of the shaded parts. Root grafting permits an integration of individuals within a stand of some forest tree species. When this occurs neighbouring trees may compete more intensely because one individual can draw on the combined root systems, or individuals may support each other.

Chapter 8

A part of a plant may meet interference from another part of the same genet, from a different genet of the same species or a member of a different species. The interactions and both ecological and evolutionary consequences of these different neighbour relations are affected by the spatial arrangement and the proportions of the neighbours as well as by the growth form of the plants. The development of individuals in mixed or pure populations can be partly predicted from a knowledge of detailed spatial relationships, provided that germination and establishment are synchronous. If the times of establishment of two neighbours differ this can have a profound effect on their relative performance and overwhelm other influences such as differences in growth rate, embryonic capital and spatial arrangement.

The growth of mixed populations may be studied by additive or substitutive experiments. The results of the latter define four contrasting forms of interference between plants and make it possible to detect which pairs of species make demands on the same limiting environmental resources or avoid making such demands. Groups of species may be studied in all possible combinations of two in order to extract generalizations about general and specific ecological combining ability. There is an important sense in which the neighbours of a plant are part of its environment and species may be ranked according to the stability of their performance in the face of the different environments provided by other species as neighbours.

Chapter 9

Mixed populations may involve forms that are so similar in reaction to one another that they partition the environmental resources simply in proportion to their representation in the mixture. More commonly there is differentiation between the forms that permits one to gain at the expense of the other's growth, reproduction and/or numbers of survivors. Model populations have been studied in which one form succeeds at the expense of another at a rate independent of frequency. Frequency-dependent situations also occur and may permit two forms (e.g. species) to grow in self-stabilizing mixtures. Difference in nutrition, the timing of growth or sensitivity to different causes of mortality may permit such frequency-dependent interaction. There are close similarities between the results of "competition" between animals (when the results are measured as changes in number) and plants (when results are measured as changing dry weight).

Chapter 10

Plants may interfere with the distribution of neighbours by depleting limited resources. An effect will occur only when the depletion zone created by one plant includes the zone available to another. Resources (or supply faction) are exhaustible and contrast with conditions (quality faction) such as temperature that are not exhaustible. The nature of a resource (light, water, nutrients, O_2 and CO_2) determines how a plant may affect a neighbour's growth; the diffusion of light, gases and nutrients are at very different rates and determine how far away from an individual depletion effects are sensed. It also determines which resources are easily studied. Light, water and nitrates are probably the three resources most commonly involved but the interaction between resource factors makes it unrealistic to isolate any one resource as that for which "competition" occurs: the extent of interference below ground is not well understood.

Chapter 11

Plants may modify the environment of their neighbours by changing conditions,

reducing the level of available resources or by adding toxins to the environment. All
of these effects can be shown quite clearly to operate in experiments with artificial
populations, but there are probably no examples of plant interactions in the field in
which the mechanism has been clearly and unambiguously demonstrated. Interactions dominate most situations that have been analysed in the field and the search
for unique factors of competition may not be very sensible. Death or a reduced
growth rate are often attributable to "competitive" effects, but interpretation of
competition in the field may depend on the recognition of much more specific
symptoms. There is a need for a more rigorous philosophy in the search for cause
and effect, perhaps like that adopted by pathologists and defined in Koch's
postulates.

Chapter 12

Most animals that eat plants take only parts and the remainder of the plants may
regenerate. Removal of leaves or roots from individual plants in a population may
damage their position in a competitive hierarchy and reduce their reproductive output. However, some compensation for lost leaves may occur in increased assimilation by the remaining leaves. Damage by animals to the cambium of trees or to the
apical meristems may have effects far greater than the fraction of the plant body
that is eaten, particularly when the effect is to change the morphology of the plant.
When a plant loses tissues to a predator the loss of mineral nutrients in the tissue
may be more important than the loss of energy involved.

Chapter 13

The unreliability or seasonality of a food source puts constraints on the life cycles
of herbivores. Monophagous specialists may require life cycles that are tightly synchronized with the life cycle of the prey. Monophagy usually involves tight evolutionary integration of both predator and prey and classic predator—prey cycles may
be hidden by rapid coevolutionary processes. A monophagous predator probably
rarely causes extinction of the host plant. Polyphages are probably less efficient in
their use of food and polyphagy may lead to local extinctions among food plants
rather than regulation. A critical issue is whether the feeding habits of a polyphagous
predator are apostatic and if so at what prey density the predator changes its feeding
image. Palatability is a difficult feature to measure and define but relative palatabilities clearly play a large part in determining the effects of a polyphage in a plant
community. The search range of a predator may itself determine the limits of abundance of a food plant and seed dispersal takes on a special significance in this context.

Chapter 14

Grazing animals may determine the relative abundance of different species in a habitat. Part of this role is due to selection between the plants on offer in a pasture, both between species of plant and within species. Different seasonal patterns of grazing can produce very differently composed swards. Many of the elements in the plant—animal interaction are sufficiently well understood for effective computer models to be made that predict changes in the populations of both the sward and the animal. The influence of grazers is in part due to defoliation but trampling and the deposition of dung and urine are important. A grazing animal tends to be a diversifier of a grassland community, creating locally different microenvironments for seedling establishment and the subsequent growth of plants, and continually initiating regeneration cycles on a small scale within the community.

Chapter 15

Fruits, and particularly seeds, are choice foods and both are often heavily predated. Fruit predation is associated with dispersal and attractive extra-ovular structures may be regarded sometimes as payment for dispersal. Predation of embryos is a loss to the reproductive capacity of the parent plant; it may be reduced by saturating the predator's needs by synchronous glut output of seeds, and by dispersal which increases the search time of the predators. The nature of the predator's activities, in particular its habits of search, may determine the distribution of seeds that remain to germinate and form new plants. Levels of seed predation may have no influence on the population dynamics of the food plant if the predators remove only what would be doomed to die from other later density-dependent processes. The predation of fruits may be encouraged by non-synchronous reliable output. Thus the selective pressures acting on the reproductive habits of plants may be quite different according to whether the predator disperses or destroys the seeds.

Seeds tend to be temporarily available food sources and birds, mammals and ants may store caches of seeds and extend the period of availability. The volume of seed cached will tend to exceed average need, with the result that an excess is commonly buried and plants establish from the buried stores. Coevolution of predator and food plant may have occurred, for example between squirrels and pines, but there are many aspects of predator behaviour that seem to have no coadaptive interpretation

Chapter 16

Many of the generalizations about animal predators on plants can be extended to plant pathogens. There is a close parallel between the plant niches colonized by insects and by fungi. There is no comparable process to the search of prey by an animal. The passive nature of dispersal of pathogens (unless an animal carries the pathogen) has the effect of making distance an even more critical determinant of infection than it is of predation. Dispersal of pathogens in time occurs because of the persistence and longevity of resting spores. Coevolved specialization of races of host and pathogen is well developed and microevolutionary interaction may complicate the ecological effect of disease.

Chapter 17

The most striking effects of animal predators on populations of plants occur after introductions, e.g. in biological control of weeds. Convulsions in the numbers of predators and prey are often recorded after such an event, but subsequently numbers tend to stabilize at levels determined, for example, by the search range of the predator and by coevolutionary changes in predator and prey. The absence of frequent damage does not show that a predator is unimportant in regulation: where a predator regulates a plant population the activity may only rarely be seen, and the regulating process may only be exposed after perturbation.

Chapter 18

The categories of annual and biennial plant life cycle are arbitrary divisions in a continuum of growth strategies. Short life with precocious reproduction can permit a rapid rate of population increase provided that seeds germinate in the season following their production. In practice, most short-lived plants produce a high proportion of dormant seeds that represent a sacrifice in the potential capacity for population increase. The seeds of some sand dune annuals are exceptional in this respect and appear to depend entirely on last year's seed production for this year's population. Among annual plants can be found species with high reproductive output and high mortality rates and others (*Avena fatua*, *Vulpia membranacea*) in which seed output is low and the mortality risk is also low. In one example (*Vulpia membranacea*) in which the causes of mortality have been analysed in detail, density-dependent control of seed output coupled with density-independent mortality is sufficient to regulate the population within the limits normally found in the field. Annuals include forms with indeterminate growth habits in which the end of life is determined by a harsh season and forms with a determinate habit in which an episode of flowering is controlled by external stimuli and followed by death. The biennial habit is an extension of such a determinate, monocarpic system extended over several years and it is probably rare in nature for biennial species to complete their life cycle within 2 years.

Chapter 19

The population dynamics of perennial herbs has been studied in a small number of species of grassland and woodland herbs. Long-term detailed mapping offers the only safe way to extract demographic measurements. Both survivorship and depletion curves show remarkable smoothness over years suggesting that year-to-year variations in the environment are less important than might be imagined. There is sometimes clear seasonal variation in the risk of death, and the risk tends to be greatest during the phases of most active growth of the survivors.

The expectation of life of genets in a pasture or woodland may be very long and some individuals persist for many decades in the midst of a continual flux of new recruitment and death. It is not uncommon for populations to be dominated by genets that, although long-lived, die without flowering. In natural communities the juvenile (non-flowering) periods are often greatly extended compared with the performance of the same species in experimental gardens.

Chapter 20

A population of trees can be described by its age-structure or its size structure. There are great dangers in assuming that size and age are closely related. A population may develop towards a characteristic size-structure irrespective of its age structure.

The behaviour of a population is the summation of the behaviour of its individuals and the most informative studies have followed the fates of individuals. Demographic studies show that most forests have an age-structure reflecting past disasters: there is probably no forest that has achieved a balanced age distribution. The interpretation of forest structure commonly looks forwards towards what it is becoming — towards some equilibrial or climax state; it is suggested here that historical interpretation is usually more realistic. A population is more effectively explained as the outcome of past events than as a stage in development towards some end.

In some species, natural root grafting allows physiological integration of genets within a population — a converse of the fragmentation of genets by the decay of interconnections.

Shrubs pose special demographic problems: the age-structure of above-ground shoots may diverge sharply from that of below-ground parts. As with clonal herbs, the demography of clonal shrubs is intimately determined by the morphology (architecture) of the growth system.

Chapter 21

Seed production may be interpreted as an alternative to vegetative vigour and long life. Physiological information generally supports the idea that there are limited resources available to a growing plant that are allocated between seed products and other activities that are of greater selective advantage in competitive environments. The reproductive regime of a plant (its seed production) may involve allocation of limited resources between many small or fewer large seeds. It is unrealistic to believe that reproductive capacity is an adaptation to the rigours of the environment. Strategies of growth and reproduction need to be interpreted as compromises between conflicting specific adaptabilities towards the general compromise in strategy that maximizes fitness.

Chapter 22

The progeny produced by an organism at different stages in its life history make different contributions to leaving descendants to subsequent generations. In expanding populations precocious reproduction enhances the intrinsic rate of natural increase and precocious progeny have high "reproductive value" — provided that the death risk is evenly distributed through the life of the parent or increases with age. Different strategies e.g. iteroparous reproduction, delayed monocarpy and infinite life may be fitter in populations that experience most selection when these populations are in decline. Long juvenile periods and intermittent seed production characterize species in advanced succession — K species. In advanced succession it can be envisaged that the last seed produced by a parent before death may sometimes have the highest reproductive value. Physiological and genetic potential for quite profound changes in life-cycle strategy exist within plant populations.

Chapter 23

The biological diversity of a community has ecological and evolutionary meaning only in so far as it is sensed by the individuals within it: an imposed anthropomorphic scale may be irrelevant. Diversity exists as the somatic and phenotypic variation within a genet, differences of age and between genotypes of a species as well as diversity at the species level. Differences between organisms determine whether they can persist together as neighbours or engage in an exclusive struggle for existence. The form and life cycle of a plant determine how it senses the heterogeneity of the environment, both in space and time. Some growth forms enforce clonal monotony on a community, others maximize the genet's experience of a variety of neighbours.

The branches in food chains are essentially unstable and evolved biological specialization is required to give them stability. Predators may permit a vegetational diversity that disappears when the predator is removed, but this is not a general rule. It depends on the relative aggressiveness and palatability of the plants involved.

Chapter 24

Natural populations characteristically contain genetic diversity, even when there is close inbreeding or apomixis. This can be interpreted as balanced polymorphism or sometimes as a transient polymorphism in communities subject to repeated hazards and cyclic phenomena or to the disruptive selection imposed by the variety of interspecific neighbours. Natural environments are heterogeneous both in space and time and this makes for dilemmas in evolution under natural selection. Life in a successional environment or a mosaic habitat may force a variety of solutions such as genetic and somatic polymorphism, plasticity, speciation, heterozygotic advantage. An evolutionary process driven by effects on individual fitness cannot be expected to maximize performance in groups, such as is required in agriculture and forestry.

Contents

The Natural Dynamics of Plant Populations

Plants, Vegetation and Evolution

To W. K.
Teacher and Friend.

1

Experiments, Analogies and Models

A biology of populations is a study of the numbers of organisms. It seeks to answer questions about the differences in the numbers of organisms from place to place and from time to time. It is concerned with the life cycles of organisms because a population cannot be adequately described without taking into account that it may include young and old, big and small, male and female. It is concerned with the direct action of the physical environment which may influence population growth and individual survival. It is concerned with the stresses caused by the growth of populations — when too many organisms chase too few resources or when one organism eats another. It is concerned with the selective processes which, favouring some individuals against others, lead to ecological and evolutionary change.

The logical starting point in a discussion of population biology must be a reference to Malthus whose "An Essay on the Principle of Population" was the real beginning of a science.

> Through the animal and vegetable kingdoms, nature has scattered the seeds of life abroad with the most profuse and liberal hand. She has been comparatively sparing in the room and nourishment necessary to rear them. The germs of existence contained in this spot of earth, with ample food, and ample room to expand in would fill millions of worlds in the course of a few thousand years. Necessity, that imperious all pervading law of nature, restrains them within the prescribed bounds. The race of plants, and the race of animals shrink under this great restrictive law.
>
> (Malthus, 1798)

Malthus realized that, given unlimited resources, the size of a population increases in a geometrical progression (1, 2, 4, 8, 16 . . .). Because the resources of the environment are always limited, such a geometric progression cannot continue indefinitely and some sort of ceiling of numbers is ultimately reached or approached. Malthus' main concern was of course to make the point that overpopulation is a social evil, but the impact of his work was much wider: both Darwin and Wallace read the essay and realized that overpopulation in nature gave to populations which were innately variable, the necessary surplus of individuals to be eliminated in natural selection. Moreover, as Darwin particularly emphasized, overpopulation itself provided a powerful driving force in natural selection in the ensuing struggle with neighbours for limiting resources.

An important early development in population theory was the work of Verhulst who attempted a formal mathematical description of population growth in a restricted environment (Verhulst, 1839). Verhulst's work was almost completely ignored until Pearl and Reed (1920) independently derived essentially the same formula, the Verhulst or Verhulst-Pearl logistic equation. The form in which this is commonly given for a set of specified conditions is

$$\frac{dN}{dt} = rN \left(\frac{K - N}{K} \right)$$

The rate of growth of the population = The potential rate of increase of the population per unit of time x The degree to which the population N falls short of its maximum attainable value, K.

The integrated form of the logistic equation is given as

$$N_t = N_0 e^r \left(\frac{K - N}{K} \right) t$$

This equation has had great influence and continues to do so partly because it emphasizes two parameters, r and K, which are potentially measurable and which together define ideal population behaviour. The parameter r, commonly called the "intrinsic rate of natural increase" defines a potential rate of increase for a population under given environmental conditions — a measure of the rate at which it will expand provided that it is unimpeded by a shortage of resources or any of the other inhibiting consequences of population growth. There is a growing tendency to talk and write about r-phases in the life of a population — phases when it explodes with near-exponential growth after a disaster or after a new colonization into an unexploited environment. The term r-species is also used to describe those species whose populations spend most of their time in exponential recovery from disasters or successive invasions.

Similarly, K defines an upper limit or saturate value to which a population may develop and which it may not exceed — this may be determined in a Malthusian manner by a limiting supply of consumable resources or by any other activity of the population which puts a ceiling on its further growth: e.g. limitations of space, territorial behaviour, toxin production. Thus one may write of K-phases in the life of a population — periods when it is prevented from further expansion as some consequence of its own size — and of K-species, whose populations spend a large part of their time under stress from the presence of neighbours.

Much experimental effort has been put into simulating the phases of population growth, particularly with micro-organisms and insects. Reasonable fits to the logistic curve can be obtained provided that the right organism is chosen and the conditions are carefully defined (e.g. the growth of *Paramecium* growing in a limited volume of nutritive medium: Gause, 1934). However, even in such rarified experimental models, departures from the ideal curve of population growth are commoner than good conformity (Feller, 1940; Sang, 1950). Lag phases at the start of the exponential growth period and over-shooting of the theoretical K-asymptote are common and the life cycles of most organisms involve periodicities of reproduction so that the growth of their populations occurs in steps or jumps rather than as a smooth population growth curve. The logistic curve remains central to population biology, not because it describes how populations behave but because it provides a standard base of ideal behaviour against which the reality can be judged and measured.

Experimental models of plant population with continuous growth

There are few plant populations that are at all suitable for studying a process of continuous population growth, not only because most plants have life cycles and therefore discrete jumps of population size, but because the fixed rooted habit of most plants prevents them from moving about in search of the limited resources and escaping from their neighbours. The result of clumped distributions is that some members of a population are under density stress even though unexploited resources may be present nearby. The choice of a suitable plant as a model for continuous population growth is therefore limited to free-living algae such as *Chlorella* or free-floating aquatic plants such as the duckweeds (*Lemna* spp.) or the water ferns (*Azolla* and *Salvinia*). These are all very odd and exceptional plants.

A good early example of a population study using an alga was made with cultures of *Scenedesmus* by Roach (1928). A continuing input of energy for growth was provided in the form of light (this alga can use carbohydrate instead) and four different levels of light intensity were obtained by the unsophisticated but effective means of placing the cultures at varying distances from the light source. The cultures were grown in a mineral salts medium at a constant temperature in a water bath. The experimenter chose to measure the bulk of alga per cubic centimetre as the parameter of population growth. The growth rate of the population is expressed as the slope of the line on the graph relating the logarithm of the bulk of alga per millilitre to the passage of time. Population growth was exponential over the first 6—7 days at all light intensities (i.e. *r*-phase unimpeded growth) but the rate declined after this time at the higher light intensities when the bulk of alga exceeded a critical value. The alga showed different growth rates in the four light environments (Fig. 1/1). In a longer-term experiment with the alga *Chlorella* three distinct phases could be recognized: (i) an exponential phase of growth; (ii) a phase of declining relative growth rate as density began to impede multiplication; and (iii) a phase of relative stability in numbers (Fig. 1/2) (Priestley and Pearsall, 1922).

Among higher plants, the duckweeds *Lemna* spp. and the aquatic ferns *Salvinia* and *Azolla* make good experimental models. Individual fronds of *Lemna* spp. produce lateral buds which themselves become fronds, and in most species break free from the parent and themselves continue to bud. The growth of populations can be followed by counting fronds or by measuring the change in their dry weight (Clatworthy and Harper, 1962). If a population is started by inoculating one or a few fronds into a beaker of a suitable culture solution placed under light it starts into an exponential growth phase. This phase continues provided that the mineral culture medium is continually replenished *and* the fronds are continually harvested to maintain low density in the culture. In this way the intrinsic growth rate of a species under specified conditions can be measured (Fig. 1/3). There are obvious differences in the unimpeded population growth rates of the four species: populations of the fastest growing species, *Lemna minor*, doubled in 0.88 days and of the slowest, *Salvinia natans*, in 1.4 days.

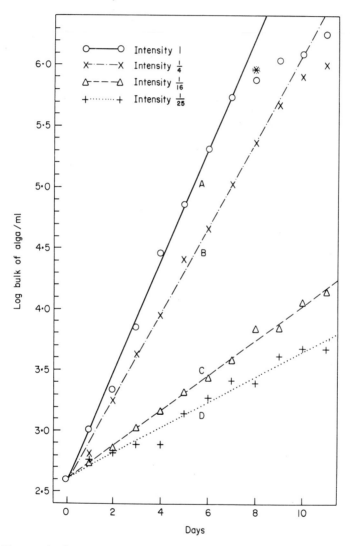

Fig. 1/1. The growth of cultures of the alga *Scenedesmus* in a mineral salts medium at different light intensities. (From Roach, 1928)

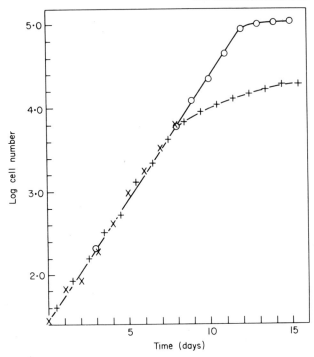

Fig. 1/2. Population growth curve for the alga *Chlorella* in culture. This alga forms clumps and so experiences the effects of density when there are still available resources in the medium. The initially exponential growth rate declines after 8 days but if the clumping is prevented by shaking the exponential rate continues for four more days. (From Pearsall and Bengry, 1940) x and + unshaken, o shaken

The intrinsic rate of natural increase is of course a function of a specified environment. The population growth rates of *Lemna polyrrhiza* and *L. minor* were measured in media of different concentration of phosphate (KH_2PO_4) and under two light intensities (Fig. 1/4). The population growth rate of *Lemna minor* proved to be remarkably stable over a 64-fold range of phosphate concentrations — whereas that of *Lemna polyrrhiza* fell off steeply with increasing concentration so that it became negative (moving towards extinction) at the highest concentration. It must be stressed that *these measurements were made under conditions in which the density of fronds had no influence on the growth rate* (a faster rate of harvesting had no effect on the growth rate, nor did more frequent replenishment of the culture solution). The experiments measured a density-independent property of the populations — their un-

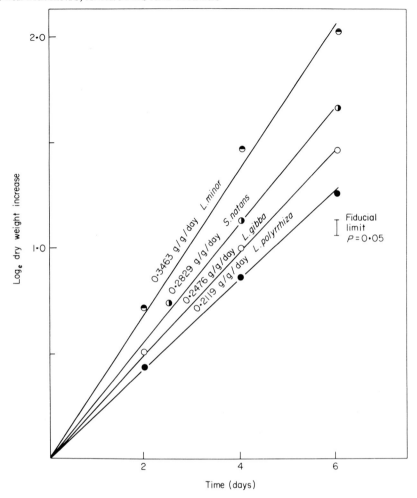

Fig. 1/3. Exponential growth of *Lemna* spp. and *S. natans* in uncrowded cultures at 25°C. (From Clatworthy and Harper, 1962)

impeded growth rate. This population parameter need tell us nothing about the rate of growth of the species in an environment of limited resources; an environment in which the presence of neighbours reduces the quality of the environment for growth.

When populations of *Lemna* spp. or *Salvinia* are allowed to continue growth without harvesting, fronds accumulate and the growth rate slows down. This happens even when the culture medium is replenished so frequently that more rapid replenishment does not change the situa-

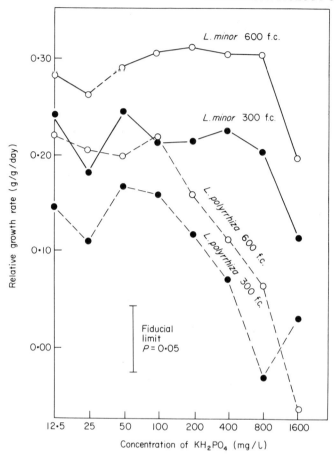

Fig. 1/4. The relative growth-rates of *L. minor and L. polyrrhiza* at 25° C. under light intensities of 300 and 600 f.c. in Steinberg solution containing various concentrations of KH_2PO_4. (From Clatworthy and Harper, 1962)

tion. The fronds quickly become crowded and form a mat in which some fronds shade others. These crowded populations begin to resemble plants rooted in soil because they cannot move freely. As the frond mat thickens the whole population becomes a hierarchy: an upper layer of rapidly budding fronds shades a zone of more slowly growing forms and, beneath this, the lowest fronds of the mat cease to form daughters, become pale and die, sinking as detritus to the bottom of the culture vessel.

The growth curve of the whole population can be divided into three fairly clear phases (Fig. 1/5). There is no lag phase and the populations start directly into exponential growth — Phase I. During this phase the increase in the population per unit time is a function of the capital in-

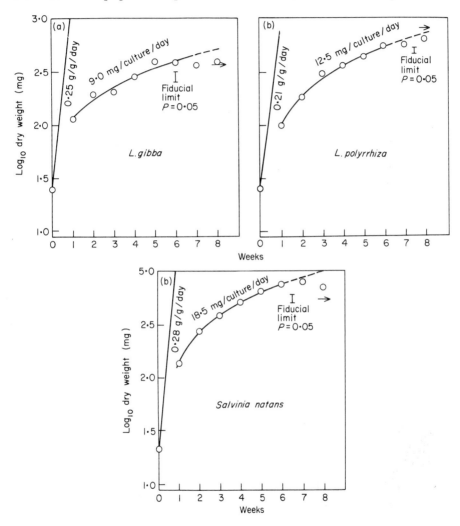

Fig. 1/5. The growth in dry weight of populations of *Lemna* spp. and *S. natans* in self-crowding cultures. The straight line plotted in each figure (Phase I) represents the intrinsic growth-rate of uncrowded cultures. The curve in each figure (Phase II) is fitted on the hypothesis that the growth made in weeks 2, 3, 4, and 5 is a constant weekly increment per culture. Data are plotted on a logarithmic scale so that variances are homogeneous throughout. Fiducial limits are calculated from the error between replicates. The arrow → indicates the final yield after 12 weeks in a separate experiment under similar conditions. (From Clatworthy and Harper, 1962)

vested, i.e. the number of fronds present. This is the phase of unimpeded growth. It can be expressed as fronds per frond per day or as grams per gram per day.

After a few days the population becomes crowded and resource limited. The growth rate declines steadily and becomes dominated by the limited environment ("resources") of the culture vessel. During this Phase II each vessel gains a constant addition of fronds per unit of time. The growth of the population is now determined by the "size" of the environment, not by the size of the capital invested. Thus, if the inoculum is regarded as an investment, one can regard Phase II *either* as a period of steadily declining compound interest *or* as a period of constant simple interest. Just as each species has a characteristic growth rate in Phase I so each has a different rate in Phase II (Table 1/I).

Table 1/I

Population growth rates of some aquatic plants in culture

Species	Relative growth rate (g/g/day)		Simple growth rate (g/beaker/day)		12 weeks "stable" population (g/beaker)	
	Phase I	(Rank)	Phase II	(Rank)	Phase III	(Rank)
L. minor	0.35	(1)	14.7	(2)	603	(2)
L. gibba	0.25	(3)	9.0	(4)	355	(4)
L. polyrrhiza	0.21	(4)	12.5	(3)	776	(1)
S. natans	0.28	(2)	18.5	(1)	524	(3)

During Phase II it seems likely, although there is no formal proof, that the growth of the cultures is limited by the amount of light falling on them. During the whole of this phase the mat of fronds has been steadily increasing in thickness. After about 8 weeks the populations reach a nearly steady state (Phase III): new fronds are continually recruited at the top and are balanced by the death of fronds from the dark underside of the mat. Each of the species has a characteristic value of biomass that it achieves in the "stable" phase (Table 1/I). There is no obvious correlation between the growth rates in the different phases.

An interesting feature of these growth curves is the tendency for the populations at the end of Phase II to overshoot the Phase III value. This suggests that the feedback process by which crowding regulates the growth rate is not perfectly efficient in damping the approach to an asymptote (K), as would be the case in an ideal Verhulst-Pearl model.

Superficially *Lemna* and *Salvinia* appear ideal species for simple population models but they are awkward in practice. The smooth growth of populations hides real life-cycle events (in particular ageing and senescence) which might not be expected in such a superficially uneventful growth system. A frond of *Lemna* has a characteristic length of life, even in exponentially growing populations. The life expectancy of a frond depends on the temperature (a frond lives 1.8 times as long at 20°C as at 30°C) but not on the light intensity (Ashby *et al.*, 1949; Ashby and Wangermann, 1951).

During its life a frond produces a succession of daughters at a rate dependent on both light intensity and temperature. Each successive daughter produced by a mother frond is smaller than its predecessor so that the last frond produced before the mother dies is often very small indeed. Successive daughters from a mother frond are not only smaller but also produce fewer daughters (though at the same rate). This process of "ageing" is counterbalanced by a rejuvenation in which fronds that are small because they arose late in the life of the mother always produce daughters larger than themselves. The smaller the frond the greater is the increase in area of its daughters. In the case of very small fronds this process of rejuvenation occurs over several generations until the maximum frond size is reached. In this way an "age" distribution and a frond size distribution are built up and maintained in an exponentially growing culture. No one appears to have studied the ageing–rejuvenation cycle in self-crowding cultures.

In nature the free-floating aquatics face the recurrent hazards of unfavourable seasons which are spent either as turions (starch-packed fronds which sink into the mud and overwinter there) or as seeds (or spores in *Salvinia*); flowering and seed formation are rare in many of the species of *Lemna* and these phases of the life cycle are omitted when the species are used as simplified laboratory models of population growth. However, even the ageing–rejuvenation cycle brings an unwelcome complexity into the experimental modelling of simple populations: a population of as simple an organism as *Lemna* contains members that behave in different ways. An important feature of the exponential growth of populations is that the age structure reaches a stable distribution – the "stable age distribution" (see e.g. Williamson, 1972). In all of the experiments on the population growth of *Lemna* species described in this chapter the individual experiments were started with a group inoculum taken from a long-established culture which had been maintained in exponential growth. It is easy to see how the very existence of an ageing—rejuvenation cycle would upset measurements of the intrinsic rate of natural increase started from a single frond or from a culture which had a specialized age distribution.

The behaviour of mixtures of species growing together

The Malthusian concept of the remorseless growth of populations to limits imposed by their resources implies that the pressures of population growth may reach over from one species to another if the species have resources in common. This was strongly expressed by Darwin (1859) in "The Origin of Species". He emphasized the mutual interference between taxonomically closely related species:

> As the species of the same genus usually have though by no means invariably, much similarity in habits and constitution, and always in structure, the struggle will generally be more severe between them, if they come into competition with each other, than between species of distinct genera. . . One species of charlock has been known to supplant another species; and so in other cases we can dimly see why the competition should be most severe between allied forms which fill nearly the same place in the economy of nature; . . .

The theoretical and experimental study of the growth of populations of two species in each other's presence has been an enormously stimulating area ever since Darwin's work. Theoretical developments had an abortive start. An extraordinary paper both in its vision and its sophistication was given by Carl Nägeli to the Maths-Physics Section of the Bavarian Academy of Science in 1873 under the title "On the displacement of plant forms by their competitors" (Nägeli, 1874). In it he considered various mathematical formulations of the effects of two species growing together, on each other and on themselves; he produced models in which only one of a pair of species survived and displaced the other and also model situations in which pairs of species formed a stable mixture. The models took account of the life cycles of plants, their seed production, mortality and longevity. Nägeli included in his models concepts of density dependence, and even frequency dependence. He anticipated the classic treatments of the problem by Volterra (1926) and Lotka (1926) by more than 50 years — and his contribution was so much in advance of its time that it appears to have been wholly ignored (Harper, 1974). It was a book by Gause (1934), "The Struggle for

Existence", that combined the mathematical treatment of interactions between populations, stemming from the work of Lotka and Volterra, with simple laboratory experiments and started a flood of studies on inter-action within mixed populations of animals. The Verhulst-Pearl equation can be written for two species, N_1 and N_2, including in each equation a contribution from the influence of the second species:

$$\frac{dN_1}{dt} = N_1 r_1 \frac{(K_1 - N_1 - \alpha N_2)}{K_1}$$

and

$$\frac{dN_2}{dt} = N_2 r_2 \frac{(K_2 - N_2 - \beta N_1)}{K_2}$$

The factors α and β describe how many of each of the species are equiva-lent to one of the other in saturating its population limits. There are four formal solutions to these paired equations depending on the values of α and β relative to the ratios of K_1 and K_2.*

(i) If $\alpha > \dfrac{K_1}{K_2}$ and $\beta > \dfrac{K_2}{K_1}$ then only one of the pair of species will

persist in a mixture and the "winner" will be determined by the starting proportions.

(ii) If $\alpha > \dfrac{K_1}{K_2}$ and $\beta < \dfrac{K_2}{K_1}$ then species 1 will be eliminated from

the mixture.

(iii) If $\alpha < \dfrac{K_1}{K_2}$ and $\beta > \dfrac{K_2}{K_1}$ then species 2 will be eliminated from

the mixture.

(iv) If $\alpha < \dfrac{K_1}{K_2}$ and $\beta < \dfrac{K_2}{K_1}$ then both species persist together in an

equilibrium.

*A more generalized model (including that of Lotka and Volterra as a special case) has been proposed:

$$\frac{dN_i}{dt} = r_i N_i (1 - (N_i/K_i)^{\theta_i} - \alpha_{ij} N_j/K_i)$$

The value θ describes any asymmetry in the growth curve of the single species (Gilpin and Ayala, 1973). This gives a better fit to experimental data with *Droso-phila* and could usefully be studied with a variety of other organisms.

There are a great number of interesting elements in this argument. Perhaps the most important is that it attempts to formulate (as α or β) the role of species in the life of each other. It gives an interpretation of successional processes: the interference that the presence of one species exerts in the population regulation of another. It gives, in solution (iv) a formal description of the type of interaction between two species that permits them to live together and so provides a model for species diversity in nature. The four solutions are independent of the values r_1 and r_2, i.e. the result of mixing two populations is independent of their intrinsic rates of natural increase. Moreover, the concepts of K_1 and K_2, the population limits for two species, provides the starting point for formal definition of the "niche". A number of later workers have written about "Gause's law" which has been variously phrased, e.g. "two species with the same ecology cannot coexist", or "two species with the same niche requirements cannot form steady state populations in the same region" (Hutchinson and Deevey, 1949).

A large number of experiments have been made with animals to "test the theory". In practice, as Slobodkin has pointed out, the experiments do not so much test the theory as test the organisms, so that if two species which appear to have the same ecology persist together in a uniform experimental environment this finding does not disprove "Gause's hypothesis"; rather it proves that the organisms do not have the same ecology or that the environment was not uniform! Of course, just as Verhulst's law of population growth seldom fits exactly to experimental data, so Gause's treatment of two species living together is based on the Verhulst law and may only describe ideal situations. The uniform environment is extremely difficult, if not impossible, to create experimentally and it becomes increasingly difficult to envisage two species that have "identical ecology". The value of "competition" theory is that it focuses the design of experiments about succession and diversity in nature onto mechanisms by which species interact, whether to produce stable equilibria or by elimination of one species by another. The very idea of a uniform environment suggests the ecological significance of environmental heterogeneity; the very idea of species with "the same ecology" poses the question: what kinds and magnitudes of difference between species lead to mutual exclusion or balanced cohabitation? These are questions that have dominated much of the work of population biologists between 1935 and the present day.

Studies of the behaviour of mixed plant populations in continuous growth systems have been made using *Lemna* species and *Salvinia*. The inoculation procedure was the same as in the single species cultures described earlier. Three pairs of species were grown in culture: (i) *Lemna minor* + *Lemna polyrrhiza*; (ii) *Lemna gibba* + *Lemna polyrrhiza* and (iii) *Salvinia natans* + *Lemna polyrrhiza*. Each pair of species was represented by 24 cultures and each was inoculated with the equivalent of 25 mg dry weight of each species. At weekly intervals three cultures were sacrificed to a destructive harvest in which the species were separated and the number of fronds counted and weighed. Destructive harvesting is a regrettable necessity in this type of experiment with floating aquatic plants. The process of population growth results in the formation of a mat of fronds in which the spatial position of a frond in the mat determines its success in trapping light and the fronds could not be counted or weighed without disturbing their arrangement. The procedure in experiments with some small animals is much simpler, and in similar experiments with the flour beetles *Tribolium castaneum* and *T. confusum* it is easy to pass the flour through a sieve at intervals, count the beetles and then return them to the culture vessels (Park, 1955) but in the experiments with duckweeds each harvested culture has to be a sacrificed replicate.

The results of a set of such experiments are summarized in Fig. 1/6, a, b and c and Table 1/II. In mixtures of *Lemna minor* and *L. polyrrhiza*

Table 1/II
The relative performance of some aquatic plants in culture,
measured in different growth phases

(a) Intrinsic rate of natural increase (r) or RGR; Phase I.
L. minor > *S. natans* > *L. gibba* > *L. polyrrhiza*
(b) Simple growth rate:Phase II.
S. natans > *L. minor* > *L. polyrrhiza* > *L. gibba*
(c) "Stable" population:Phase III.
L. polyrrhiza > *L. minor* > *S. natans* > *L. gibba*
(d) Success in mixed cultures.
$\frac{S.\ natans}{L.\ gibba}$ > *L. polyrrhiza* ≥ *L. minor*

---

there was no clear indication after 8 weeks that one species was succeeding at the expense of the other. The ratio of the dry weights of the two species shifted rather erratically between harvests and there was considerable variation between replicates at the same harvest. It appears that these two species are fairly evenly matched in a struggle for existence. In Park's experiments with *Tribolium*, although inevitably only one species survived from two species mixtures, the struggle was sometimes protracted and not resolved for 400 days! After 56 days *Lemna minor* appeared to be less healthy than *L. polyrrhiza* and would probably have slowly disappeared from the mixture.

Mixtures of *Lemna polyrrhiza* and *Lemna gibba* moved steadily towards dominance by *L. gibba* and mixtures of *L. polyrrhiza* and *Salvinia natans* moved even more rapidly towards dominance by *Salvinia*. The

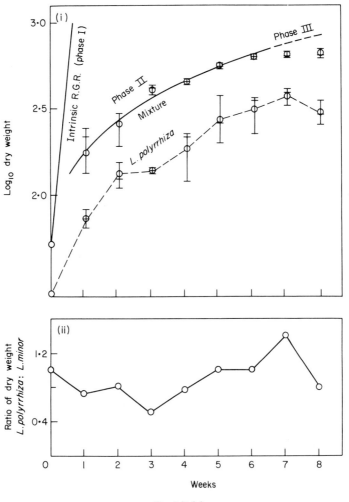

Fig. 1/6 (a)

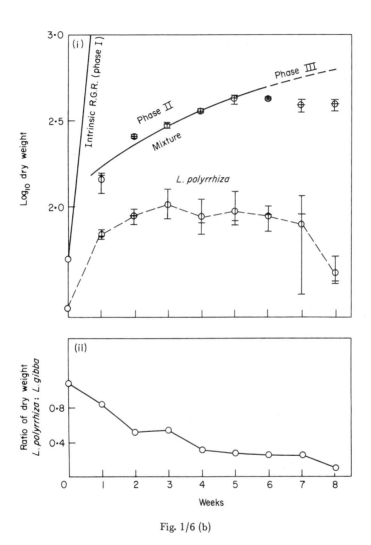

Fig. 1/6 (b)

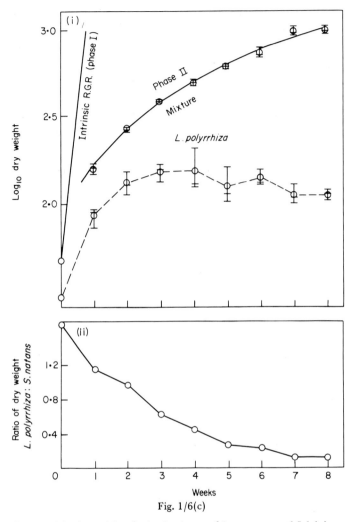

Fig. 1/6(c)

Fig. 1/6. The growth in dry weight of mixed cultures of Lemna spp. and Salvinia
    (a) *L. polyrrhiza* and *L. minor*
        (i) The dry weight of the mixture of species and of one of the component species
            (*L. polyrrhiza*) in mixed culture. The growth curve is divided into three phases
            (I, II, III) as in Fig. 1/5.
        (ii) The ratio between the two species in mixture based on their dry weights.
    (b) *L. polyrrhiza* and *L. gibba.*
        (i and ii as in (a))
    (c) *L. polyrrhiza* and *Salvinia natans.*
        (i and ii as in (a))
(From Clatworthy and Harper, 1962.)

relative success of the species towards each other is shown in Table 1/II. The comparisons were incomplete: the potentially very interesting struggle between *Lemna gibba* and *L. minor* could not be followed because these two species are morphologically so similar in Phase I and early Phase II of their population growth that they cannot be separated. No single parameter of population growth in pure culture was an efficient predictor of which species would succeed in mixtures. The intrinsic rate of natural increase would have predicted rapid success for *L. minor* in mixtures with *L. polyrrhiza* but this was in fact the most closely balanced of the mixtures studied. The Phase II growth rate in pure cultures would have predicted that *L. polyrrhiza* would succeed at the expense of *L. gibba*; the reverse was the case. The "stable" Phase III population size would have predicted that *L. polyrrhiza* would "win" against *L. gibba* and *Salvinia*; in fact *L. polyrrhiza* "lost" in a struggle for existence with both these species!

Neither theoretical nor experimental models of the growth of populations in isolation are sufficient to predict the outcome of a struggle for existence between them. The observer is forced back onto aspects of the natural history of the species to explain their interactions. The success of *Salvinia natans* in mixture with *Lemna polyrrhiza* seems to depend on the growth form of the two species. *Salvinia* produces its new fronds in the air and then lowers them onto the water surface. New fronds of *Lemna* are formed at the water surface and *Salvinia* is therefore bound to overtop and shade *Lemna*. The success of *L. gibba* in mixtures with *L. polyrrhiza* seems to depend on the habit of *L. gibba* of developing aerenchyma in its fronds when they are crowded. Their buoyancy then ensures that the fronds of *L. gibba* remain high in the frond mat and are not often overtopped by *L. polyrrhiza*. The mixture of *L. gibba* and *L. polyrrhiza* is interesting in another context — that of the productivity of mixed communities. The mixture of the two species moves towards dominance by the species that produces the smaller standing crop!

The whole plant as a model population

Species of *Lemna* and *Salvinia* are very unusual sorts of plants; they happen to be convenient (like *Paramecium* and *Drosophila* among animals) for illustrating some fundamental principles of population biology. The growth of a population of fronds from a mother frond is of course the growth of a clone. *A clone of* Lemna *is a single plant that falls to pieces as it grows!* Conversely, it may be useful to regard a single more "normal plant", such as an oak tree, as a population of units that remain attached together. It is a tenet of classical morphology that the higher plant is composed of a repetitive branched system of units each consisting of a segment of stem, a leaf and its axillary bud (see e.g. Arber, 1950). The whole shoot system is a population of such units which may gain a unity from the possession of a common root system. Even this unity is lost in the clonal growth of plants such as the grasses or the strawberry when independent adventitious root systems develop from the vegetative buds and old connecting parts of the shoot system die and rot away.

The branched root system is also in a real sense a population, a constantly changing number of foraging units. Both the root and the shoot of the higher plant grow exponentially by increasing the number of parts; the growth rate characteristically declines even in an isolated plant, as the parts make demands on limited resources available for the whole: leaves on the same shoot begin to shade each other. Growth of the individual may approach an asymptote as flowering and fruiting occur in an annual or as the respiratory burden of accumulated support tissue comes to balance with the assimilatory activity. All of the phases of development of a *Lemna* culture are mirrored in the life of a single terrestrial flowering plant. When one plant grows close to another, the growth rates of one or both may be affected, its branching and its number of structural units may be reduced and we see the influence of one level of population, the number of neighbouring plants, on the development of another level of population, the number of structural units that make up each individual plant.

The plant as a population of parts

The idea of a plant as a population of parts is not a trite analogy but important for understanding the population biology of plants. There are fundamental parallels between the population biology of animals and plants but the differences are sufficiently great that it can be dangerous to generalize from one kingdom to another without recognizing the differences. The population-like structure of the individual plant is among the most important of all these differences.

The individual fruit fly, flour beetle, vole, rabbit, flatworm or elephant is a population at the cellular but not at any higher level. Starvation does not change the number of legs, hearts or livers of an animal but the effect of stress on a plant is to alter both the rate of formation of new leaves and the rate of death of old ones: a plant may react to stress by varying the number of its parts. An individual plant of the annual *Chenopodium album* may, under stress of nutrient deficiency or if grown at high density, flower and set seed when only 50 mm high, but, given more ideal growing conditions, may reach *ca* 1 m in height and produce 50 000 times as many seeds as its depauperate stressed counterpart. Some of the difference between the two plants is determined by differences in the birth rate of new parts, some by differences in the death rate of old parts; a relatively tiny part of the variation is in the size of the parts themselves.

The population-like structure of a higher plant is emphasized by the way in which plants are classified. Almost the whole of the information used by taxonomists in the description of plants involves features of the leaves, flowers, seeds etc., i.e. the modular units of construction. The form of the whole is taxonomically almost irrelevant because it is so enormously variable. The reliable repeatable unit is the constructional module, not the whole.

The fundamental equation of population biology is:

$$N_{t+1} = N_0 + \text{Births} - \text{Deaths} + \text{Immigrants} - \text{Emigrants}.$$

Immigration and Emigration have little meaning for the whole plant as a population of parts but the main Birth − Death element in the equation is readily applied to the leaves or some other modular unit of structure. The plant (quite unlike a higher animal) has an age structure in its organs just like any more conventional population. The age structure of growing plants of *Linum usitatissimum* is illustrated in Fig. 1/7.

For some purposes the population dynamics of the parts of plants may be more useful than the dynamics of the individual whole plants in a community. A zoologist wanting to know about the food supply for a predator may be much more concerned with the expectation of life of a leaf than of the plant that bears it. Figure 1/8 shows the results of treating a pasture as a population of leaves, ignoring entirely any concern with the number of plants, genotypes or tillers present. Individual batches (cohorts) of leaves were marked on emergence and their fate was followed at intervals of time. This gives survivorship curves for the leaves in a pasture and emphasizes the relatively short life of the photo-

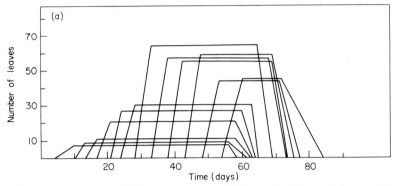

Fig. 1/7. The age structure of the leaves on a plant of *Linum usitatissimum*. The population of leaves is shown as successive cohorts that expand during three day periods. (From Bazzaz and Harper, 1976)

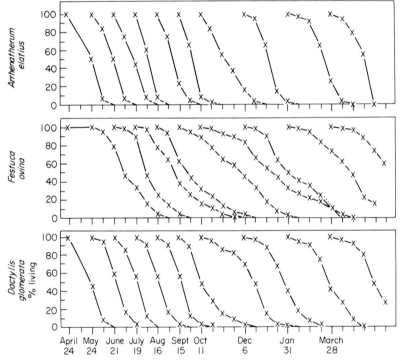

Fig. 1/8. Survivorship curves for cohorts of leaves appearing in a pasture at different periods of the year. (Reproduced from unpublished data of P. Williamson by kind permission)

synthetic unit, the difference between the expectancy of life of the leaves of different species and the changing death risk through the seasons. This study was made by a zoologist interested in the food supply of molluscs.

A leaf has a life history, a changing pattern of behaviour from birth (as a primordium on a meristem) to death from senescence or some environmental hazard. During its life it changes from being an importer and consumer of resources to being an exporter. Figure 1/9 shows the

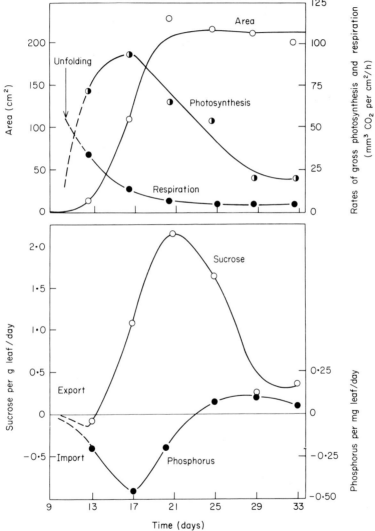

Fig. 1/9. The life cycle of a leaf. (a) Changes in area, rate of photosynthesis and of respiration of the second leaf of cucumber. (b) Changes in the rate of import and export of sucrose and phosphorus by the second leaf of cucumber. (From Milthorpe and Moorby, 1974 after Hopkinson, 1964. Reproduced by permission of the Oxford University Press)

life cycle of a cucumber leaf as a consumer and supplier of carbohydrate and of one nutrient element, phosphorus. A rootlet also has a life history though it is a less tidy unit to study (or count!) than a leaf.

The fact that the leaf with its axillary bud is the unit of construction favoured by morphologists and physiologists does not mean that it is necessarily the module of organization appropriate for all descriptions of a plant as a population of parts. The stem and root system must be included in any full descriptive system. One way of doing this is the so-called pipe model (Shinozaki *et al.*, 1964) of plant growth in which the module of construction is regarded as a leaf plus its dependent linked fraction of the stem and including an appropriate fraction of the root system. This model was designed with trees in mind and is conceptually elegant though very difficult to use in practice. The parts of a plant are interrelated on a scale that makes such a fasces of independent pipes quite unrealistic as a physiological model. Other modules are used in different contexts. The agronomist concerned with the growth of grass has for long used the tiller — the branch unit — as a very convenient countable module. Practical considerations necessitate the use of the tiller as a module because it is usually quite impossible to count the number of genetic individuals in a grassland community. For the apple grower, the relevant module of structure will be the shoot system — the number of long and short shoots. This measure goes much further in predicting the behaviour of an apple orchard than does the number of trees.

The units of birth and death may be modules that include several leaves (in *Taxodium distichum*, for example, the leafy shoot is the obvious module of construction in the canopy as it is shed as a unit). A whole system for the description of tree architecture has recently been suggested that is based on the concept of a modular structure. The module is the repeating unit of form and usually also of death (Hallé and Oldeman, 1970; Oldeman, 1972). Figure 1/10 shows a characteristic "logistic-shaped" growth curve for the population of modular units that comprises a single tree of *Rhus typhina*.

The "ramet" is the unit of clonal growth, the module that may often follow an independent existence if severed from the parent plant. In many higher plants, clonal growth occurs by the formation of a population of parts of what was originally one seedling. The ramets are then the effective units that are readily counted in the field (e.g. for straw-berries, or creeping buttercup, *Ranunculus repens*). In other cases the useful measure is the number of leaves, for example in the cases of bracken fern (*Pteridium aquilinum*) or white clover (*Trifolium repens*). The module appropriate for a population study will vary with the organization of the species to be studied.

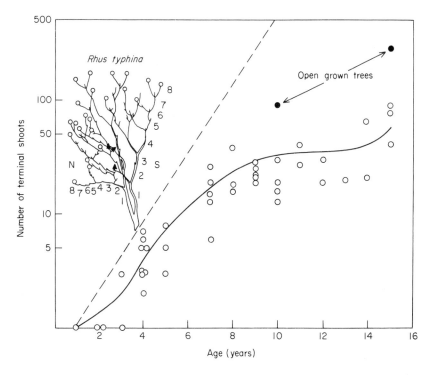

Fig. 1/10. Tree growth as the development of a population of modules. The figure is calculated for *Rhus typhina*. Each point on the graph represents the number of terminal growing points on a tree of a particular age. The broken line indicates the number of shoots expected if each growing shoot gives rise to two new shoots in the next growth period. The terminal meristem in this species aborts. (From J. White, unpublished)

The existence of two levels of population structure in plants makes for difficulties, but the problems are much greater if their existence is ignored. One of the strongest reasons why a population biology of plants failed to develop alongside that of animals was that counting plants gives so much less information than counting animals. A count of the number of rabbits or *Drosophila* or voles or flour beetles gives a lot of information: it permits *rough* predictions of population growth rates, biomass and even productivity. A count of the number of plants in an area gives extraordinarily little information unless we are also told their size. Individual plants are so "plastic" that variations of 50 000-fold in weight or reproductive capacity are easily found in individuals of the same age. Clearly, counting plants is not enough to give a basis for a useful demography. The plasticity of plants lies, however, almost entirely in variations of the number of their parts. The other and closely related reason why plant demography has been slow to develop is that the clonal spread of plants and the break-up of old clones often makes it impossible to count the number of genetic individuals.

The problems are great: they can be regarded as insoluble and a demography of plants an unattainable ideal, or they can be ignored with a certainty of serious misinterpretation, or they can be grasped and methods, albeit crude, developed to handle the problem. Parts of this book attempt to face the problem; in other parts the problems scarcely arise.

The way in which the problem can be faced is to accept that there are two levels of population structure in plant communities. One level is described by the number of individuals present that are represented by original zygotes. Such units represent independent colonizations. Such individuals will be called *genets* (Kays and Harper, 1974). An individual genet may be a tiny seedling or it may be a clone (e.g. *Holcus mollis*) extending in fragments over a kilometer (Harberd, 1967). The number of genets in a population of plants is formally equivalent to N in the population dynamics of most animal populations. Each genet is composed of modular units of construction — the convenient unit may be the shoot on a tree, the ramet of a clone, the tiller of a grass or the leaf with its bud in an annual — and the number of units of construction per genet is termed η. The number of modular units in a plant population is then described by $N\eta$ and variations in the behaviour of a population of plants may occur through changes in N or η or some combination of the two. In practice, because the modular units are relatively constant in size, compared with the variation in size of a genet, the weight of a genet roughly parallels the number of its modular units. Simple allometric relationships occur between leaf weight, total plant weight and stem diameter (Ando *et al.*, 1962; Kittredge, 1944). There are many occasions when it is more convenient to use units of weight than units of construction.

This approach involves deliberately avoiding the term vegetative reproduction and in this book this is the only place in which the phrase is used.* There are strong reasons for discarding such a widely used phrase: the most important concerns the way in which we interpret the evolution of plant form. The unit of variation that results from the sexual process and that displays mutation and recombination is the developing product of the zygote. A clone of *Trifolium repens*, even though it may be fragmented and extend in patches over 18 m (Harberd, 1963) is the phenotypic representation of one genotype. The fact that it has grown very large and penetrated various parts of a grassland habitat is just a proof that it represents a very successful genotype. It is successful in the same way as a single tree that has emerged into the canopy and extends its branches over and among those of other trees. The ability of some plant species to form fragmented phenotypes of a single genotype is just one of the variety of successful ways of playing the game of being a plant.

The ability of the modules of many plants to form their own roots gives them a chance to tap water and nutrients that are unavailable to plants with a root system that radiates from a single point. It is a habit like the twining or tendrilled habit that confers a special advantage on a plant structure; it is dangerous to regard it as in any way the equivalent of a reproductive process. A. S. Watt (1940, 1943, 1947), in his classic analyses of the bracken fern (*Pteridium aquilinum*) which spreads far by a complex branching system, likens a bracken plant to a tree laid on its side. This analogy is very real and it may be helpful to divide plant growth into two categories: (i) that which has been dominated in its evolution by selective pressures to attain height (and shade neighbours) which leads inevitably to the woody habit; and (ii) that which has been dominated by pressures to expand laterally (and reach limited water and nutrients) — this leads equally strongly towards the lateral branching, node rooting habit of clonal species. Trees like *Populus tremuloides* gain the best of both worlds, with woody tall systems and clonal growth by forming suckers.

The treatment of an individual plant (genet) as a population in its own right has led to the development of techniques of growth analysis in which the growth of a plant is treated as an exponential process. Growth is described as a continuous process and measured as the "rela-

---

* The distinction made here between "reproduction" and "growth" is that reproduction involves the formation of a new individual from a single cell: this is usually (though not always e.g. apomicts) a zygote. In this process a new individual is "reproduced" by the information that is coded in that cell. Growth, in contrast, results from the development of organized meristems. Clones are formed by growth — not reproduction.

tive growth rate" (RGR).

$$\text{RGR} = \frac{\log_e W_2 - \log_e W_1}{t_2 - t_1}$$

The RGR is conventionally measured in terms of dry weight, perhaps unfortunately, because what is then examined is the resultant of the processes of birth and death of parts and these underlying realities are disguised or never measured. It can be argued that deeper understanding of the plant growth process itself will come from adapting a yet more complete population-type analysis to the growth of the individual plant. Certainly the logistic function can be applied rather easily to the growth of single plants (e.g. Fresco, 1973). The conventional techniques of growth analysis are considered further in Chapter 10.

Most animals have a growth pattern that is quite unlike that of plants and except at the fundamental level of cell number there is no useful sense in regarding a *Drosophila* or a vole as a population of parts. However, the populational structure of a genet is common in some animal forms, most notably the hydroids, corals and their allies. The population biology of *Hydra* is an almost exact parallel to that of the duckweeds: an individual genet grows by budding, the buds separate and bud in turn. A population of free-living units represents a single original zygote: one *Hydra* genotype enters a struggle for existence with others as a phenotype that is fragmented. In colonial hydroids the fragmentation does not happen and a more or less organized sedentary colony is composed of repeated modular units that are the products of a single zygote. A vertical section through a coral reef is comparable with a profile through a forest. In the corals and colonial hydroids the population dynamics have, like higher plants, two distinct aspects of population growth: multiplication of the number of genets (or zygotes) and multiplication of the number of modules that compose a genet. The full population biology of such animals would need, as with plants, to take account of changes both in η and in N.

A somewhat similar and very striking analogy with the population biology of plants is that of colonial Hymenoptera. Bees, wasps, ants and termites have growth systems involving the dispersal of fertilized females (cf. developed embryos in plants) and growth of a colony of modules. There is a division of labour between the modules (workers, soldiers, nurses, etc.) comparable to the specialization of leaf modules as carpels, stamens, etc. in the higher plant. There are, of course, important differences (e.g. in the cytology of the modules in the Hymenoptera), but the parallels are striking. The population biology of ants, like that of a higher plant, has to take into account N (the number of genets or zygotes) and η (the number of modules developing from each genet); the effects of density or other adverse elements in the environment may be seen *either* in a small number of colonies *or* small colonies *or* some combination of both effects.

A diagrammatic model of a plant population

The structure of most of this book is built around a diagrammatic model of population behaviour (Fig. 1/11). This describes the sequence of events that determines the success of a seed in leading to the production of more seeds. A plant is only the means by which a seed produces more seeds (modifying Spencer's aphorism that a hen is only the means by which an egg produces more eggs). The model starts with dispersed seed

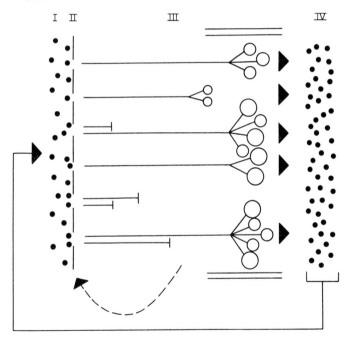

Fig. 1/11a. Elements of the population dynamics of a monocarpic (semelparous) plant.
  I = the bank of seeds in the soil
  II = the recruitment of seedlings (the environmental sieve).
  III = the phase of growth in mass and in number of modular units.
  IV = the terminal phase of seed production.
  (Adapted from Harper and White, 1971)

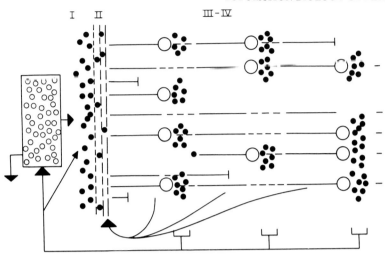

Fig. 1/11b. Elements of the population dynamics of 1 cohort (one generation) of a polycarpic (iteroparous) plant. Symbols as in Fig. 1/11a. (From Harper and White, 1971)

— the seed rain which forms an inoculum. This may be stored in a dormant state forming a living "bank" of quiescent plants in or on the soil (I). From this "bank" seedlings may be recruited, depending on the physical conditions that they experience (II). The population of growing seedlings, sifted from the numbers and variety in the soil bank, makes demands on resources of the environment which may be insufficient to allow vigorous growth. Death or a depauperate condition may result. The survivors are shown in the model as branched systems (modules), potentially branching exponentially but constrained within resource limits (III). During the development of the plants from an episode of recruitment there may be a feedback on subsequent processes of recruitment from the bank — the growth of a plant changes its own environment and that of its successors. The growth of the plant is seen as culminating in (Fig. 1/11a), or including (Fig. 1/11b), phases of seed production. The seeds are dispersed (IV) and enter the cycle again.

The sequence of events in the model is not wholly neat and tidy but summarizes the essential features of most plant life cycles (and some colonial animals as well). Predation can be readily envisaged as acting at any point in the life cycle, as also can pathogenic activities and the physical hazards of the environment; the model allows for the presence of two or more species. The model is designed to focus attention on organisms as life cycles and is a basis for the comparison of strategies either visually (Fig. 19/19) or as matrix models (Chapter 19).

Dispersal, Dormancy and Recruitment

2

Dispersal: The Seed Rain

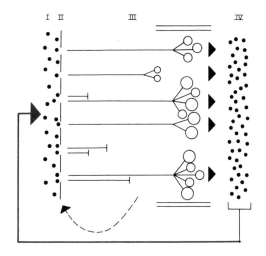

The flux of seed (or other propagules) into and out of a unit of habitat determines the *potential* population of that habitat. Most areas of land and water are continually invaded by an inoculum of propagules from vegetation elsewhere, spores, seeds, invading suckers, rhizomes etc. If the areas are already colonized by vegetation they serve also as a source of inoculum, providing propagules which fall in their own area and of inoculum which becomes spread to other areas. The inoculum present in an area (the potential vegetation) is the resultant of this flux. Clearly any attempt to discuss the dynamics of plant populations must concern itself with the parameters of such a flux (Fig. 2/1).

The dispersal of propagules has been a favourite area of natural history, and there is a vast anecdotal literature (Ridley, 1930; van der Pijl, 1969). Few theoretical studies have been made until very recently (e.g. Gadgil, 1971; Levins, 1968; MacArthur and Wilson, 1967). Superficially there appear to be two distinct contexts in which one can consider dispersal: (i) that of the expanding range and increasing population

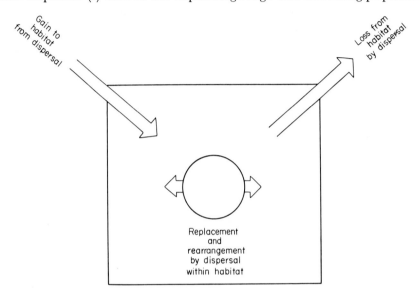

Fig. 2/1.

size of an invading species into a new area, island or continent and (ii) as part of the process by which an established and stabilized population maintains itself within the ever-shifting "islands" that constitute the pattern within established vegetation. In reality these two situations are not contrasting, but parts of the same system.

Vegetational succession depends on the fugitive properties of plants. All successional species are doomed in their present habitats, and their continued survival depends on escape and establishment elsewhere. The frequency of such events depends on the life cycle and length of tenure of a place in the succession, and on the frequency and distance at which new sites suitable for colonization become available. An epibiotic outbreak, e.g. the spread of *Opuntia* in Australia, is a rare, large-scale event, but of same fundamental nature as the common, small-scale shifting occupancy of sites by a species in its native vegetation.

The significance of dispersal in the life cycle of a plant must there-fore depend on the spatial and temporal heterogeneity of its environ-ment. Just as the dispersal habit of a species reflects the selective advan-tages that dispersive qualities have conferred in the past (ultimate factors), the present dispersal characteristics of the species contribute to determining its present ecological range and population size (proximate factors).

The dynamics of a plant population with specific dispersal parameters (Fig. 2/1) depend on the rate at which it is eliminated from colonized areas, the frequency of appearance of new areas suitable for coloniza-tion and the dispersive characteristics that determine how quickly pro-pagules from "dying" areas find and exploit new understocked habitable sites. Dispersal may directly affect population size over the potential range of a species. The population size of a species may be low because:

(i) The areas suitable for its occupancy (e.g. availability of sites for germination) are few.

(ii) The habitable areas are separated by distances that are great relative to the dispersibility of the species.

(iii) The carrying capacity of the habitable sites is low.

(iv) The time for which these sites are suitable is short relative to the rate of dispersal of propagules.

(v) The habitability of a site is of short duration (e.g. successional displacement is rapid).

(vi) Colonization and full exploitation of the carrying capacity of a habitable site is slow.

(this interpretation derives largely from the treatment of dispersal by Gadgil, 1971)

> In spite of the heavy toll of death exacted during dispersal a very strong ten-dency to disperse obtains in a very large number of groups of plants and animals. In a very general way, the factor favouring evolution of dispersal would be the chance of colonizing a site more favourable than the one that is presently in-habited. . . . An organism should disperse if the chance of reaching a better site exceeds the expected loss from the risk of death during dispersal or the chance of reaching a poorer habitat. In many cases a mixed strategy of a proportion of the organisms staying on in the same habitat while the rest disperse could be the most advantageous strategy.
>
> (Gadgil, 1971)

Attempts to determine the fittest dispersal strategies for specified en-vironmental regimes have so far proved too complex to be handled other

than through numerical experiments on digital computers. The effort
involved is, however, justified as much by the intellectual rigour needed
to specify the problems and state them clearly as in the solutions to the
problems. A series of numerical experiments suggest that optimal dispersal depends both on the variability of habitable sites in space and on the
variability of the carrying power of these sites in time.

The very idea of dispersal as an element of natural history which can
be quantified and studied numerically immediately focuses attention
on the desperate poverty of hard quantitative information. Classic
surveys of seed dispersal (Ridley, 1930; van der Pijl, 1969) are far more
concerned with the rare events of long distance colonization than with
quantifying the day-to-day happenings in more normal ecological situations: are more concerned with the precise and subtle variations in form
that aid dispersal than with the ways in which these variations affect
populations of plants.

The formal meaning of "disperse" is given by the Oxford English
Dictionary as: "(1) to scatter in all directions; to rout. (2) to spread
about; to send to or station apart at, various points. (3) To divide, dis-
part. (4) To distribute from a source or centre. (5) To spread about; to
diffuse. (6) To dissipate." (The partial synonym "disseminate" has its
etymology in the sowing of seeds.) The implication of all these definitions is that something that was concentrated is spread about more evenly.
In fact, with the exception of weed seeds sown in contaminated crop
seed, seed dispersal usually achieves strikingly uneven distribution!

Characteristic seed dispersal patterns in flowering plants; mechanisms not involving animals

The seed falling on a unit area of land might be expected to be a
function of several variables: (i) the height and distance of the source of
the seed, (ii) the concentration at the seed source, (iii) the dispersibility
of the seed (its weight, possession of wings, plumes, etc.) and (iv) the
activity of distributing agents (e.g. wind direction and velocity).

In the case of wind dispersal one might expect some simple function
to apply to the relationship between the density of dispersed seed and
the distance from the source. Most of the relevant data come from
studies of weeds and of forest trees. The weed studies have been made
on isolated plants or from the edge of dense stands. The forest examples
are mainly from studies of isolated trees, or of dispersal from the edge

of forest clearings. All of the studies therefore represent plants in essentially invading phases. The nature of the dispersal process from individuals within a more or less continuous stand is difficult to study because
the seeds from the individual must be labelled to distinguish them from
other seeds of the species dispersed in the same area. This problem is
soluble and in a classic study of the distribution of pollen from a tree of
Pinus coulteri, Colwell (1951) fed radioactive tracers at high dosage to
an individual tree; the pollen produced was marked sufficiently for individual grains to be recognized and distinguished from others dispersed
in the forest (Fig. 2/2a). The same procedure can be followed for seed
dispersal, by using paints, dye or isotopes to mark seeds on an inflorescence before dispersal (Mortimer, 1975; Watkinson, 1975).

The relationship between seed number and distance from a stand of
Senecio jacobaea was studied by Poole and Cairns (1940) in a New
Zealand pasture (Fig. 2/2b). Of all seed shed, 60% was found around the
base of the plants; only 0.39% of the total seed was distributed further
than 4.6 m from the edge of the plot containing the plants, 0.08% of
the seed was distributed more than 9 m, 0.02% more than 18 m and
0.005% more than 36 m.

A very similar-shaped curve, but measured over much greater distances, was obtained for the passage of seeds of *Tussilago farfara* across
uncolonized open areas of land in the newly reclaimed Polders of the
Zuider Zee (Bakker, 1960) (Fig. 2/2c).

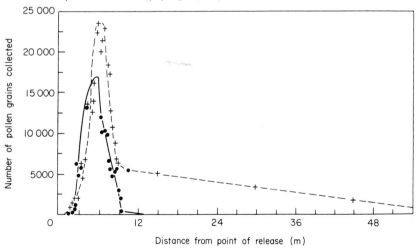

Fig. 2/2a. The distribution of pollen from a tree of *Pinus coulteri* within a forest stand. (From
Colwell, 1951)

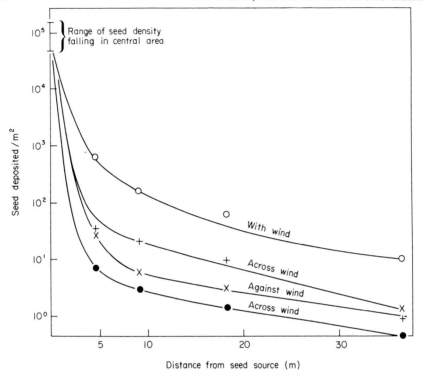

Fig. 2/2b. The distribution of seed of *Senecio jacobaea* from a dense population *ca* 20 m square
measured in relation to the direction of the prevailing wind. Eye-fitted curves. (From
data of Poole and Cairns, 1940)

Seed fall of *Pinus echinata* was measured (Yocom, 1968) by laying
seed traps at right angles from the sides of a clearing within a forest. The
trees were *ca* 21 m high and the clearing 238 m x 257 m. The densest
seed fall was at the forest edge and in a band 10 m wide just within the
margins of the forest. 50% of the seed fell within 20 m of the forest
edge and 85% had fallen within 50 m to give a seed rain of 3.9–7.7 seeds
per m² at that distance. A similar study (Roe, 1967) made on the dispersal of Engelmann spruce (*Picea engelmannii*) into forest clearings
showed a remarkably regular relationship between the logarithm of the
density of seed deposited and the distance from the forest edge; again
the densest fall of seed was just within the forest margins (Fig. 2/2d).
Turbulence is important in keeping seed airborne and seeds tend to fall
in zones of locally reduced turbulence; this, at least in part presumably

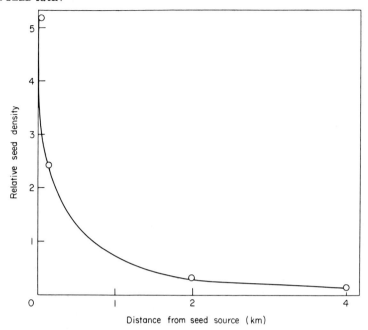

Fig. 2/2c. The distribution of seeds of *Tussilago farfara* from a dense stand across newly reclaimed polders, Holland. (From data of Bakker, 1960)

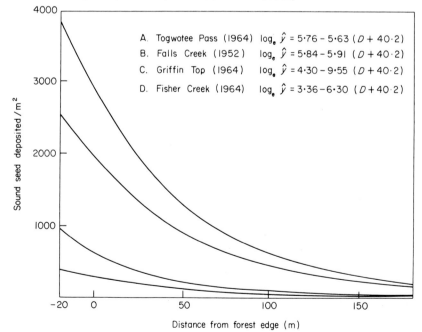

Fig. 2/2d. The distribution of seeds of *Picea engelmannii* from the edges of forest clearings in
Utah, U.S.A. The calculated regressions are shown. (From data of Roe, 1967)

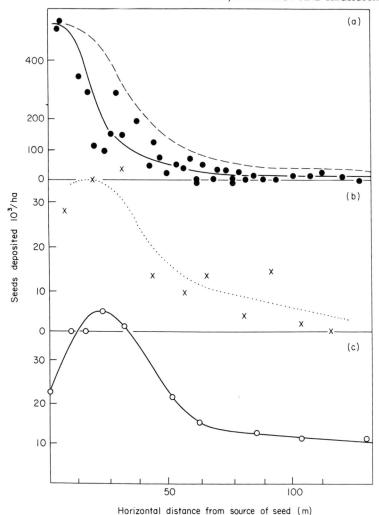

Fig. 2/2e. The distribution of seeds of *Eucalyptus regnans* from trees *ca* 75 m high.
  (a) From the edge of a dense forest. Measurement taken at 55° from the direction of
      the prevailing wind. The dashed line is an estimate in the direction of the prevailing
      wind.
  (b) From the edge of a sparse forest.
  (c) From an isolated tree.
  (From Cremer, 1965)

accounts for the occurrence of the densest seed rain just within the
margin of a forest.

The pattern of seed dispersal from forests margins may differ from
that from isolated trees. In *Eucalyptus regnans*, the pattern of dispersal

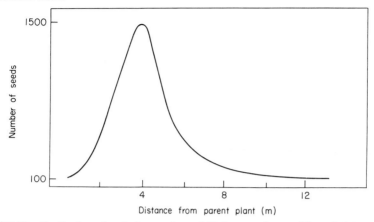

Fig. 2/2f. The distribution of seeds from a plant of *Verbascum thapsus*. (From Salisbury, 1961)

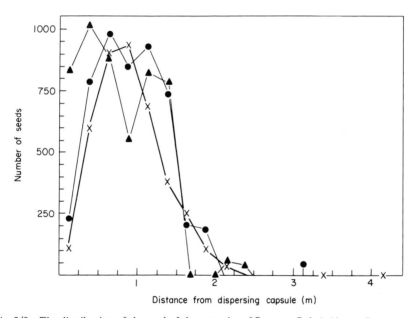

Fig. 2/2g. The distribution of the seed of three species of Papaver, *P. hybridum* ▲ *P. argemone* ●
and *P. dubium* x. The capsules were held each at the height characteristic of its
species and exposed to the same variations in wind velocity. (From Salisbury,
1942)

from isolated trees showed the densest rain of seeds occurring at 30 m
from the source, whereas the dispersal curve from the edge of a forest
of the same species showed the densest seed rain at the forest margins
(Cremer, 1965) (Fig. 2/2e).

The dispersal of seed by wind depends on (i) how fast the seed falls through air, measured as its terminal velocity, (ii) the height at which the seed is released and (iii) the speed and turbulence of the wind between the ground and the point of release (Cremer, 1965). The terminal velocity of seed of *Eucalyptus regnans* is of the order of 3.5 m/s (Gilbert, 1958) and of Douglas fir (*Pseudotsuga douglasii*) 1–1.5 m/s (Isaak, 1943) – the fir is the better dispersed. Dispersal of seeds by wind may be in four essentially different ways, three of which depend on specialized structures attached to the seed or fruit and so represent an expenditure of energy and resources – dispersal has a cost.

The dispersal of dust seeds

The flight Reynolds numbers for spores, pollen and the tiny dust seeds such as those of *Monotropa* and many orchids are usually so small that movement in air is dominated by viscous forces. Although they are much heavier than the volume of air they displace, dust seeds and spores falling vertically under gravity in still air have very small terminal velocities (of the order of 100 mm/s). Such seeds are therefore very sensitive in their trajectories to fluctuations of wind and convection currents. Similar velocities can be obtained in minute as well as in large eddies of air, so that suspended movements become possible in an environment that appears to be tranquil. When overall convection velocities in excess of these terminal velocities are sustained for some length of time (as in normal heating and cooling of the atmosphere in day and night) dust seeds can easily be carried to very high altitudes and vast distances of travel become possible or indeed the rule (Burrows, 1975a, b).

The dust seeds depend for the detail of their movement on the detail of the airflow around them and if the air flow is known the trajectories can be defined accurately. All that is required for this form of dispersal is that the seed be very small and very light. This clearly puts constraints on species that employ such a dispersal method, since embryo size and food reserve must be minimal. It is intriguing that all seeds with this dispersal mechanism have anomalous methods of seedling nutrition: they are total parasites or depend on a mycorrhizal association (see Chapter 21).

The dispersal of plumed seeds and fruits

Seeds in this group move at much higher Reynolds numbers and aero-

dynamic drag occurs, more because of the inertial forces in the flow
than because of viscous forces (Burrows, 1973). The rate of fall of these
seeds depends on structures such as the pappus or plumes on the seed or
fruit coat. The effect of a pappus on seed dispersal in the Compositae is
shown in Figs 2/3a and b. The terminal velocity of a propagule in still
air is crudely related to the ratio between achene weight and pappus
weight (Fig. 2/3b) but much more closely related to the ratio of pappus
diameter to achene diameter (Fig. 2/3a). This implies that the reward
for expenditure is not a simple cost measurable as dry weight involved,
but the actual design of the pappus is also critical. The very open pappus
form of *Hypochaeris radicata* leaves it with nearly twice the terminal
velocity of *Erigeron acer* although the two species have about the same
ratio of pappus diameter to achene diameter (Sheldon and Burrows,
1973). The effects of changes in the terminal velocity on the distance
of dispersal in a specified wind and convection speed are shown in Fig.
2/4a and the dispersal curves are calculated for two different heights of
seed release. Height of release has an overwhelming role in determining
distance of dispersal (Fig. 2/4b). For a member of the Compositae in
which dispersal is favoured in natural selection, there is a variety of
evolutionary pathways that can produce the same effect:

(i)   to decrease the weight of the seed;
(ii)   to increase the ratio of pappus structure to achene;
(iii) to improve the drag efficiency of the pappus structure;
(iv)   to release the seed higher.

The flower stalk of the dandelion (*Taraxacum officinale*) is very
plastic and elongates particularly after flowering: this can be interpreted
as a very effective way of enhancing the role of the pappus in seed dispersal (Small, 1918). For the plumed seeds and fruits the distance of
dispersal is strongly influenced by the humidity, particularly in species
where the pappus is hygroscopic (Fig. 2/4b).

The dispersal of seeds with a wing and a more or less concentrated central mass

A seed with a wing and with the weight of the seed concentrated in a
central area is perhaps the only arrangement that can give stable flight
over a substantial distance from the point of release *in still air*. This, the
long glide, depends on perfect geometry, and slight variations due for

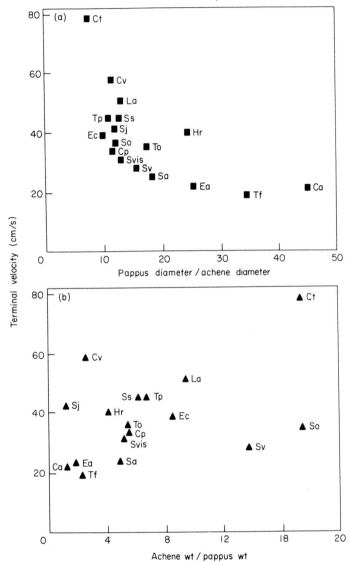

Fig. 2/3a. The relationship between the pappus diameter: achene diameter ratio and the terminal velocity of the achene—pappus units of selected Compositae. Note: *Hypochaeris radicata* has a very open pappus form.

Fig. 2/3b. The relationship between the achene weight: pappus weight ratio and the terminal velocity of the achene—pappus units of selected Compositae.
Ca, Cirsium arvense; Cp, C. palustre; Ct, Carduus tenuiflorus; Cv, Carlina vulgaris; Ea, Erigeron acer; Ec, Eupatorium cannabinum; Hr, Hypochaeris radicata; La, Leontodon autumnalis; Sa, Sonchus arvensis; Sj, Senecio jacobaea; So, Sonchus oleraceus; Ss, Senecio squalidus; Sv, S. vulgaris; Svis, S. viscosus; To, Taraxacum officinale; Tf, Tussilago farfara; Tp, Tragopogon porrifolius.
(From Sheldon and Burrows, 1973)

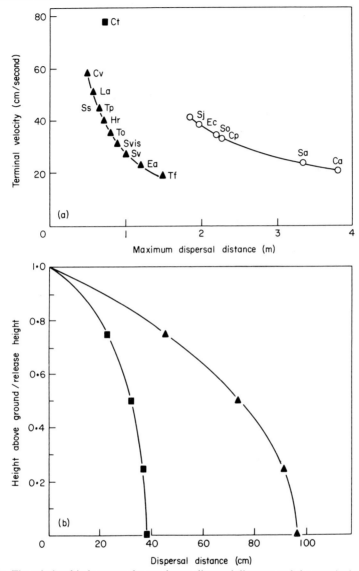

Fig. 2/4a. The relationship between the maximum dispersal distance and the terminal velocity
of the achene—pappus units of selected Compositae. A wind speed of 5.47 km/hour,
convection speed of 3.05 cm/second and the ground boundary layer are considered.
The plant height of each species (30.5 cm, ▲; 61.0 cm, ■; 91.5 cm, O) is taken as the
mean height, to the nearest foot (30.5 cm), as given by Clapham, Tutin and Warburg
(1962).

Fig. 2/4b. The maximum dispersal distance possible by achene—pappus units of *Senecio vulgaris* stored at 0% (▲), and 75% (■) humidities. A wind speed of 5.47 km/hour; a
convection speed of 3.05 cm/second; plant height of 30.5 cm; and the effect of the
boundary layer are considered. K (0%) = 1.25, K (75%) = 0.13.

(From Sheldon and Burrows, 1973)

example to uneven ripening of the wing can result in extremely (apparently) erratic movements. Wind can increase the distance of glide and the aerodynamics of seeds of this type are not unlike aeroplanes in unpowered flight (Burrows, 1973; 1975a and b). This type of seed is best known among lianes and trees of tropical forest. Within a still forest environment the glide may be the only reliable way of escaping from the parent tree unless animal dispersal agents are employed.

Winged seeds and fruits which rotate when falling

A perfectly symmetrical winged seed can fall in a perfectly straight line in still air. Dispersal depends on wind and on the initial angle of release: slight asymmetries of structure of the wing can have large effects on the pattern of motion. The critical element in the rotating system of seed dispersal is that (like a glider) it can develop lift. The flight paths are extremely subtle and forecasting depends on six degrees of freedom! (Burrows, 1975a, b). It may well be that many of the fine variations between even closely related species (e.g. *Acer* spp.) have real meaning in their dispersal biology.

Groups of species caught in different evolutionary pathways may face the same selective forces, e.g. a selection pressure to disperse. The ancestral constraint (the archetype) may determine in which direction an evolutionary response occurs. It is unlikely that an *Acer* will evolve a pappus, or a *Taraxacum* a wing. For all the systems of wind dispersal, increasing the height of release brings immediate rewards in enhanced dispersal; otherwise wind dispersal is increased by subtle modifications within a limited number of aerodynamic systems. All of these except the dust seed system involve expenditure of energy and resources that might have been used in making more seeds or in some other evolutionary modification of form. The study of aerodynamics of seed dispersal begins the quantification of an aspect of natural history and is just one case in which this reveals a very subtle interaction of costs and benefits.

When light unwettable seed lands on water, it tends to become concentrated at the edges of the water. Achenes of *Tussilago farfara* have been observed speeding at 40 m/min across water surfaces to the banks of the water edges on the Dutch Polders (Bakker, 1960). The edges of lakes and water-courses become depositaries for extremely dense seed populations.

Where a seed is dispersed by ejaculation from a capsule, the height of the capsule plays a large part in determining the distribution pattern of the seeds. The tall biennial mulleins (*Verbascum* spp.) which have no specialized mechanism to force seed out of the capsule, give a seed distributional pattern with the highest density of dispersed seeds at a distance of 3 m from the parent plant, a zone of relative seed poverty immediately under the parent, and a rapid fall-off in seed density with distance beyond 3 m (Fig. 2/2f) (Salisbury, 1961).

Many of the leguminous trees of tropical rain forests possess large seeds with no obvious dispersal mechanism (e.g. species of the genus *Mora* of which the seeds may weigh 0.4 kg). The dispersal rate of *Mora* spp. is extremely low and the range of some of these species may still be expanding from the sites in which they originally evolved (Beard, 1946)! Seeds which are not wind dispersed or ejaculated often have some linkage with animal behaviour and this may result in very specialized patterns of distribution.

The role of animals in the dispersal process

Where an animal is involved in seed dispersal, the often very species-specific feeding habits, territorial and migratory behaviour, introduce a complicated element into the pattern of distribution. The animal may eat and digest seeds (a loss to dispersal) or eat fruits and pass seed undamaged in faeces (a possible gain to dispersal). It may transport seed passively when its specific territory, range of movement or place of grooming may determine how far the seed is moved and in what sort of site it is deposited. The animal may collect and store seed, burying it in local aggregations from which localized very dense seedling populations may ultimately emerge.

There are important distinctions between those birds that possess the type of hard gizzard which destroys seeds (e.g. the turkey *Meleagris meleagris*; the hawfinch (*Coccothraustes coccothraustes*) and those in which the gizzard is soft (e.g. the thrush allies, *Turdus* spp.) and, although soft fruit parts are digested, seeds pass in a viable state into the faeces. Division of labour between these two types of predator sometimes occurs in their treatment of a fruit like the rose; redwings, fieldfares and thrushes (*Turdus musicus, T. pilaris* and *T. viscivorus*) eat the fleshy receptacles and swallow the achenes, but these pass through the digestive tract and are voided in a viable state in the faeces. The achenes may then in turn be eaten by the hawfinch (*Coccothraustes coccothraustes*) which harvests them from the faeces (Mountfort, 1957).

A bird feeding on fleshy fruits will often not swallow the seeds, but will eject them from the bill. In the case of the mistletoe, *Viscum album*, the individual sticky seeds are rubbed off the bill, sometimes into crevices in bark in which they will germinate and establish but sometimes in totally inappropriate places, e.g. on telephone wires. Many birds eject pellets of several seeds as a single unit: the English robin (*Erithacus rubecula*) may eject such pellets; one pellet contained 20 raspberry seeds, 2 seeds of whitecurrant and 3 seeds of redcurrant (Lack, 1965). The effect of this type of behaviour is to concentrate seeds in small groups and the behaviour pattern of the animal may result in the faeces or pellets themselves being locally concentrated. Seeds of the yew (*Taxus baccata*) are dispersed by many of the thrush allies; these have been observed feeding on the yew trees, but perching on beech trees in the neighbourhood while they deposited faeces. Individual droppings contained 6—12 yew seeds which eventually fell or were washed to the ground under the beech trees (*Fagus sylvatica*) (Mountfort, 1957). Similar localized seed droppings have been found from birds feeding on *Podocarpus* species in New Zealand (Beveridge, 1964). Faeces of pigeons contained up to 100 seeds of *P. dacrydioides* or 12 seeds of *P. spicatus*. After feeding, the birds flew to other trees, often of different species, and dropped dense deposits of faeces largely composed of sound podocarp seeds. (A single dropping from a bird captured in the forest yielded 178 *Fuchsia* seedlings.) In the same area in which such active dispersal was occurring there was also massive seed predation; in one month up to 40% of the seed of *Podocarpus totara* was destroyed by parakeets.

Unravelling the dynamics of the seed rain is an important and relatively neglected, though difficult field of research. Presumably seed storage by animals, by which a number of seeds are placed in small groups, is generally a disadvantage in seedling establishment, producing conditions in which the seedlings will enter into a very intense struggle for existence with their close neighbours. Such aggregation is, however, almost always found where seed is buried by animals for subsequent retrieval.

> In recently cut ground (on Wicken fen) several clumps were seen of tightly packed *Rhamnus frangula* seedlings of about 10 cm high. Dr A. S. Watt suggested that these might be mouse stores, and digging showed at once that this was indeed what they were. Collections of stones, buried and forgotten or deserted

had germinated together, piercing the few centimeters of peat from the mouse run to the surface. There were 30–50 seedlings in each small clump.

(Godwin, 1936)

The larger ruminants distribute seeds in local concentrations — seeds that are eaten with herbage may pass unharmed through the digestive tract to be deposited in faeces. Hansen (1911) recorded that a cow grazing on a weedy field consumed in a day 89 000 seeds of *Plantago* spp. and 564 000 seeds of *Matricaria chamomilla* of which 85 000 and 198 000 respectively were voided at 58% and 27% germination capacity. These voided seeds were of course concentrated in the localized patches of dung within which they remain dormant for longish periods. Few seeds will germinate in dung until it has thoroughly decomposed, though seeds of *Nitraria* germinate readily in the fresh faeces of the emu (Noble, 1975).

Ants play a curious part in the distribution of seeds. The harvester ants (Myrmicineae) gather seeds from many species of plants growing near their nest and store them in underground chambers (Trevis, 1958). The stored seed is eaten by the ants and fed to their young. If seeds germinate in the nest the seedlings are carried out and deposited at the edge of a clear, vegetation-free, courtyard which normally lies around the nest. Some of these seedlings may succeed in establishing: the seeds have been collected from a wide area and deposited in a very narrow area. The vast majority of the collected seeds are eaten or decompose within the underground storage caverns. If any significant seed dispersal is achieved by such predation, it must be exceedingly rare, and at enormous cost to the plant. The harvester ant *Veromessor* collects as many as 15×10^6 seeds per acre per year in the Sonoran desert, about 8% of the total seed present. The harvesting is not random and in one study (Trevis, 1958), 90% of the seed collected was found to be from three genera: *Oenothera, Mentzelia,* and *Malvastrum.* (The great majority of the seeds in the area were of *Plantago* species which tended to be ignored.) The effect on the seed population remaining to germinate therefore was highly selective. Other ant species, however, seem to play a more positive role in plant population dynamics, particularly those species which have developed an attracting mechanism such as elaiosomes (Ridley, 1930). These elaiosomes or oil bodies are eaten by ants but the seed itself is not eaten, though the seeds are often carried about. Most of the studies of plant—ant interactions have been concerned with the morphology of elaiosomes (e.g. Berg, 1975) or the domestic economy of the ants (Briese, 1974). There seems to have been no serious attempt to study the role of seed-eating ants in the domestic economy of the plant populations. That one can imagine a biological advantage to the plant in an animal—plant interaction is no proof that there is an advantage; inability to spot a functional relationship is no reason for denying one. Ant—plant relationships are so complex that it may require cooperative studies between botanists and entomologists to unravel the mutual benefits and damage. It should not be difficult to perform the experiment of removing ants from local areas of vegetation and following changes, if any, in plant establishment.

The role of man in seed dispersal

The part played by man in the widespread dispersal of seeds from continent to continent as well as locally has been well recorded, mainly in relation to the invasion of new continents or islands by species introduced by colonizing man (Salisbury, 1961; Elton, 1958). Even locally, the dispersal of seed by man may create focal points for the development of plant populations. Areas of the Irish peatlands which have been cleared to depths of 3 m or more of their peat, leaving a virtually sterile substrate, are colonized along pathways where seed has been carried on agricultural implements or boots. Populations have subsequently spread from these pathways to become important weeds in these "new" habitats.

The role of man in maintaining a continuous seed rain is particularly developed in agriculture. During the act of crop harvesting, many weed seeds are harvested and, during subsequent preparation of the seeds for resowing, weed seeds may persist together with the crop seed, and subsequently be sown with it. A repeating mixed population of crop and weed is thus created and maintained. Various attempts have been made to break this vicious circle by seed-cleaning techniques of various types from primitive winnowing machines to sophisticated modern machinery (Horne, 1953). These controls have been so effective in some cases that a species like *Agrostemma githago*, common in British arable land in the early twentieth century, is now almost extinct. This annual weed produces large tuberculate seeds which are retained in the capsule until the grain crop is threshed. The spiny tubercles make it easily removed from a cereal grain sample, an Achilles heel in the life cycle of the species. It remains an important component of weed populations in those parts of Europe where primitive agricultural practices continue.

Attempts have been made nationally and internationally to control the seed rain of weeds that may be spread with crop seeds (Horne, 1953; Rollin and Johnston, 1961). A typical regulation is that seed may not be sold (or the weed seed presence must be declared) if the seed of the agricultural crop contains more than 1% by weight of a weed infestation. Nevertheless a cereal crop sown at 30 kg/ha (a normal sowing rate for a small grain cereal), and containing a 1% infestation by weight of weeds, would imply sowing *Galium aparine* at 52 seeds/m², *Polygonum convolvulus* at 27.7/m², and *Raphanus raphanistrum* at 24.5/m². A 1% weed infestation in a pasture grass seeds mixture sown at 2.9 kg/ha would involve sowing *Cerastium vulgatum* at 185 seeds/m², *Plantago major* at 92/m², *Holcus lanatus* at 74/m² or *Rumex crispus* at 12/m² (Horne, 1953). Not only is such a seed rain deposited in an area that has normally been cultivated and prepared for plant growth but the seeds of the weeds are sown together with the crop, usually drilled in and nicely buried for optimal germination and establishment.

The movement of seed after dispersal

Seeds do not necessarily remain where they land after dispersal. The achenes of composites may blow along the ground until they accumulate near obstacles or fall down crevices. Soils that have cracked after drought provide just such crevices and the seeds of composites that fall into the cracks are often held in position by the pappus which remains at the soil surface with the achene dangling into the crack. Seeds of other species have specialized mechanisms which cause them to move horizontally or to bury themselves (*Avena fatua* and *Erodium* spp.). Most information about such post-dispersal movements is anecdotal but in a series of experiments Mortimer (1974) measured the distances moved by marked seeds in the field on four different types of soil surface. He sowed seeds of *Plantago lanceolata*, *Dactylis glomerata*, *Holcus lanatus*, *Arrhenatherum elatius* and *Festuca rubra*. Any seed that moved to the edge of the 0.5 m square quadrats was returned to the centre and its movements were again followed. The soil surfaces had been prepared deliberately to give a range of surface roughness which was measured as the log of the variance of surface height (see Chapter 5). The roughest soil gave a value of 3.04 on this roughness scale and the smoothest surface 0.27. Figures 2/5a–c show the movements of seeds on the soil as

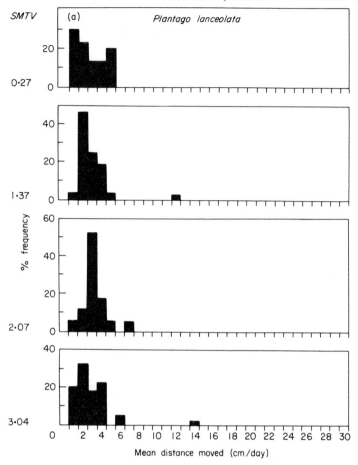

Fig. 2/5a. The movements of seeds on soils of different roughness. The values are expressed as a frequency distribution of movements per seed per day. Soil roughness is measured as SMTV = the variance of the height of the soil surface. (From Mortimer, 1974)

frequency distributions of movement per seed per day. As might be expected, the seeds moved furthest on the smooth soils and tended to remain in position, trapped in crevices, on the rough soils. The species differed sharply: *Plantago lanceolata* which has mucilaginous seeds scarcely moved, whereas *Arrhenatherum elatius* which has a hygroscopic awn was very mobile, moving up to 37 cm per day. *Festuca rubra* had intermediate behaviour and typified the other grasses. On the smoother soils no seeds remained where they landed. Clearly sticky seed traps

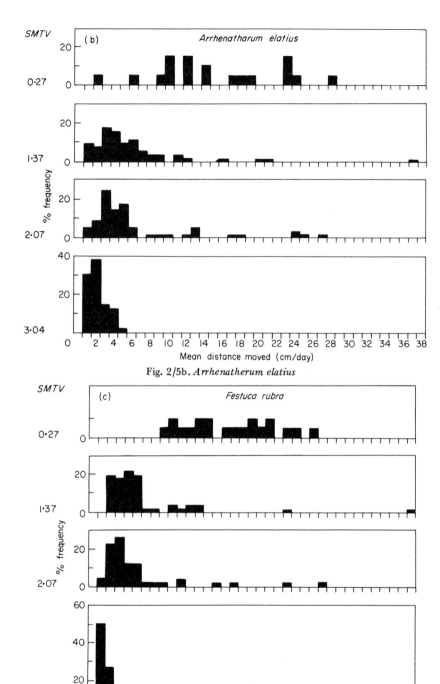

Fig. 2/5b. *Arrhenatherum elatius*

Fig. 2/5c. *Festuca rubra*

would have given a misleading picture of the real extent of seed dispersal. Movements would undoubtedly have been much less except that animals such as ants played a role.

General relationships between seed density and distance from source

Pathologists concerned with the spread of epidemics have been interested in drawing generalizations from a variety of anecdotes. The spread of an infection in a crop has many analogies in the invasion of a habitat by a colonizing plant (see e.g. Chapter 20 for an account of the invasion of land by *Pinus taeda*). For epidemics of crop diseases van der Plank (1960) has found it useful to relate the logarithm of infection density to the logarithm of distance from source (Chapter 16). This has been done for 11 examples of seed dispersal in Fig. 2/6 where the graphs are shown alongside inverse square law and inverse cube lines. Van der Plank has shown that the slope of the line relating disease intensity to distance gives a great deal of information about the way in which an invasion will spread. Generally if the slope is that of the inverse square law or steeper, a population will spread into colonizable territory as an advancing front — "a horizon of infection". The steeper the slope the more sharply defined is the infection front. Dispersal curves with slopes less than that of the inverse square law will lead to a spread-out pattern of isolated colonists which may subsequently act as foci for new infections.

The seed dispersal curves of Fig. 2/6 show a wide variety of slopes. The commonest pattern approaches a cubic relationship (e.g. *Papaver dubium* (8), *Dactylis glomerata* (10), *Dipsacus sylvestris* (a), all of which are species with no special adaptations to wind dispersal. *Picea engelmannii* (5) in one of its sites also followed this pattern. The herbaceous species in this group often dropped very few seeds close to the parent but outside this zone the decline of seed density with distance was always steep.

Poverty of seed fall in the immediate neighbourhood of the parent seems to be characteristic of plants in isolation (cf. *Eucalyptus regnans* (3 and 4), though this was not the case with *Dipsacus sylvestris*.

The species with specialized mechanisms of wind dispersal generally (*Senecio jacobaea* (1) and (2) *Tussilago farfara* (11) and *Picea engelmannii* in site 6) would tend to colonize not as advancing fronts or a horizon of colonization but as isolated individuals over a great distance.

The examples in Fig. 2/6 suggest that isolated plants are unlikely to

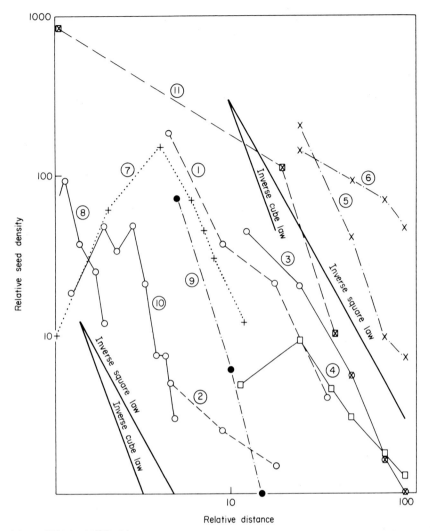

Fig. 2/6. The relationship between seed number and the distance of seed dispersal. The data are plotted on log/log co-ordinates to show the relationship between patterns of seed dispersal and the inverse square and inverse cube laws.

(1) *Senecio jacobaea* — from the edge of a dense population in the direction of prevailing wind (see Fig. 2/2b)
(2) as (1) but against prevailing wind (see Fig. 2/2b)
(3) *Eucalyptus regnans* — from the edge of a dense forest (see Fig. 2/2e)
(4) as (3) but from an isolated tree
(5) *Picea engelmannii* — from edge of a forest clearing (A in Fig. 2/2d)
(6) as (5) — from B in Fig. 2/2d
(7) *Verbascum thapsus* — from an isolated plant (see Fig. 2/2f)
(8) *Papaver dubium* — from an isolated plant (see Fig. 2/2g)
(9) *Dipsacus sylvestris* — from isolated plants (Werner 1975)
(10) *Dactylis glomerata* — from isolated infructescences (Mortimer, 1974)
(11) *Tussilago farfara* — from the edge of a dense population (see Fig. 2/2c)

form horizons of colonization but that dense stands will colonize as an advancing front (cf. 5 and 6); also that spread in the direction of the prevailing wind may be as a horizon but against the wind is more likely to be as scattered individuals. These generalizations are made more as hypotheses than as evident truths. There are far too few studies of dispersal for valid generalizations to be made at this stage but van der Plank's approach to the problem of epidemic spread provides a stimulating model against which new observations can be compared and new studies designed. What, for example is the shape of the dispersal distance curve for animal-dispersed seeds?

The seed rain as a process in time

The dispersal of seeds is usually seen as an area phenomenon and is studied in two dimensions, but there is also a depth and a time dimension to dispersal. Seed that lands on a soil surface may stay on the surface or be distributed down a soil profile. This depth aspect of dispersal is considered in Chapter 4.

The time element in the intensity of a seed rain derives in part from differences in the time at which seed is ripened and in part from differences in the time of retention of ripe seed before it is dispersed. The optimal times for flowering and for seed germination are often in very different seasons: the seed may be released as it ripens, held on the plant and then released in a sudden burst or in a slow sequence over a long period. Not much attention has been given to the biological significance of the timing of seed release. It is clearly under genetic control, for one of the regular consequences of the domestication of plants is that seed is retained by the parent plant and not dispersed.

Many of the species characteristic of disturbed habitats, particularly horticultural weeds, possess a long flowering period and a correspondingly long period of the year when seed is ripened and released, and in these species the seed is usually released immediately it is ripe. In contrast the weeds of arable land usually concentrate the period of flowering and seed production within a narrower season, and in some cases require the act of harvesting and threshing for full release of the seeds. These are weeds that have evolved a dispersal habit parallel to that of the crop and maximize the chance of dispersal with the crop seed. Within a single community the seed rain varies in its species composition through the year. This is illustrated for five grasses in a single meadow plot in Fig. 2/7. Both flowering and seed production of *Poa annua* and *Senecio vulgaris* may occur at all months of the year in Britain. These are inhabitants of very unpredictable environments (see Chapter 21); others, for example *Taraxacum officinale* and *Bellis perennis*, have a peak of flowering and seed production but the process continues at lower intensity over a period of 6—8 months, and are plants from

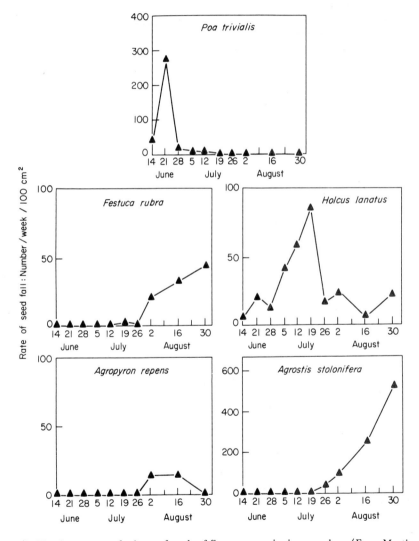

Fig. 2/7. The time course of release of seeds of five grass species in a meadow. (From Mortimer, 1974)

environments of higher predictability. Generally, in northern temperate regions, the later successional species tend to have narrower flowering and seed ripening seasons, often under precise photoperiodic control. One consequence of this is that in the season of seed production there is a sudden richesse of food for predators which may have been limited by shortage of food at other times of the year.

That seed may be ripened during a relatively short period does not necessarily imply that it is released and dispersed in a seed rain in a correspondingly short period. Figure 2/8 shows the spread over time of

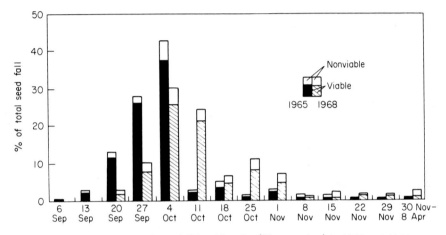

Fig. 2/8. The seasonal pattern of seed fall in white pine (*Pinus strobus*) in 1965 and 1968. (From Graber, 1970)

seed release in white pine (*Pinus strobus*) in two years of heavy seed fall. Other pines, e.g. *Pinus radiata, P. cembra*, may retain seed within the cones on the tree for 5 or more years with no loss of viability (Fielding, 1965) before seeds are released. In moist, cool climates a great proportion of the cones is retained on the tree. After a forest stand has been destroyed by fire the cones open quickly and liberate seed. There is clearly adaptive value in such behaviour because most cones and seeds on the ground are destroyed during fires (Fielding, 1964).

Seed retention by the parent plant can be significant on a shorter time scale. The annual species of poppy which are common arable weeds in Britain, include species such as *Papaver rhoeas* with small seeds and large pores in the capsule and which liberate almost all the seed within a day or so of the capsule ripening (Harper, 1966). Other species, such as *Papaver argemone*, have larger seeds, small capsule pores and a

lengthy period of seed release; seed may still be in the capsules the following year. The timing of seed release in the poppies is not related to the
time of seed germination. A clue to the significance of seed retention
and dispersal may be that the cultivated poppy *Papaver somniferum* has
lost the property of dehiscent capsules and the seeds are retained until
man harvests and threshes them. Unfortunately for the farmer, birds
find a capsule full of seeds a choice food supply and bird damage is a
major hazard in the cultivation of this crop. *Papaver rhoeas* escapes
predation because the capsules are empty almost as soon as they are
ripe. *Papaver argemone* retains seed but the capsule is spiny and birds
do not attack this species. Occasionally spineless forms of *P. argemone*
occur and these are then taken by birds. It seems likely that the predation by birds may account for some of the dispersal characteristics of
the poppies.

The selectional dilemma of whether to disperse or retain seeds is neatly illustrated by comparing the hazards to the seed crop on the plant
and on the ground. In a stand of white pine in Maine, of an average of
804 cones per tree 44 were killed by insects on the tree (white pine cone
beetle, *Conophorus conipenda* and Cone Moth, *Dioryctria abietella*),
and squirrels cut and destroyed 29 cones. The remaining 731 cones
released seed and these were then almost all eaten on the ground by
blackbirds (*Euphagus carolinus*) in September and October. Voles and
mice were also thought to collect the seeds (Graber, 1970). Natural
selection might well be expected to optimize the timing of seed fall in
relation to the chance of survival — from seed held to seed dispersed
may be "out of the frying pan into the fire". In some situations release
of all seed in a short time may oversaturate predators' demands and be
the optimal strategy for leaving survivors (see Chapter 15). In some
situations a steady slow release of seed may be optimal, maximizing
the chance that some seed lands on the ground during a few favourable
days and escapes predators by quick germination.

Selection of dispersal characteristics

In a discussion of the selective advantages of dispersal mechanisms it is
desirable to be able to quote evidence that the behavioural and morphological characters associated with dispersiveness are heritable and can be
selected. Some of this evidence comes indirectly from the change of
plants under domestication. Dispersal of seed is clearly a quality selected

against by man in harvesting crop plants. He only harvests what has not been dispersed. The loss of shattering properties in the Gramineae cultivated by man as cereal crops, and the loss of capsule dehiscence mechanisms in cultivars of *Linum usitatissimum* and *Papaver somniferum* are indirect evidence that dispersive qualities can be selected against.

Evidence that dispersiveness can be favoured by selection is less easy to find. Sakai (1958) was able to select fugitive qualities in *Drosophila*, the ability to escape from crowded cultures. It would be interesting to attempt a comparable selection programme with plants. In Poole and Cairns' (1940) study of dispersal in *Senecio jacobaea* they were able to show that the mean seed weight of seeds which were dispersed furthest was less than that of the seeds which fell close to the parent. This evidence is incomplete without a demonstration that seed size in *Senecio jacobaea* is heritable, but in general seed size has relatively high heritability (see Chapter 21).

Perhaps the most beautiful example of the evolution of a seed dispersal habit is that of *Camelina sativa* in which local races have evolved with dispersal habits that ensure that the seeds stay with the crop plant, flax or linseed. Divergence in the weight of seed and its degree of winging can be directly related to the seed behaviour of the crop in the winnowing machines (Stebbins, 1951).

The senses and mobility of animals permit search for food, homes and safety. Techniques of direct search and escape are barred to the higher plant. Instead plant dispersal is by crude scattering or involves the animal. In most cases the animal claims its reward by eating seeds, or is "paid" in elaiosomes, fleshy fruit walls, etc. The animal can search for the plant, but the plant cannot search to escape. The number of seeds left to germinate after the animals (dispersers and non-dispersers alike) have taken their fill has to be explained in terms of the animals' behaviour (see Chapters 13 and 15). It is relevant in all this complexity that the average plant, like the average animal, leaves on average just one descendant!

3

Dormancy

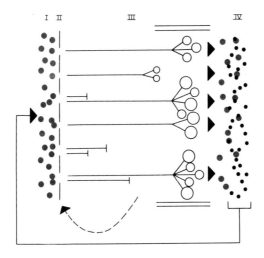

In the life cycles of most organisms there is a stage of rest, a period when metabolism runs slowly. The resting period usually occurs during unfavourable conditions when the organism opts out of its normal active life and enters a phase more resistant to environmental hazards.

Among the variety of animal species dormant stages can be found in a variety of positions in the life cycle. In the larger mammals there is usually a daily sleep period and in small mammals there is often a seasonal period of dormancy. Rest stages in insects may be found at every stage of the developmental cycle: diapause occurs in eggs, larvae, pupae or in the imaginal stage.

In plants the dispersal phase is usually dormant. Dispersal, particularly by wind, is easier if the structures have little weight and seeds are usually dehydrated before they are dispersed. The dehydrated systems have very slow metabolism. Dormancy in plants has been defined as "a state in which viable seeds spores or buds fail to germinate under conditions of moisture, temperature and oxygen favourable for vegetative growth" (Amen, 1968). The phase of dormancy may be a seasonal phenomenon in the life of a perennial plant — a rhythm of meristematic activity in which periods of active leaf production alternate with periods of rest. Such a rhythm occurs even in the relatively constant environments of tropical forest where the rhythm may not be strictly seasonal and where different shoots on the same plant may sometimes be in different phases of the cycle of activity and quiescence. The dormant phase may involve the loss of all aerial parts so that the unfavourable season is spent as dormant meristems underground (or under water); or it may involve the loss of leaves while dormant meristems remain more or less protected above ground (the deciduous habit); or the leaves may remain present and functional but the meristems pass through a period in which no new leaves are developed.

For perennial plants the seed is an alternative means of perennation but this role is not so critical as it is in annuals where the dormant seed is the crucial link between generations.

A phase of dormancy in the life cycle is just one way in which an organism can survive through a seasonal environment — in a sense it is a weak solution to the problem of adaptation to a changing environment because the time spent in the dormant state is time lost in capturing resources, time lost in a struggle for existence with neighbours. An alternative strategy of life in a seasonal environment is two distinct phases of growth each adapted to a specific season. Such somatic polymorphism is found in some desert shrubs: a relatively large leaf is formed in the wetter season and is abscissed and replaced by a foliage of small leaves or scales during the drier season. Such species have a seasonal cycle in the transpiring and assimilating area. Good examples of this alternative to dormancy occur in *Thymus capitatus, Poterium spinosum, Teucrium polium, Artemisia herba-alba* and *Zygophyllum dumosum* (Orshan, 1954, 1963; see Fig. 3/1).

The existence and duration of a dormant phase in the life cycle are

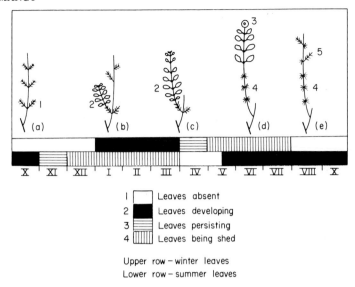

1	Leaves absent
2	Leaves developing
3	Leaves persisting
4	Leaves being shed

Upper row – winter leaves
Lower row – summer leaves

Fig. 3/1. Seasonal dimorphism in *Teucrium polium*.
  (a) The lower part of a dolichoblast developed during the previous spring bearing small partial brachyblasts (1).
  (b) and (c) One of the brachyblasts of the same branch turning into a new dolichoblast bearing winter leaves (2).
  (d) The same dolichoblast reaching its maximal height. The older winter leaves have already been shed and new brachyblasts (4) have developed in their axils. (3) The inflorescence.
  (c) All the winter leaves already shed. The upper brachyblasts slowly elongating (5).
  (From Orshan, 1963)

presumably evolved in response to (i) the probability of suffering greater hardship by continuing growth, (ii) the effort or cost of a seasonally dimorphic phenotype and (iii) the cost of producing a homoeostatic growth form that is tolerant of the whole range of environmental conditions that are experienced. There is some evidence that the costs of wide tolerance are great — the ability to function in a wide range of environments is only to be obtained by reducing efficiency at the optimum (Sacher, 1967; Levins, 1968). Extreme examples of solutions to the problem of life in a seasonal environment would be the annual habit (the extreme of solution (i) above) the somatic polymorphism of leaf habit (solution ii) and the evergreen habit (solution iii).

Any adaptation to a seasonal phenomenon requires that the behaviour of the organisms is synchronized with the changes of the seasons. The breaking of winter dormancy might be brought about as a reaction to the warming of the environment — for example seeds germinate when the temperature of the soil rises above a critical level. This is dangerous behaviour if the seasonal rhythm of the environment has unpredictabilities (environmental noise). The warming of the soil is a particularly unreliable indicator of the arrival of spring because a few days of precocious warm weather may be followed by a killing frost. Similarly in many environments the supply of water is uncertain and seeds that are triggered into germination by an early rain after the dry season may die if the next rain is delayed. If germination is clocked by some other seasonal stimulus that is reliable (e.g. photoperiod) evolutionary experience can optimize the end of dormancy to coincide with the best chance of (a) survival and (b) an adequate length of growing season.

There is a further advantage in a dormancy-breaking mechanism that depends on an external clock. If the organism reacts to favourable conditions when they appear, it will always be too late because development takes time. A plant that is transferred from full light to shade develops a changed structure of leaves and an altered ratio of leaf area to dry weight (Blackman and Wilson, 1954) but while this new foliage is being produced the plant is in an unadapted state. Triggers to development which predict a changing environment will generally be more efficient than those that are themselves the changed conditions. There is therefore "no necessary relation between the physical form of the signal and the response evolved. . ." (Levins, 1968).

The solar clock has been adopted as the external stimulus for many of the developmental events associated with dormancy. The disadvantage of such clocked prediction is that the organism may fail to be in the optimal stage of development in a precociously good season but in compensation it is guarded against the hazards of the late lethal season. How far it is guarded against the latter may be selectively built into the clocking response. Presumably the optimal clock time for a response (such as germination) is determined selectively by the reward of leaving a few more descendants in the good seasons and the cost of being caught in the wrong state in a bad season.

The less seasonally regular is the environment, the less efficient is the solar clock at predicting appropriate development. Thus desert rainfall or the melting of a snow cover though crudely seasonal, are often highly erratic in time. The occurrence of a forest fire or a hurricane may make conditions suitable for plant growth where it was unsuitable before, but these are events unpredictable from year to year. Similarly a seed buried in the soil may be too deep for it to establish a seedling successfully and events which may bring it to the surface are again unpredictable from year to year. In such situations an external clock controlling dormancy is valueless and the breaking of dormancy is usually a direct response to the favourability of environmental conditions.

In an environment in which favourable growth conditions are seasonal, dormancy is usually clocked by solar rhythm. This may be called *"seasonal dormancy"*. When there is only a small seasonal element in the occurrence of favourable conditions (or their occurrence is spatially rather than temporally variable) dormancy tends to be both imposed and released by the direct experience of the unfavourable or favourable conditions *"opportunistic dormancy"*. The populational consequences of *"seasonal"* and *"opportunistic"* dormancy are quite different — the former is predictive, the latter responsive. Opportunistic dormancy can often be imposed on seeds with a seasonal germination if they fall in an unfavourable environment.

It may be said that "some seeds are born dormant, some acquire dormancy and some have dormancy thrust upon them" (Harper, 1957). The three categories represented by this statement may be called innate, induced and enforced dormancy. *Innate dormancy* is the condition of seeds as they leave the parent plant and is a viable state but prevented from germinating when exposed to warm, moist aerated conditions by some property of the embryo or the associated endosperm or maternal structures. *Induced dormancy* is an aquired condition of inability to germinate caused by some experience after ripening. *Enforced dormancy* is an inability to germinate due to an environmental restraint — shortage of water, low temperature, poor aeration, etc. These categories of dormancy are not completely tidy but are useful particularly in a populational context. Induced and enforced dormancy are "opportunistic".

Innate dormancy

Much of the vast literature on the subject of innate dormancy is part of plant physiology, and is a sophisticated and well reviewed subject. Innate dormancy can itself be conveniently subdivided into four main types.

Incomplete development

The process of growth of an embryo to a stage fit for the germination process to occur, may not have been completed while the embryo was still borne on the parent plant; what is shed is morphologically incomplete. An example of this type is found in *Heracleum sphondylium* where the development of the embryo continues at the expense of extra-embryonic reserves for several months after the seed is shed. This process imposes a necessary time lag between dispersal and germination.

Control by a biochemical trigger

A biochemical process may need to be stimulated before the germination process can begin. Often this trigger to the germination process is a seasonally related stimulus which can switch on the germination process at an adaptively appropriate time of year. Photoperiodically operated triggers act through modifications of the phytochrome system (see discussion by Black, 1969). Seeds of *Betula pubescens* require light and long days for successful germination after the seed is shed (Black and Wareing, 1955). In a normal 24 h cycle with 4 h of light and 20 h of darkness about 30% germination was obtained, but in a rhythm of 20 h of light and 4 h darkness about 90% germination occurred. Although light is required, the length of the dark period is also important and when seeds were exposed to either 24 h or 8 h periods of light separated by different lengths of dark period, the percentage germination declined progressively with increasing dark period. Even very short periods of light were sufficient to stimulate germination, provided that the dark period was not long. In the field, where only variations within 24 h cycles occur, the significance of this particular trigger system is presumably that few seeds of *Betula pubescens* germinate successfully in short autumn days.

The sensitive and subtle light dependency of *Betula pubescens* is further complicated by a temperature dependence; at 20°C the photoperiodic effect is lost and a high rate of germination is maintained in both long and short days, even after a single exposure to light. Moreover, if the seed has been chilled, the light requirement is lost. Variations on the *Betula* theme have been found in a number of species, for example a wide range of dicotyledonous herbs (Isikawa, 1954).

Other stimuli to germination in seeds with innate dormancy include temperature effects, for example chilling or a fluctuating temperature. In many cornfield weeds of the genus *Papaver* a diurnal fluctuation between 10° and 30°C breaks the dormancy of a large proportion of the seeds. Alternations between 20 and 30°C or between 10 and 20°C are less effective, but usually break the dormancy of a few seeds in the population. An alternation between 10 and 30°C is not likely to occur in the upper layers of the soil in British cornfields until late April or May and so fixes the time of germination of the majority of the seeds (McNaughton, 1960).

It is important to notice that in most experiments on the breaking of innate dormancy, the treatment does not produce an "all or nothing" effect. Even in unfavourable photoperiods, a proportion of the seeds of *Betula pubescens* germinated. There is clearly a spectrum of requirements by the seeds in a single sample which may reflect different genotypes, different maternal influences and perhaps different ages and ripening conditions. The physiologist may often stress the precision of a dormancy-breaking mechanism, whereas to the population biologist the "error" or "noise" in the process may be more interesting.

Various germination stimulants are effective in laboratory practice and may cast light on the mechanisms of biochemical control of germination but seem to have little relevance to field conditions. Treatments with gibberellins or thiourea come in this category. However, one commonly used laboratory stimulant to germination, the nitrate ion, may be more relevant. The nitrate concentration of the soil solution often rises quite sharply as the soil temperature increases in spring (Russell, 1962). It may be that the stimulation of germination of seeds of *Chenopodium album* and a number of other species by nitrate is involved in the timing of field germination.

The removal of an inhibitor

There is no very clear division between the triggering of a biochemical process which may destroy a germination inhibitor and the physical leaching, or removal of the source of, an inhibitor. It is, however, convenient to regard the action of chemical inhibitors of germination which may be leached or destroyed by external agents as distinct from a breakdown process which occurs within the tissues of the seed. Substances that inhibit seed germination *in vitro* have been isolated from a very large number of seeds and fruits (Evenari, 1949), particularly from the fleshy fruit coats or receptacles of the Rosaceae: the presence of an inhibitor can often be detected simply by leaching seed samples with water when germination may proceed rapidly, but germination is inhibited if the leachates are added back to the seeds.

The physical restriction of water or gas access

The presence of an impermeable or relatively impermeable seed or fruit
coat may prevent water uptake by seeds and so prevent germination
until physical damage has been done to this barrier. The seed dormancy
of *Avena fatua* can often be broken simply by pricking the pericarp.
This type of dormancy is particularly common in species that inhabit
sand dunes where the abrasion is by the movement of sand. Scarification
is also a common requirement by the seeds of alpine plants (Amen,
1966). Seeds that require abrasion tend to break dormancy at different
times rather than in a sudden flush.

The genetic control of innate dormancy

Innate dormancy often appears to be under strict and simple genetic
control though sometimes modified by a maternal effect (not surprisingly, in view of the large maternal contribution to the endosperm and the
totally maternal origin of the ovary). In *Nicotiana tabacum* the need for
light to promote germination was found to be dominant to light-indiffer-
ence. Reciprocal crosses all required light but there were differences in
sensitivity influenced by the maternal parent (Honing, 1930).

The wild ancestors of most commercial crops bear seeds with innate
dormancy which has been lost in the process of domestication
(Schwanitz, 1957), though a degree of seed dormancy has been conserved as protection against germination in wet weather while the seed is
still in the inflorescence. Wild populations of *Papaver rhoeas* bear *ca* 97%
of dormant seeds but cultivated forms normally have no dormancy.
Among seed populations of this poppy collected in the wild *ca* 3% of
the seeds germinate readily direct from the capsule. These seeds often
germinate in the field in autumn and are normally killed by winter frosts
or by cultivations. If they survive the winter they form overwintering
rosettes of leaves and develop to very large plants the next spring.
Autumn germinated plants of the closely related *Papaver dubium* produce about 21 times as many seeds as the commoner spring germinating
forms (Arthur, 1969; Arthur *et al.*, 1973). The presence of the non-
dormant forms in the wild populations is a genetic polymorphism, i.e.
two alternative genetic forms each proving its fitness in a different
"patch" in a temporally uncertain environment.

Further evidence of genetic control of dormancy in *Papaver* comes
from interspecific crosses. *Papaver dubium* and *P. lecoqii* yield 0.3—

1.3% germination without pretreatment but the interspecific hybrid gives 81.0–98.7% germination under the same conditions if *P. lecoqii* is the female parent. Similarly *P. argemone* x *P. apulum* produced non-dormant seed with *P. argemone* as female parent although both species produced mainly dormant seed (Harper and McNaughton, 1960).

Nicandra physaloides possesses one pair of iso-chromosomes. The presence or absence of an isochromosome determines whether a seed will germinate readily (2n = 20) or is dormant (2n = 19). A botanic garden stock of this species would have disappeared had it not been for the appearance of some of the chromosomally deficient types which had survived dormant and buried for 28 years (Darlington and Janaki-Ammal, 1945).

Genetically controlled dormancy polymorphism is probably the exception rather than the rule among plants but another example illustrates its possible significance. In *Spergula arvensis* a seed coat character is genetically controlled and associated with germination. Three seed forms are found in nature: a homozygous papillate form bearing about 120 papillae, a homozygous smooth coated form and a heterozygote in which the seeds bear 60 papillae (New, 1958, 1959, 1961). The seed coat is of course maternal and an individual plant bears seeds all of the same form. There is a definite cline in the proportions of the forms across the British Isles, the non-papillate forms being more frequent in the north. A similar cline occurs across Europe. The papillate seeds germinate more readily than the smooth at 21°C and the reverse is true at lower temperatures.

Somatic polymorphism and innate dormancy

"Somatic polymorphism" implies the production of seeds of different morphologies or behaviour on different parts of the same plant — not a genetic segregation but a somatic differentiation. It represents an allocation of different fractions of the seed output to different ends. In *Xanthium* species, for example, the seeds are borne in pairs, so that a large and a small seed are joined and dispersed together. The dormancy-breaking requirements of the two seeds are different, with the result that at least 12 months normally separate the germination of the two seeds in each dispersal unit. This is an obvious insurance strategy and if one year provides a lethal environment there is a second batch of seeds waiting for the next season. Similarly in *Avena fatua* and *A. ludoviciana*, the grains borne on different parts of the individual spikelets have different germination requirements. This was demonstrated by putting paint spots on the different grain before sowing and following their pattern of germination: perhaps the first use of "mark—recapture" techniques in a study of plant populations. In *Avena fatua* var. *septentrionalis* the first grain of the spikelet lacks dormancy and the remainder have deeply dormant seeds (Thurston, 1957).

As its name implies, the composite genus *Dimorphotheca* bears within a single capitulum two sizes of achene, the larger borne by the disc florets and the smaller by the rays. Again there is a separation of germination time as there is also in *Synedrella* in which the smaller achenes are produced by the disc florets (Salisbury, 1942). Similar differentiations between seed size and behaviour are common in the Compositae, e.g. *Calendula, Bidens, Crepis*; the very structure of a capitulum lends itself to diversification of the flowers and seeds borne within it. In *Gymnarrhena* cleistogamous and exposed flowers produce different sized seeds. The cleistogamous flowers are produced below the soil and bear 1—3 large achenes (6.5 mg) which lack a pappus. The open flowers produce many small seeds (0.37 mg) bearing a pappus. This is one of the most extreme of the divided strategies that somatic polymorphism implies. Moreover, in this case the allocation of resources to the two seed types is plastic — in seasons of low rainfall only cleistogamous flowers are produced and dispersal is sacrificed to the need to leave a few large "safe" seeds sown close to the parent (Koller and Roth, 1964). That the large seeds do indeed represent a safe investment is shown by withholding water or giving only limited daily irrigation. Mortality is then concentrated almost entirely in the plants from small seeds (Table 3/I).

Many species of the Chenopodiaceae show somatic polymorphism in their seeds. Individual plants of *Chenopodium album* may produce four different types of seed in two colour categories (black and brown) and two seed coat categories (reticulate and smooth). The brown seeds are thin-walled and larger than the black and germinate quickly, even at low temperatures, when they are dispersed. The black seeds require a cold experience or the supply of nitrates before they will break dormancy (Williams and Harper, 1965). The seedlings from brown seeds are commonly killed by winter cold or in agricultural cultivations but, if they survive, produce very large plants with a higher reproductive output than the plants from black seeds which germinate in spring. There is a cline in the proportions of brown and black seeds produced by plants

Table 3/I

The death of seedlings of *Gymnarrhena micrantha*
during drought. (Koller and Roth, 1964)

(a) water was withheld for various lengths of time
and then irrigation was resumed.

	Duration of interruption (days)			
	1	3	5	7
Dead seedlings from aerial achenes	7	17	21	24
Dead seedlings from subterranean achenes	0	0	0	9

(b) varied amounts of water were added to the
soil daily, 20—60 g water/400 g soil.

	Water content (ml per container)			
	20	25	45	60
Dead seedlings from aerial achenes	13	10	1	3
Dead seedlings from subterranean achenes	3	2	0	0

across the British Isles which strongly suggests that the polymorphism is
selectively favoured. It is rare for the brown seeds to exceed 3% of the
total produced by a parent and most of these are among the first seeds
to be ripened by the parent. Probably the ratio of the brown to black
seeds is environmentally governed as in other genera such as *Halogeton*
and *Atriplex*. In *Atriplex heterosperma* small black seeds are produced
early in the season followed by large brown seeds and in *Halogeton*
brown seeds are produced during long days but when the plants are
transferred to short days black seeds are produced. Again there is a
difference in the dormancy-breaking requirements of the different
colour morphs (Frankton and Bassett, 1968).

Seed size polymorphisms are found in the Cruciferae, e.g. *Aethionema*
which produces indehiscent 1-seeded and dehiscent many-seeded siliquas
(Zohary, 1962). Somatic seed polymorphism appears to be largely
restricted to relatively short-lived, fugitive species, particularly weeds.
Presumably in a long-lived perennial the risk of a poor season for seed-

ling establishment is relatively unimportant. Most of the examples of
the phenomenon are found in the four families Compositae, Chenopodiaceae, Gramineae and Cruciferae, in all of which the floral structure or arrangement of the inflorescence lends itself to a division of
labour. It is difficult to find any phenomenon in the animal kingdom
which corresponds to somatic polymorphism except perhaps the
differentiation of castes in the colonial Hymenoptera.

Somatic polymorphism permits a degree of sensitivity in adjustment
of the proportion of morphs which is lacking in genetic polymorphisms.
The proportion of morphs can be altered directly by environmental
influences on the seed parent and the allocation of resources to the
two seed types can be subtle and directly responsive to environmental
change. Moreover some genetic control is possible — but control of a
continuum of possibilities, not just a few alternatives. The proportion
of disc and ray florets in the Compositae can be changed by selection
and has of course been so changed by plant breeders interested in
producing decorative forms in the Compositae. However, in a somatic
polymorphism the proportion of, for example, ray and disc florets
produced on a plant derived from sowing a seed from a ray floret is
the same as the proportions obtained by sowing a disc achene. This
gives the polymorphism a degree of buffering against sudden selective
forces that may temporarily favour one or the other morph.

Seed and dormancy polymorphisms are so common, at any rate
amongst weedy species, that it becomes dangerous to ascribe to a
species any one germination requirement. Very commonly the seed of
a species is tested in a variety of dormancy-breaking regimes until one
is found which stimulates as many as say 60% to germinate; this is
then described as the characteristic pattern for that species. It is rare
to ask questions about the remaining 40%: are they alive; what is
their dormancy-breaking requirement? This problem is illustrated by
some experiments on the germination of *Rumex crispus* which is often
described as bearing light-demanding seeds. In a comparison of the
seeds from different plants in the same site and from different sites,
three laboratory environments were used: (i) darkness and a constant
temperature; (ii) darkness and alternating temperature; and (iii) light
and alternating temperature. Although most seeds from all plants germinated in regime (iii) the progeny of individual plants varied enormously (Fig. 3/2) in their ability to germinate in darkness or at constant temperature. A little of this variation could be ascribed to the

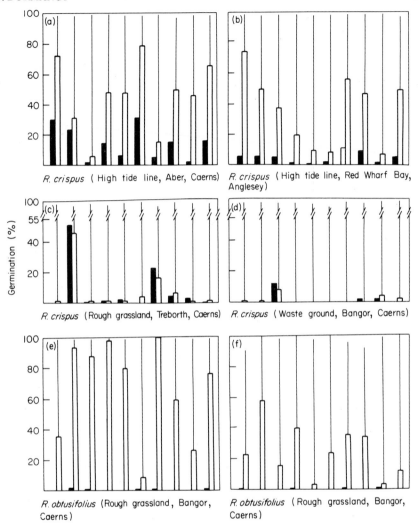

Fig. 3/2. The percentage germination of seeds from ten different plants in each of six different habitats tested under contrasting germination conditions. Vertical lines: Light, alternating temperature; Black bars: darkness, constant temperature (20°C); White bars: (a–d) darkness, alternating temperature (10–20°C), (e) and (f) darkness, constant temperature (20°C) interrupted by one brief exposure to daylight. (From Cavers and Harper, 1966)

ripeness of seed or its position on the plant but the majority of the variation was between plants within a small habitat sample and the variation was greater between plants than between habitats (Cavers and Harper, 1966). This species has no one characteristic germination response. Similar diversity has been found in *Calotropis procera* between the seeds of phenotypically similar plants (Sen *et al.*, 1968). Of course the general tendency to study the behaviour of bulk samples of seed completely hides this type of phenomenon.

Levins (1969) has an interesting treatment of the population biology of dormancy which is based on the concepts of environments as fine or coarse grained. The grain describes the way an individual experiences the heterogeneity of the environment. The heterogeneity may be good and bad seasons, lethal or favourable events, good or bad years. In a fine-grained environment the individual is exposed to the various environmental factors in small doses, "short term fluctuations or wanderings through many patches of an environmental mosaic. In the limiting case each individual experiences the same range of environments in the same frequencies and there is no uncertainty." This would be true of the variations in climate between years for a long-lived tree species which in a lifetime would sample most of the possible range of variation. "In a coarse-grained environment, each individual spends its whole life, or a long period at least, in a single alternative, either because the environment varies from generation to generation or because the spatial patches of the environmental mosaic are large."

Analyses of optimal strategies in a heterogeneous environment disclose several different types: (i) If the alternative environments (say the different types of weather from year to year) are not too different in the way they affect a particular biological system the optimal strategy is a single type best suited to some intermediate environment.

(ii) If the alternative environments are very different and the environment acts in a fine-grained way the optimum is a single type specialized to the more common environment. If the frequency of the alternative environments changes, the optimal form may shift abruptly.

(iii) If the environments are very different and act in a coarse-grained manner, the optimum strategy is polymorphism. Seed dormancy in the fugitive annuals of disturbed habitats and many desert species adapted to uncertain rainfall commonly represents this type of compromise.

Of course in wholly homogeneous environments dormancy has no rôle and in so far as homogeneity is ever found in nature, some of the mangroves of tropical maritime communities best exhibit this state. The processes of seed development and germination are continuous — the seedling falls from the parent directly into a predictable environment and simply continues growth.

Enforced dormancy

A state of seed dormancy may be maintained in or on the soil or submerged in water by the absence of necessary conditions for early seedling growth — a shortage of water, an unsuitable temperature or an unfavourable atmosphere. Such seeds may be said to be enforced in dormancy.

Carbon dioxide narcosis is commonly thought to be one of the factors responsible for enforcing a state of dormancy on seed buried in soil. In practice it is not easy to determine CO_2 tensions at the level that might be important, i.e. within the seed coat and in contact with the embryos. Equilibrium tensions in the pore spaces of the soil are probably irrelevant, for the respiring seed will be a source of CO_2 and the highest concentrations are likely to occur within the seed itself. When seeds are exposed to varied partial pressures of CO_2 the external concentrations needed to inhibit germination are high compared to those commonly found within soil pores, e.g. $0.1-1.0\%$ in forest litter (Brierley, 1955), $0.1-1.6\%$ in arable and pasture soils (Russell, 1950).

Lowered oxygen tension has also been held responsible for the dormancy of buried seeds and there are a number of laboratory demonstrations that severe oxygen starvation may maintain a dormant condition (Fig. 3/3). Again, however, measured values in soil pores only very exceptionally fall below 18% (Russell, 1950) and it is difficult to make the requisite accurate estimates of O_2 tensions within the seed coat. On present evidence, gaseous inhibition of germination seems insufficient to account for the massive accumulation of viable but dormant seeds found under most vegetation (see Chapter 4).

Induced dormancy

Induced dormancy implies that a seed has, through some experience after release from the parent, acquired a dormancy which was not innate and which does not require continued enforcement — the seed has changed its dormancy state. The classic study of induced dormancy is the work of Kidd (1914) and Kidd and West (1917) who studied the role of CO_2 narcosis on the seed of several species. They found that such narcosis was normally reversed or removed on exposure of the seed to normal air (i.e. dormancy had been enforced) but in *Brassica alba* the carbon dioxide experience left the seeds in a dormant condition requir-

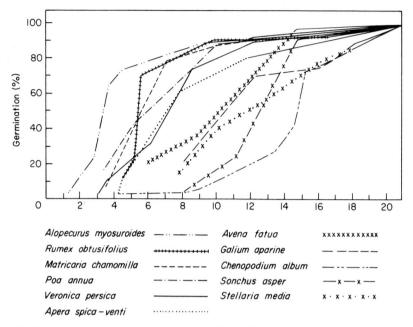

Alopecurus myosuroides	— ·· — ·· —	Avena fatua	xxxxxxxxxxxx
Rumex obtusifolius	++++++++++++++	Galium aparine	— — — —
Matricaria chamomilla	— — — — —	Chenopodium album	— ·· — ·· —
Poa annua	— · — · — ·	Sonchus asper	— x — x —
Veronica persica	————————	Stellaria media	x · x · x · x · x
Apera spica – venti	················		

Fig. 3/3. The germination of seeds of various weeds in relation to the partial pressure of oxygen. (From Müllverstedt, 1963)

ing special treatment such as the removal of the testa or drying and re-wetting before the seeds would germinate.

Kidd studied induced dormany as a field phenomenon by sowing seeds in bags over decaying grass. He dug pits 45 cm deep, 60 cm square, and at the base of the pit he packed fresh grass to a depth of 7.5 cm. The pit was then filled with soil. A control pit with no grass was also dug and refilled with soil, but without grass. Seed of *Brassica alba* was sown at 7.5, 15 and 22.5 cm depth in both pits. In the control pit the seed germinated 100% and vigorously at all depths. In the pit containing decaying grass only 3 of the 25 seeds sown at 7.5 cm sprouted and at the deeper layers no germination occurred. The ungerminated seed from the deep layers remained dormant after removal from the soil and eventually germinated intermittently unless the testa was removed when the dormancy was immediately broken. The carbon dioxide concentration in the soil over the decaying grass varied from 12.4 to 18.8%, which was well within the range that induced dormancy in the laboratory experiments but exceptionally high for normal soils.

Dormancy may be induced by a period of intense drought. This phenomenon is particularly well developed in the Leguminosae and depends on the hilum acting as a hygroscopically activated valve. When the atmosphere is dry the hilum valve opens and allows the loss of water but in a humid atmosphere it closes and prevents the gain of water (Fig. 3/4). The embryo therefore dries progressively to a value equilibrating with the driest environment it has experienced. Dormancy of such "hard" seeds is only broken by scarification which makes the seed coats permeable to water (Hyde, 1954).

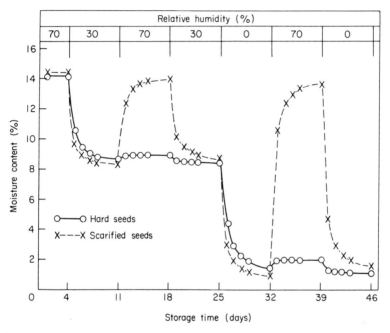

Fig. 3/4. Changes in moisture content occurring in white clover seeds transferred successively to chambers of different relative humidity. (From Hyde, 1954)

An important contribution to the understanding of induced dormancy was made in an elegant field experiment of Wesson and Wareing (1967, 1969). They studied the buried viable seed in a pasture which had a history of arable cultivation up to 6 years previously. The top 2 cm of soil was removed and soil samples were taken at various depths, down to 10 cm. The critical feature of this study was that the samples were taken in the dark — previous investigators studying the germination of buried seed had taken their samples during the daytime. By this pro-

cedure Wesson and Wareing ensured that during the sampling process
the buried seed did not receive any light which might stimulate germination. The soil samples were divided into 3 kg samples, sifted under weak
green light, spread into thin layers and placed in light or total darkness.
The results are shown in Table 3/II. From the soils exposed to light,

Table 3/II
The germination of seeds from soil samples taken beneath a 6-year-old
pasture. The soils were sampled in the dark and subsequently spread in thin layers
and placed in light or darkness. (Wesson and Wareing, 1967)

3 kg samples of soil. Counts of seedlings made after 28 days.
Means of 6 replicates.

Date of sample	Treatment	Grasses	Dicots
Nov. 1964	Light	14.0	7.0
	Dark	0.5	0.5
Aug. 1965	Light	19.0	7.0
	Dark	0.6	0.5
Oct. 1965	Light	9.2	11.0
	Dark	0.6	1.0

seedlings of 23 species of dicotyledons and many unidentified species of
grass appeared. In the darkness only three species germinated and the percentage emergence of these was much reduced compared with their behaviour in the light. In a second experiment holes were dug in the field 70
cm square to depths of 5, 15 and 30 cm. The operation was performed in
complete darkness or under weak green light and the holes were then
covered either with sheets of asbestos to exclude light, or with glass.
Counts of seedlings were made in quadrats 0.25 cm square at the base
of the pits and the results are shown in Fig. 3/5. Clearly the access of
light was sufficient to stimulate germination. The access of oxygen and
opportunity for CO_2 to diffuse out of the exposed soils were obviously
improved whether the holes were covered with asbestos or glass and the
experiment therefore offers some indirect evidence that it was not gas
relations in the soil that maintained the buried seed in a dormant state.

After 35 days the asbestos cover was removed from the holes and
replaced by glass and there was then a rapid emergence of seedlings.
Most of the seeds that required light for germination in these experiments were from species whose seeds do not have a requirement for

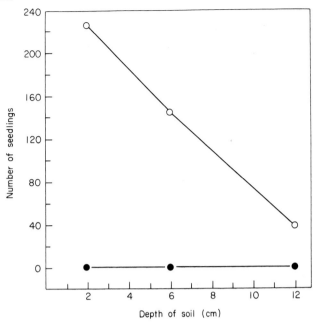

Fig. 3/5. The number of seedlings emerging at the base of pits dug in the field to three depths. ● light excluded ○ light admitted. (From Wesson and Wareing, 1967)

light when fresh. Somehow burial had induced a requirement for light. Cold treatment may induce a light requirement in the seeds of *Stellaria media* and this may be one way in which seeds shed in the autumn acquire a light-dependent condition by spring time. Another possibility is that seeds shed from the parent and falling under vegetation may be thrown into a dormant state by the excess of far-red wavelengths in the light that penetrates the foliage. Such seed then requires red or white light to break the dormancy and by now the seed may have become buried. Wesson and Wareing (1969) buried seed in soil in a laboratory experiment and showed that burial inhibited germination to an extent that depended on the depth and water content of the medium. The dormancy could be prevented by artificially aerating the medium, which suggested that the inhibitor was gaseous and might arise from the seeds themselves — it was not CO_2.

Patterns of seed dormancy in natural communities

Groups of species which live in the same area must possess common en-

vironmental tolerances which enable them to survive the hazards of the
area. There may however be different strategic solutions which accom-
plish the same tolerances. Groups of species living in the same area must
also differ in ways which permit them to escape from an exclusive
struggle for existence. It is interesting to examine the ranges of dormancy behaviour found in some contrasting environments.

A desert environment

Went and his colleagues in the Earhart Laboratory, Pasadena made a
detailed study of the germination physiology of species from the Mojave
and Colorado deserts (Went, 1957). They recognized four groups.

   Group 1. A group of species limited in occurrence to the aftermath
of summer storms (July—August) which are relatively infrequent, occurring once every 5—20 years. Species germinating after such storms include *Bouteloua barbata, B. aristidoides, Pectis papposa, Euphorbia
micromera, Portulaca oleracea* and most of the species of desert
shrubs. Germination of this group of seeds can be obtained in the laboratory in well wetted soil maintained at temperatures between 26 and
30°C.

   Group 2 includes species which germinate after winter rains, e.g. *Gilia
matthewsii, G. aurea* and many other species of this genus, *Oenothera
palmeri, Nama demissum* and *Mimulus bigelovii.* Laboratory germination of this group can be obtained in well wetted soil at temperatures
between 10 and 20°C.

   Group 3. Species which germinate in the field usually after November
rains. At altitudes of 1000 m they may germinate in October and even
earlier in the season at greater heights. This group contains *Oenothera
deltoides, O. cordifolia, Lupinus sparsiflorus, Abronia villosa, Salvia
columbariae* and *Phacelia campanularia.* In the laboratory these species
germinate at temperatures intermediate between Groups 1 and 2.

   Group 4 includes species which have no specific timing and seedlings
are found in the fields at almost any time of the year, e.g. *Palafoxia
linearis, Datura discolor* and *Cucurbita palmata.* This poorly timed
group may suffer calamities in the field and the last two species are
often seen killed by frost, something not seen in plants of the other
groups.

   Within groups the mechanisms which ensure precise but *opportunistic*
germination are often different. In *Euphorbia* spp. and *Pectis papposa*

water-soluble inhibitors are present in the seeds and rainfall of sufficient duration washes them out. *Baeria chrysostoma* is very sensitive to inhibition by salt concentrations which are common on the surface of desert soils. Heavy rains lower the osmotic pressure in the upper soil layers and allow seeds of this species to germinate. *Avena* species are delayed in germination by the attached lemma and palea which hinder the entry of water into the grain. Light rains are ineffective in permitting these seeds to germinate, since the germination process is sufficiently slow that water from a slight rain has been lost before germination has properly started. *Cercidium aculeum* possesses an impermeable seed coat and this is scarified by sand and gravel during storms; populations of this species therefore tend to be restricted to storm-washed gravel and sand deposits that accumulate after storms. *Physalis* spp. and *Lycium* spp. bear pulpy fruit coats which contain germination inhibitors. Only when the soil has remained moist for sufficient time for the fruit coat to rot does germination occur.

All of the mechanisms for breaking dormancy observed in this desert area are opportunistic, reacting directly to the environment becoming favourable and apparently not clocked in relation to solar time. Water is involved in one way or another, sometimes with a temperature limitation which imposes a degree of seasonality. The essential hazard of this environment for seedlings is the gross unpredictability of the rainfall.

A sub-alpine environment

There are apparently some general correlations between the type of seed dormancy and the altitudinal zone. Amongst alpines scarification seems to be the commonest dormancy-breaking mechanism (Amen, 1966), and at higher elevations a greater proportion of the species required cold moist stratification than in lower areas (Mirov, 1936). But within any one zone, species are usually present which bear non-dormant seeds as well as other species with different dormancy types. A study of the germination requirements of freshly collected seed in two subalpine areas in Colorado showed seven species lacking dormancy, four requiring scarification, two requiring moist cold stratification and four species with more complex needs (Pelton, 1956).

A humid tropical environment

The humid tropics are not areas of unvarying climate continuously favourable for plant growth, although the seasonal fluctuations may often be much dampened compared with temperate regions. Periodicities of growth are common, especially in the extension of new shoots. However, there is often a marked non-synchrony between the behaviour of neighbouring plants of the same species in a population, which suggests that there is little environmental adaptation in the periodicity (see discussion by Longman, 1969). There is relatively little reported research on the germination requirements of the trees of tropical rain forest, and a survey would be of the greatest interest.

This book is not the appropriate place to attempt a general review of dormancy, and detailed reviews are available in Lang (1965), Vegis (1961, 1964), Hartsema (1961), Amen (1968) and Woolhouse (1969) which consider mechanisms as well as adaptive significance. The examples and categories considered here are chosen for their relevance to the behaviour of populations, in particular in the ways that dormancy determines what is available and when, for recruitment into seedling populations. Clearly much of seed and vegetative dormancy is just part of the rhythmic adaptation of organisms to temporal rhythms in the environment. However, a large part of innate dormancy and all of enforced and induced dormancy are in a real sense a strategic alternative to dispersal. Dispersal, if successful, places seeds in places where the resultant plants succeed better than if they had remained by the parent, while opportunistic dormancy is successful if the plants from dormant seeds succeed better by waiting for a suitable environment to appear than by germinating immediately. Most plant communities are successional and each species is doomed to local extinction; the two strategies of "escape to somewhere else" or "wait until the right habitat reappears" are alternative ways of meeting the deterioration of a local habitat.

4

The Seed Bank

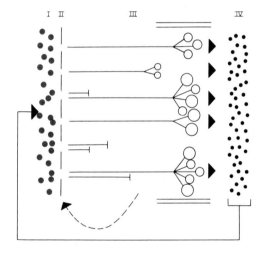

In most habitats occupied by higher plants, the number of individuals present as dormant propagules vastly exceeds the numbers present as growing plants. This situation is largely unique to the plant kingdom. Hibernation and diapause in animals usually represent a purely seasonal pattern of behaviour; only in the most primitive animals, for example eelworms, does a vast dormant population accumulate over the years. The store of seeds buried in soil (the seed bank) is composed in part of seeds produced on the area and partly of seeds blown in from elsewhere

(Fig. 4/1). Seeds normally arrive on the soil in a dormant condition. They may then need to await stimuli or conditions before they can germinate or they may have further dormancy imposed upon them by the experiences they meet on or in the soil. Seed in a dormant state in the seed bank can be regarded as a "deposit account". In addition it is convenient to recognize and distinguish a "current account", a temporary stage in which the only hindrance to immediate germination is a shortage of water and a favourable temperature. The current account derives (a) from the direct dispersal onto an area of seeds whose germination requirements are simple or whose stimulus requirements have already been met, and (b) by recruitment from the dormant seed bank, that is

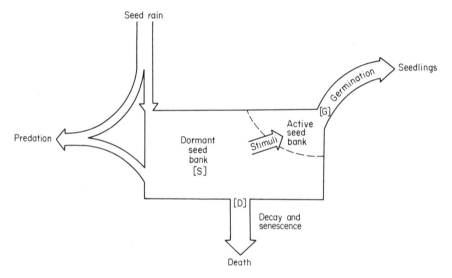

Fig. 4/1. Diagrammatic flow chart for the dynamics of the population of seeds in the soil. The symbols G, S and D refer to the model of Cohen (see text)

withdrawal from the deposit account. There is a two-way flow between the two accounts. Seeds in the seed bank are continually added to by the seed rain and thus represent a record of past as well as present vegetation growing on the area and nearby. It represents a source from which new vegetation may quickly arise if the existing stand is destroyed.

The numbers of buried and viable seeds within the soil can be estimated by taking samples and maintaining them at a favourable temperature, stirring at intervals and counting the seedlings as they emerge (e.g. Roberts, 1968; Brenchley, 1918; Brenchley and Warington, 1930;

Milton, 1939; Livingstone and Allessio, 1968). Alternatively in a very laborious procedure, the seeds in a soil sample can be counted directly under a dissecting microscope (Olmsted and Curtis, 1947) and in some cases it has proved possible to extract seeds from the soil by flotation in concentrated salt solutions and to follow this by measurements of seed viability by germination tests or by tetrazolium vital staining (Malone, 1967).

The first serious attempts to estimate the seeds in a seed bank were made on arable soil at Rothamsted by Brenchley and Warington (1930, 1933, 1936). Their experimental site was the unique Broadbalk field which had been planted to winter wheat every year from 1843. Soil was sampled to a depth of 15 cm on areas 10 cm x 7.5 cm; 140 such samples were taken in each year of the study. The soil was washed through sieves to concentrate the seeds which, with the remaining soil, were placed in porous earthenware pans. As seedlings emerged from the soil they were counted, identified and removed. The soil sample was thoroughly turned over every 3 months and vigorously cultivated every 6 weeks. The estimates of the seed bank were of a staggering magnitude, *ca* 39 000 seeds/m² representing 47 species. The true list of species was almost certainly greater than this as seedlings of certain genera, e.g. *Papaver*, and some grasses, were not determined to specific level, and were grouped in generic categories. More than two-thirds of the whole seed population was of *Papaver* species (27 800 seeds/m²). These estimates are certainly underestimates as the sample plots included some cultivation treatments which had depressed the weed population. It is interesting that major differences in the weed floras of differently manured plots were reflected in the buried viable seed flora, a clear indication that seed dispersal was not of sufficient magnitude to mask the local plot effects. Parts of the experiment included areas which were fallow (regularly cultivated and hoed). Two years of fallow produced changes in the size of the seed bank of individual species, some (e.g. *Polygonum aviculare, Scandix pecten-veneris*) fell over 2 years to 8 and 5% of their previous densities in the seed bank. Other species, particularly those with very rapid growth and seed production, completed their life cycle between cultivations, and increased their representation in the seed-bank. (The effect of fallowing for 2 years was to reduce the total seed-bank by only 6%, but the fallowing treatments were not sufficiently frequent to prevent short-lived weeds from completing their life cycles between cultivations (Brenchley and Warington, 1945).

The seed bank in an arable soil is of course a special case, because cultivations are continually stirring and inverting the soil profile, burying seeds that have been distributed on the soil surface, and bringing up to the surface seeds previously buried. However, comparable seed densities appear in analyses of the soil under permanent grasslands.

Essentially the same method of soil sampling was used by Chippindale and Milton (1934) and Milton (1936, 1939), for grasslands in Wales, though they used raw soil samples for laboratory germination tests rather than a sample prepared by Brenchley's rather cumbersome sifting procedure. Milton also sampled to greater depths (35 cm) and separated his cores into layers to give a depth profile of seed distribution. As might be expected, the depth profile shows a sharp fall-off in seed density (Fig. 4/2a–3) though occasional seeds were found at depths of 22–23

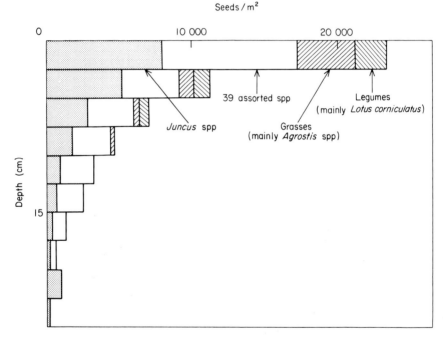

Fig. 4/2a. The depth distribution of the buried viable seed population in the soil under a marshy grassland (Frongôch, Wales). (From data in Chippindale and Milton, 1934)

cm, even in the soil under pastures which had never been cultivated within the historical record. In some undisturbed situations, seeds of small-seeded species may be found even deeper than in Milton's experi-

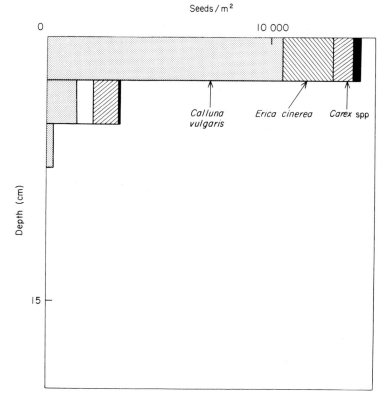

Fig. 4/2b. The depth distribution of the buried viable seed in the soil beneath a *Calluna vulgaris* association (Ponterwyd, Wales). (From data in Chippindale and Milton, 1934)

ments. *Striga asiatica*, a very small-seeded hemiparasite, extended down a soil profile of an undisturbed soil to depths of 150 cm and probably beyond (Robinson and Kust, 1962) (Fig. 4/2g).

In the natural grasslands of central Wales, the depth distributions found by Milton were regularly of high abundance in the top 2.5 cm of soil, with a rapid decline in density with increasing depth. Just as in Brenchley's arable land, the seed banks under pasture were often dominated by a single species or genus, and the depth distribution of this one species tends to overshadow the different patterns of less abundant species. Thus under a natural pasture, dominated by the grass *Nardus stricta*, seeds of *Calluna vulgaris* represented more than half the total population, and 78% of this was in the topmost 2.5 cm of the profile. In contrast, four species of *Juncus* were only meagrely represented

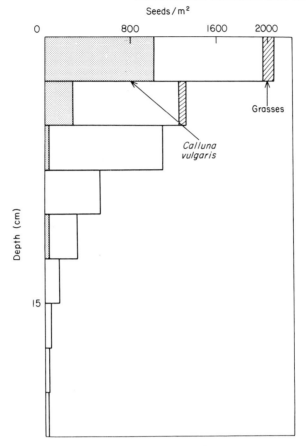

Fig. 4/2c. The depth distribution of buried viable seed in the soil beneath a hill pasture dominated by *Nardus stricta* (Ponterwyd, Wales). (From data in Chippindale and Milton,
1934)

in the top 2.5 cm and existed at greatest densities in a zone 7.5—12.5 cm
below the surface. On occasions, *Juncus* species dominated the seed
bank down the whole profile (Fig. 4/2a).
   The most remarkable features of these results are the extreme variations between species in the number of viable seeds that they leave in
the soil and the absence of any close correlation between the surface
vegetation and the seed flora of the soil beneath. For example in the upland areas examined by Milton, very high counts of buried seeds of
Calluna vulgaris and *Erica cinerea* were found, although these species

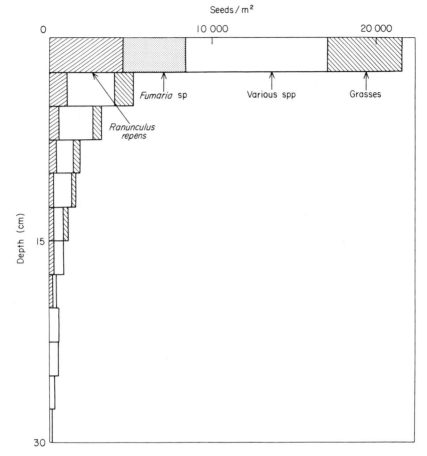

Fig. 4/2d. The depth distribution of buried viable seed beneath a floristically rich hay meadow (Wallog, Wales). (From data of Chippindale and Milton, 1934)

played only a very small part in the vegetation of the area. All of the areas examined by Milton were grasslands, yet the proportion of grass seeds present in the soil-bank was astonishingly low. Under a pasture dominated by *Molinia caerulea* only 0.3% of the seed bank was from this species. When grass seeds were present in significant quantity, they were usually *Holcus lanatus*, *Poa trivialis*, *Poa annua* and *Agrostis* spp. and on rare occasions approached 1 kg of seed/ha. Where a grassland had a history of past arable cultivation, the seeds of arable weeds were often present, even when the last arable practice was more than 70 years before.

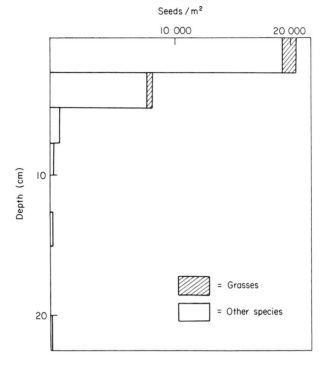

Fig. 4/2e. The depth distribution of buried viable seed in the soil beneath a natural hill pasture dominated by *Molinia caerulea.* (From data of Chippindale and Milton, 1934)

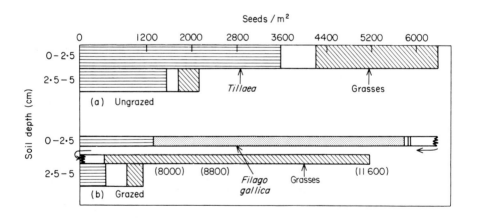

Fig. 4/2f. The depth distribution of seeds in Californian bunch grass communities. (From Major and Pyott, 1966)

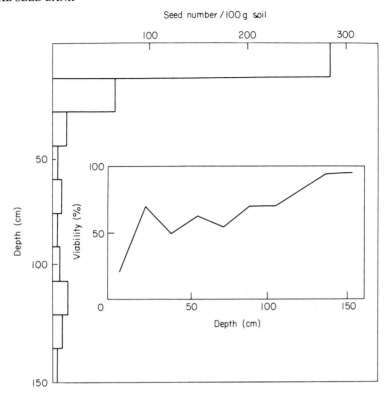

Fig. 4/2g. The depth distribution of buried seed of *Striga asiatica* and (inset) variation in viability with depth (lakeland sandy soil, North Carolina). (From data of Robinson and Kust, 1962)

In general, in the British uplands the largest seed banks occur where the soil conditions are acid and/or waterlogged. These may in themselves be conditions which favour seed storage, but in addition, many of the characteristic species of acid, wet habitats (*Juncus* spp., *Calluna vulgaris*, *Agrostis* spp.) produce a very high density of seeds.

In areas dominated by perennial grasses, the seed bank may be quite small. In Russian meadows dominated by *Bromus inermis*, Rabotnov (1956) obtained 280–2450 seeds/m². Where the species composition was more varied, as in a *Geranium pratense* type grass meadow, higher figures (*ca* 17 000) are obtained. The seed production from perennial grasses is low compared with many annual species and is also rather unpredictable, depending in part on the season, and in part on the timing of grazing activities.

Where annual species dominate communities, as in California bunch grass sites, relatively high values are found in the seed-bank, and these are even greater where the land has been grazed (Major and Pyott, 1966) (Fig. 4/2f). It is not altogether clear why the presence of grazing animals increases the seed bank. In part it may be that seeds are enforced in dormancy in the faeces, in part that the opening of the habitat by tight grazing allows annuals with high seed production to enter and contribute to the seed bank. The grazing animal may also increase the seed bank by trampling, creating conditions in which seeds retain greater viability within the soil. Even local and temporary disturbance, allowing one season's growth of an annual species, may be reflected in an enormous relic of this event within the seed bank. In the native prairies of the Great Plains where rather low values of 300—800 seeds/m^2 are commonly found (Lippert and Hopkins, 1950), locally disturbed areas may have seed banks of up to 20 000 seeds/m^2, primarily composed of a single species, e.g. *Sporobolus.* This can also happen in more frequently disturbed and cropped systems, and some of the highest density seed banks have been found in horticultural soils where buried viable seed populations as great as 157 000 seeds/m^2 reflect a single short-lived species with high seed production which has recently played a purely temporary part in the vegetation (Roberts, 1968).

Classic studies of the buried seed flora of maturing and mature woodland communities have been made in the "old field" successions of Japan and New England where the withdrawal of agriculture has allowed secondary successions towards developing woodland. Three stages in the Japanese succession are illustrated in Fig. 4/3 — also showing seasonal changes in the depth distribution of viable seeds. In a New England study at Harvard Forest by Livingstone and Allessio (1968), soil samples were taken for analysis of the buried viable seed from 16 sites representing different stages in the old field succession. Buried seed populations ranged from *ca* 1250—5000 seeds/m^2. These values are lower by an order of magnitude than those found in arable soils or in annual grasslands, but correspond reasonably well with the numbers of seeds under permanent grasslands of the kind studied by Milton. Some of the results of the Harvard Forest study are shown in Fig. 4/4. The stability of the buried seed flora is remarkable when seen against the profound changes in the species constituting the actively growing ground cover. Seeds of the species of the late successional stages and climax do not make an appearance in the seed bank until very late: they appear first under a

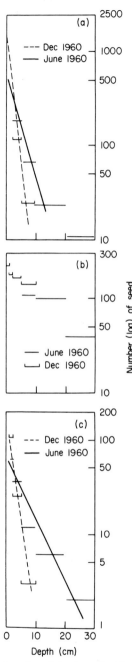

Fig. 4/3. Changes in the distribution of seeds in the soil with depth and over time in three stages of an old field succession in Japan. a. *Erigeron* stage, b. *Imperata* stage c. *Pinus* stage. Note the differences in scale. (From Numata *et al*, 1964)

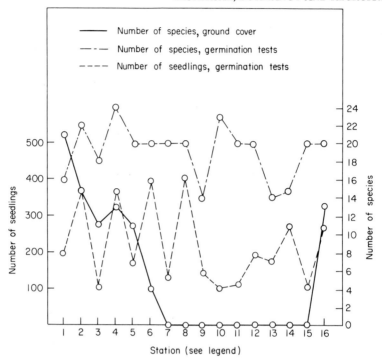

Fig. 4/4. The buried viable seed population in the soil beneath various stages in the old field succession in Massachusetts. The 16 stations were chosen to represent a time sequence in the
succession from first year of the succession (Station 1) through to 70-year-old white
pine dominated forest (Station 16). (From Livingstone and Allessio, 1968)

37-year-old stand of white pine (*Pinus strobus*).

In some hardwood and softwood forests in Maine the buried seed
population diverges remarkably in species composition from that of the
vegetation; much of the vegetation (at least 15 species in one woodland)
was from species that were not present in the vegetation and many
species present in the vegetation were not represented by seed in the
soil (Olmsted and Curtis, 1947). A unique feature of this study was that
estimates of the seed bank were made by direct observation of the sifted
fractions of the soil using dissecting microscopes. Germination tests were
then made of the extracted seed after a cold treatment. There was a very
great disparity between the values obtained by the two methods, implying either that a vast number of seeds remain in the soil dead but un-
decomposed and readily recognizable, or that the germination test was
inappropriate to break the dormancy of the supposedly inviable seeds.

Seed densities of $122-2850/m^2$ were obtained by the direct count method, values very similar to those obtained by other workers in woodlands, but the extraordinarily low densities of $5-173$ viable seeds/m^2 are difficult to reconcile with most other records (Table 4/I).

The presence in the seed bank under a woodland of seeds of species not present in the vegetation may imply that they are relics of previous vegetation or that they have been carried in as part of the seed rain from other areas. The presence of abundant seed of *Rubus* spp. in forest soils when the species is absent from the vegetation probably represents bird-dispersed seed. Its presence in the seed bank accounts for the extreme rapidity with which this genus enters the successional phases of felled woodlands (Oosting and Humphreys, 1940). The seeds of tree species are usually quite uncommon in the seed bank, though seeds of *Betula*, essentially a successional genus, may occur. Part of the reason may be that the larger tree seeds are so heavily predated and decompose rapidly but another factor is the delayed reproduction of climax tree species which may be present in the vegetation for 20 or more years before they produce seed.

The buried seed population of mature or climax communities generally contains a living, though dormant and heavily biased record of the past vegetational history of the succession. In general, the species of early successional phases contribute more to the buried population than do the dominants of the more mature phases. The pioneer species of succession commonly lie dormant — ready, as it were, to initiate the next new succession after some disaster (e.g. fire, felling, hurricane or herbicide) has destroyed the mature system. In a sense there is a circular argument here; species which have pioneered the succession are strongly persistent in the soil-bank and so appear as pioneer species in the next succession on the area. Do they persist because they are pioneers or do they become pioneers because they have persisted? Similarly it may be asked of the climax species, do they appear late in the succession because their seeds do not persist well in the soil-bank and have poor dispersal mechanisms, or does their failure to persist in the soil-bank and their low dispersibility in some way reflect a necessary biological property of mature stand species?

The vertical movement of seeds in the soil

The ways in which the buried seed population acquires its depth distri-

Table 4/I
Some examples of buried seed populations

Area studied	Vegetation	Seeds/m^2	Author
Maine, U.S.A.	White pine (70 year stand)	1000 (173)	Olmsted and Curtis (1947)
	White pine (80 year stand)	320 —	Values without parentheses are direct counts by microscopic examination. Values in parentheses are the result of germination tests
	Spruce, fir (30 year stand)	2850 (49)	
	Red pine plantation (24 year stand)	532 (5)	
	Beech, yellow birch, sugar maple (110 year stand)	1000 (91)	
	Beech, yellow birch, sugar maple (50 year stand)	218 (42)	
	Sugar maple (150 year stand)	122 (11)	
Massachusetts, U.S.A. Harvard Forest	Stages in old-field succession; see Fig. 4/3	1250—5000	Livingstone and Allessio (1968)
Prairies of Great Plains, U.S.A.	Native prairie	300—800	Lippert and Hopkins (1950)
	Disturbed areas (mainly *Sporobolus*)	20 000	
Meadow grass-land, U.S.S.R.	*Bromus inermis* dominant	280—2450	Rabotnov (1946)
	Geranium pratense, grass meadow	16 980	
Bunch-grassland, California, U.S.A.	Ungrazed area (top 5 cm only)	8230	Major and Pyott (1966)
	Grazed area (top 5 cm only)	12 200	

Table 4/I (continued)

Area studied	Vegetation	Seeds/m^2	Author
Arid grassland, Saskatchewan Canada		4000—15 000	Budd *et al.* (1954)
Sown grassland (U.K.)	Newly sown leys	4940—18 800	Champness (1949) Champness and Morris (1948)
Lowland grassland (U.K.)	Managed for hay	38 000	Chippindale and Milton (1934)
Arable land (U.K.)	Continuous wheat	34 100	Brenchley and Warrington (1933)
	Mainly cereals	56 500	Roberts (1958)
	Vegetable crops	1600—86 000	Roberts and Stokes (1966)
	Continuous wheat (27 800 *Papaver* spp.)	40 000—75 000	Brenchley and Warington (1930)

bution are rather obscure. Some seeds, particularly small ones, undoubtedly move down the soil profile of loose-textured soils with percolating rain water; this has been successfully imitated under laboratory conditions (R. Livingstone, pers. comm.). The very deeply buried seeds of *Striga asiatica* (Fig. 4/2g) almost certainly passed down with rain water. A few seeds possess self-burial mechanisms, e.g. species of the genera *Hordeum, Triticum, Avena, Erodium* but these are clearly in a minority among seed types and are mainly weedy species of disturbed habitats. The depth distribution of weed seeds in arable land is of course partly explained by the soil inversions involved in cultivation.

In undisturbed vegetation some burial of seeds occurs by the piling of successive layers of leaf litter on the soil surface after seed fall, the burying action of soil cast up by earthworms (Darwin, 1882) and moles, and by the caching activities of birds and rodents. Some seeds may presumably also fall down earthworm burrows, the cavities left after root decay and cracks caused by drying—wetting cycles in the soil.

Earthworms actively carry some seeds down into their burrows, which may extend 7—8 ft from the surface. Darwin (1882) reported

> I found at Abinger in Surrey two burrows terminating in similar chambers at a depth of 36 and 41 inches, and these were lined or paved with little pebbles, about as large as mustard seeds; and in one of the chambers there was a decayed oat grain with its husk. Hensen likewise states that the bottoms of the burrows are lined with little stones; and where these could not be procured, seeds, apparently of the pear, had been used, as many as fifteen having been carried down into a single burrow, one of which had germinated. We thus see how easily a botanist might be deceived who wished to learn how long deeply buried seeds remained alive, if he were to collect earth from a considerable depth, on the supposition that it could contain only seeds which had remained long buried.

Most deeply buried seeds appear not to germinate until and if they are brought near to the surface and the frequency with which this happens is very difficult to assess. Probably most of the seed bank is doomed to die and decay *in situ*. Events which bring soil from depths and expose it at the surface of the soil are likely to be rare outside agriculture, although within agriculture such constant churning of the soil-bank is the norm. Some burrowing animals such as moles, gophers and earthworms (McRill, 1974) bring soil and seeds from various depths to the surface and in woodlands and forest the falling tree brings up soil from deep in the profile. It is doubtful whether many other forces in nature disturb the accumulating soil bank of seeds; only these specialized and local events bring a fragment of the population to the surface. Man, the ploughman, has imposed a churning action on what are predominantly downward-moving seed burial systems and thereby has suddenly favoured those species whose decay is slow in the seed bank.

Losses from the seed bank other than by predation

Various classic attempts have been made to determine the rate of loss of seeds from a soil-bank due to old age, fungal attack, decay and germination, the most famous being the experiments of Beal started in 1879 (Darlington, 1931, 1951; Darlington and Steinbauer, 1961) and of Duvel started in 1902 (Duvel, 1902; Goss, 1924; Toole and Brown, 1946). In both Beal's and Duvel's experiments seeds of various species were buried in containers, and left for various periods of time. At intervals these caches were dug up and tested for viability.

In Beal's experiment, seeds from a variety of species found near Michigan were placed in pint bottles of sand and buried. They included seeds of 20 herbaceous species and seed of *Quercus rubra* and *Thuja occidentalis*. At the first examination of a sample of these buried seeds 5 years later, the tree seeds had lost viability, but many of the herbaceous species continued viable, and 50 years later seeds germinated of *Rumex crispus, Oenothera biennis, Verbascum blattaria, Brassica nigra* and *Polygonum hydropiper.*

In Duvel's experiment, seed of three groups of species was buried, representing crops characteristic of the area, seeds of weeds not native to the area, and seeds of weeds that were native to the area. The first to lose the power to germinate were crop seeds and the last were the seeds of weeds native to the area.

Viable seeds of *Nelumbo nucifera* have been reliably dated at 150–250 years old (Exell, 1931) and another radiocarbon dated record for *Nelumbo* sp. (a species similar to *N. nucifera*) gives an age of 1040 ± 210 years for seeds that were still viable (Libby, 1951). Similar techniques show that the seed of *Chenopodium album* and *Spergula arvensis* may remain viable in the soil for 1600 years (Ødum, 1965). A number of generalizations emerge from the very scattered literature on seed longevity in the soil. (i) Long-lived seeds are characteristic of disturbed habitats. (ii) Most long-lived seeds are annuals or biennials. The biennials in particular appear frequently in soil samples taken from soil in dated archaeological sites (S. Ødum, pers comm.). (iii) Small seeds tend to have much greater longevity than large ones and the very large seeds have very short lives, e.g. the nuts of trees. (iv) Aquatic plants may have great seed longevity and the conditions in mud below water may inhibit decay. (v) Seeds of mature tropical forest species have very short lives.

Differences in longevity are well known for species in dry storage (Roberts, 1960) and there are genetic differences between strains of the same species (e.g. in *Zea mays,* Haber, 1950; Lindstrom, 1942).

Old age in itself can lead to loss of viability: the seed of crop species stored under ideal conditions in a commercial seed store steadily lose viability (Roberts, 1960). In the absence of fungi and seed-destroying pests, seed viability declines as mutations accumulate during the period of storage, and irreversible biochemical changes take place which lead to death (Figs 4/5, 4/6).

In the presence of soil pathogens, the process of death may be acceler-

ated. Their role is mainly known in the case of crop species, though
Tadros (1957) found that fungal decomposition of seeds (or very young
seedlings) in the soil could account in part for the different floristic
composition of the natural vegetation on serpentine and non-serpentine
soils. Seed of *Emmenanthe* spp. which failed to survive on unsterilized
serpentine soil, survived and germinated successfully on sterilized soil.
Just how complicated the process of seed destruction by pathogens may
be is illustrated in a series of experiments on crop plants (Leach, 1947).

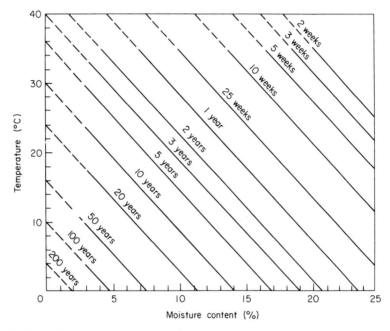

Fig. 4/5. The influence of age and storage conditions on the loss of viability of stored wheat
   grains. The data are expressed as isochrons for the "half-life" of the seeds under varied
   combinations of temperature and humidity. The broken lines indicate theoretical and
   probably unrealistic extrapolations. (From Roberts, 1960)

The seed of spinach was rapidly rotted by species of *Pythium* within a
temperature range of 12–20°C, but not at low temperatures, e.g. 4°C.
Rotting by *Rhizoctonia* did not occur at temperatures below 12°C and
was rapid at 20°C. Water melon seed was not damaged by *Pythium* or
Rhizoctonia at 35°C, but these fungi became steadily more lethal as
temperatures fell below this value. Seeds of the pea (*Pisum sativum*)
were strongly attacked within the range 12–25°C, and in sugar beet

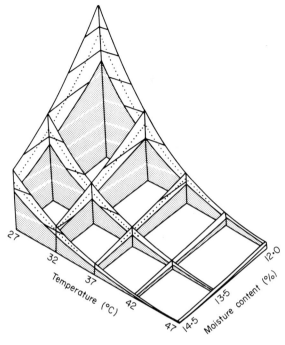

Fig. 4/6 A three-dimensional diagram of the theoretical relationship between temperature, moisture content and viability in rice (cv. 'Toma 112'). The vertical scale represents time and the horizontal lines indicate 100 day intervals. The dotted line represents the half-life and the unshaded areas above and below this represent the standard deviation of death points in time. (Roberts, 1961)

from *Rhizoctonia* between 16 and 30°C and from the seed-borne *Phoma* between 4 and 20°C. A similar differentiation between seeds of different species is produced by temperature in the seed rotting of *Zea mays* and *Triticum vulgare* by *Gibberella zeae* (Dickson, 1923, 1928).

An impression of the subtlety with which the environment interacts with the soil microflora to determine seed survival is shown in experiments on the fate of maize grains introduced into soil by deliberate sowing. Seed sown at intervals out of doors at Oxford, England, from early spring to mid summer showed steadily increasing success in avoiding death in the soil; but seed sown in the cold early spring, or in the hot dry mid summer, when germination was delayed by the cold or the drought, suffered heavy mortality (Fig. 4/7). Seed sown in the late summer germinated readily but as the soil cooled down in the autumn

more and more of the seed rotted before germination. This seed rotting
was largely prevented when the seed was treated with fungicides, strongly indicating a pathogenic cause for the seed destruction (Harper *et al.*,
1955). At low temperatures there was a consistent increase in seed
death with increasing soil moisture contents. Even when soil was sampled

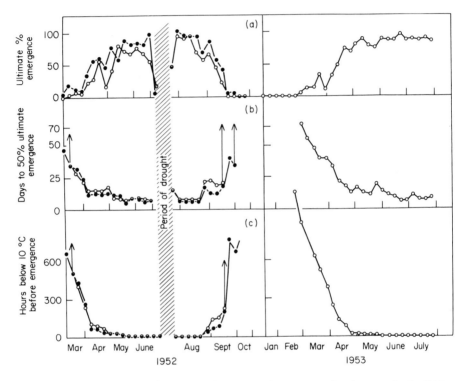

Fig. 4/7. The influence of sowing date on the success of emergence of maize sown in the field.
Open and closed symbols refer to different varieties. (From Harper *et al.*, 1955)

from the field and maize sown in it in the laboratory, maintained at a
constant temperature of 8°C, and the mortality was corrected for variations in water status, there remained drifts in the "pathogenic potential"
of the soil, indicating that the ability of the soil microflora to cause seed
decay is something that itself is constantly changing. The most favourable environment for seed rotting appears to be one that is just not
suitable for germination, or permitting it only extremely slowly (Harper,
1955). Maize grains lose viability most quickly at temperatures or moisture contents just too low for germination (Fig. 4/8).

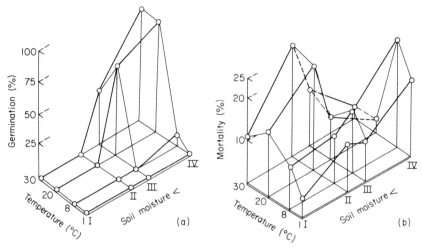

Fig. 4/8. The influence of soil moisture and temperature on (a) germination and (b) mortality of maize grains in soil (cv. 'White Horsetooth'). (From Harper, 1955)

When seeds are placed in the soil bank in which rotting processes occur, the rot does not affect the population as an all-or-nothing phenomenon; there is instead a steady reduction in the number of seeds that remain viable depending on the severity and length of the adverse period. Thus, when seed rotting occurs in the field its effect may be to eliminate a species or simply to change its numerical representation in the seed bank. Changes in the environmental conditions that affect the critical factors of temperature and water content of the soil will be reflected in changes in the density of viable seeds that remain in the seed bank. When for example the soil temperature is deliberately modified by some cultural practice, it may indirectly operate on the seed–pathogen interaction to alter the density of a species that germinates from that soil. Soil temperature may be altered by changing the colour of the soil surface or by altering the slope and aspect of the soil and so affecting the amounts of solar radiation absorbed. When maize was sown on pyramids of soil with north, south, east and west facing slopes of various inclinations or in soil of which the surfaces were differently coloured, one effect of the treatment was to change the frequency with which seeds were rotted and so the density of the resulting stands of plants (Figs 4/9, 4/10).

Measures of the overall rates of loss from seed banks have been made only on some horticultural soils (Roberts, 1970). In disturbed and in undisturbed soils a log-linear decay of a seed bank occurs with the passage of time if new seed entries are prevented. This implies that the seed population within the soil-bank has a continuous and constant death

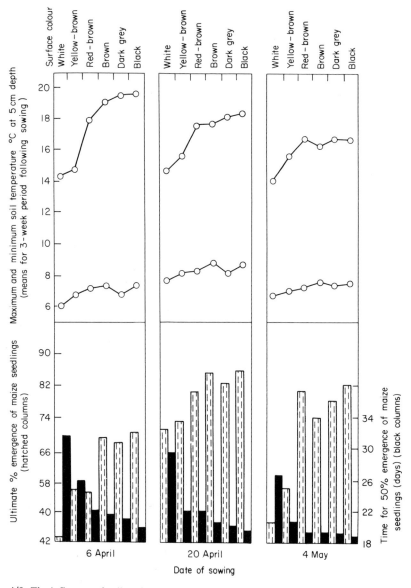

Fig. 4/9. The influence of soil surface colour on soil temperature and indirectly on the mortality and germination speed of maize sown in the soil. (From Ludwig and Harper, 1958)

risk. This death risk, which in the case of horticultural weeds implies a
half-life of about 1 year, differs somewhat between the species that
have been intensively studied (Fig. 4/11). It would be of great interest to
compare decay rates in this way for other species in other communities,
excluding the seed rain from areas of land in for example forest and permanent grassland and determining the rate at which the seed bank of
different species decays when new additions have been prevented. This
approach would be ecologically much more meaningful than that in the
original experiments of Beal and Duvel in which the seed was buried

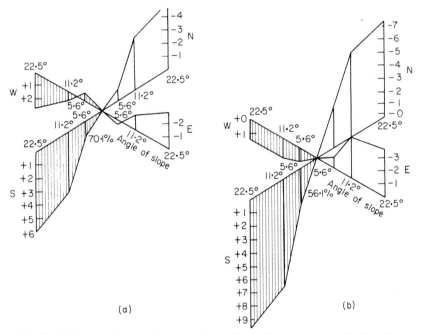

Fig. 4/10. The influence of slope and aspect of the soil on soil temperature and indirectly on
the success of sown seeds in evading rotting in the soil. In each figure the slope is
expressed outwards from the point of intersection of the compass lines. The point
of intersection represents horizontal ground on which 70.1% of the sown seeds produced seedlings in 1953 and 56.1% in 1954. The vertical scales represent the deviation
from this horizontal land value. Data summarized over seven sowing dates from March
30th to May 11th. Experiment at Oxford, England. (From Ludwig et al., 1957)

under the artificial conditions of enclosure in sealed jars and in sand
rather than in the natural soil system. Germination itself is of course a
source of continuing loss of seeds from the seed bank and the act of

cultivation, which by bringing seeds to the surface stimulates germination, appears to reduce the seed bank more rapidly than is the case in an undisturbed soil (Roberts, 1970).

Optimal dynamics of the seed bank

The population dynamics of the seed bank can in some senses be divorced from consideration of the whole life cycle of the plant population; the seed bank has dynamic properties of its own. Cohen (1966) attempted to model the parameters affecting the size of the seed bank in such a way that optimal strategies might be calculated for different theoretical environments. In Cohen's model the following parameters were introduced:

S = Number of seeds present, i.e. the size of the seed bank

Y = The number of seeds produced per germinated seedling, a random variable depending on environmental conditions that in this model are assumed to be independent of population density (in practice they will often be density dependent).

In a sense the parameter Y represents the return to the seed bank on the investment that is represented by the fraction of the bank that germinates.

G = The fraction of the total seeds present in the bank that germinate in each year (the fraction of the capital of the seed bank that is invested in germination).

D = The fraction of the seed bank that decays each year. This represents a depreciation of uninvested capital.

P_i = The probability associated with Y_i.

The relationship between the size of the seed bank in year T and its size in year $T + 1$ can then be written:

$$S_{t+1} = S_t - S_t \cdot G - D \cdot (S_t - S_t G) + G \cdot Y_t S_t$$

Intuitively it would appear that decreasing D, the fraction that decays each year, and increasing Y, the number of seeds produced per germinating seedling, increases the growth rate expectation of the seed bank. Cohen then poses the specific question, how do variations in G influence the expectation of growth of the seed bank at any combination of D, Y and P_i. Figure 4/11 shows the long-term expectation of the growth rate

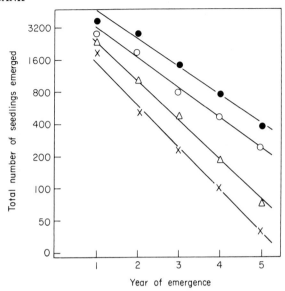

Fig. 4/11. The number of seedlings emerging in successive years from a horticultural soil from which seed immigration was prevented. ● = *Thlaspi arvense,* △ = *Stellaria media,* ○ = *Capsella bursa-pastoris,* x = *Senecio vulgaris.* The data represent the results of experiments started in 1953, 54 and 55 and the data have been combined. (From Roberts, 1964)

of the seed bank plotted against the fraction of the seeds germinating each year, i.e. *G*, for a variety of abstracted environmental conditions. These environmental conditions are:

(a)   two different values of P_i, a probability of successful reproduction of 0.1 or 0.8;

(b)   two values of *Y*, the number of seeds produced per germinating seedling, 5 or 500,

(c)   two values of *D*, the fraction decaying each year, 0.1 or 0.8

A number of interesting conclusions emerge. If there is a high probability of successful reproduction ($P = 0.8$, $Y = 500$) the optimal strategy is for a high fraction of the seeds to germinate each year and the proportion decaying is almost irrelevant. However, if there is a high probability of low reproduction ($P = 0.8$, $Y = 5$) the rate of decay is irrelevant if most of the seeds germinate each year, but becomes very important if there is dormancy. In contrast, if there is a low probability of reproduction, whether it be at a high or low rate, a high germination percentage is disastrous and so is a high decay rate.

Cohen's model also predicts that there should be a strong correlation between the fraction of the seed that germinates and the fraction of the ungerminated seed that decays; and also a correlation between the fraction that does not germinate in any given year and the probability of the total or near failure of the germinating seeds to produce a seed crop.

The bud bank

Although the buried viable population of plants in the soil is usually thought of as a reserve of seeds, other dormant meristems may accumulate in considerable numbers as bulbs, bulbils and buds on rhizomes, corms and tubers. A single node from the rhizome of *Agropyron repens* may, after one season's growth, leave as many as 825 dormant rhizome buds overwintering (Haddad, 1968). Such a reserve of dormant buds which may reach densities of $620/m^2$ (Fail, 1964) is in some critical ways different in its biological consequences from a reserve of seeds. In the first place, because such buds have been produced clonally, they represent the growth of single genets whereas a buried seed population conserves as yet untested genotypes. Secondly, vegetative buds may be maintained dormant by forces of correlative inhibition so that some degree of integrated control of the breaking of dormancy is possible. In *Agropyron repens* buds may be relieved of correlative inhibition by breaking up the rhizome or by damage to shoots which have broken dormancy and emerged. After such damage, previously dormant buds may become active. The weed *Agropyron repens* may be controlled by regular cultivations which break up the rhizomes and destroy growing shoots and so progressively remove the buried bud population from its imposed dormant state, but this can be a long process, and is testimony to the stabilizing effect that such dormant populations may have in a hazardous environment.

The length of life of dormant buds in the soil varies considerably depending in part on the rate of rotting of the parent organ and its food reserves — buds borne on rhizomes do not appear to survive after the rhizome has rotted. Rhizomes and their buds in *Agropyron* have a life rarely exceeding 3 years. There are many dormant buds on the corms of *Ranunculus bulbosus* which may be stimulated into growth if the main shoot is destroyed; but if the main shoot grows normally these buds are lost with the parent corms which rots before it is 1 year old. In contrast the long-lived rhizomes of for example bracken (*Pteridium*

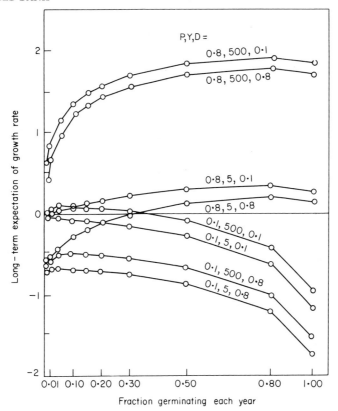

Fig. 4/12. Theoretical relationships between the long-term expectation of growth of the seed bank, and the fraction of the seed bank germinating each year. Calculated for: various probabilities of a seed producing a seed-bearing plant (*P*); a low or high seed output (*Y*); and different seed decay rates (*D*). (From Cohen, 1966)

aquilinum) and *Iris* spp. bear correspondingly long-lived buds. Vegetative buds in the form of bulbs and bulbils usually include food reserves within the dormant bud tissue and survive long after the parent organs have decayed.

Of course many plants pass the adverse season as perennating underground organs (hemicryptophytes, geophytes, cryptophytes) but this is simply a repeated cyclic seasonal pattern, not a long-term storage of meristems. However there may on occasions be a stored bank of dormant vegetative plants which may remain unobserved above ground for several years. Delphinium plants mapped by Epling and Lewis (1952) were found to lead a curious perennial existence in which individuals interposed years or groups of years in a dormant underground state with years of active growth and flowering. These phases of dormancy and growth were not synchronous for different plants within a population.

A most remarkable example of vegetative dormancy comes from the regrowth of stumps of Coolibah trees (*Eucalyptus microtheca*) which remained dormant for 69 years after felling and, after a storm in 1974, 60% of the stumps produced vigorous new shoots (Hughes, 1974).

The soil is not only a reserve of dormant stages of higher plants, but is also a reserve and living record of disease organisms. This may be just as important in the population dynamics of higher plants. Agricultural experience shows that most plants grown as crops accumulate specific disease floras and faunas, and that in the absence of chemical control, crop rotations are required to control epiphytotics. Potato cultivation encourages a build-up in the soil of such species as *Synchytrium endobioticum* and *Spongospora subterranea*; similarly wheat cultivation encourages build-up of *Ophiobolus graminis, Cercosporella herpotrichoides* and *Puccinia* species. Lucerne (*Medicago sativa*) builds up Fusarium wilt and many Cruciferae build up populations of *Plasmodiophora brassicae.* In some cases these pathogens persist in the soil as slow-growing or resting mycelium but in many cases as resistant long-lived spores. It seems probable that an entirely similar disease accumulation occurs in the soil under more natural vegetation but there has been little serious study of this possibility. If the accumulation of specific pathogens in the soil bank occurs in nature, it may need to be taken into account in interpreting successions, population fluctuations and local movements of species in the mosaic of natural vegetation.

5

The Recruitment of Seedling Populations

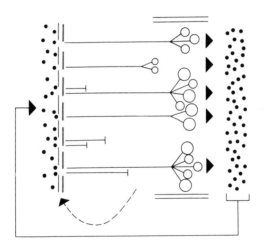

The establishment of a seedling from a seed involves a series of precise deterministic events within an environment in which the scale of hetero-geneity is determined by the size of the seed. The presence or absence and the density of a seedling population depends not only on the avail-ability of seed but on the frequency of "safe sites" that provide the precise conditions required by a particular seed. Great subtlety in the size, shape and dormancy-breaking requirements of seeds combined with an equal heterogeneity in the environment at the appropriate scale of size is sufficient in itself to determine different species compositions of established seedlings from the same seed input. It will usually be impossible after seedling growth to discover what were the direct causes of density, pattern and composition of a plant population. Correlations made between the distribution and abundance of adult plants and environmental features are extremely unlikely to reveal real causes because the scale of environmental heterogeneity that is relevant and readily studied for the mature plant is of quite a different order from that which determines the behaviour of an individual seed.

From among the vast numbers of seeds present in the soil and arriving on the surface through dispersal, only a tiny fraction germinates to give seedlings. In a crude sense, the numbers of seedlings appearing can be thought of as a function of the number of "safe sites" offered by the environment. A "safe site" is envisaged as that zone in which a seed may find itself which provides (a) the stimuli required for breakage of seed dormancy, (b) the conditions required for the germination processes to proceed and (c) the resources (water and oxygen) which are consumed in the course of germination. In addition a "safe site" is one from which specific hazards are absent — such as predators, competitors, toxic soil constituents and pre-emergence pathogens. A population of a particular species may be absent from an area *either* because it offers no "safe sites" even though seed is abundantly present, *or* because seed is absent although safe sites are abundant. Different densities of seedlings may result from different seed densities deposited in a uniformly "safe" environment or a different frequency of safe sites when seed is abundant.

The development of a seed to a seedling depends on conditions which are immediately localized to the environment of that seed and seedling and the development of a population of seedlings depends on each individual finding itself in an appropriate set of conditions. At the scale of size of a seed the physical environment is exceedingly heterogeneous: not only in the biblical sense that some seeds fall on stony ground, but also in the sense that a worm cast is a mountain to a mustard seed, a fallen leaf is a shade from light (or from the eye of a predator (Shaw, 1968b)), a raindrop is a cataclysm (Clements, 1964). It is important to think about developing plant populations in these terms and at the scale of environmental heterogeneity relevant to individual seeds — otherwise the real events that make for life or death of the individual become lost in the statistics of vegetation, and real causes become lost in generalizations about correlations.

In the flow diagram which heads this chapter, the recruitment of a seedling population from the available seed is represented by a sieve — an environmental lattice of safe and unsafe sites. Much of this chapter is concerned with experimental demonstrations of the subtlety with which the sieve operates, a subtlety which depends on very species-specific requirements for seed germination and extremes of environmental variation at the scale of size of individual seeds.

Germination of seeds on artificial substrates

A very convenient demonstration of an environmental sieve can be made by sowing seeds onto the surface of a sintered glass plate. Such an inert flat surface may be provided with water from beneath and, if the sinter is mounted in a funnel, the water supply may be fed through a manometer so that it is maintained at constant tension. The tension of water in the pores of the sinter determines the shapes of the water menisci in the pores. It is relatively easy to set up a series of such germination substrates at a range of tensions from 0 to 200 cm of water.

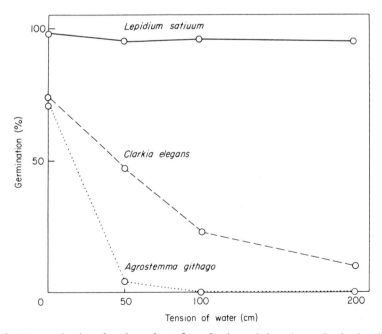

Fig. 5/1. The germination of seeds on the surface of a sintered glass plate maintained at different water tensions. (From data in Harper and Benton, 1966)

Seeds sown on the surface of such a sinter may or may not germinate
and discrimination between species may be brought about by varying
the water supply to the plate (Harper and Benton, 1966). Figure 5/1
illustrates such an experiment, in which seeds of *Lepidium sativum,
Clarkia elegans* and *Agrostemma githago* were sown. All three species
germinated freely on the wettest plates, but *Agrostemma githago* was
prevented from germinating by the slightest increase in tension.
Lepidium sativum germinated freely over the full range of water supply
whereas the germination of *Clarkia* decreased regularly as the tension
was increased. In such a highly simplified environment, water supply
acted as an environmental sieve, discriminating between the germination
of the three species and between the seeds of a species.

It is especially interesting that, even in such a uniform environment,
the particular conditions of water supply did not determine an "all or
nothing" environment but rather, operated to affect the probability that
a seed would germinate. Increasing water tension reduced the chance
that a seed would germinate and this chance was affected differently for
the three species. Thus on a sinter at 50 cm tension, 47% of the seeds of
Clarkia elegans germinated, at 100 cm only 23% germinated, and at 200
cm tension 10% germinated. Clearly some sort of heterogeneity either
in the seed sample or in the environment caused the water supply to
discriminate between seeds that did or did not germinate. The critical
determinant appears to be whether the contact made between seed and
sinter was sufficient to allow the seed to absorb water faster than it lost
water to the atmosphere. A critical balance between these two processes
is required if the seed is to reach a water content sufficient for germination. In such an environment, seeds that do not germinate may often be
persuaded to do so simply by rolling them over so that the micropyle is
nearer to the surface of the sinter. Alternatively, almost all the seeds,
even of a very sensitive species like *Agrostemma,* will germinate if
evaporation is prevented by covering the mouth of the sinter funnel
with polythene.

A comparison of a range of species is shown in Table 5/I. The larger
seeds are the most sensitive to water availability in this environment (cf.
species in Group 4) as might be expected from the ratio between the
seed surface available for water loss and the zone of contact that such
seeds make with the sinter. Seeds with rough or tuberculate surfaces are
similarly extremely sensitive. In contrast, seeds with mucilage, which
markedly increases the seed—sinter contact and the zone of water flow

The percentage germination of the seeds of various species sown on sintered glass plates at controlled water tensions

Species and seed weight (mg)	Standard (filter papers in petri dishes)	Conditions of test — Sintered glass plates under tension			
		0 cm	50 cm	100 cm	200 cm
Group 1 Seeds with copious mucilage					
Lepidium sativum (0.0027)	100	98	95	96	95
Camelina sativa (0.0012)	99	98	96	96	94
Group 2 Seeds with less copious mucilage					
Linum usitatissimum (0.0059)	96	92	88	84	73
Plantago major (0.0002)	49	46	30	30	30
Sinapis alba (0.0074)	92	96	73	55	24
Group 3 Tuberculate seeds					
Reseda alba (0.0005)	35	45	8	7	5
Agrostemma githago (0.028)	70	71	4	0	0
Melandrium rubrum (0.0006)	57	39	0	0	0
Group 4 Seeds with smooth testa — no mucilage					
Clarkia elegans (0.0004)	81	73	47	23	10
Brassica napus (0.0038)	97	87	9	0	0
Pisum sativum (0.23)	98	11	0	0	0
Vicia faba (0.66)	100	0	0	0	0

From Harper and Benton, 1966.

to the seed, are remarkably insensitive to variations in water supply. Indeed if seeds of *Agrostemma githago* are provided with a mucilage coat their germination becomes largely independent of tension.

This type of experiment shows that even in a controlled, "homogeneous" environment the numbers of seeds that are recruited into a germinating population are determined by the individual properties of each seed, how it lies, what its shape is, what contact it makes with its substrate, how fierce is the evaporating power of the atmosphere. This complex of factors determines which seeds germinate and which do not, discriminating not only between species but also between individuals of a species. In this way the density of a seedling population may be determined.

The interpretation of the distribution and abundance of plants in nature seldom takes into account the operation of such a "sieve" if only because, by the time a plant is growing, the peculiarities of the microsite surrounding its seed have long disappeared. Evidence for the operation of such sieves in nature has to come from deliberate observation of marked seeds in the field and this is obviously very difficult except for large seeds. Such evidence is, however, available from experiments of Watt (1919) with acorns (*Quercus robur*). The acorns lose water when they lie on the soil surface and the amount of water that they accumulate varies with the amount of contact between the acorn and the soil. "This . . . shows the value of even contact with the soil in lowering the net loss, while an atmosphere of greater relative humidity assists still more in that direction." Watt showed that acorns failed to germinate on soils "characteristically flat, almost devoid of vegetation, without humus and of a light sandy nature" whereas germination was successful "whereever there is a light covering of humus over the acorns" or "in depressions in the soil" or indeed where "the necessary environment is provided to inhibit the drying process". He was able to control the success of acorn germination experimentally by altering the extent to which individual acorns were buried and exposed to the atmosphere.

Watt's experiments were done in the drier east of Britain. In the wetter west the oak *Quercus petraea* germinates quite well on the surface of the soil, provided that it is not eaten by predators. In the west the safety of a germination site is primarily determined by whether it hides the seed from the eye of a predator (Shaw, 1968a, b); (see also Griffin, 1971, for a description of oak establishment in California).

Slight variations in the physical state of a substrate may affect the numbers of seeds that germinate successfully. This can again be demonstrated using sintered plates to control the water status of a simulant soil. Slate dust may be obtained in various particle sizes and a layer, placed on a sinter, serves as a germination substrate. Slate dust of different particle sizes gives "soils" of different porosities. As the amount of water held in a soil is a function of the pore size and also of the water tension within the soil, it is possible to create on sinters, "soils" of the same water content at different tensions (by using different particle sizes) or at the same tensions but at different water contents. Moreover, soils that are in the process of being dried and soils in the process of being wetted possess different water contents at the same tension, and correspondingly, different tensions at the same water content. It is therefore possible to vary both soil water content and tension independently by using different particle sizes and both wetting and drying cycles. The results of an experiment are shown in Fig. 5/2a and b. Clearly the number of seeds recruited into the germinating population varies continuously with variations in either tension or water content. The size of the recruited population can be altered by *either* a slight change in soil particle size, *or* a slight change in water tension, *or* a slight change

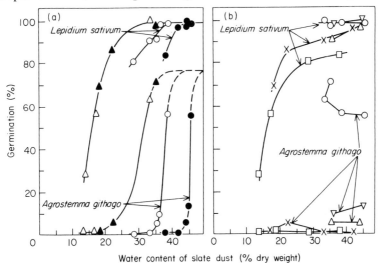

Fig. 5/2. The relationship between ultimate percentage germination of seeds of *Lepidium sativum* and *Agrostemma githago* and the water contents of Slate Dusts I (△, wetting cycle; ▲, drying cycle) and Slate Dust II (○, wetting cycle; ●, drying cycle) at controlled tensions on the wetting and drying cycles. The curves, fitted by eye, connect points on: (a) similar surfaces, and (b) at the same moisture tension (○, 0 cm; ▽, 25 cm; △, 50 cm; x, 100 cm; □, 200 cm). (From Harper and Benton, 1966)

in soil water content. The opportunity that these subtleties offer in explaining the fine scale of environmental hospitality in the field are clearly enormous.

The seed on the soil

Although laboratory experiments in controlled systems are instructive in showing how complex seed–substrate interactions may occur, a different level of experimentation is necessary to establish the link between this type of study and "real" field phenomena. A group of such studies has been made in which seed has been sown·on deliberately prepared natural soils. Seeds of *Bromus rigidus* and *B. madritensis* were

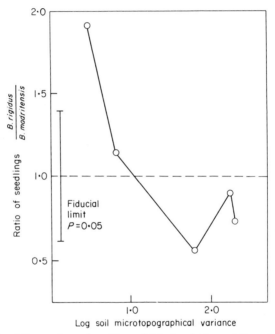

Fig. 5/3. The ratio of the numbers of seedlings of *Bromus rigidus* and *B. madritensis* establishing from an equiproportioned mixture sown on soils of varied microtopography. The soil microtopography is expressed as the logarithm of the variance of the height of the soil surface. (From Harper *et al.*, 1965)

sown on Yolo silty loam soil which had been prepared in pots to give a range of surface microtopographies (Harper *et al.*, 1965). The soil was obtained from the field in a sun-dried state and broken by hand to give different sized clods *ca* 5, 3—5, 2—2.5 or <1.25 cm diameter. Pots were filled with soil of a specific clod size and in addition some pots of the finest soil were heavily watered and allowed to dry out for 3 weeks to give a hard-capped, cracked soil. Onto these soils seeds of the two grasses were dropped in equiproportioned mixture. The pots were watered at weekly intervals with the equivalent of 2.5 cm of rain and germination of the seedlings was followed. A crude measure of surface roughness was made by lowering a bank of ten needles onto the soil surface and determining the heights of the tops of the needles when the points all touched the soil. The variance of these heights gave a measure of the soil microtopography.

The ratio of the numbers of the two species germinating on the soil surfaces was strongly affected by the microtopography (Fig. 5/3), because subtleties in the microsites on the different surfaces gave different opportunities for germination of the two species. The seedling population on the smoothest soil was predominantly of *Bromus rigidus* and on the roughest soil *B. madritensis* predominated. It was not difficult to see how this happened; the grain of *B. madritensis* is small and light with a curved awn, whereas that of *B. rigidus* is larger and bears a straight needle-like awn. Grains of *B. madritensis* falling on the smooth soil surface tended to roll onto their convex surface with the embryo pointing impotently into the air. On the rough surfaces the grains tended to curl around the soil clods and make good contact with the soil surface. In contrast, many of the long straight grains of *B. rigidus* fell and lay across soil clods with the embryo suspended in the air; on the smooth soil the grains of *B. rigidus* lay flat on the ground making good contact with the soil and its water. Occasionally grains of *B. rigidus,* falling like darts, entered cracks and then germinated quickly. If the soil surface had been in a plastic condition *B. rigidus* might more often have exploited its dart-like properties in this way.

A more critical experiment was made in the same way using seeds of *Brassica napus* and *Chenopodium album*. Microtopography had a marked effect on the relative establishment of the two species (Fig. 5/4), and again the different behaviour of these two species on the soil surfaces could be seen in the way in which they came to rest after sowing. The seeds of *C. album* are tuberculate and rough and tended to remain on the side of a soil clod at the position of first landing. The smooth round seeds of *B. napus* roll down into crevices. In a similar experiment made with two spherical-seeded species, *Brassica napus* and *Raphanus sativa*, a difference in diameter (1.5 and 2.5 mm respectively) was sufficient to differentiate their behaviour on soil surfaces (Fig. 5/5) (White, 1968).

All of these experiments emphasize a somewhat unusual aspect of seed size and shape as well as a neglected element of environmental variation in determining the size and quality of plant populations. It is not then surprising to find a number of aspects of seed morphology which become more meaningful under this type of analysis. The existence of seed polymorphisms, for example tuberculate and non-tuberculate seeds (which are common in the Caryophyllaceae) or of small and large seeds (common in the Compositae) becomes readily interpretable as an adaptation to life in an environment that is heterogeneous at the scale relevant to seedling establishment. A particularly neat example of precise adaptation in seed—substrate contact comes from a study of

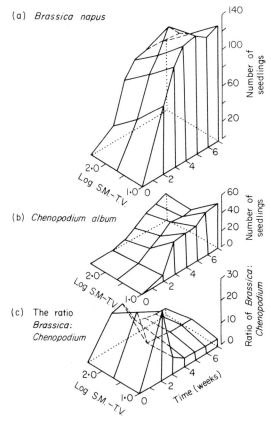

Fig. 5/4. The influence of soil microtopography on the establishment of populations of *Chenopodium album* and *Brassica napus* from seed. The microtopography of the soil is expressed as the logarithm of the variance of the height of the soil surface. (From Harper *et al.*, 1965)

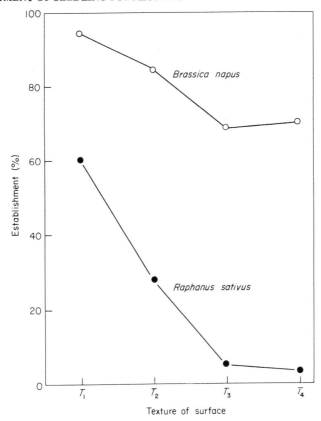

Fig. 5/5. The influence of soil microtopography on the establishment from seed of *Brassica napus* and *Raphanus sativus*. The varied soil surfaces were obtained by using different particle sizes of vermiculite: T_1 = 5.5–9mm, T_2 = 5.4–2.3 mm, T_3 = 3.1–2.1 mm and T_4 < 2.1 mm. (From White, 1968)

Blepharis persica. The seeds of this species bear unicellular hairs on the testa. When the seeds fall on wet ground and touch the soil with their flat sides, hydration and swelling of the hairs raises the seeds to an angle of 30–45° to the soil surface and brings the micropyle into contact with the soil. This stage takes only a few minutes. The tips of the hairs become mucilaginous and later, as they dry, form connections between the seed and the soil which help seedling anchorage and root penetration. If the integuments which bear these hairs are removed, root penetration is poor, even if the seeds are deliberately oriented with the micropyle towards the soil (Gutterman *et al.*, 1967)!

Seeds of *Panicum turgidum* have a convex and a flat side. Germination is much improved if the seeds land flat side downwards (Koller and Roth, 1963). An extraordinary role is claimed for the seed shape of *Eusideroxylon zwageri* in which the seeds are the shape and size of golf balls. This species is said to be confined in its distribution to the bottom of valleys because the seeds always roll downhill (Witkamp, 1925), a macrotopographical extension of the microtopographical effect already described for *Brassica*!

Structures which confer dispersibility on a seed may also influence the places and positions in which the seed lands. The rigid pappus of many Compositae illustrates this clearly. The achenes of *Taraxacum* bear stiff pappus filaments on a beak projecting from the achene and come to lie on the soil surface at an angle. There is some evidence that this is an adaptively superior angle (Sheldon, 1973). The pappus was removed and the achenes were deliberately placed at various angles to the soil surface; superior germination was obtained with the achenes at the angle at which the pappus would have supported them (Fig. 5/6). In

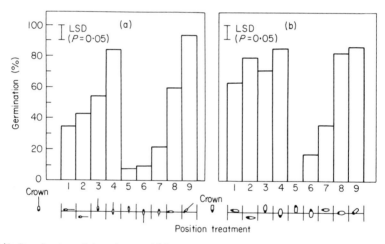

Fig. 5/6. Germination of the achenes of (a) *Taraxacum officinale* and (b) *Sonchus oleraceus* sown in different positions on a water-supplying substrate. (From Sheldon, 1974)

other genera of the Compositae there are different fates for the pappus after the act of dispersal is over: it may absciss quickly (*Cirsium* spp.), open and close hygroscopically, altering the achene's position on the soil (*Taraxacum officinale*) or collapse and become mucilaginous (*Senecio* spp.). These differences seem bound to interact with soil micro-sites in determining which seeds germinate where. (A general review of the significance of the shapes and sizes of seeds is given by Harper *et al.*, 1970.)

Small-scale variations in the environment — "safe sites"

Terms such as micro- or small depend for their meaning on the eye of the beholder — a rift valley is microscale relief to a continental geographer; a 2 cm pebble is microscale relief in the environment of an oak tree but macroscale relief in the environment of an acorn; a 1 mm sand grain is microrelief in the environment of an acorn but macroscale to an orchid seed. So far in this chapter attention has been concentrated on scales of environmental heterogeneity of much the same order as the size of the seeds: it is these that are the environmental sensors in the germination phase. Several experiments have been directed to examine the effects of a slightly larger scale of heterogeneity, sufficient to include small populations of seeds and seedlings.

Seed of *Plantago lanceolata*, *P. media* and *P. major*, was sown on an artificial seed bed of 150 x 150 cm area and 30 cm depth overlying a free-draining layer of gravel. The seed bed was prepared from sterilized horticultural compost (John Innes No. 2), sifted to give a fine crumb soil to a depth of 5 cm from the surface (Harper *et al.*, 1965). The whole area was divided into 4 replicate blocks each containing 9 plots and was sown with seed of the three species uniformly at a density of 1666 seeds per m². The nine plots in each block were treated in different ways. (1) A hole 12.5 x 12.5 cm and 1.25 cm deep was made by pressing a square of glass onto the soil and then removing it. (2) A similar hole was prepared 2.5 cm deep. (3) A sheet of plate glass 12.5 x 12.5 cm and 0.6 cm thick was placed horizontally on the soil surface and allowed to remain there. (4a) A sheet of plate glass 12.5 cm long and 3.75 cm high was inserted vertically into the soil to a depth of 1.25 cm leaving 2.5 cm projecting vertically above the soil aligned from north to south; (4b) as (4a) but aligned east to west. (5) A thin-walled wooden box 12.5 x 12.5 cm without top or bottom and with sides 3.75 cm high was pressed 1.25 cm into the soil. (6) A similar wooden box with sides 2.5 cm high was pressed 1.25 cm into the soil. (7) A similar box with sides 1.25 cm was pressed into the soil leaving the top of the box flush with the soil surface. (8) A control plot received no special treatment. The arrangement of the experiment is shown in Figs. 5/7 and 5/8a

The experiment was set up out of doors and received only natural rainfall. Seedlings appeared in flushes after rain; and the positions of

Fig. 5/7. Experimental layout for the study of the effects of small alterations imposed on the soil surface on the germination and establishment of *Plantago* species (cf. Fig. 5/8).

these were mapped and they were then removed. Three separate flushes of seedlings were mapped in this way and the results are shown in Figs 5/8b, c and d and Table 5/II.

To a very remarkable extent the various treatments differentiated between the three species sown on the area. Seedlings of *Plantago lanceolata* emerged in great numbers from the deeper of the two hole treatments and around the little wooden boxes. 23.5% of all the emerged seedlings of this species were from the deep holes and 44.8% from the immediate neighbourhood of the little wooden boxes. In striking contrast 46.7% of all the seedlings of *Plantago media* emerged under or around sheets of glass lying flat on the soil surface and only 21.9% in the environment of the little wooden boxes. In the control areas only 8.3% of the seedlings of *Plantago lanceolata* appeared and 10.5% of the seedlings of *Plantago media*, but 27.9% of the seedlings of *Plantago major* appeared in these control areas. The latter species was remarkably unresponsive to treatments. No attempt was made to determine the variations in the physical factors of the environment that were caused by the treatments. Indeed it is questionable whether sufficiently sensitive instrumentation is available for the purpose, but the nature of the responses suggest strongly that soil-water relations were amongst the

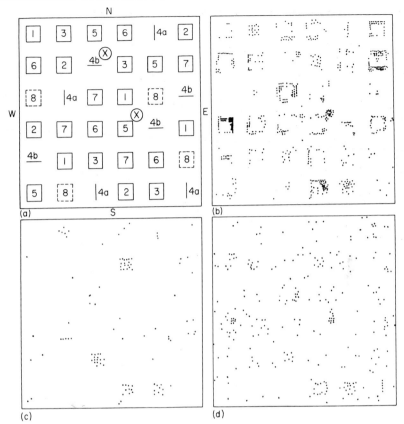

Fig. 5/8. (a) Plan showing the distribution of various types of objects placed on a soil surface sown with seeds of *Plantago* species (see text and Fig. 5/8). x = worm casts. (b) The distribution of seedlings of *P. lanceolata* in relation to objects and depressions on the soil surface. Combined records of 1 May and 11 June 1959; all seedlings were counted and removed on each date. (c) As (b) for seedlings of *P. media*. (d) As (b) for seedlings of *P. major*. (From Harper *et al.,* 1965)

critical variables introduced by the treatments.

   Not only does this experiment reveal <u>the very fine scale of the discrimination that can occur between species</u> — but it also makes the important point that the abundance of a particular species is governed by the frequency of occurrence of particular microenvironments. If we had wished to increase the number of seedlings of *Plantago lanceolata* in the experimental area we should have created more holes or inserted more wooden boxes. For *Plantago media* we could have increased the population density by laying down more sheets of glass.

Table 5/II
The emergence of seedlings of *Plantago* species in relation to the
distribution of various modifications in the soil surface

(From Harper *et al.*, 1965)

Treatment	Percentage of total emergence for each species		
	P. lanceolata	*P. media*	*P. major*
1. Hole 1.25 cm deep	7.3	4.8	8.0
2. Hole 2.5 cm deep	23.5	11.4	18.5
3. Glass sheet flat on surface	9.6	46.7	11.6
4a. Glass sheet vertical N–S	2.3	2.9	3.3
4b. Glass sheet vertical E–W	3.8	1.9	2.2
5. Box 2.5 cm above surface	10.6	3.8	4.7
6. Box 1.25 cm above surface	11.0	4.8	8.7
7. Box 0 cm above surface	23.2	13.3	12.0
8. Control – no treatment	0.4	0.0	3.3
9. Seedlings not associated with treatments 1–8 (43.5% of total sown area)	8.3	10.5	27.9
Total number of seedlings emerged	689	105	276

All seedlings found within the treated area (12.5 × 12.5 cm) and including a 2.5 cm broad surrounding border were regarded as associated with a treatment. Figures are means of four replicates accumulated over two sampling dates, 1 May and 11 June 1959.

During the course of this experiment two earthworms emerged from below the prepared seed bed and deposited casts on the soil surface (x in Fig. 5/7a). A mass of seedlings of *Plantago lanceolata* developed around these casts. This species is among those that are ingested by worms and pass through the gut without harm (McRill and Sagar, 1974). Among the variety of environmental heterogeneities deliberately introduced into this experiment, this naturally occurring event in some intangible way increased the meaning of the whole experiment – natural events producing the same order of reaction as an experimental treatment imposed by man. It was another unpredicted element in an experiment that had originally stimulated the design of the *Plantago* experiment. Drills of seed of *Plantago lanceolata* had been sown in arable land to raise young plants for a physiological study. Emergence and establishment had been poor except where pigs, which had escaped from a nearby field, ran across the plot. Excellent seedling establishment occurred in the footprints.

Small-scale events like the pressing of a hoof have repercussions in localized plant behaviour, but only rarely can the behaviour be traced back to the causal event. During a period when a population of *Ranunculus bulbosus* was under fortnightly census in a permanent grassland (see Chapter 19) cattle walked across the plots when the ground was very wet. The imprints of their footprints were mapped. In the following spring the distribution of seedlings was mapped and is shown in Fig. 5/9: the seedlings neatly followed the hoof outlines though there was no

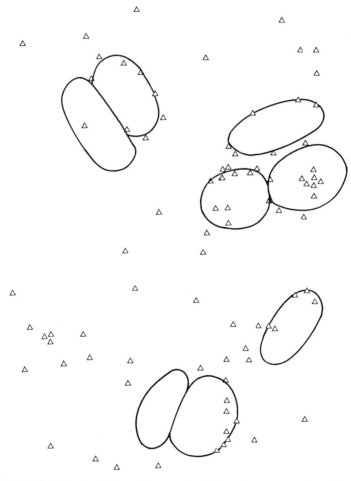

Fig. 5/9. The distribution of seedlings of *Ranunculus bulbosus* in a permanent grassland in relation to cattle hoof marks made in autumn when the ground was wet. (From J. Sarukhán, unpublished)

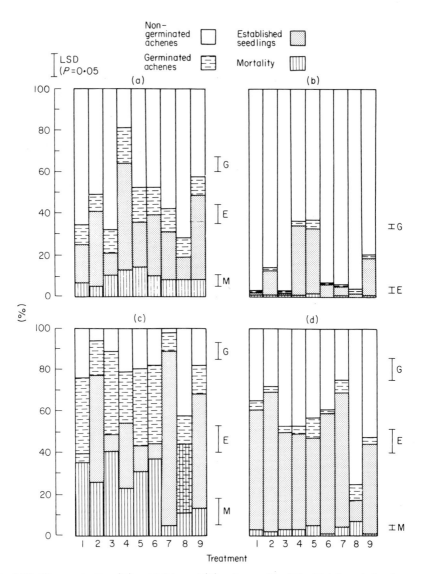

Fig. 5/10. The germination (G), establishment (E) and mortality (M) of (a) *Senecio jacobaea* and (b) *Leontodon autumnalis*, and (c) *Tragopogon porrifolius* and (d) *Erigeron acer*, in relation to a variety of micro-environmental modifications on a soil surface.
(1) Control, the finely particulate soil receiving no modifications.
(2) A wooden cross with arms 20 x 5 cm and 0.5 cm thick, placed on the soil surface.
(3) The soil compacted at 0.4 kg/cm².
(4) A transparent polythene pyramid, placed on its apex, the base measuring 20 cm² and the sides lying at an angle of 22.5° to the soil surface.
(5) The same, but of black polythene.
(6) An open wooden box, base 20 cm², 5 cm deep and with walls 0.5 cm thick, placed on the soil surface.
(7) A hole 20 cm² x 2 cm deep.
(8) A flat-topped uncompacted mound, base 20 cm², 2 cm high above the general soil level.
(9) Twenty white plastic strips 7 x 1 cm, randomly distributed over the plot.
(From Sheldon, 1974)

longer any trace of the footprints.

A rather more complicated experiment was carried out (Sheldon, 1974) to determine the effects of various obstructions and treatments of a soil surface on the germination of four species of Compositae, and the results are shown in Fig. 5/10.

The species specificity of the process of seedling establishment is also emphasized in a comparison of the biology of some *Ranunculus* spp. which are perennial herbs of grassland. The distribution of mature plants of three species, *Ranunculus repens, R. acris* and *R. bulbosus,* is related to the drainage pattern of the ground and in the typical ridge and furrow

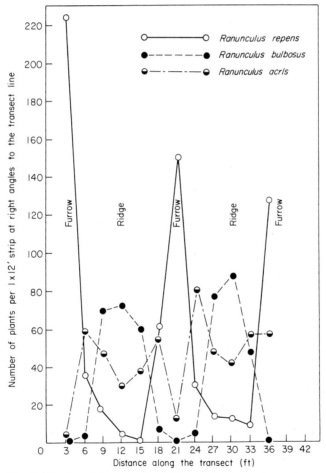

Fig. 5/11a. The distribution of three species of *Ranunculus* along a transect across ridge-and-furrow grassland. (From Harper and Sagar, 1953)

grasslands of the English Midlands *R. bulbosus* tends to occupy the ridges, *R. repens* the furrows and *R. acris* the intermediate zone (Fig. 5/11a). When seeds of these three species were sown on soil in pots which were maintained waterlogged, free-drained or at an intermediate level the recruitment of seedlings in the different water regimes tended to parallel the field distribution (Fig. 5/11b). It is an interesting feature of this experimental result that although the position of the water table affected the three species differently it did not do so as an all-or-nothing effect. The position of the water table altered the probability that a seed would produce a plant: it changed the frequency of "safe sites".

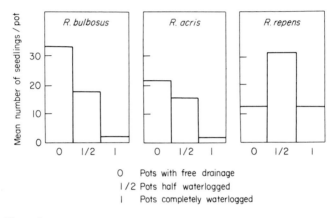

O   Pots with free drainage
1/2 Pots half waterlogged
I   Pots completely waterlogged

Fig. 5/11b. The early establishment of three species of *Ranunculus* sown in pots with the water table maintained at different levels (Sowings were made on June 26th 1953 and counts made on July 28th 1953). (From Harper and Sagar, 1953)

The safety of a germination site is determined not only by the necessary elements that it provides for seedling establishment but by the hazards that it excludes. One of the most acute hazards for a tiny seedling is desiccation at a time when the root system is scarcely established. For a plant adapted to mires or marshes the water supply may be particularly critical during short periods of drought. *Narthecium ossifragum* is one such mire inhabitant and its seeds were sown on a sloping soil surface with a maintained water table. The establishment of seedlings is shown in Table 5/III. Very early in seedling development the populations were restricted to a narrow zone close to the water table. Another population of seedlings was subjected to desiccation for 3, 14 or 48 h. Table 5/IV shows the recovery after this experience. As the seedlings grew older the death risk from such an experience rapidly declined.

Table 5/III
The influence of water table on the establishment of populations of
Narthecium ossifragum from seed
(From Summerfield, 1973)

Number of days on slope	Depth of water table below soil surface (cm)					
	1	3	6	9	12	15
21	0	0	0	0	0	0
31	8	4	0	0	0	0
37	72	28	0	0	0	0
45	76	32	4	0	0	0
49	85	46	4	0	0	0
55	88	46	4	0	0	0

Table 5/IV
The effect of varying periods of desiccation on the survival of
seedlings of *Narthecium ossifragum* of various ages
(From Summerfield, 1973)

Developmental stage at the time of drought	Percentage survival after drought		
	3 h drought	14 h drought	48 h drought
1. Radicle just emerged (2–3 days old)	0	0	0
2. 1st leaf 1–1.3 cm; root 1.0 cm (7–9 days old)	100	84	0
3. 3–4 leaf stage (35–40 days old)	100	100	100

There is as much arbitrariness involved in defining "germination" and "establishment" as there is in putting limits to other phases of plant development. A seed that is going to germinate may begin to show marked changes in the ultrastructure of its cells within 30 min of wetting, but external evidence that a seed is going to germinate may await the emergence of a radicle. In some species (e.g. *Hyoscyamus niger* (Ødum, 1965)) the testa may indeed crack and a radicle protrude when a dead seed is wetted, so emergence of a radicle may not be reliable evidence of germination. In other species (e.g. *Trillium* spp.) a radicle may emerge in the first season after germination and the aerial shoot may be delayed by a year. In practice it is normal to measure germination in a laboratory experiment by the number of seeds that have developed an emerged radicle. When seeds are sown in soil and the emergence of the radicle cannot be observed "germination" is commonly measured as the number of seedlings emerging from the soil surface. Such observations are better recorded as "seedling emergence" as the numbers will represent all the seedlings that germinated (in the laboratory sense) minus those that died in the pre-emergence stage. A convenient later phase in the recruitment process may be called "establishment": when the seedlings have expanded a photosynthetic surface and are theoretically capable of pursuing an existence independent of their seed reserves.

Most of the "sifting" qualities of the environment that have been illustrated concern seeds that are exposed on the surface of the ground. If seeds are buried, many of the extreme hazards involved in absorbing and losing water are avoided. At the same time burial isolates the seed from light and thus from one of the most important germination stimuli. So many species bear seeds that have or acquire a light requirement for germination that surface germination must be extremely common (see Chapter 4). However, there are many species which will germinate in the dark, including the seeds of most crop plants and the latter are usually sown in, rather than on, soil to maximize germination. Other, wild, species possess burial mechanisms which force the seed into the soil, e.g. the twisting awn of *Avena fatua* and of *Erodium* spp., and although these self-burying mechanisms are usually considered as seed dispersal devices (Ridley, 1930) it is more appropriate to regard them as comparable to the mucilage of *Lepidium* or the hairs of *Blepharis* which maximize water uptake and minimize water loss by the germinating seed.

Recruitment from buried seed populations

Although a seed that is buried beneath the soil surface or hidden below a fallen leaf may be buffered against rapid water loss to the atmosphere, an extra work load is involved in comparison with a surface-germinating seedling in the need to extend the potentially photosynthetic area above the soil surface or above the sheltering object. There has been a great deal of agronomic research aimed at determining an optimal depth of sowing for different crop plants (Black, 1959; Heydecker, 1956). One of the most detailed studies was made by Black (1956a) and Black *et al.*

(1963) with *Trifolium subterraneum*. Seed samples of the variety 'Bacchus Marsh' were selected with mean seed weight of 3, 5 and 8 mg and these were sown at three depths 1.3, 3.1 and 5 cm. Depth of sowing affected the time taken for seedling emergence from the soil, all seed sizes produced emerged seedlings from 1.3 cm depth on the 4th day, but the smaller the seeds the later their seedlings emerged from the deeper sowings. Indeed seedlings from the smallest seed were not capable of elongating the hypocotyl to more than 3.7 cm. The corresponding degree of hypocotyl elongation from the larger seed sizes was 5.2 and 6.7 cm. Resources needed for hypocotyl extension are translocated from the cotyledons and the deeper the seed is sown the greater is the proportion translocated. The seedlings of deeper sown seeds therefore have less food store in the cotyledons at the time of emergence, but apparently the photosynthetic efficiency is not impaired and the vigorous growth of the seedlings depends on the area of the cotyledons, not on their weight.

Other evidence for the effects of depth of sowing on seedling recruitment comes from studies of *Linum usitatissimum* in which there is variation in seed size both between and within cultivars (Harper and Obeid, 1967). The major effect of depth of sowing was on the time taken for seedling emergence — from shallow sowings the first seedlings to emerge were from the smallest seeds — from deeper sowings the seedlings from large seeds emerged first (Fig. 5/12). There was a very close correlation between seed size and length of the hypocotyl (Fig. 5/13).

These experiments all suggest that only seeds with extensive seed reserves are likely to be able to form established seedlings from deeply buried seed. Correlated with this is the general absence of a light requirement for the germination of large seeds. As pointed out earlier there are likely to be fewer problems in the water relations of small than of large seeds on the soil surface. The various properties of seed size, the problems of water uptake, the response to depth of sowing and the extent of light sensitivity in germination are probably parts of a coadapted syndrome.

A large fraction of the buried viable seed in natural and agricultural soils remains dormant. What fraction of this population germinates at depths from which seedlings fail to emerge and so die is obscure. The evidence from the experiments of Wesson and Wareing (see Chapter 3) suggest that the buried seed population has in large part acquired a light requirement and is only recruited into a seedling population after a soil

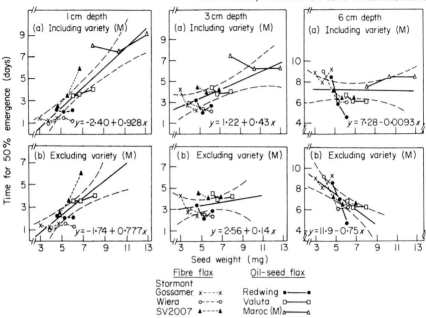

Fig. 5/12. The relationship between seed weight and velocity of seedling emergence from 1-,
3-, and 6-cm depths of sowing. The three seed size categories of each variety are
shown connected by a line. Fiducial limits ($P = 0.05$) of the regressions are shown
as broken curves on either side of the linear regression line. % Germination of M
('Maroc') was poor and unreliable — the calculations are shown with and without
'Maroc'. (From Harper and Obeid, 1967). Reproduced from Crop Science Vol. 7
pages 527—32 by permission of the Crop Science Society of America.

disturbance which exposes the seed to light. If this is generally true,
experiments on depth of sowing may need to be interpreted with much
care. Experimentally sown seed will normally receive light while it is
being sown and start into germination no matter how deep it is sown,
provided it receives appropriate temperature and water supplies. A corresponding naturally dispersed seed landing on and being incorporated
into the soil when the temperature is low or the soil is dry may not germinate and may then acquire a need for light, so that it remains dormant
even when temperature and water supplies again become favourable.

Little is known for natural communities about the relationships between the seed reserve in the soil and the fraction that is recruited into
seedlings each year but there is some information from the weed populations of arable crops. The weed flora of a crop of winter wheat represented only 4% of the viable seed population in the top 10 cm of soil

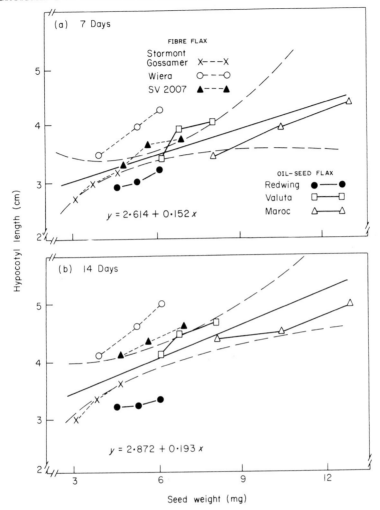

Fig. 5/13. The relationship between seed weight and hypocotyl length (a) 7 and (b) 14 days after sowing. (From Harper and Obeid, 1967). Reproduced from Crop Science Vol. 7 pages 527–32 by permission of the Crop Science Society of America.

(Barallis, 1965) and similar values have been found by Kropáč (1966) in arable fields in Czechoslovakia. Both a dense stand of *Matricaria recucita* (Chancellor, 1965) and an infestation of *Alopecurus myosuroides* (Naylor, 1972) represented 1–4% of the buried seed population.

It is usually quite uncertain whether a seedling population is recruited evenly from all age groups in the buried seed population. An attempt to answer this question for *Alopecurus myosuroides* was made (Naylor, 1972) using a technique of mark and recapture. When natural seed dispersal was occurring, seed which had been marked with fluorescent paint was sown on the soil surface. Such paints in various colours can be obtained as aerosols and a fine spray of paint droplets is applied to the floral membranes that surround the grains of this species. In the following season after normal ploughing, discing and sowing of the crop, an infestation of *Alopecurus myosuroides* developed in the crop and on excavation it was possible to recognize which of the seedlings came from the deliberately sown seed. Using this technique Naylor estimated that between 62 and 71% of the populations of *Alopecurus myosuroides* in two sites was derived from seed less than one year old. In a study of the field germination of naturally sown seed of *Trifolium subterraneum* (Donald, 1959) 92.1% of the seedlings appeared in the first year after sowing, 6.3% in the second year and 1.0, 0.52 and 0.07% in the third, fourth and fifth years. These data emphasize that populations, even of annual plants, have age distributions if the age of the plants is measured from the time of zygote formation instead of from the time of germination. The greater part of such populations is apparently composed of "young" individuals with a progressively lower representation of older age groups. The different age groups recruited into the growing population have experienced different histories and probably different selective forces. Seedlings that emerge from long-buried seed will have escaped selection by recent events and may be genotypically different from those derived from younger seeds. This must have the effect of buffering the whole population against rapid genetic change due to temporary selective forces.

Clearly if population biology advances to the stage of a predictive science, it should be possible to use the parameters of (i) seed rain, (ii) soil bank, (iii) recruitment rate and (iv) expected mortality to predict the size of seedling populations. An attempt has been made to do this for populations of *Alopecurus myosuroides* in arable fields in south and east England. Soil samples were taken from a number of fields in which this species had been present as a weed. The samples of soil were taken to a depth of 2.5 cm as it was known that 90% of the populations of this species were derived from seed in the top 5 cm of the soil. The soil samples were placed in germination trays (flats) in a glasshouse, watered regularly and the seedlings of *A. myosuroides* emerging were counted. Calculated regressions between these estimates of the potential weed density and the actual field densities in the various fields the following year gave remarkably good predictions. The "Weed Predictive Index"

accounted for 97% of the variance in field density of the weed in the
following April and 94% in May (Naylor, 1970a, b).

The compaction of soil influences the chance of a buried seed producing an emerged seedling: this was elegantly demonstrated for seed of
Medicago sativa sown at different depths in soil that was afterwards
compacted by the application of different weights to the soil surface
(Triplett and Tesar, 1960). Seedlings often fail to emerge from soil that
has become "capped" by an impervious clay layer after the impact of
raindrops. Whether such capping prevents germination (by impeding gas
flow to the seeds) or emergence because of the physical impediment is
not clear. Physical impediment is however the more likely explanation
because groups of seedlings seem able to break through such capped soil
more readily than isolated individuals: physical force seems to be needed.
The physical impediment discriminates between species. In the case of
the onion (*Allium cepa*) there is a very good correlation between seed
weight and successful emergence from depth whereas the radish (*Raphanus sativa*) gives a very poor correlation. The shape of the onion seedling
is ideally formed to penetrate the soil whereas that of the radish offers
more resistance and the larger the cotyledons the greater is the resistance
(Hewston, 1964).

Many (perhaps most?) seeds in nature germinate on the soil surface.
The texture and "strength" of the surface may then restrict penetration
by the radicle with the result that a successfully germinated seed dries
up and dies (Stolzy and Barley, 1968). Seedlings of different species
differ in the ability of their radicles to penetrate the soil surface: in a
controlled comparison of four grassland species the order of effectiveness of penetration into a compacted surface was *Trifolium subterraneum* > *Medicago sativa* > *Phalaris tuberosa* > *Lolium perenne* (Campbell and Swain, 1973). On sodium clays the surface layer often forms a
hard crust — very small applications of gypsum are sufficient to change
the texture of this surface layer from an inhospitable to a hospitable
environment for germination (Davidson and Quick, 1971).

Recruitment depending on the interaction of many factors

A demonstration of the way several factors can interact in determining
the recruitment of seedlings of species in nature comes from a study of
the two junipers *Juniperus communis* and *J. virginiana* in New England
(Livingstone, 1972). *Juniperus communis* characteristically establishes
by the side of exposed stones in old field sites. Cattle do not graze or

trample close to the stones. Seed of *J. communis* is eaten by red squirrels and birds. The cedar wax wing (*Bombycilla cedrorum*) depulps the seeds and eats the kernels but the American robin (*Turdus migratorius*) eats and digests the pulp and excretes the seeds. The robin sits on the stones when defaecating. A count of the number of droppings on stones and control "non-stone" areas showed 196 on stones compared to 14 on the control areas. The faeces contained many seeds of *Juniperus*: 1410 seeds of *J. communis* were deposited on stones and only 53 on the control area. The seeds are washed off the stones into crevices in the soil and frost heaving of the soil in winter creates a crevice 2–3 mm wide around each stone. The water run-off from the stones into the crevices and the protection they offer from evaporation ensure that the seed spends the summer experiencing the warm *moist* stratification that is required by *J. communis* but not by *J. virginiana*. This sequence of events results in recruitment of *J. communis* around stones.

Chemical factors affecting the recruitment of seedlings

The process of germination makes few or no demands on the consumable resources of the environment apart from the need for oxygen and water. Most seeds (with the exception of those of some parasites and species like *Monotropa* and orchids that form endotrophic mycorrhizal associations) carry a sufficient supply of nutrients to support the developing seedling well beyond the stage at which it becomes photosynthetically independent. Indeed the large seeds of some forest trees, e.g. *Quercus*, carry enough mineral resources to support the whole of active growth during the first year. The chemical factors in the substrate are more likely to influence recruitment by toxic action than by failure to provide essential nutrients. This may occur when evaporation exceeds precipitation (particularly when irrigation is applied) and sodium chloride accumulates at the surface of the soil. This characteristically develops in a patchy fashion and inhibition of germination may then occur in the patches of salt. Seeds germinate in such zones only after heavy rains when the salt concentration at the surface is temporarily lowered (Went, 1957).

In an experiment in which seed of *Plantago* species was deliberately introduced as a seed rain into a variety of habitats from which these species were normally absent, germination and seedling emergence occurred in most communities (woodlands, heaths, acid grasslands) but in very acidic sites, although a radicle started to emerge, this quickly blackened and the seedlings died before cotyledons were expanded (Sagar and Harper, 1960). Clearly this is a case in which recruitment was directly affected by soil conditions — probably chemical. It is, however, not particularly easy to discriminate between direct toxic action of soil chemical conditions and indirect effects due to soil-borne pathogens which are themselves determined in distribution by the chemical conditions (see Chapter 4).

An important attempt to determine the direct role of soil chemistry in the recruitment of a seedling population was made in a "seedling bioassay of some Sheffield soils" (Rorison, 1967). A variety of species was studied: two calcifuges — *Deschampsia flexuosa* and *Digitalis purpurea*; two calcicoles *Erigeron acer* and *Scabiosa columbaria*; and three intermediate or more widely tolerant species — *Rumex acetosa*, *Betonica officinalis* and *Urtica dioica*. Seed was sown on the surface of soil samples taken from the field and maintained moist in closed plastic boxes in controlled environment chambers. The soils varied from podsols to mull-rendzinas with a range of pH from 3.5 to 7.4 and from 1% carbonate to 38.6%. The progress curves of germination and seedling survival are shown in Fig. 5/14. "The most striking differential responses came from *Urtica dioica* and the two calcicoles, *Scabiosa columbaria* and *Erigeron acer*. Here germination was most rapid and most complete on the calcareous soils and progressively slower and less complete on increasingly acid soils." The percentage germination of *E. acer* for example rose steeply from a start on day 3 to about 90% by day 10 and 95% by day 16 for seeds on the three calcareous soils and on a control garden soil. On the acid soils there was a somewhat delayed start and a less steep rise to a plateau at 68% germination; on the still more acid soil germination only reached 34% and on the most acid of all there was no germination until day 5 and a peak at only 10% on day 7. Even *Digitalis purpurea*, a calcifuge, suffered in speed and amount of germination on the most acid soil. These appear to be clear-cut examples of chemical discrimination between species at the seedling recruitment stage. It is however remarkable that with such a deliberately chosen wide range of species and of soil types no species was precluded from germinating on any soil; only relative, not absolute success was affected.

The seedling bioassay experiment appears to show that discrimination between species by chemical properties of the substrate is only important for extremely specialized species on extreme substrates. Water, light and a suitable temperature are the major determinants of whether

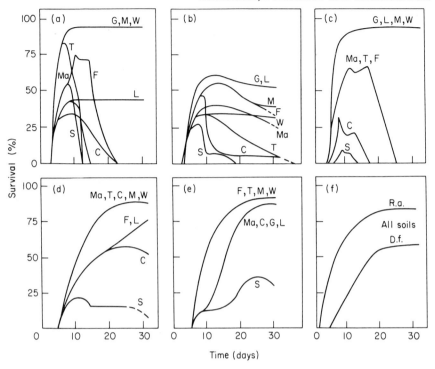

Fig. 5/14. Seedling survival of (a) *Urtica dioica,* (b) *Scabiosa columbaria,* (c) *Erigeron acer,*
(d) *Betonica officinalis,* (e) *Digitalis purpurea,* (f) *Rumex acetosa* and *Deschampsia
flexuosa,* on nine experimental soils. Individual soils are represented only when
they gave rise to a distinct survival curve. S = millstone grit, pH 3.5; F = ditto pH
4.1; Ma = edale shale pH 4.1; C = loessic drift over carbonif. limestone pH 3.7;
T = coal measure shale and sandstone pH 3.8; G = ditto pH 6.1; L = magnesian limestone pH 7.2; M = carboniferous limestone pH 7.0 and W = ditto pH 7.4. (From
Rorison, 1967)

a seed embarks on the hazardous life of being a seedling. Once the
radicle of the germinated seedling starts to explore the soil system a
whole sequence of new hazards appears, and the nature of the substrate
now eliminates whole populations of species (Fig. 5/14) and changes the
survivorships of those that do persist.

The present and preceding two chapters have taken what might be
called a "Gleason" approach to the population biology of plants. ". . .
every species of plant is a law unto itself, the distribution of which in
space depends upon its individual pecularities of migration and environmental requirements. Its disseminules migrate everywhere, and grow

wherever they find favourable conditions. . . . It grows in company with any other species of similar environmental requirements, irrespective of their normal associational affiliations. The behaviour of the plant offers in itself no reason at all for the segregation of definite communities" (Gleason, 1926). In fact, order and organization appear in populations because of feedback from existing populations to new recruits and because these recruits themselves interact. The presence of vegetation has profound influences on the nature of the environmental sieve.

The feedback from developing vegetation and its influence on the recruitment of seedlings

As vegetation develops on an area of land there are major changes in the environment at and beneath the soil surface which alter its hospitality as a seed bed. In particular a shading canopy reduces the light intensity and changes its quality at the soil surface, reduces the rate of direct evaporation from the surface and creates a zone of generally higher humidity and more equable temperature. In addition falling leaves, and later their decomposition products, change the physical and chemical properties of the soil itself and may cover and bury seeds that have fallen on the surface. All of these changes alter the nature of the environmental sieve that selects those seeds that will germinate. This feedback is shown as an arrow on the flow diagram that heads this chapter. One form of feedback from growing vegetation is the induction or enforcement of dormancy in the seed population. Rising carbon dioxide concentrations in soil due to root respiration and the decomposition of litter may reduce the number of seeds that germinate (Chapter 3) and many claims have been made that growing vegetation produces toxins which inhibit seed germination (Chapter 11). Some of these processes are considered in other parts of this book and only a few examples are treated in detail in the present chapter.

It is difficult to find good field evidence of germination of any species being promoted by a vegetation cover. What few observations have been made in the field generally show reduced germination. Only in laboratory studies has it been possible precisely to separate effects on germination from effects on subsequent survival (or on post-germination, but pre-emergence, mortality). A series of experiments (Foster, 1964) on the establishment of daisies (*Bellis perennis*) in grassland illustrates the ways in which an existing sward affects recruitment. Six simulant grazing treatments were applied to a grass sward: (A) no defoliation — the grass was allowed to grow tall and dense; (B) a single defoliation — the exclosure had been erected at the end of April and the cut was taken 15 weeks later; (C) the sward was defoliated 3, 9 15 and 23 weeks after exclosure — every time the grass reached a height of 15 cm it was cut back to *ca* 2.5 cm; (D) as (C) but cut 2, 4, 12, 17 and 23 weeks after exclosure — this treatment in particular kept the sward short in May; (E) the sward was kept continuously short by cutting 2, 4, 6, 9, 12, 15,

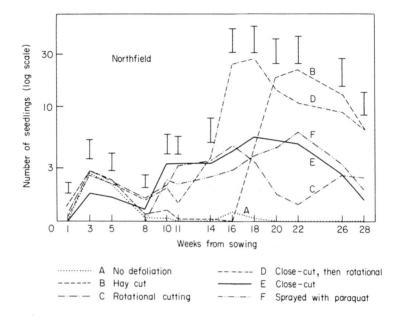

Fig. 5/15. The establishment of seedlings (per 30 cm square) of *Bellis perennis* from seed sown into a pasture. The pasture was subject to a variety of experimental treatments — see text. Fiducial limits (*P* = 0.05) are shown for each sampling date. (From Foster, 1964)

17, 20 and 25 weeks after exclosure. In addition a "bare" treatment (F) was imposed in which plots were sprayed with the herbicide Paraquat (dipyrillidium chloride) which killed all above-ground vegetation. Seeds of *Bellis* were scattered onto plots treated in these various ways, at 50 or 500 seeds per 30 cm square quadrat. Patterns of seedling emergence are shown in Fig. 5/15. In all treatments a small flush of seedlings appeared in the first 3 weeks after sowing — but these rapidly died in both the undefoliated and once defoliated treatments (A) and (B). Seedlings continued to appear in the frequently cut and the grass-killed plots.

Most striking was that a great flush of seedlings appeared 14—18 weeks after the start of the experiment in treatments (B) and (D). These specific sward treatments created excellent conditions for germination of seed that had been lying dormant, and because the two flushes were not synchronous, the flushes cannot be ascribed to a period of particularly favourable macro-climate. In these two treatments the cut swards allowed much light to reach the ground surface at the time just preceding a flush of germination. The flush immediately followed the cut in the once cut-treatment (B) and the frequent May defoliation (D) produced a very open sward. The data are summarized in Fig. 15/16 for the

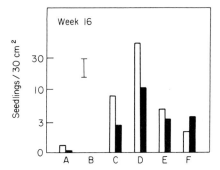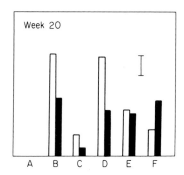

Fig. 5/16. The establishment of seedlings of *Bellis perennis* from seed sown into a pasture to which various management treatments had been given — see text. Fiducial limits are shown at $P = 0.05$. □ = no fertilizer; ■ = N.P.K. fertilizer applied. (From Foster, 1964)

number of seedlings present at week 16 and 20, week 16 preceding the seedling flush of (B) and week 20 being after that flush. This figure also shows the impact on the seedling population of applying an N—P—K fertilizer. In all treatments in which the grassland sward was present applying fertilizer resulted in poorer seedling recruitment and survival; yet when the grasses were absent (in the Paraquat plots) the same fertilizer treatment increased the seedling population of *Bellis*. Where the fertilizer stimulated grass growth this reduced the hospitality of the sward to *Bellis* seedlings.

Very similar results were obtained from experimental sowings of *Rumex acetosa* and *R. acetosella* in hill grasslands (Putwain *et al.*, 1968). However the frequent failure of seeds to germinate under swards is not really understood. The germination of seeds of *Bellis* is promoted by light and hindered by raised concentrations of CO_2, but both effects are relatively slight (Foster, 1964). Changes in the composition of light after it has passed through a leaf canopy may be one of the critical factors hindering the recruitment of seedlings under vegetation.

The phytochrome system which controls the light sensitivity of germination in many species involves an interconversion of two forms of phytochrome. Red light (*ca* 660 nm) causes the formation of the active form of phytochrome (P_{fr}) which induces germination and the induction can be reversed by far-red light (*ca* 730 nm). These processes were largely studied on seeds of lettuce, of one particular variety 'Grand Rapids'. The development of this elegant piece of plant physiology was made largely by Flint and McAlister (1935), Evenari *et al.* (see Evenari, 1965) and Borthwick *et al.* (1954). The natural significance of this process remained obscure until it was pointed out (Vezina and Boulter, 1966) that far-red light predominates over red under leaf canopies and it was suggested (Cumming, 1963) that this was probably important in the germination of seeds of *Chenopodium*. It was then shown (Black, 1969) that "leaves of *Tilia* act as a perfect far-red filter which can be used to demonstrate red—far-red reversibility in Grand Rapids lettuce"; peak transmission through these leaves is at 730 nm. A similar series of observations by Taylorson and Borthwick (1969) showed that light filtered through leaves of tobacco, maize and soybean also inhibited germination of the seeds of a number of weed species. Further studies in this area are much needed; a feedback mechanism which prevents seeds from germinating under established vegetation while allowing them to remain viable would seem to be of high adaptive value.

Feedback processes from established vegetation onto the seed bed beneath have many other complexities which are illustrated in a beautiful experiment made by Hermann and Chilcote (1966). They were concerned to explore the factors that hindered the establishment of new stands of Douglas fir (*Pseudotsuga douglasii*) after clear-cutting or forest fires. They laid out 90 1 m² quadrats in 1 ha site from which Douglas fir had been clear-cut and burned 3 years previously. They spread different seed-bed materials to a depth of 2.5 cm on the plots: (i) charcoal as thoroughly charred chips; (ii) litter — a mixture of unburned small twigs, needles, leaves and dried moss; (iii) light burned soil — mineral soil mixed with ashes; (v) an unburnt control; (v) hard-burned soil — transformed by fire to bright red cinders; and (vi) sawdust from newly cut logs of Douglas fir. The plots were sown with seeds of Douglas fir scattered broadcast on the plots at 500 seeds/m².

Variations in shade were superimposed on these treatments by erecting cheesecloth-covered frames to reduce light intensity to 75% and 25% of full daylight and control plots received full daylight. Emerged seedlings were recorded and marked every week. Surface temperatures were measured with thermocouples and some very striking temperature differences appeared. During the summer, the surface of the charcoal soil exceeded 60°C for 370 min on some days and never for less than 270 min. The duration of these high temperature experiences fell in the order charcoal > litter > light burn > no burn > hard burn > sawdust. On the surface of the sawdust the comparable figures were 160 min above 60°C (longest) and 90 min (shortest) per day. The effect of light shade was to reduce the maximal temperature on the seed beds by *ca* 9°C. The heavy shade largely prevented temperature differences between the seed beds from developing and the temperatures, even during a hot spell in July—August ranged between 43 and 49°C.

The seed beds dried out at different rates, and most rapidly beneath unburned or severely burned soil: shade hindered the development of differences in the rate of drying out between the different seed beds.

The seed had been treated with a rodent repellent and there was no significant loss of seed by predation. There were however very large differences in the number of seeds germinating to give countable seedlings. Both seed-bed and light regimes produced significant effects (seed bed $P < 0.01$, light $P < 0.05$) but there was no significant interaction (Tables 5/V and 5/VI).

Table 5/V

The mean numbers of seeds of *Pseudotsuga douglasii* forming
seedlings per m^2 on six different types of seed bed

Sawdust	Litter	Light burn	No burn	Hard burn	Charcoal
84	108	130	133	243	401

Table 5/VI

The mean numbers of seeds of *Pseudotsuga douglasii*
forming seedlings per m^2 under three light regimes

Percentage of full sunlight		
100%	75%	25%
6274	4054	5166

From Hermann and Chilcote, 1966.

The differences in germination of Douglas fir on the various plots were followed by differential mortalities. 62% of all the seedlings emerging on the plots died between March and November in the first year, the major causes being destruction by animals, heat injury and "damping-off". The damage by animals during the germination period was due to nibbling of seedlings by mice and larger seedlings were taken by deer and rabbits. The mice preferred heavy shade for their activities: mortality from this cause was only 13% on the open plots, but increased to 31% in the light shade and 57% in deep shade. It is not difficult to see the ways in which a patchy distribution of a ground cover of plants (instead

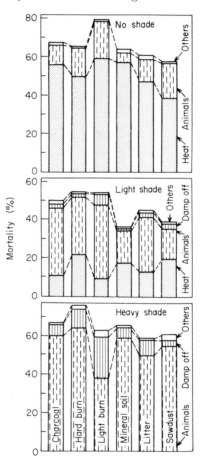

Fig. 5/17. The causes of mortality among seedlings of *Pseudotsuga douglasii* developing on six different substrates and under three shading regimes in a clear-cut forest in Oregon, U.S.A. (From Hermann and Chilcote, 1965)

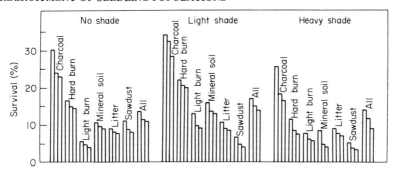

Fig. 5/18. The number of seedlings of Douglas fir still alive after the first, second and sixth growing season as a percentage of the seeds sown on a variety of prepared soil surfaces on a forest floor under three shading regimes. (From Hermann and Chilcote, 1965)

of muslin) can offer similar preferred feeding grounds for small shy mammals. In this way an indirect feedback would occur on seedling establishment (see Chapter 11).

Damping-off of seedlings was primarily due to species of *Fusarium* and *Rhizoctonia* and only 0.1% of the seedlings died from this cause on the open plots, but this hazard affected 3% of the seedlings on the half shaded and 9% on the most shaded plots. A large number of seedlings were lost from heat damage, most of this loss occurring on the unshaded plots (53%), with no loss from this cause on the most deeply shaded plots.

The role of the different mortality factors is shown as percentage of germinated seedlings in Fig. 5/17. What is highly remarkable about this set of data is that, despite the wide difference in the causes of mortality between treatments, the *overall percentage mortality* in the different treatments varies within such narrow limits (Fig. 5/18). This strongly suggests that there is compensation between the causes of mortality. Up to a limit, if a seedling is not killed by one agent it will be killed by another.

The Effects of Neighbours

6

The Influence of Density on Yield and Mortality

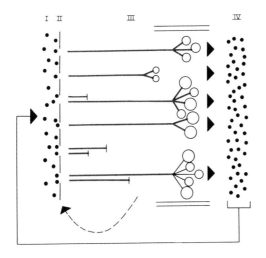

The presence of a plant changes the environment of its neighbours and may alter their growth rate and form. Such changes in the environment, brought about by the proximity of individuals, may be called "interference", a blanket term which does not define in any way the manner in which the alterations in environment are produced and includes neighbour effects due to the consumption of resources in limited supply, the production of toxins, or changes in conditions such as protection from wind and influences on the behaviour of predators. Higher plants react

to stresses of density by plastic responses as well as an altered risk of
death. The population-like structure of an individual plant fits it admirably to respond to stresses by varying the birth rate and the death rate
of its parts, leaves, branches, flowers, fruits, rootlets, etc. Such variation
is impossible for most animals except as the analogous variation of the
number of individuals in a coral or sponge or the number of colonial
Hymenoptera in a colony. Some resemblance to the plastic reaction of
higher plants to density is found in fish, in which density stress may be
reflected in the size that individuals attain, the time required to reach
maturity and the number of eggs produced. However, such animals have
nothing comparable to the birth and death of parts that is so characteristic of the growth of a plant.

Most experimental studies of the reactions of plants to density have
been made on stands of single species — the logical starting point for
approaching any understanding of the stresses present in the more
normal species-diverse communities of nature. Many of these studies
have been made by agronomists or foresters to discover ideal densities
for crop or forest yields.

The influences of density on growth

A typical yield/density response for a plant population is illustrated in a
study of *Trifolium subterraneum* made by Donald (1951). He sowed
seeds in pure stands over a wide range of density to give seedling densities from 6 to 32 500 per/m^2, a range likely to include that of any artificial or natural seeding of this species. Seven densities were included in
the experiment and plants were harvested after 62, 131 and 181 days.
The results, like those in most comparable experiments, refer only to
the above-ground parts of the plants and are shown in Fig. 6/1a and b
as the yield of dry matter per unit area.

The weight of a population at the time of sowing is the total weight
of the embryos and the relationship between "yield" and density is of
course at that time perfectly linear. However, with the passage of time
the population departs from this linear relationship and the yield per
unit area becomes independent of the number of seeds sown over a very
wide range of densities. Plants at high densities meet stress from neighbours early in their development, whereas plants at lower densities do
so only when they have grown larger. Variations in sowing density are
therefore very largely compensated by variations in the amount of growth

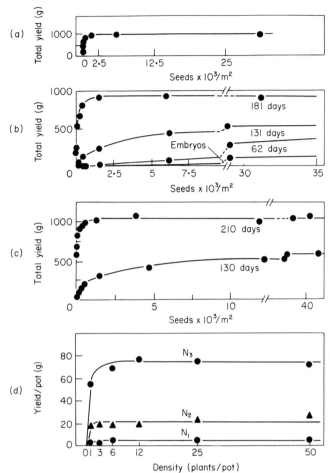

Fig. 6/1. Some relationships between yield of dry matter per unit area and the density of seeds sown.
   (a) *Trifolium subterraneum* at the post flowering stage
   (b) *Trifolium subterraneum* at various stages in development (note the break in the scale of density)
   (c) *Lolium loliaceum* at two growth stages.
   (d) *Bromus unioloides* at three levels of nitrogen fertilization.
   (From Donald, 1951)

made by individual plants. At the time of the last harvest (181 days) changes in mean plant weight exactly compensated for changes in density over the range from 1500 to 32 500 seedlings/m². In a similar experiment involving *Lolium loliaceum* essentially the same picture was obtained (Fig. 6/1c) and this type of relationship between density and dry matter production has been found to hold for a wide range of species. It has been called the "law of constant final yield" (Kira *et al.*, 1953).

In its initial growth from seed the yield of a population is determined by the number of plants present (the investment) but eventually the resource-supplying power of the environment comes to dominate the rate at which the members of the population grow and ultimately sets the limit to yield, irrespective of plant density. The population then behaves more and more as an integrated system — reacting independently of the number of its individuals — the behaviour of the plant becomes subordinated within that of the population (Harper, 1964).

Any influence that slows down the rate of plant growth might be expected to delay the onset and reduce the intensity of density stress between plants. This is shown in an experiment (Clatworthy, 1960) in which *Trifolium repens* and *T. fragiferum* were sown at a range of densities in pot culture and maintained *either* freely supplied with water but freely drained *or* with the water table maintained close to the surface *or* with the water table at the soil surface. Waterlogging inhibited plant growth but under freely drained conditions the plants grew rapidly and showed a marked plastic response to density (Fig. 6/2a and b). With impeded drainage the plants were small, even at low density, and the intensity of interference between them was reduced. Under fully waterlogged conditions the plants were small and stunted and the mean weight per plant was quite unaffected by density. In this experiment the restriction on growth was a supra-optimal supply of water. Similar experiments have often been made in which a resource has been applied at sub-optimal levels. For example, Donald (1951) sowed *Bromus unioloides*, at a range of densities, in pots of soil provided with three levels of nitrogen fertilizer: (i) no nitrogen; (ii) 150 mg nitrogen as NH_4NO_3 applied in three equal doses; and (iii) 700 mg nitrogen per pot as NH_4NO_3 applied in five equal doses during the growing season. The effects of these treatments on the final yield of the populations are illustrated in Fig. 6/1d. The law of constant final yield again applied to dry matter production per unit area but the more nitrogen was added, the higher was the plateau yield. The yield of a population was then independent of the number of plants present but dependent on the level of supply of nitrogen. As the nitrogen resources of the soil in this experiment were almost completely exhausted by the growth of the grass, nitrogen was probably in limiting supply and responsible for determining the plateau yield.

The presentation adopted for the data in Fig 6/1a, b, and c and 6/2a

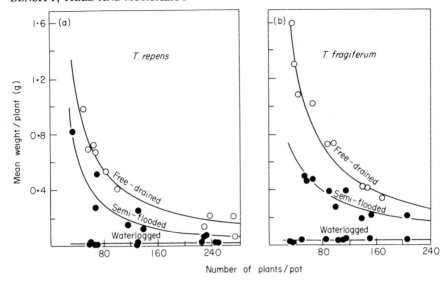

Fig. 6/2. The effect of plant density upon the mean dry weight of plants of (a) *Trifolium repens* and (b) *T. fragiferum* grown in pure stands under three water regimes. (From Clatworthy, 1960)

and b is not wholly satisfactory for the accurate comparison either of species or of environments and a considerable literature has built up concerning the most appropriate ways of analysing yield/density responses. Kira *et al.* (1953) suggested plotting the log of mean plant weight against the log of density. This treatment has been applied to Donald's data for *Trifolium subterraneum* in Fig. 6/3. The relationship between mean plant weight and density on log scales is represented by a horizontal line which slides upwards with the passage of time. As the plants grow and interfere with each other a segment of this line at highest plant density becomes inclined and the whole density range is now represented by two intersecting lines, one horizontal at the mid and low density range and the other inclined at high densities. Over the range of densities at which the line remains horizontal there is no interference between individuals. As growth proceeds this zone of no interference becomes progressively shorter and eventually disappears; the inclined section of the line increases in slope, approaching 45°, and extends to lower and lower densities. The slope of the inclined line increases from 0 and approaches a value of 1 when yield has become independent of density.

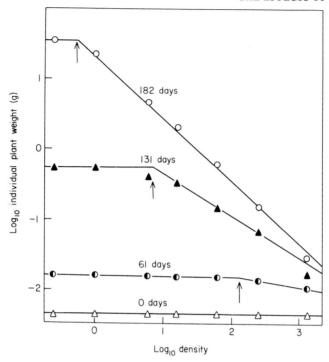

Fig. 6/3. The influence of density on the weight of plants of *Trifolium subterraneum* at successive dates from sowing. Data obtained from Donald (1951) and presented according
to the log/log transformation of Kira *et al.* (1953). The arrows show the density at
which plants of different ages begin to affect each other's growth.

There is a special elegance about this way of expressing the data but
it creates some difficulties. In particular two linear regressions have to
be used for what is a continuous density sequence and the sharpness of
their transition is almost certainly an artefact of the form of presentation.

Later, Shinozaki and Kira (1956) noticed that there was a linear
relationship between the reciprocal of mean plant weight and density
(Fig. 6/4). This linearity ("The Reciprocal Yield Law") has now been
established for a very wide range of plant species (see also Holliday,
1960 for other crop species). It is open to the criticism that unless properly weighted, very small plants at high densities contribute disproportionately large amounts of information to the regression analyses.
However properly weighted, the equation $\frac{1}{w} = Ad + B$ may be used to fit
a curve to a graph of log mean plant weight plotted against log density

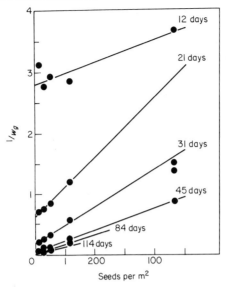

Fig. 6/4. The relationship between the density of sown seed and the reciprocal of mean plant weight in an experiment with soya beans. (From Shinozaki and Kira, 1956)

(see Fig. 6/5a and b). This treatment gives an excellent fit to the data, and has been applied with considerable success to the density response of many species, implying that a single rather simple rule underlies the growth of plant populations and providing a convenient generalization for further experimental tests.

An experimental comparison of the yield/density relationships of closely related species was made between *Bromus rigidus* and *B. madritensis* in pot culture (Harper, 1961). These two annuals of mediterranean climates commonly occur together on Californian range lands. The relationship between density and mean weight conforms well to the reciprocal yield law of Shinozaki and Kira (Fig. 6/6). The seed capital of the two species is very different, *B. rigidus* having markedly larger embryonic capital and endospermic reserves and a higher absolute growth rate than *B. madritensis*. In the period up to the first harvest, when the plants were growing relatively unstressed by density, populations of *B. rigidus* made considerably greater growth than those of *B. madritensis*. As the plants grew and density stress became an increasing force, the development of the two species populations became constrained within the same limits so that at the second and third harvests the yield per

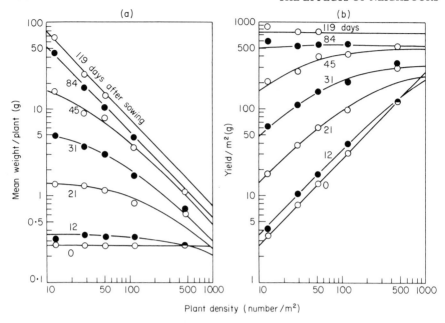

Fig. 6/5. The relationship between sowing density and yield of soybeans. Note logarithmic scales. The fitted curves are reciprocal yield equation $\dfrac{1}{w} = Ad + B$

(a) data for mean yield per plant
(b) as (a) but for yield per m².
(From Shinozaki and Kira, 1956)

unit area became independent of which species was present. This experiment included an equi-proportioned mixture of the two species and this also approached the same yield as the species in pure stand. The yield per pot became independent of density, of species, and of species combinations. The whole experiment included treatments involving added nitrogenous fertilizer and the effect of this was to increase the yield per pot *irrespective of the species, combination or density* (Fig. 6/7).

Although the reciprocal yield law has been found to apply to the garden turnip (*Brassica napus*), the Azuki bean (*Phaseolus crysanthos*), the carrot (*Daucus carota*) and a 13 year-old stand of pines *(Pinus densiflora)* which had been sown at a range of densities when 1 year old, there are some conspicuous exceptions to the law. At very high densities a stand of plants is often composed of such thin, weak individuals that they collapse (lodge) and the growth rate then declines sharply. Diseases

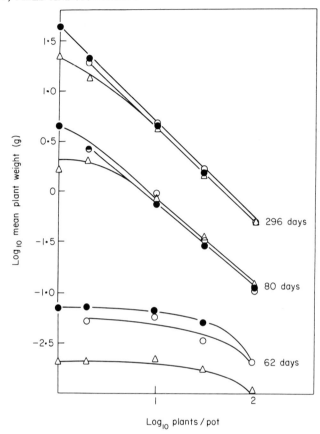

Fig. 6/6. The relationship between the density of sown seed and mean yield per plant in populations of *Bromus rigidus* and *B. madritensis* in pure stands and in a mixture of the two species.

● = *B. rigidus*.

△ = *B. madritensis*

○ = mixture of the two species — the mean weight per plant is shown taking the two species together.

(Adapted from Harper, 1961)

may develop in high density stands and a large number of individuals may then die. If this death occurs late the remaining living plants may have insufficient time to exploit the resources that are freed by the death of their neighbours. An interesting example of departure from the reciprocal yield law is the reaction of peach seedlings (*Prunus persica*) to density in experiments in a glasshouse and outdoors. The outdoor population showed the characteristic linear response between the reciprocal

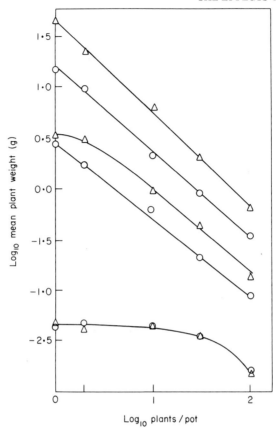

Fig. 6/7. The relationship between the density of sown seed and mean yield per plant in popu-
lations of *Bromus rigidus* and *B. madritensis* at two fertilizer levels.
△ = Low fertility
○ = High fertility
Data are the means of the yields of the two species (cf. Fig. 6/6).
(Adapted from Harper, 1961)

of mean plant weight and plant density but the glasshouse populations
behaved very differently (Hirano and Kira, 1965). There had been
previous claims that the peach root produces autotoxic exudates (e.g.
Proebsting and Gilmour, 1941) and Hirano and Kiro favoured the view
that the density effect of peaches in their glasshouse experiments could
be accounted for in this way. However, it seems odd that the effect was
not found when the plants were grown in the open and that a more
normal yield/density effect could be obtained by applying fertilizers to
the plants grown under glass.

Plant-to-plant variation

In most studies of the relationship between density and plant weight, changes in *mean* plant weight are reported. Such means are obtained by sampling and weighing a population of plants and dividing the weight by the number of plants present (or by the number of plants sown at the start of the experiment). This of course obscures any plant-to-plant variation. When plants are weighed individually it is found that populations growing under density stress have a skewed distribution of plant weight. The skewing of the frequency distribution increases both with the passage of time and with increasing density; an example is shown in Fig. 6/8 from an experiment in which *Linum usitatissimum* was sown at 3 densities and harvested at 3 stages of development (Obeid *et al.*, 1967). The frequency distribution of plant weights became strikingly skewed particularly by the time of the last harvest: at this time a hierarchy of individuals had established with a few large dominants and a large number of suppressed plants. Koyama and Kira (1956), showed that such skewed distributions represent stages in the development of log-normal distributions, and the log transformed data from the *Linum* experiment has a normal distribution. The development of such a hierarchy is not just an artifact of experimental studies of single species, similar log-normal distributions of plant weight have been found in a number of natural populations (Ogden, 1970), see Fig. 6/9. It is obviously dangerous to assume that average plant performance represents the commonest type of plant performance. Apparently the commonest type of plant found in experimental and natural populations is the suppressed weakling. Similar hierarchical development of plants in even-aged stands has been shown by Stern (1965), for *Trifolium subterranean*, and the log-normal distribution of plant weight appear to be the regular form of frequency distribution of plants in density-stressed populations. Even in the absence of density stress, populations may ultimately develop log-normal frequency distributions (Koyama and Kira, 1956) — density stress exaggerates and accelerates this trend.

The place that an individual occupies within the hierarchy of a plant population seems to be largely determined in the very early stages of plant development. The weight of an individual is a function of (a) its starting capital (the embryonic weight + some fraction of the endospermic reserves), (b) the relative growth rate of the genotype of the individual in the environment provided, (c) the length of time for which this

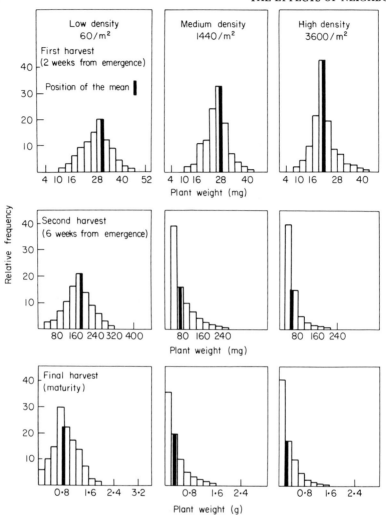

Fig. 6/8. The frequency distribution of plant weights within populations of *Linum usitatissimum* sown at three densities and harvested at three stages of development. (From Obeid *et al.*, 1967). Reproduced from *Crop Science* 7, 471–473, by permission of the Crop Science Society of America.

growth rate is continued and (d) restrictions on the rate or time of growth imposed by the presence, character and arrangement of neighbours in the population. An attempt was made to distinguish between the roles of these sources of variation (Ross and Harper, 1972). Populations of *Dactylis glomerata* were grown from seed randomly dispersed on the

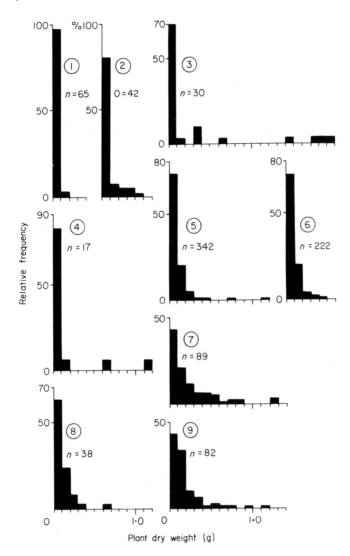

Fig. 6/9. The frequency distribution of individual plant dry weight in some mixed annual weed populations in an arable field in N. Wales. n = density of individuals per 0.5 m^2 approx. 1. Gramineae — mostly *Poa annua*; 2. *Atriplex patula*; 3. *Polygonum aviculare*; 4. all other species; 5. *Stachys arvensis*; 6. *Stellaria media*; 7. *Spergula arvensis*; 8. *Senecio vulgaris*; 9. *Polygonum persicaria* and *P. lapathifolium*. (From Ogden, 1970)

surface of soil in seed trays, at a mean density of *ca* 3 seeds/cm². An area of approximately 100 cm² was marked out in the centre of each tray and photographed to give an accurate record of the position of each seed. Each day the trays were examined for seedling emergence within the marked area and, when a group of approximately 10 plants had emerged since the last observation, each of the plants in that group was appropriately marked and its emergence class recorded. In this way all the seedlings were allocated to emergence classes. After 7 weeks' growth, each plant was harvested. 95% of the variance in individual plant weight was accounted for by the regression on the number of days, t, from emergence to harvest of the members of that group ($w = 0.26t - 2.02$ ($P < 0.001$)). The size that the plants in each emergence group would have attained if they had not suffered interference from neighbours and if their weights had been simply a function of the time allowed for growth is shown as a broken line in Fig. 6/10. The weights of later emergence groups fell well below this line and the last group to emerge con-

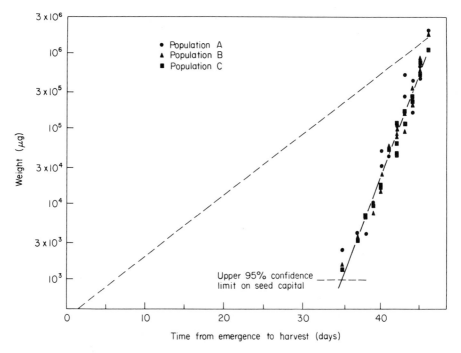

Fig. 6/10. The influence of emergence time on the dry weight per plant of *Dactylis glomerata* in a dense population – for explanation see text. (From Ross and Harper, 1972)

sisted of plants which had barely exceeded their embryonic capital even 35 days after emergence.

Plants in such an experiment may be grouped according to the time at which an individual emerges relative to the rest of the population. The position in the emergence ranking that in individual would have occupied in a population in which a hundred seeds had been sown is called the "percentage emergence ranking". When the mean log plant weight for each emergence group is plotted against the percentage emergence ranking for that group, a highly significant linear relationship is obtained (Fig. 6/11). This suggests that the amount of growth made by an individual is more directly determined by its order in the sequence of emergence than by the actual time at which it emerges. *The advantage which an early emerging seedling gains is far greater than can*

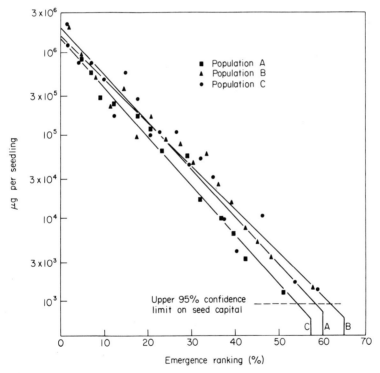

Fig. 6/11. The influence of the time of emergence of a seedling of *Dactylis glomerata* on its growth in a dense population. The time of emergence is expressed as an emergence ranking — all of the seedlings in the population being classified according to the time at which they emerged *relative to the rest*. (From Ross and Harper, 1972)

be accounted for merely by the greater time that it has been allowed to grow. The advantage must be due, at least in part, to the capture of a disproportionate share of the environmental resources by the individuals that emerge early and a corresponding deprivation of those that emerge late.

The way in which such pre-empting of resources (space capture) occurs was examined by growing plants under "restricted" and "unrestricted" conditions. Plants in the "unrestricted" treatment were grown from a single seed sown in the centre of a 7.4 cm diameter pot. Plants in the "restricted" populations were also grown from a single seed, sown in a small bare zone of 2.1 cm radius: from the edge of this zone to the edge of the pot further seeds were sown at a density of 2.5/cm^2. The growth of individual seedlings was followed by a non-destructive harvesting method. Plants growing in the "restricted" and "unrestricted" conditions initially maintained the same growth rate but those in the "restricted" population then departed from the "unrestricted" growth rate and adopted a new, but still exponential, growth rate

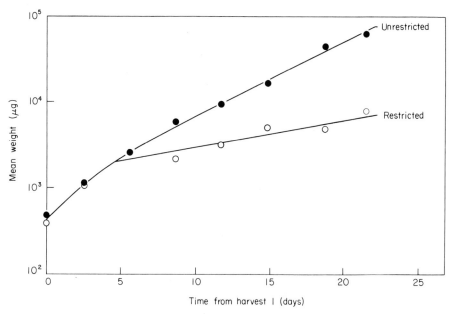

Fig. 6/12. The growth of seedlings of *Dactylis glomerata* grown without neighbours (unrestricted) compared with seedlings grown in the centre of a bare area surrounded by a circle of other seedlings at a radius distance of 2.1 cm (restricted). (From Ross and Harper, 1972)

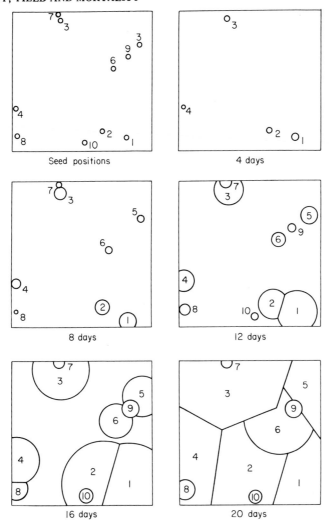

Fig. 6/13. Diagrammatic model of the pre-emption of space (= resources) by developing seedlings. (From Ross, 1968)

which they then maintained for at least 3 weeks (Fig. 6/12). Such behaviour is consistent with the view that the growth made by a plant is determined early in its life by the capture of space (or the resources implied by that space).

A model illustrating the effect of the timing of seedling emergence on space capture can be envisaged in which 10 randomly placed seedlings

are all assigned the same relative growth rate and the emergence of the
10 seedlings is spread over 10 consecutive days. The area of the space
which each seedling pre-empts is presumed to be proportional to the
weight of the seedling. A plant is assumed to stop growing when its
potential space is completely captured by neighbours — for example, in
Fig. 6/13 the potential space of plant 7 has been captured by plant 3. The
sequence of diagrams in Fig. 6/13 shows the areas theoretically occupied
by seedlings 4, 8, 12, 15 and 20 days after emergence of the first seedling.

Ross's experiments and models (see also Ross, 1968) suggest that the
timing of emergence of seedlings in a population is of far more importance than their relative spatial position, and in randomly distributed
populations he was not able to extract a significant proportion of plant
to plant variance by regressions on the distance from nearest neighbours
or on the weights of neighbours within encircling zones. He attempted
experimentally to vary the space available to a plant and its proximity
to neighbours by sowing seeds of *Dactylis glomerata* at random at a density of 3.3 seeds per cm^2 on trays of soil but, by using templates during
sowing, left empty circular zones within the sown area (see Fig. 6/14).
He allowed one seedling to grow in each empty zone but it was placed at
different distances from the edge of the zone. He also varied the size of
the empty zones. The space available had a more critical effect on seedling growth than the position of the seedling within that space (Fig.
6/15). The weight of a seedling developing in a bare zone was a function
of the cube of the radius of that zone; as a plant exploits a volume of
space above and below ground this cubic relationship is perhaps not sur-

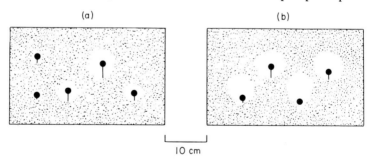

Fig. 6/14. An experimental arrangement of plants to determine the influence of the proximity
    of neighbours.
    (a) seeds sown in the centres of bare areas surrounded by dense populations — the
    radius of the bare areas is varied.
    (b) as (a) but the bare areas are of the same radius and the test seedlings are sown at
    different distances from the edge.
    (From Ross and Harper, 1972)

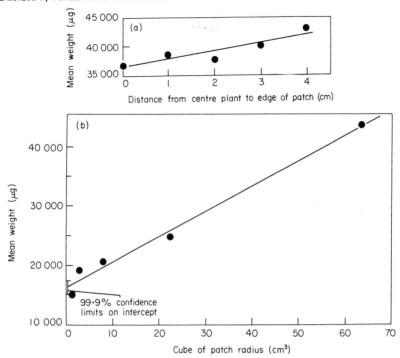

Fig. 6/15. The influence of the proximity of neighbours on the development of seedlings of
   Dactylis glomerata.
   (a) The relationship between seedling growth and its distance from the edge of a
   bare patch.

   $w = 36.44 + 1.308$ (d)   (see Fig. 6/14b)

   (b) as (a) but showing the relationship with the cube of the radius of the bare patch
   $w = 16.298 + 0.42r^3$ (see Fig 6/14a)
   (From Ross and Harper, 1972)

prising (although it is usually more convenient to describe plant density
in terms of the area rather than the volume that a population occupies).
In a stand of carrots (*Daucus carota*) sown in close rows (Mead, 1966)
20% of the variation in individual plant weight could be accounted for
by the differences in area available to each plant. The "available area"
was calculated by joining the perpendicular bisectors of the lines between each plant and its neighbour in the same row and in adjacent
rows and taking the area of the resulting polygon as the available area.
Further evidence that individual neighbours in a plant population may
affect each other's growth comes from a study (Yoda *et al.*, 1957) of
dent corn (*Zea mays*) grown in rows. There was a clear negative cor-

relation between the weights of individuals and their first, third and fifth neighbours, and a positive correlation with their second and fourth neighbours. This suggests that within a row of plants, once a difference between two neighbours has been triggered, it is progressively exaggerated.

Although the frequency distribution of plant weights in a population tends to be strongly skewed this is not necessarily true of plant height. The height of individuals in populations of annual plants (Koyama and Kira, 1956) and trees (Kuroiwa, 1960) is more or less normally distributed or even negatively skewed (i.e. skewed in the opposite direction to the weight frequency). The difference between the frequency distributions of weight and height is shown for a natural stand of *Elatostema rugosum* (Fig. 6/16). This situation suggests that at least some of

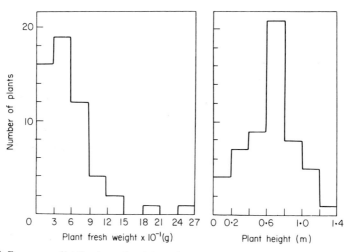

Fig. 6/16. Frequency distributions of weight and height in a natural stand of *Elatostema rugosum*. (From Ogden, 1970)

the abundant low weight individuals in dense populations tend to maintain their height, struggling into the canopy that is created by the few large plants. In populations of *Helianthus annuus* L. (Hiroi and Monsi 1964), the suppressed plants are lank and etiolated with a relatively increased proportion of non-photosynthetic to photosynthetic tissue (stem to leaf) (Ogden, 1970). In natural vegetation, the frequency distributions of height are much more complex and "plant communities are so varied in composition and structure that it is difficult to generalize.

In herbaceous monocultures, however, the changes in the distribution of weight and height frequencies with the passage of time indicate that in a juvenile population most of the individuals reach into the canopy, but as the plants grow larger, fewer and fewer individuals do so". (Ogden, 1970).

The growth rate of a young plant is greatest if it (a) establishes before its neighbours, enabling it to pre-empt resources, (b) is well separated from its neighbours and (c) has weak neighbours. The hierarchy of size seen in the log-normal distribution of plant weights seems to derive, at least in part, from a series of small but cumulative differences in space and time amongst the individual. The sharper distinctions that occur between individuals of different species ensure that when they grow together a hierarchy becomes even more clearly defined. It is likely that intra specific variations — e.g. somatic and genetic seed polymorphisms will sometimes give bimodal or multimodal hierarchies.

The influences of density on mortality

When plant populations are grown at high density, some individuals often die. Such mortality has often been disregarded in analyses of the relationship between yield and density: yield is commonly expressed in relation to the density of seeds sown or the density of seedlings established at the start of an experiment and changes in density are ignored.

Two categories of mortality can be recognized, density-independent and density-dependent. The term "density-dependent" was used by Smith (1935) to describe an increasing risk of death associated with an increasing density of animal populations. This remains general usage, though Haldane (1953) pointed out that it is perhaps more proper to distinguish between those density effects that hinder population growth and can thus act as regulators and those that may increase population growth and lead to instability — negative and positive density-dependence. The term "self-thinning" has been used for density-dependent mortality in plant populations (Harper et al., 1962; Yoda et al., 1963) and is contrasted with "alien-thinning" where mortality in one species can be ascribed to the stress from the density of an associated species.

In practice, density-independent and density-dependent mortality are not easy to separate. The mortality risk to a seedling from being hit by a raindrop or hailstone might be thought to be density-independent. Presumably the risk of being hit is independent of density but whether a

seedling dies after being hit is a function of its size and vigour, both of
which are strongly affected by density. The mortality risk of being eaten
by a predator might be thought to be largely density-independent but
the vigour with which some predators search for a particular prey
declines as the prey becomes less abundant and they may desert a particular source of food when the work involved in finding it becomes exces-
sive. Wood pigeons (*Columba palumba*) cease to predate when the density of a food falls to a level at which the birds can no longer search
quickly enough to pick up a sufficient quantity. The risk of predation is
then less at low seed densities than at high (see Chapter 16). This is also
true of the birds' predation on the leaves of white clover (*Trifolium
repens*). Limited search ranges of some phytophagous insects may ensure that epidemics break out only on dense populations, but
where inter-plant distance is high the plants may escape attack (see
Chapter 13).

Density stress changes the form of plants and will therefore usually
change the way in which they respond to a hazard — few causes of mortality are likely to be wholly free of some influence of density, but there
are examples. When the cornfield annual *Agrostemma githago* was grown
in experimental field plots at a range of densities from 1076 to $10760/m^2$
there was no change in the chance that a seed would produce a mature
plant; the mortality risk was constant at 77% at all densities, and all of the
variation in density was absorbed in plastic responses in plant weight,
branching, capsule number, etc. However in the presence of a constant
density of a second species, beet (*Beta vulgaris*) or wheat (*Triticum
sativum*) the mortality of *Agrostemma* became strongly density-dependent
(Fig. 6/17), (Harper and Gajic 1961).

Occasionally population density may actually enhance seedling establishment, though such positive density effects usually involve only the
early stages of germination and establishment. In experiments with
Rumex crispus and *Rumex obtusifolus* two densities of seed were sown
in soil at widely different water regimes. When the conditions were adverse for seedling establishment, doubling the sowing density more than
doubled the number of seedlings. However when conditions favoured
seedling establishment, doubling the number of seeds resulted in less
than a doubling of seedling numbers (Harper and Chancellor, 1959).
Positive density-dependence has been found in the germination of seeds
under artificial conditions, for example in petri dishes (Knapp, 1954);
seed of *Trifolium subterraneum* germinates more readily when sown

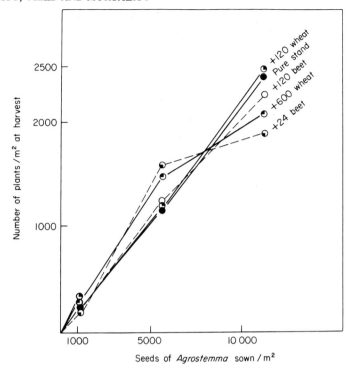

Fig. 6/17. The relationship between number of seeds sown and number of mature plants produced by *Agrostemma githago* in pure stand and in the presence of wheat and sugar beet. (From Harper and Gajic, 1961)

densely in petri dishes, apparently because accumulated respiratory CO_2 stimulates germination. Such an effect is unlikely to be significant under field conditions but other factors may act to produce the same result: for example when a soil has developed a capped surface (e.g. a clay soil drying after heavy rain) seedlings often emerge in clusters, suggesting that a group acting together may be able to break a crust that an isolated seedling cannot penetrate (Chapter 5). Such evidence of positive density-dependence in plant populations of a single species is very much the exception — most density responses are negative, reduced plant size or an increased death risk.

A mortality risk that increases with increasing density has regulating properties. It is a negative feedback which acts to constrain population size within narrower limits than the range of starting densities. It acts against unrestrained population increase. It is a buffer that can maintain populations more constant than would be produced by natural variations in seed production and dispersal.

An example of density-dependent mortality comes from an experiment in which seeds of species of *Papaver* (cornfield poppies) were sown at densities ranging from 0 to 3140/m². The highest density of mature plants obtained was *ca* 750/m². The chance of a seed producing a mature plant declined with increasing density (Fig. 6/18). Such density-dependent mortality is well known in forest nurseries where it is usually associated with pathogenic activity (e.g. Gibson (1956) showed that *Rhizoctonia* was responsible for killing a higher proportion of seedlings in overcrowded plots).

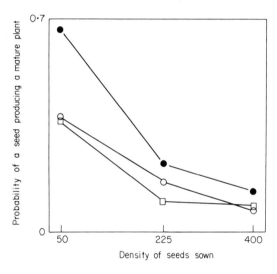

Fig. 6/18. The relationship between the density of seed sown and the chance that a seed will produce a mature plant — summarizing the behaviour of five species of *Papaver* in three differently shaded parts of an experimental garden. (From data in Harper and McNaughton, 1962)

Japanese workers have made the most thorough studies of density-dependent mortality in plants, in particular at Osaka City University (Kira, Shinozaki, Yoda, Ogawa and others) and at Tokyo (Monsi, Hiroi and others). Some examples of the relationships between sowing density and population density over time are shown in Fig. 6/19a, b and c. The data take a characteristic form; mortality occurs at high but not at low densities; the mortality continues with time and is not a sudden happening at a particular growth stage; the higher the density the sooner do

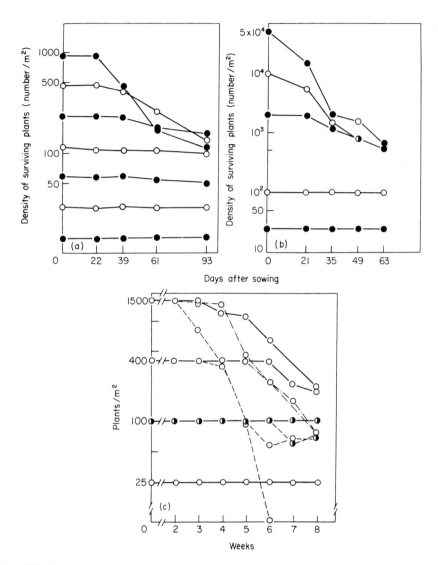

Fig. 6/19. Survivorship curves for populations of annual crop plants sown at a range of densities.

(a) Soybean (*Glycine soja*); (b) Sesame (*Sesame vulgaris*).
(From Yoda *et al.,* 1963)

(c) Sunflower *(Helianthus annuus)*——full light intensity
— · — · 60% light and — — — — 23% light. (From Hiroi and Monsi, 1966)

the first deaths occur. The severity of the self-thinning is affected by environmental conditions: populations of *Helianthus annuus* thin more rapidly if the light intensity is low (Hiroi and Monsi, 1966): some populations with very high starting densities of sunflower, grown at low light intensity, thinned to extinction within 6 weeks.

The time sequence of the thinning process can be seen clearly by plotting initial and surviving densities as the axes of a graph (Figs 6/20a and b). Soon after seedling emergence the numbers of plants are nearly linearly related to sowing density — mortality up to this point is nearly density-independent. As the plants grow, thinning at the higher densities leads to an asymptotic population density and the asymptote itself falls as the time passes. Sometimes fewer survivors remain from very high than from intermediate sowing densities.

The time trend of the self-thinning process can also be analysed by following the change in the log of plant numbers. A linear relationship between log numbers and time implies a constant exponential decrease — a risk of death to the individual which, like the decay of an isotope, remains constant over time. A hollow curve (Deevey type III) is obtained if the risk of early death is higher than that later in life. Self-thinning data are shown in Fig. 6/21a and b for three species: *Erigeron canadensis*, an early successional cornfield weed, and two crop species, *Fagopyrum esculentum* and *Sesame vulgaris*. The survivorship curves tend to be concave at first and then approach linearity. The number of such studies is still too small to allow wide generalization but they suggest that, after a seedling period in which the mortality risk is high, a nearly constant and lower risk is maintained for the rest of the life of the populations.

In natural self-thinning stands of fir near Leningrad, Sukatschew (1928) noticed that the residual density after self-thinning was greatest on thin, poor soils and that lower densities, though of bigger trees, were found on the deeper richer soils (Table 6/I). He designed an experiment to examine this phenomenon under controlled conditions. He sowed seeds of an annual, *Matricaria inodora*, at two densities in fertilized and unfertilized soil. Mortality occurred in the populations and was density-dependent, the risk of a plant dying being greater at the higher density. At both densities, however, the mortality risk was greater in the fertilized soils (Table 6/II). This result is at first sight surprising although it clearly paralleled his field observations on the firs. Apparently, the application of fertilizers increased the rate of growth of the experiment-

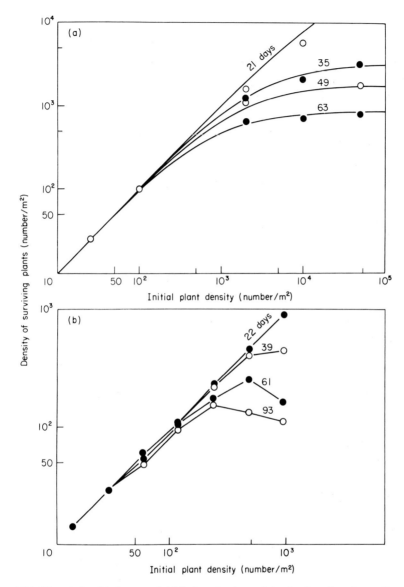

Fig. 6/20. The relationship between initial plant density and the density of survivors after various periods of time.
(a) Buckwheat (*Fagopyrum esculentum*)
(b) Soybean (*Glycine soja*).
(From Yoda *et al.*, 1963)

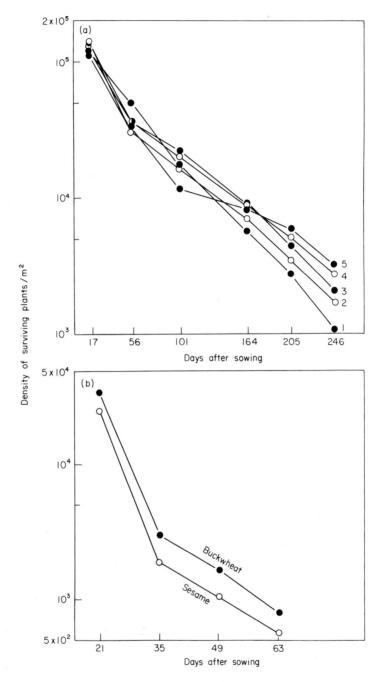

Fig. 6/21a. The time trend of self thinning in field populations of *Erigeron canadensis*. 5, 4, 3, 2, and 1 are plots receiving *in*creasing levels of fertilizer. (From Yoda *et al.*, 1963)
b. The decline in the asymptotic density with time in populations of buckwheat (*Fago-pyrum esculentum*) and sesame (*Sesame vulgaris*). (From Yoda *et al.*, 1963)

Table 6/I
The density of trunks of fir in Leningrad forests on five
natural soil types. The populations were self-thinning.
(From Sukatschew, 1928)

Types of soil condition	At age 20 years		At age 60 years	
	Predominant trunks	Oppressed trunks	Predominant trunks	Oppressed trunks
I (best)	5600	—	1300	640
II (poorer)	5850	—	1600	680
III (poorer)	6620	—	1950	650
IV (poorer)	7480	—	2280	720
V (worst)	8400	—	2780	760

Table 6/II
Density dependent-mortality in *Matricaria inodora* at 2 levels of
soil fertility. (From Sukatschew, 1928)

Dense culture (3 cm interplant distance)	% decrease in numbers
Non-fertilized soil	5.8
Fertilized soil	25.1
Mid-density culture (10 cm interplant distance)	
Non-fertilized soil	0.0
Fertilized soil	3.1

al plants with the result that there was greater population pressure or
density stress and under these conditions more plants died.

A particularly important study of self-thinning was made by Yoda *et
al.* (1963) who showed that the rate of growth of individuals in a population and their death risk were correlated and dependent on fertility.
They followed the development of dense populations of a number of
species through an annual growth cycle, taking destructive harvests at
intervals so that both the mean plant weight and the numbers of survivors could be determined on replicate plots. When the log of mean plant
weight (i.e. of the survivors) was plotted against the log of the density
of survivors, the values for successive harvests were found to lie around a
line of slope of *ca* −1.5. If all the populations at all the sampling dates had
equal weights per unit area a slope of −1.0 (45°) would have been obtained.

A slope of -1.5 implies that while the number of individuals present in the population is decreasing, the weight of the population as a whole increases; the rate of growth of individuals more than compensates (and is probably responsible for) the fall in numbers. This relationship between mean weight and density of survivors can be written $w = cp^{-\frac{3}{2}}$. A number of examples of such self-thinning are shown in Fig. 6/22a and b.

It is useful to compare the self-thinning graphs of 6/22 with the yield/ density graphs of Figs 6/3 and 6/4. In Fig. 6/22 changes are recorded in populations that have been started at *one* constant density and the graphs show the changes in numbers and mean plant weight with the

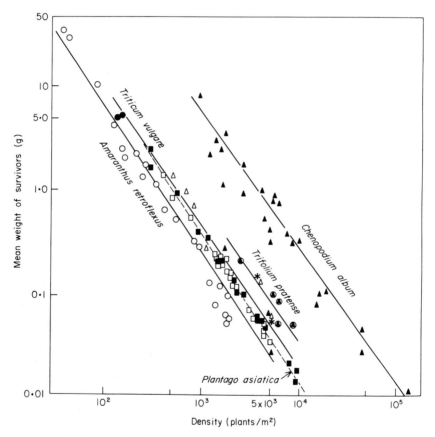

Fig. 6/22a. Changes in plant density and in mean plant weight with the passage of time. Data for *Chenopodium*, *Amaranthus*, and *Plantago* from Yoda *et al.*, (1963); data for *Trifolium* and *Triticum* from Harper and White (1970) after data of Black and of Puckeridge.

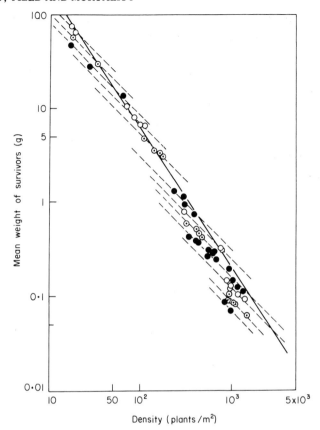

Fig. 6/22b. As Fig. 6/22a but *Ambrosia artemisiifolia elatior.* Each group of points along a
       dotted line represents plants at the same stage of growth and on soils of the same
       fertility level. (From Yoda *et al.,* 1963)

passage of time. In contrast, Figs 6/3 and 6/4 show the changes in plant
weight that occur over a range of sown densities. In Figs 6/22a and b
the regression lines represent time courses, whereas in Figs 6/3 and 6/4
different stages in time are represented by different curves. Moreover in
Figs 6/3 and 6/4 the performance of populations was expressed in relation to the starting density and mortality was ignored.

An especially detailed study was made (Yoda *et al.,* 1963) of a homogeneous stand of *Erigeron canadensis* — a species that occurs naturally
in the second year of old field successions in Japan. Field plots were
cleared of existing vegetation and seeds of *Erigeron* were distributed as
evenly as possible at a rate of $1-2 \times 10^5$ seeds/m^2 on an area of

sandy, infertile soil on which there was already a marked fertility gradient. The existing gradient was exaggerated by applying a mixed NPK
fertilizer on a sequence of plots in the ratio 5:4:3:2:1. The density of
seedlings was subsequently determined by placing small quadrats 10 cm
by 10 cm within the plots and the average dry weight of the plants was
determined at each harvest by sampling 100 random individuals from
each plot. The results show clearly the "Sukatschew effect", i.e. self-
thinning was most intense on the high fertility plots (Fig. 6/23). Taking

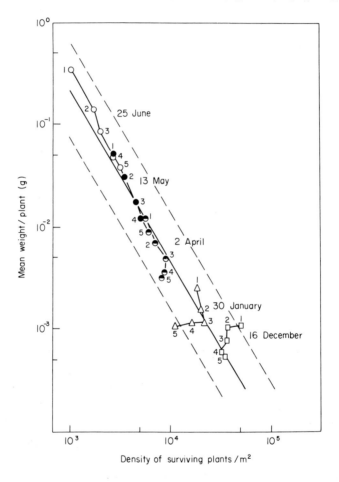

Fig. 6/23. Changes in numbers and individual plant weight of *Erigeron canadensis* with time.
The numbers 1, 2, 3, 4 and 5 represent a gradient of decreasing fertility. (From Yoda
et al., 1963, redrawn with calculated regression and 0.95 confidence limits — the
first harvest, Nov. 7, is omitted).

together the data for the five fertility levels and the five dates of harvest, a plot of log mean plant weight against log density of survivors gives a series of points which are clustered along a common line. The fitted regression for all the data (except the first harvest) has a slope of -1.66. This is 3° greater than and does not differ significantly from a slope of 1.5. During the 9 months of the experiment the populations declined dramatically from an average density of 122 400 seedlings/m² to *ca* 1060 plants/m², so that the adult population represented less than 1% of the seedling population. The data present a vivid picture of a population of rapidly growing plants from which individuals are continually being lost in the self-thinning process, a population that is dynamic both in numbers and the size of individuals. Increased fertility stimulated the rate of growth of survivors and increased the number of plants dying.

A similar experiment was made (White and Harper, 1970) using seed of commercial varieties of *Raphanus sativus* and *Brassica napus* — species chosen because they could be relied on to give rapid, even germination. Seeds were sown at a density of 180 per 20 cm diameter pot and after 2 weeks the number of seedlings was counted and reduced to 150 per pot (*ca* 4.8×10^3 seedlings per m²). Harvests were taken $6\frac{1}{2}$, 13 and 17 weeks after sowing — the plants did not flower during this period. The relationship between the mean weight per plant and the surviving density is shown in Fig. 6/24. The calculated regression for all species, mixtures, fertility levels and harvest times is $\log w = 2.28 - 1.45 \log p$ or $w = 190 \, p^{-1.45}$. This is obviously again closely similar to the 3/2 power law of Yoda *et al.* and again the greatest thinning occurred on the most fertile soil. The species scarcely differed in their relationship to each other though *Raphanus* suffered slightly greater mortality than *Brassica*.

The 3/2 power law has been tested mainly with annual plants but there is evidence that it holds true for forest trees as well. The empirically derived thinning tables used by foresters to optimize timber production conform rather well with the 3/2 thinning rule. Management tables have been produced for a number of different forest tree species on different soil types. They provide advice on the thinning regimes which, with the initial planting densities commonly used, provide the greatest growth in girth consistent with maximal production of timber volume per unit area. Forestry management tables are concerned with timber volumes rather than with mean plant weight and so the log of the mean volume per tree has been plotted against the log of the surviving density after the thinning regime recommended for each species

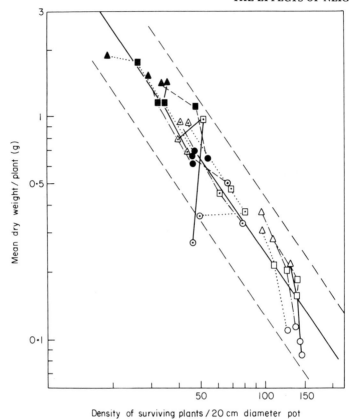

Fig. 6/24, Changes in numbers and mean plant weight with time of *Brassica napus* ──── and
Raphanus sativus · · · · in pure stands and in mixtures ─ ─ ─ and ─ · ─ · ─. The confidence limits are given for *P* = 0.05.
○ = low fertility, □ = medium fertility, △ = high fertility.
○, □, △ = first harvest, ⊙, ▣, ◮ = second harvest, ●, ■, ▲ = third harvest.
(From White and Harper, 1970)

(Fig. 6/25a and b). There is again a striking linear relationship but with
gradients ranging from −1.72 to −1.82 that are considerably steeper
than expected under natural self-thinning, but it must be remembered
that in these tables it is useful volume of timber that is the variable, not
the mean weight of the individual, and this may be sufficient to steepen
the regression slope.

   The thinning lines for the various species shown in Fig. 6/25a and b
are quite distinct. The mean volume expected at a particular density
differs from species to species. These differences are themselves of con-

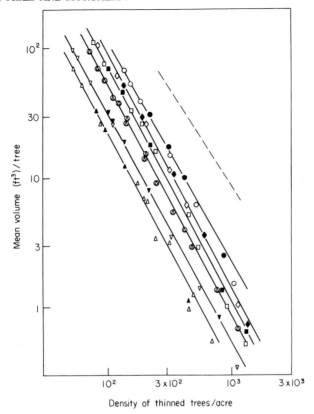

Density of thinned trees/acre

Fig. 6/25a. The relationship between the density of trees and their mean volume in fully stocked
stands up to 100 years after planting. Calculated from data in Forestry Management
Tables (Bradley *et al.*, 1966). For each species two or three sets of yield class data (Y.C.) are
given — these represent sites of different yielding ability and are essentially comparable with the fertility levels in Figs 6/22b and 6/23. Noble fir, Y.C. 240 (o); Y.C. 160 (●);
western hemlock, Y.C. 260 (◊), Y.C. 180 (♦); Sitka spruce, Y.C. 260 (□), Y.C. 200 (■);
Scots pine, Y.C. 160 (△), Y.C. 100 (▲); European larch, Y.C. 140 (▽), Y.C. 80 (▼); syca-
more/ash/birch (joint tables), Y.C. 120 (△), Y.C. 80 (▲). The data for the separate yield
classes lie along common thinning lines, whose gradients range from −1.72 to −1.82.
A slope of gradient −1.5 is shown at right for comparison. (From White and Harper,
1970)

siderable interest. Does the parameter c which defines the position of a
species on the graph describe some quality of canopy shape which
affects the degree of mutual interference between individuals? Among
the species represented in Fig. 6/25a and b the highest volume per tree
at any given survival density is obtained from *Abies nobilis* — a tree of

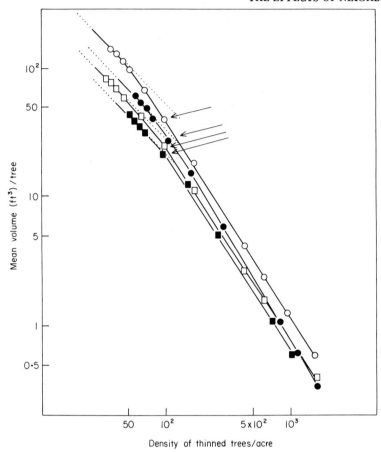

Fig. 6/25b. Relationships between density and mean volume per tree for beech and oak in fully stocked stands for up to 150 years of age. The data have been recalculated from Forestry Management Tables (Bradley *et al.* 1966) which predict yields obtained in accordance with the thinning practices advocated. Two sets of yield class data are given for each species: beech, Y.C. 100 (○), Y.C. 60(●); oak, Y.C. 60 (□), Y.C. 40 (■). At points of maximum mean annual increment (indicated by arrows for each yield class) the slopes change from a gradient of −1.5 to −1. (From White and Harper, 1970)

markedly pyramidal shape. The order in which the species fall by the same criterion is: *Abies nobilis* (strikingly pyramidal) > *Tsuga heterophylla* (tapering crown) > *Picea sitchensis* (broadly pyramidal) > *Pinus sylvestis* (crown often flattened and irregular) > *Larix decidua* (irregularly pyramidal) > sycamore, ash and birch (deciduous and round-crowned). This order suggests a topological series and that tree geometry is important in determining thinning relationships.

The forestry thinning tables for different "yield classes" give points that lie on the same thinning lines — variations in soil quality have the same effect as in the experiments with annuals (Fig. 6/23) of moving points up and down a common thinning line, not shifting the position of the line.

The yield/density relationship for long-lived stands of oak (*Quercus robur*) and beech (*Fagus sylvatica*) have slopes of −1.48 to −1.62 (Fig. 6/25b), values much closer to the ideal slope of −1.5. However, in the oldest stands the gradient changes to a value of −1.0. In these old forests the total volume of timber per unit area has become constant — any thinning that is practised permits the remaining trees to increase in size but only sufficiently to compensate for the volume removed at thinning. Further examples of the 3/2 power law in natural and artifically thinned populations can be found in White and Harper (1970).

The mechanism of the self-thinning phenomenon is only dimly understood. The plants that are most likely to die in natural (as in forest) thinning are the smallest and "weakest". The clearest direct evidence that this is the case comes from an experiment made on populations of *Trifolium subterraneum* by Black (1958). This species produces a rather wide range of seed sizes even from a single plant. Samples of large and small seeds were sown in pure stands and in equiproportioned mixtures all at high density. Self-thinning occurred in the seedling population after about 40 days' growth. The populations from large seeds suffered more rapid mortality than those derived from small seeds (Fig. 6/26). The size of the seedlings is closely related to the size of the seeds and the real growth rate is a function of the seedling size, particularly cotyledon area. Apparently the faster-growing, larger, and more vigorous seedlings produced a density stress amongst themselves which resulted in a greater mortality than in the populations of the same density but of the smaller and slower growing seedlings. When small and large seeds were sown together, the mortality was concentrated almost exclusively amongst the plants derived from the small seeds: the stress of density was absorbed by the death of the smaller members. Not only did the plants from large seeds come to dominate the canopy of the mixed stands but at the end of the experiment (after 120 days) the plants from the small seeds were intercepting only *ca* 3% of the total light intercepted by the canopy of the mixture (Fig. 6/27). It is not possible to make the rigid argument

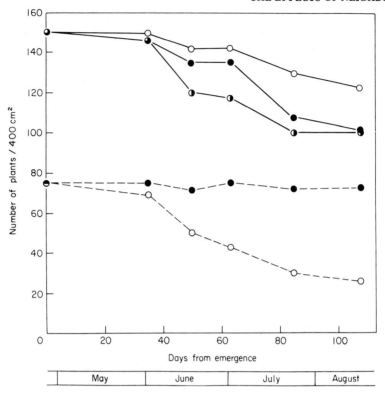

Fig. 6/26. Changes in plant numbers in populations of *Trifolium subterraneum* derived from large seeds = ●, small seeds = ○ and mixed swards = ◑ and also for both components within a mixed sward = ● — — — ● and ○ — — — ○. (From Black, 1958)

that the plants that died did so because they were starved of light, but this seems the most probable direct or indirect cause.

If the self-thinning process involves the progressive elimination of the weakest individuals it immediately becomes pertinent to ask questions about the nature of the hierarchy of size or vigour that develops in a growing population. It has already been pointed out that plant populations growing under density stress develop a log-normal distribution of weights (Figs 6/8, 6/9) and that this may arise because of differences in "space occupancy" at very early stages in establishment. Self-thinning removes the smaller individuals from the population and tends to stabilize the degree of skewness.

Many of the weaker individuals in a population extend their foliage to the top of the canopy but do this by means of long, spindly stems and a proportionately greater respiratory burden. Their net assimilation rates may therefore be expected to be lower than that of the dominant plants in the canopy. It may be that the self-thinning process involves changes in the net assimilation rates of the individuals and that the plants that die are those with very low or negative net assimilation rates.

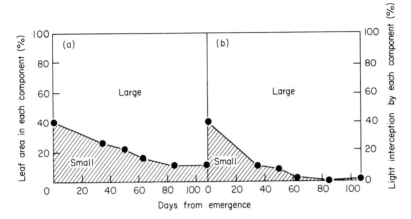

Fig. 6/27. The percentage of the leaf area (a) and the percentage of the light interception (b) of plants of *Trifolium subterraneum* from large and small seeds grown in a mixed sward. (From Black, 1958)

It would be of the greatest interest to determine the frequency distribution of net assimilation rates among the individuals of a self-thinning population.

Considerable support is given to the above interpretation of the self-thinning process by some data obtained by Hiroi and Monsi (1966; analysed by White and Harper, 1970). Sunflowers (*Helianthus annuus*) were grown at starting densities of 1600 and 400 plants/m² and shaded so that they received 100, 60 or 23% to full daylight. The numbers of survivors and their mean weight were measured in replicate plots over a period of 10 weeks. The self-thinning lines closely followed the 3/2 power law, but were displaced and the gradients slightly reduced at the lower light intensities (Fig. 6/28). The lower the light intensity, the more rapid was the thinning with the result that (i) at any given density of survivors the plants were heavier if the light intensity was higher and (ii) at any given weight of plants the number of survivors was greater at

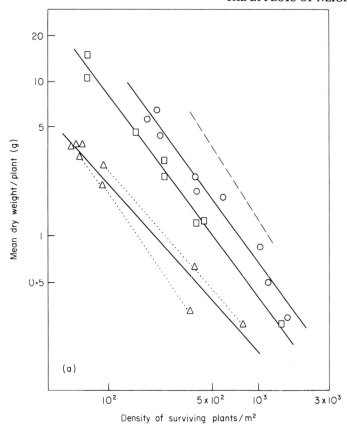

Fig. 6/28. Changes in numbers and mean plant weight in *Helianthus annuus* populations grown at 100% (○) 60% (□) and 23% (△) light intensity over a period of 10 weeks. (From White and Harper, 1970 calculated from data of Hiroi and Monsi, 1966)

high than at low light intensity. Such a reaction to light contrasts completely with the reaction of plant populations to nutrient supply. Changes in the level of fertility had the effect of moving points up or down a common self-thinning line whereas changes in the light intensity altered the position of the line (the value of c). These observations are wholly compatible with the view that self-thinning involves the death of individuals whose net assimilation rate falls below some critical value: shading of the population increases the proportion falling in this category.

Most experiments on self-thinning have been made under conditions in which light might be expected to be in limiting supply within canopies. It is not difficult to envisage comparable self-thinning processes occurring where other supply factors dominate the chances of survival.

Mortality in experimental populations of single species is a steady process in which, after the seedling phase, the risk of death does not change appreciably with age. It appears that one of the consequences of the growth of individuals is to generate a density stress which continues to expel a more or less constant proportion of the survivors in unit time. The same effect extends to control the death rates of the parts of individual plants.

Grasses, like most clonal plants, have a clearly defined modular structure in which the genet develops as a population of tillers. Each tiller is itself a population of leaves. Under stress, leaves may die or whole tillers or whole genets. A series of populations of Lolium perenne were sown at densities ranging from 330 to 10 000 seeds/m² under three light intensities (full, 2/3 and 1/3 of full daylight). As seedlings emerged they were loosely marked with a cotton loop so that the later developing tillers could be related to the original seedling even if the connections between the parts of the genets disappeared. Seven harvests were made over 180 days and records were taken of the number of surviving genets, the number of surviving tillers, the weight per genet and the weight per tiller (Kays and Harper, 1974). The results are shown in Figs 6/29, 6/30 and 6/31. Figure 6/29 shows that during the course of the experiment there was a continual death of genets. The higher the density the sooner was the onset of mortality. The denser populations had lost 60—70% of their members after 180 days. During the last 60 days all the populations were losing genets at about the same rate so that, although mortality started at different times at different densities, the half-life of a population rapidly became independent of the starting conditions. During the period when genets were being lost from the populations, the number of tillers was changing independently and after 180 days the number of tillers per unit area was almost independent of the starting density — there was no significant difference in tiller density per plot after 130 days in any one light regime. The populations were adjusting the effective density of tillers per plot throughout the experiment and the number present at any time was the consequence of (i) differential mortality of genets, (ii) differential birth rates of tillers per genet and (iii) differential death rates of tillers.

The populations at the three light intensities behaved differently (Fig. 6/30a, b and c) but in all cases the three processes (i), (ii) and (iii) contributed to bring the mean density of tillers per unit area to a value that

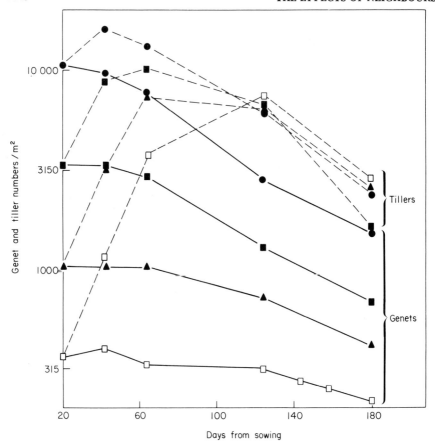

Fig. 6/29. Changes in the density of tillers – – – and of genets——— in populations of *Lolium perenne* sown at a range of densities. (From Kays and Harper, 1974)

was characteristic of the light regime but quite independent of the start-ing density of seeds. In the most shaded regime the tillers were smaller and their density was lower than in full light.

Figure 6/31 shows the relationship between the density of surviving genets at the various sampling times and the mean weight per genet. The populations grew and self-thinned as if constrained by a 3/2 thinning law. However, at the lowest light intensity the thinning line approached a slope of 45°, i.e. the growth rate of the genets was just sufficient to compensate for the effects of thinning. This exactly parallels the results from the sunflower experiments (Fig. 6/28).

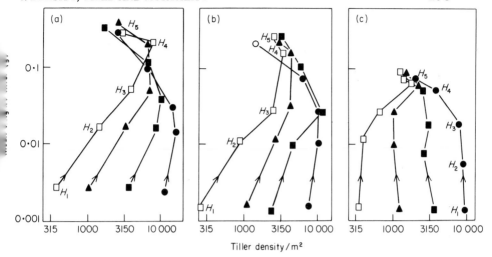

Fig. 6/30. The relationships between tiller density and mean weight per tiller in populations of *Lolium perenne* grown at three light intensities (a) 100%, (b) 60% and (c) 30% of full daylight. H_1, H_2 etc. are successive harvests. (From Kays and Harper, 1974)

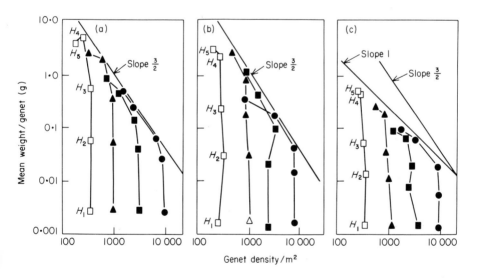

Fig. 6/31. The relationships between the density of genets and the mean weight per genet in populations of *Lolium perenne* grown at three light intensities (a) 100%, (b) 60% and (c) 30% of full daylight. H_1, H_2 etc are successive harvests. (From Kays and Harper, 1974)

Of course the behaviour of populations of a single species grown in pots or field plots is far removed from the complexities of natural systems. In particular, in such experimental studies extremes of water stress can usually be avoided that may be critically important in the field. The relevance of the experiments to the interpretation of natural vegetation is comparable to the relevance of a test tube study of an enzyme to the understanding of the functions of cells or the study of a root segment in a manometer to understanding the activities of a root system on a whole tree. The relevance comes largely from the act of faith that complex systems may be understood by breaking them down into parts and then reassembling them. This is not to deny that the whole is more than the sum of its parts but an act of faith that the remainder consists of interactions that are themselves appropriate objects of study. The parts of population biology that are discussed in this chapter establish that growth and death are interrelated at the population level in a sufficiently integrated fashion for formal mathematical relationships to be derived. Some of these may be species-specific but many have generalities far beyond the species level. The generalizations that emerge provide an appropriate base from which to extend into study of the greater complexity.

7

The Influence of Density on Form and Reproduction

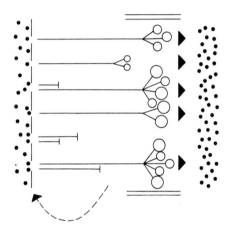

Higher plants are plastic in size and form: this plasticity derives mainly from their population-like structure (see Chapter 1). Much of the fine genetic control that determines the difference between species is dependent on the form of the repeating units of construction, particularly leaves and flowers. The size and form of these units is usually tightly canalized and changes only fractionally over widely varied environmental conditions. In contrast, the number of the units and thus the size of the whole plant varies greatly both with age and with conditions. Thus while zoologists commonly use aspects of the size of whole organisms,

195

for example length and height, as distinguishing taxonomic characters, such features are of remarkably little taxonomic value in plants. To the population biologist, however, variations in the number of parts is of great interest because it is this variation that largely determines the reproductive activity of the genet and thus its potential contribution to the next generation.

The proximity of neighbours can profoundly affect the development of individual plants and their plastic development is one of the powerful density-reactive mechanisms that contribute to regulate the reproductive output of a population. If each type of organ represented a constant fraction of the plant, irrespective of plant size, this chapter would simply repeat the generalizations made about density and plant yield made in Chapter 6 — but this is not the case. Although the dry weight of a plant population tends to compensate more or less perfectly for variations in density, the parts are not all altered to the same extent. A density-stressed individual is not simply a miniature version of its vigorous low-density counterpart.

During the growth of plants under density stress the allocation of assimilates between different structures becomes proportionately altered. In particular the ratio of seed to total dry matter changes so that at high density reproductively inefficient plants are often developed: for example in populations of *Agrostemma githago* grown at high density many of the plants that completed a full annual cycle of growth died without having formed capsules, or formed capsules which contained no seeds. Such plants are vegetative eunuchs, consuming resources but leaving no descendants. In agricultural crops the developmental consequences of density are important because the optimal density for a particular product, seed, latex or a storage root may be different from the optimum density for the production of dry matter. The optimal density for grain production by maize is almost always lower than the optimal density for the production of leaf for fodder (Fig. 7/1).

Various attempts have been made to describe algebraically the relationships between density and yield of a particular plant part. Shinozaki and Kira (1961) showed that the total yield of dry matter per unit area of a plant population could be described by the reciprocal equation

$$w^{-1} = A\rho + B,$$

where A and B are constants, w is yield per plant and ρ is density. A modified form of this equation gives a more universal fit to population

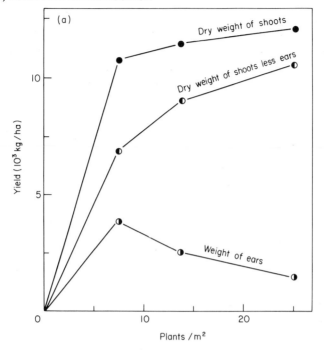

Fig. 7/1. The relationship between sowing density and the yield of ears, shoots less ears and total shoot weight for populations of maize (*Zea mays*). (From Harper, 1961)

data including specified parts:

$$w^{-\theta} = A\rho^{\phi} + B$$

(Bleasdale and Nelder, 1960). For most purposes ϕ can be taken as 1 (Bleasdale, 1966; 1967) so that the equation then becomes

$$w^{-\theta} = A\rho + B.$$

When $\theta = 1$, the equation describes the common form of density response found for total yield and when θ is <1 fits those situations in which there is a clear optimal density for the production of a plant part such as the bulb of the onion, the tap-root of beet and seed. It had been shown by Kira *et al.* (1953) that there was a linear relationship between the logarithm of total yield per plant and the logarithm of a plant part so that

$$\log w_{total} = \log K + \alpha \log w_{part},$$

or

$$w_{\text{total}} = K w^{\alpha}_{\text{part}},$$

where α and K are constants. Combining this equation with that of the "reciprocal" law it can be shown that

$$\theta_{\text{part}} = \theta_{\text{total}} \times \alpha.$$

An elegant test of the goodness of fit of these various equations was made by Fery and Janick (1971) for the yield—density relations of the ear or grain and the total above ground parts of maize (*Zea mays*); an extract from their data is given in Fig. 7/2. Such an analysis of the plastic responses of plants has been made only for agricultural crops and those of horticulture, where it offers a chance to define and channel breeding programmes for crop improvement. Such analysis could be very revealing in comparisons of naturally occurring species. Are there species or ecotypes with homoeostatic allocation of assimilates between the various organs ($\theta = 1$) and others with unstable allocation ($\theta < 1$) which might be adaptively superior in certain environments?

Most of the information about the density responses of plants is descriptive rather than analytical and most of this chapter is concerned with describing for a variety of plant types, those changes in form and reproduction that are brought about by density stress. The role of density as the stress factor is emphasized in this as in other chapters because of its obvious relevance to population biology but stresses such as drought and nutrient shortage may of course also cause plastic changes in form and there is seldom any reason for supposing that the stresses created by density are qualitatively different from those caused by a shortage of supply factors that exists independently of population size.

Annual plants

A classic study was made by Clements *et al.* (1929) of the effects of density on the growth of wheat and sunflower. He chose cultivars because seed was available in large quantities and because most cultivated species lack the seed dormancy that often complicates experiments with wild forms.

Wheat was sown in pots of soil at densities of 4, 8, 16, 32, and 64 plants per pot (*ca* 40—700 seeds per m^2) and various plant parts were measured when the plants had reached maturity. The logarithms of the

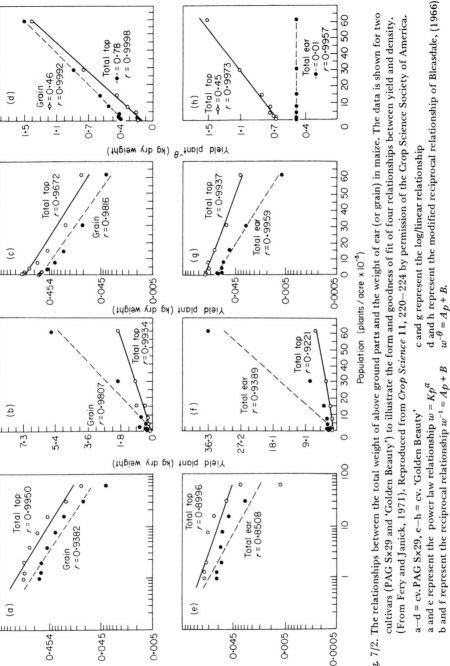

Fig. 7/2. The relationships between the total weight of above ground parts and the weight of ear (or grain) in maize. The data is shown for two cultivars (PAG Sx29 and 'Golden Beauty') to illustrate the form and goodness of fit of four relationships between yield and density. (From Fery and Janick, 1971). Reproduced from *Crop Science* 11, 220–224 by permission of the Crop Science Society of America.

a–d = cv. PAG Sx29, e–h = cv. 'Golden Beauty'   c and g represent the log/linear relationship
a and e represent the power law relationship $w = Kp^a$ d and h represent the modified reciprocal relationship of Bleasdale, (1966)
b and f represent the reciprocal relationship $w^{-1} = Ap + B$ $w^{-\theta} = Ap + B$.

various measurements are plotted against population density in Fig. 7/3. The slopes of the lines on a logarithmic plot describe the rate of change of each character with changing density and give a direct comparison of the relative plasticity of the various organs. Some structures were relatively stable, particularly height, leaf width, stem diameter and the number of spikelets borne per spike. Branching (tiller formation) in contrast was very plastic. The number of live tillers per genet fell from about 11 at low density to 1 at the highest density and was the most plastic of

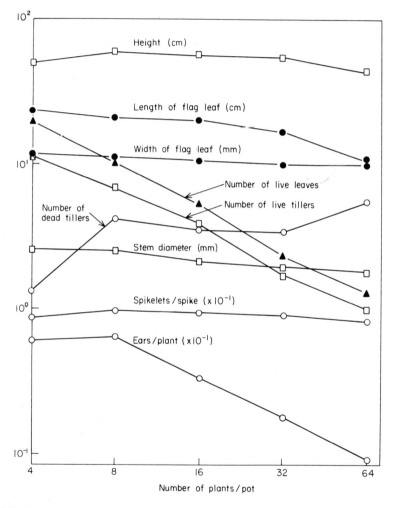

Fig. 7/3. The plasticity of the components of form and seed yield of wheat sown at a range of densities. (Drawn from data of Clements *et al.,* 1929)

all the features measured. Part of this response to density was due to a
reduction in the number of tillers formed (their birth rate) and part to a
marked increase in the number of tillers dying before the end of the
experiment (about half of the density effect on tiller number was to
reduce the number of tillers formed and half to increase the number that
died). It is an important feature of the population biology of plants that
parts of a genet may die as a result of density stress, e.g. the lower
branches of a shaded tree or whole rooted ramets in a vegetative clone.
In the annual cereals such as wheat, all tillers, whether they form an
inflorescence or not, die at the end of the first growing season or before
(unless the plants are prevented from flowering and very heavily fertilized; see Langer, 1972).

The weight of seed produced by a wheat plant is the product of three
"yield components": (i) the number of fertile tillers per plant, (ii) the
number of grains per ear, and (iii) the mean weight per grain. The comparative plasticity of the "yield components" of wheat is illustrated by
an experiment in which an 816-fold range of sowing densities was used
(Puckridge and Donald, 1967). The result was a 43.2-fold variation in
fertile tillers per plant (ears), 1.7-fold variation in grains per ear and only
1.04-fold variation in mean weight per grain. The behaviour of wheat
under density stress is probably typical of the annual grasses. The relative plasticity of the different organs reflects the growth cycle of the
species: flowering in wheat is triggered by vernalization and photoperiodic stimuli and until the plants have received the appropriate
sequence of experiences they remain vegetative. During this phase
density stress affects the number of tillers and thus the potential number
of ears. After the flowering stimuli have been received and the meristems
change to the flowering condition, the number of tillers scarcely
increases further. The stress of density that is experienced after this
point is reflected in the mortality of tillers and a reduction in the size of
the ears that have been initiated. The potential size of the inflorescence
is determined relatively early in the post-vegetative stage before the
tillers elongate telescopically and the ear becomes visible. Subsequent
density stress may cause some abortion of flowers or grains. All of these
adjustments occur before grain filling starts and it is therefore not surprising that grain size absorbs very little of the density stress.

The field bean (*Vicia faba*) has a contrasting growth form and
response to density (Hodgson and Blackman, 1957a and b); it has hypogeal germination and branches arise from the axils of the cotyledons

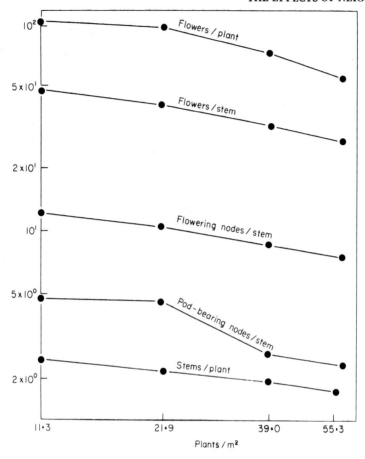

Fig. 7/4. The influence of density on the plasticity of *Vicia faba*. (Drawn from data of Hodgson and Blackman, 1957a)

below ground. These in turn may branch at or below ground level to give a multi-stemmed plant. The stems above ground are usually unbranched and bear clusters of flowers in the axils of the leaves. Flowers are continually produced through most of the growth cycle and, in marked contrast to wheat, the terminal meristems are never used up in flowering. This growth pattern has been called indeterminate because the meristems could theoretically continue to produce new leaves and flowers indefinitely. Like most annuals in the Leguminosae, *Vicia* responds to shading and to density by etiolation: the increase in height is brought about by elongation of the internodes rather than an increase in their number. The plastic reactions to density stress are illustrated in Fig. 7/4. As with wheat, part of the response to density is a reduction in the number of plant parts produced (branches, flowering nodes) and part is by abortion (death of old leaves, abscission of flowers and pods). The mean seed weight is relatively constant: a 5-fold variation in plant density produced only 0.11-fold change in seed weight (0.60—0.65 g). The relative constancy of mean seed weight over a density range in this experiment is particularly interesting because the year to year variation in seed weight is quite large. In one of the experiments repeated at the same density over 3 years the mean seed weight varied by as much as 50% (0.50—0.74 g).

The cultivated sunflower (*Helianthus annuus*) was the second species studied by Clements in his pioneer work on density responses. Most commercial varieties of sunflower bear a single large capitulum at the top of an unbranched stem, and sometimes there is also a diminutive lateral. The capitulum is developed from the terminal meristem when the plant is quite small (15—20 cm high) and in subsequent growth there is no potential for any further increase in the number of plant parts. Density stress is therefore absorbed by a reduction in the size of plant parts (leaf length, stem diameter) or by abortion. The sunflower is unusual in having highly plastic seed size — variations in density may produce at least 6-fold variation in achene weight. Thus the sunflower typifies an extreme determinate pattern of growth in which the homeostasis of seed size has been sacrificed but it is very doubtful whether this behaviour is anything but an interesting anomaly of a specialized cultivar: the wild species of sunflower behave quite differently. The wild forms of *Helianthus annuus* are branched and bear many capitula over a protracted flowering season; the effect of density on the wild forms is to reduce the numbers of branches and capitula. In the wild forms, a range of densities of 156-fold produced only a 1.25-fold variation in mean achene weight. The cultivars, bred for their curiously determinate growth habit have apparently lost the ancestral ability to regulate seed size. A number of the contrasts in the response to density of the cultivated and wild sunflowers are illustrated in Fig. 7/5 (there is a further discussion of these experiments in Bradshaw, 1965).

The corncockle (*Agrostemma githago*) is a Caryophyllaceous weed of annual cereals, branching just above ground level, growing to a height of 70—100 cm and developing a terminal cyme of flowers. The capsules contain 40+ seeds and a vigorous plant may produce 50+ capsules. The

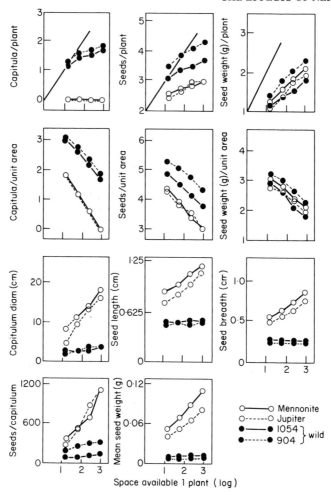

Fig. 7/5. Responses to density of *Helianthus annuus* – the wild and cultivated sunflower. (From Khan, 1967)

response of this plant to density is: (i) a reduced number of basal branches (to the extreme of a quite unbranched stem); (ii) a reduction in the size of the inflorescence (to the extreme of a single, or even no flower per stem); (iii) a reduced number of seeds per capsule (to the extreme of barren capsules); and (iv) a reduction of mean seed weight of *ca* 5% (see Fig. 7/6). The growth pattern of *Agrostemma* is monopodial and determinate – the terminal meristems are used up in flowering – and

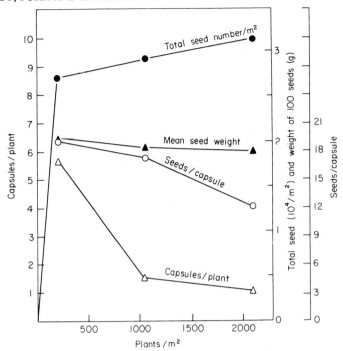

Fig. 7/6. The influence of sowing density on seed yield and the components of seed yield for *Agrostemma githago*. (From Harper, 1961)

it exemplifies the plastic response of determinate annuals in contrast to species like *Vicia faba, Veronica hederifolia, Anagallis arvensis* and other species with axillary inflorescences and continuing terminal growth.

An experimental comparison of density responses

An unique comparative study of the density responses of a group of species was made between *Bromus inermis, B. tectorum, Capsella bursa-pastoris, Conyza canadensis, Plantago lanceolata, P. major, Senecio sylvaticus, S. viscosus* and *Silene anglica* (Palmblad, 1968). Seed was sown in imitation of natural dispersal by dropping from a height of 30 cm onto bare soil at densities equivalent to 55 275, 2750, 5550 and 11 000 seeds per m². The subsequent behaviour of the various populations is shown in Fig. 7/7a, b, c and d. There was density-dependent variation in (1) the proportion of seeds that produced seedlings, (2) the

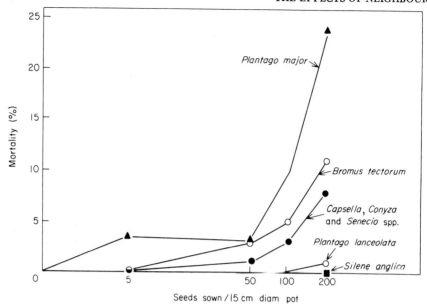

Fig. 7/7a. The percentage mortality of various species as influenced by density. (Drawn from data of Palmblad, 1968)

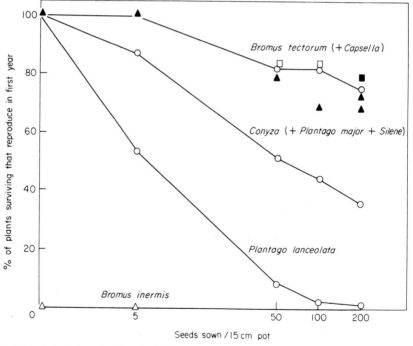

Fig. 7/7b. As 7/7a but showing the influence of density on the percentage of plants that survive and produce seed in the first year.

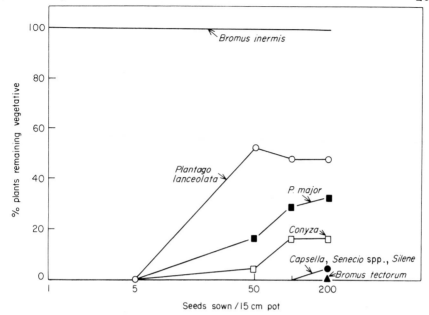

Fig. 7/7c. As 7/7a but showing the influence of density on the percentage of plants that survive to the end of the first year but remain vegetative.

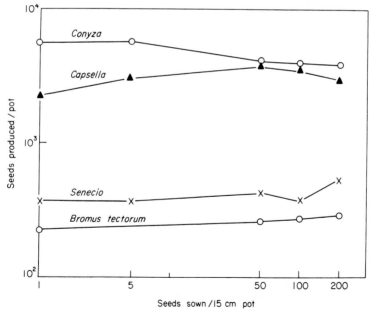

Fig. 7/7d. As 7/7a but showing the influence of density on the numbers of seeds produced per population.

proportion of seeds that grew to maturity, (3) the proportion of survivors that reproduced and (4) the proportion of survivors that remained vegetative after the first growing season. *Bromus inermis, Conyza canadensis* and *Silene anglica* reacted to density stress by density-dependent mortality in the earliest phases of germination and seedling emergence; in contrast *Senecio sylvaticus, Plantago lanceolata* and *Bromus tectorum* showed no density-dependent mortality in the seedling phase but between seedling emergence and flowering more plants died and in all species except *Silene anglica* death was density-dependent, particularly so in *Plantago major*. With the exception of *Bromus inermis*, all individuals that survived to the flowering period at the lowest density flowered and produced fruits, but the chance of an individual producing seed declined with increasing density. *Plantago lanceolata*, which had made no adjustment to density by mortality, made the greater part of its response in a reduced chance that an individual would reach a critical size for flowering in the first year.

Even at the highest density, 73—75% of the surviving plants of *Capsella bursa-pastoris* set seed. The annuals all died after flowering; only *Plantago lanceolata, P. major* and *Bromus inermis* overwintered and grew on in the second season. *Bromus inermis* does not flower in the first year unless it has received appropriate vernalization and photoperiodic stimuli. This species was maintained in culture through to the second year to examine the continuation of its life cycle. All the plants that survived to the end of the first year also overwintered successfully.

At the end of the first growing season the total seed production per pot was calculated for each species and the effect of density is summarized in Fig. 7/7d. The narrow range of variation in seed production per pot, despite the wide range of densities sown, is very remarkable. The 200-fold range of seed input was never represented by even as much as a 2-fold range of output. This set of data illustrates most elegantly the way in which density-dependent mortality and plasticity together regulate the reproductive output of a population.

Biennial plants

The foxglove, *Digitalis purpurea*, is typical of that large group of species which, when grown in favourable conditions, make vegetative growth in the first season and flower and die in the second. When conditions are less favourable, a proportion of the plants continue in the vegetative state for 2, 3 or more years: the bienniality is facultative rather than obligate. The foxglove is a fugitive inhabitant of woodland communities, developing population flushes in woodland clearings and then declining in numbers as the canopy cover becomes more complete. It is also common in hedgerows and on screes and generally in slightly acidic habitats where there is disturbance (e.g. fire) that is not annually recurring.

An experimental study of the density response of this species was made by sowing a field experiment at a very wide range of densities (30–100 000 seeds per m²). Although this range is so large, it almost certainly under-represents the range naturally occurring under dense stands. The plants were allowed to develop for 2 years before harvest and the responses to density in the second growth season (first flowering season) are shown in Fig. 7/8a.

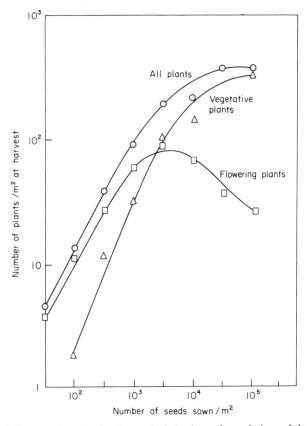

Fig. 7/8a. The influence of sowing density on the behaviour of populations of the foxglove, *Digitalis purpurea*. (From Oxley, in preparation)

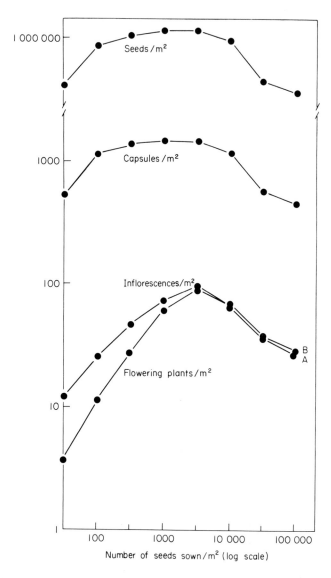

Fig. 7/8b. The relationship between the density of sowing of *Digitalis purpurea* and the components of seed yield. (From Oxley, in preparation)

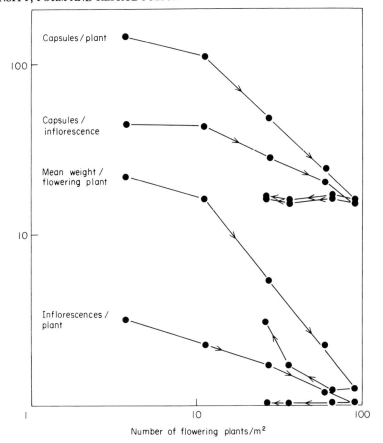

Fig. 7/8c. The relationship between the number of flowering plants per m² of *Digitalis purpurea* and various elements contributing to the seed yield of the population. The arrows indicate progressively increasing densities of seed sown — cf. Fig. 7/8b. (From Oxley, in preparation)

The chance that a seed would produce a surviving plant fell rapidly with increasing density as also did the probability that a surviving plant would flower in the second season. Of those plants that did flower in the second year, increasing density reduced both the number of inflorescences per plant and the number of capsules per inflorescence; the number of seeds per capsule and the mean weight per seed remained very constant. Just as in the group of species studied by Palmblad, density stresses were absorbed in (i) mortality, (ii) a changing proportion

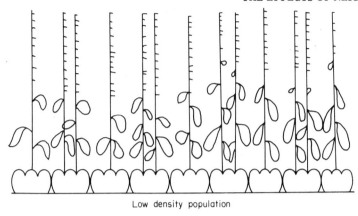

Low density population

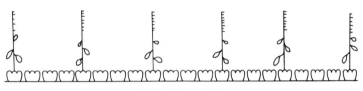

High density population

Fig. 7/8d. Diagrammatic illustration of the patterns of growth and flowering in populations of *Digitalis purpurea* at high and low densities. (From Oxley, in preparation)

of plants that flowered or remained vegetative and (iii) a reduced reproductive output of those that did flower.

The dry weight of the populations approached, as expected, a constant yield per unit area; this was made up of flowering plants plus those that remained vegetative. At low densities the greater part of the dry weight was made up of flowering individuals but at high densities it was mainly vegetative rosettes. Figure 7/8b shows the effects of density on (i) the number of flowering plants, (ii) the number of inflorescences, (iii) the number of capsules and (iv) the number of seeds produced per m². All these values rise with initial increases in density, reach a more or less stable intermediate value and subsequently decline. A 3000-fold variation in the input of seeds gave only a 3-fold range of seed output per unit area! The highest rate of seed production (*ca* 1 000 000 per m²) was obtained from sowing densities between 1000 and 10 000 per m². The lowest values of seed production were obtained from inputs of either 30 seeds per m² or 100 000 seeds per m²!

A major effect of density in populations of *Digitalis purpurea* was to increase the proportion of plants that failed to flower in the second year. During the growth of the plants it seems likely that what was initially a normally distributed range of plant sizes had become skewed and a size hierarchy had developed within the populations, particularly at the high densities. From this hierarchy the smallest individuals had been lost in the thinning process. During the second season of growth the remains of the hierarchy had formed into two fractions, one exceeding a critical size for flowering and the other not. The flowering plants elongated rapidly to a height of *ca* 1 m bearing cauline leaves and producing a high canopy; the vegetative plants remained as rosettes with no vertical extension though they sometimes reached flowering size in subsequent years. The flowering and vegetative forms within a population were as different in morphology as two quite distinct species and must play a quite different ecological role in the community. The plants that flowered escaped from the stressed environment of the overcrowded rosettes into a new, higher and relatively uncrowded environment.

One of the consequences of the change in population structure that results from density in a population of foxgloves is that the same number of flowering plants per unit area can be obtained from sowing a low or a high number of seeds. In the first case the flowering plants will be intermingled with a few rosettes whereas in a stand at high density the flowering plants grow out of a dense mass of suppressed rosettes. The same density of flowering plants may behave quite differently according to whether it emerged from a high or low density stand. This is illustrated in Fig. 7/8c and would represent a puzzling situation to the ecologist in the field faced with foxgloves flowering at the same density but with quite different form (Fig. 7/8d).

The distinction that arises in a foxglove population between flowering and non-flowering individuals is an extension of a hierarchy into a bimodal population and this may be the normal pattern of density response in monocarpic perennials. Evidence that there is a critical rosette size for flowering and that thinning affects mainly the smallest plants comes from an experiment with the teazle *Dipsacus fullonum* (Werner, 1975). In this species it appears that the chance that a plant will flower in its second or subsequent years is very remote unless a rosette of at least 20 cm diameter has been developed in the previous summer. These large plants seldom died. The risk of death was concentrated in that part of the population less than 9 cm diameter (Fig. 7/9).

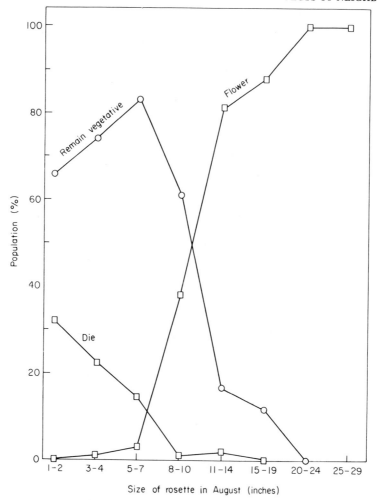

Fig. 7/9. The relationship between the size of rosettes of teazel (*Dipsacus fullonum*) and the probability of flowering in the following year. (From data of Werner, 1975)

There is virtual continuity between annual, biennial and perennial growth habits and this is particularly clear when any attempt is made in the grasses to distinguish between annual tillering forms and clonal perennials. The typical annual grass, such as wheat, branches repeatedly in its foreshortened stem to form a tussock of tillers and most of these develop independent root systems. Some of the tillers flower and die; the remainder, the vegetative or sterile tillers also die when or before the fertile tillers flower. In some species the sterile tillers overwinter and flower in the following year; in others the larger proportion of the tillers remain vegetative and in subsequent years produce new daughter tillers of which some flower and others continue to perennate and enlarge the genet. It is obviously an exercise in semantics to define a point at which branching becomes clonal growth but it is sometimes convenient to take the establishment of its own root system as the point at which a branch has become a tiller or ramet. If this definition is taken, we are forced into the position in which wheat is to be regarded as a clonal annual; although this seems odd it is not illogical.

The influence of density on a rhizomatous perennial (*Tussilago farfara*)

Tussilago farfara is an opportunistic colonizer of disturbed open ground, often invading and colonizing it at an early stage in a succession until it is overtopped by other species. It may remain a long-term occupant of habitats that are burned annually, e.g. railway banks. The seeds are wind dispersed and *T. farfara* is a primary colonizer from long distances of the newly reclaimed polders of the Netherlands (Bakker, 1960) (Fig. 2/2c). It also colonizes newly reclaimed land that has been exposed after peat has been stripped from blanket bog in Ireland.

Germination occurs in April or May, shortly after the seeds have been shed and a tap-root is quickly formed. Contractile roots develop during the first summer from the lowest nodes of the stem and rhizomes grow out from the same region. By the first autumn the main shoot may bear several large leaves and already inflorescence buds may be present. The leaves die in the autumn and a cluster of flowers is produced in the spring. The main shoot then dies. During the first growing season some of the rhizome tips grow upwards to produce ramets which are vegetative in the following season. The main shoot was the only connecting link between these ramets and after its death the genet is represented by a number of unconnected parts. Rhizomes initiated in year 1 normally produce leafy shoots (together with daughter ramets) in year 2 and flower in year 3. Each ramet is monocarpic though the genet is of course polycarpic (iteroparous). Such a life cycle and growth form contains a number of plastic elements.

Seedlings of *T. farfara* were planted into a loam—sand mixture in wooden boxes 45 x 45 x 30 cm deep at densities of 1, 4, 16 and 64 per box. During the first year of growth there was a rapid response to density, particularly in the rate of rhizome production and to a lesser extent in the mortality of ramets. The relative changes in the density of leafy shoots (vegetative plants) are shown in Fig. 7/10a. The boxes planted with one plant contained 11 ramets at the end of the first season and 13 in the second season. The population of shoots at the highest density had doubled by the end of the first growing season and then declined to near its starting density in the second season. A comparable but longer-observed experiment was made in open field plots and showed the same type of response to density stress continued into the later part of the second season (Fig. 7/10b).

The stress of high density affected flowering even more strongly than vegetative growth. In the winter after the first growing season, the dry weight of inflorescence buds per box paralleled the range of planting densities (though the 64-fold variation in plant density was reflected in only a *ca* 4-fold variation in the weight of dormant buds). In May of the second season however, the populations at the highest density produced the lowest weight of reproductive tissue (flowers + seeds + peduncles + receptacles). By the end of the second season the order of reproductive activity had been completely reversed — the greatest weight of inflorescence buds was in the boxes planted initially with just one seedling and the lowest weight of inflorescence buds was in the boxes planted at the highest density (Fig. 7/10c).

At flowering time, those plants that had developed a rosette of leaves in the previous season produced a cluster of peduncles each bearing a capitulum with a number of achenes. The seed produced by a population (i.e. per box) is the product of (a) the number of flower clusters per box, (b) the number of mature capitula per flower cluster and (c) the number of seeds produced per capitulum. These values are shown in Fig. 7/10d. Nearly every plant produced a flower cluster but there was great plasticity in the number of capitula per cluster and further plasticity in the weight of seed borne per capitulum which was more than halved by the increased planting density. The weight of seed produced per box was remarkably stable though with a clear optimal density for maximal seed production. Seed production per box varied only 1.5-fold over a 64-fold range of planting density. The production of rhizomes was very much more sensitive to density than the production of seeds and at high density the allocation to rhizomes fell to nearly $\frac{1}{3}$ of its value at low density (Fig. 7/10e).

Populations of clonal plants like *Tussilago* develop a hierarchy of size

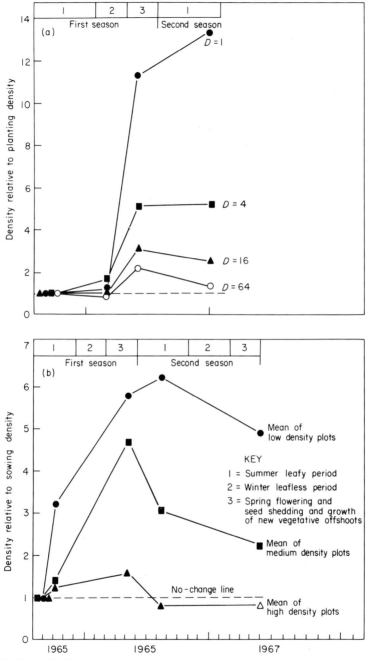

Fig. 7/10. Changes in the density of leafy shoots of *Tussilago farfara*:
(a) planted at densities of 1, 4, 16 and 64 seedlings per pot;
(b) sown at a range of densities in a field experiment. Values are shown as changes relative to the starting density. (From Ogden, 1968).

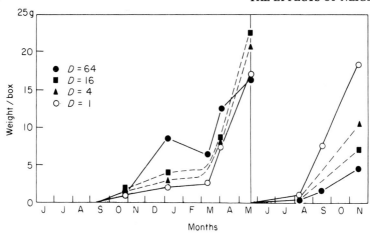

Fig. 7/10c. The influence of density on the reproductive activity of *Tussilago farfara*. The graphs show the change in the weight of reproductive structures over two seasons from seedlings planted at densities of 1, 4, 16 and 64 per box. "Reproductive structures" include flower buds, seeds, seed heads, flowers and peduncles. (From Ogden, 1968)

among the genets and the process of elimination of the smaller genets continues until a few large clones dominate the population. So few perennial herbs have been studied to determine their response to density that it is impossible to say if *Tussilago* behaves in a typical fashion. There are unusual features about this species such as the very early death of the older tissues and the high and reliable reproductive activity. It may be that in perennials that accumulate a more persistent ageing body, such as *Iris*, the effects of density are seen in a more complete inhibition of flowering.

The influence of density on the form and reproduction of trees

The tree depends for its very form on the continual accumulation of dead tissue — it is a thin layer of living sepulchre enclosing a growing corpse. A meaningful comparison of the effects of density on a tree with the effects on a herbaceous plant would require determination of the amounts of living tissue in the two cases and this is a virtually impossible task with the tree. Almost all of the information about density responses in trees is dominated by the accumulated effects over the life

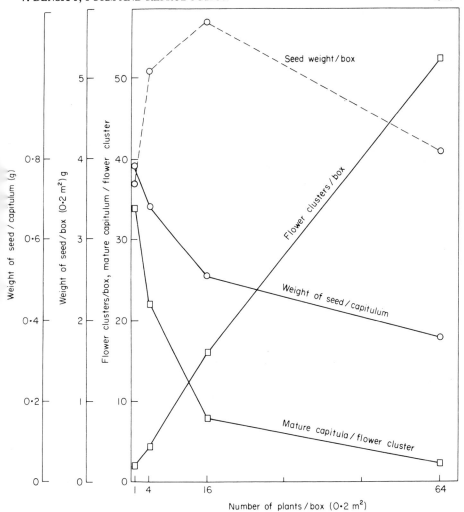

Fig. 7/10d. The influence of density on the components of seed yield of *Tussilago farfara*. (From data of Ogden, 1968)

of the individual. In fact it is difficult to find any single fully documented record of the influence of density on a stand of trees throughout their life cycle. No detailed studies of the density responses of annuals were made before the 1920s and it requires a peculiar devotion to science to plan a formal density experiment with trees that will rarely complete a life cycle in the lifetime of the experimenter. Results are so slow to

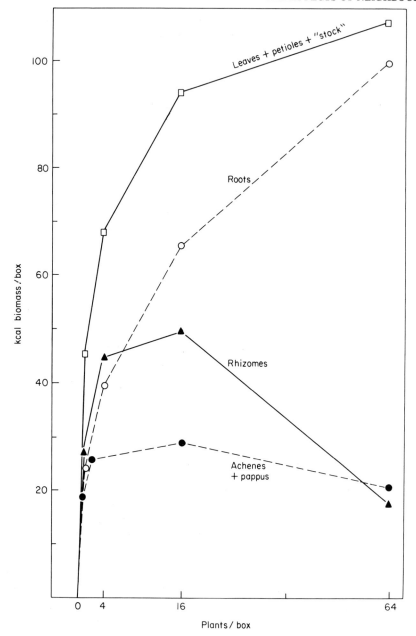

Fig. 7/10e. The influence of plant density on the partition of biomass (expressed as kcal) between various structures in *Tussilago farfara*. (From Ogden, 1968)

emerge that it is not surprising that foresters have concentrated on the sort of experiment that has direct relevance to the production of timber. However, the forester is as aware as any other practising manager of plant growth, of the profound influence that the proximity of neighbouring trees has on the form, development and the growth rate of his species. At all the stages from the seedling nursery through initial planting, various thinning and pruning stages, selective felling and, in the management of seed orchards, silviculture is concerned with optimizing density.

The kinds of measurement used for recording the growth of herbaceous plants are quite unsuitable for trees. Even height is hard to measure and parameters like dry weight are obviously useless. Consequently foresters have developed a highly specialized technology and associated vocabulary for the description of trees that is quite alien to that used by agronomists and ecologists who deal with herbs. The word "density" has itself a specialized meaning in forestry usage. To the agronomist or ecologist it is the number of plants per unit area; in forestry the word density implies some measure of the stress present within a forest that is caused by the demands made by the trees. The "density" of a forest measures not only the number of trees present per unit area but also takes account of their size and sometimes their form.

The forester's concept of "density" is described by Carron (1968):

> . . . to define in concise terms just what we mean by the density of a stand is rather difficult and perusal of the literature on stand density suggests that foresters are far from agreement on what the term implies. The basis of our concept of density lies in the fact that if we look at a forest stand we get an impression of the way the tree material fills up an air space. We think of the density of the stand in terms of the amount of tree material per unit area or space. This concept of density gives us an attribute which improves the value of our description of the stand but is valuable for practical management only if we can express it in a quantitative way . . . The most likely variables for inclusion in a measure of stand density are those which are readily measured, are intercorrelated with the other main variables of size, and in sum or average represent a stand characteristic efficiently.

So much of a tree is inaccessible that if measurements made at ground level give predictive information about the inaccessible upper parts this is clearly of enormous value; hence the very great importance that foresters attach to correlations between parts of a tree. One unique measure plays a very large part in the descriptions of forest stands, the Diameter at Breast Height (DBH), either measured over the bark (DBHOB) or under the bark (DBHUB). In an even-aged stand there is generally a good correlation between the diameter of the trunk and the height and spread of the canopy of individuals (e.g. Fig. 7/11); thus DBH summarizes a lot of the information that a forester needs to guide his management of a forest.

Measurements of forest "density" take various forms:

(i) *The stocking rate.* The term stocking is the forester's equivalent to the "density" of the agronomist (the number of trees per unit of land) and is occasionally used to describe the early phases of establishment of a plantation. It seldom plays much part in describing later phases of forest growth.

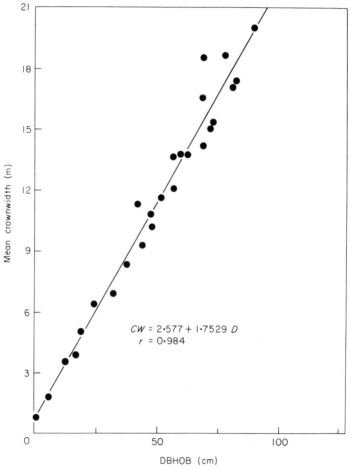

$$CW = 2 \cdot 577 + 1 \cdot 7529 \, D$$
$$r = 0 \cdot 984$$

Fig. 7/11. The relationship between crown width and stem diameter in open grown *Eucalyptus obliqua.* (From Curtin, 1964)

(ii) *The stand density index.* If the number of trees per unit area is plotted against their average diameters (DBHOB) for stands at "full density" a straight-line relationship can usually be found (Reineke, 1933). The slope of this line is the same for a number of different species although the height of the curve (the constant in the regression equation) differs from species to species. The slope and height of the curve are not much affected by the age of the trees or the quality of the site (Fig. 7/12). If, therefore, a chosen forest stand is sampled and the number of

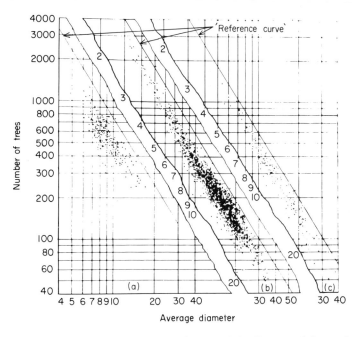

Fig. 7/12. The relationship between the density of trees and the diameter of the trunk at breast
height for: A = mixed conifer stands in California; B = Douglas fir (*Pseudotsuga douglasii*)
in Washington and Oregon; and C = Douglas fir in North California.
Diameters are given in inches and density as numbers per acre but the log–log transformation gives the data an effectively universal scale. The "reference curve" in each
graph is linear on log–log coordinates and represents the maximum number of trees
it is possible to have of a given trunk diameter. (From Reineke, 1933)

trees and their mean diameter are plotted on such a graph and the plot lies on the curve of full density for that species, the chosen stand may be said to be at full density. If the population is represented by a point lying below the reference line it is not fully utilizing "space" or resources. For comparative purposes Reineke suggested that a line be drawn parallel to the reference line through the point representing the chosen stand; the number of trees corresponding to the 10 in (25 cm) value of DBHOB is then read from the graph. This procedure makes it possible to make comparisons within and between species. Clearly formal algebraic descriptions of the relationship between the point and the reference line may also be used for such a comparison. Reineke's stand density index has been widely used in practical forestry. The major theoretical difficulty lies in choosing "full density" stands for calculating the reference line which is ideally an asymptotic limit. There are clear parallels between Reineke's index and the thinning curves described in Chapter 6 (e.g. Fig. 6/25) but in the latter there is an empirical test of full density, that which cannot be exceeded without forcing the excess individuals to die. Such a criterion is manifestly unacceptable in commercial forestry.

(iii) *Number of trees and form*. The spacing of trees has an influence on the rate at which a trunk tapers in diameter from base to apex and could therefore contribute a measure of "density". But it involves measuring diameter at different heights and is therefore a slow affair.

(iv) *Number, diameter and height* can be combined as a measure of forest "density", particularly with species or situations where height is not tightly correlated with DBHOB.

(v) *Number, diameter, height and form* (taper) could ideally be used to assess the volume of a forest stand but this again involves measuring extremely awkward features, and is too time-consuming for practical purposes.

(vi) *Percent crown closure*. The crown of a tree is the volume in which assimilation occurs; the degree to which the crowns of trees in a forest occupy the available volume or area could give a measure of density stress, always provided that light was the limiting resource affecting forest growth. The percent crown closure can be estimated as the horizontal area of crown expressed as a percentage of the horizontal ground area.

(vii) *The crown competition factor*. It is rather generally found that the development of the crown is highly correlated with DBHOB. This is an interesting statement about tree form. The fact that the precise form of the correlation differs between species (Dawkins, 1963) implies that the "form" of trees and perhaps other plants is susceptible to quantitative description. One extremely interesting measure of the stress of density in a forest is the Crown Competition Factor which uses the

linear relationship between crown width and stem diameter of uncrowded (open grown) trees as a basis from which to assess the "competition"
between trees in a crowded stand (Krajicek *et al.*, 1961). Such a factor
measures "the amount of space available to trees in a stand as compared
to the amount of canopy space that they would occupy if they were
completely open grown" (Curtin, 1964, 1970).

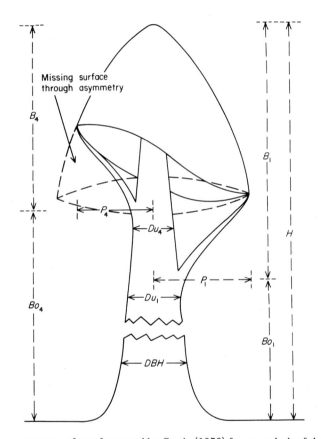

Fig. 7/13. The components of tree form used by Curtin (1970) for an analysis of the interrelationships between parts of individuals in a forest stand.

The interrelationships between the parts of a tree that constitute its
form have been studied in depth in forests of *Eucalyptus obliqua*
(Curtin, 1970). The relevant measures of shape are shown in Fig. 7/13.
A series of relationships define form and can be represented by simple
equations which are independent of the history of the stand, its age, site

quality and density:

$$D_u = 0.14 + 0.18\ D + 0.90\ BD/H'$$

$$Bo' = -1.89 + 1.09\ H - 0.86\ DuH'/D$$

$$B = -1.89 - 0.09\ H' + 0.86\ DuH'/D$$

$$Du = -1.46 + 0.49\ CW \text{ (accounting for 92\% of the variance of } Du)$$

These equations simplify to

$$CW = 1.78 + 0.67\ D + 1.28\ BD/H'$$

$$CW = 1.36 + 0.68\ D + 1.90\ BD/AH'$$

Nothing quite like these attempts to describe the form of plants seems to have been tried for plants outside forestry.

Examples of tree spacing experiments made with forest trees

One of the rare examples of a study of the direct effects of spacing on the growth of trees was made on Sitka Spruce (*Picea sitchensis*) in Northern Ireland (Jack, 1971). A stand had been established in 1949 and the experimental treatments started in 1960 when the "crop" was *ca* 4 m in height with a mean DBHOB of *ca* 6 cm and with the side branches nearly touching. The trees were at that time *ca* 1.83 m apart (*ca* 3000 stems/ha) and the young trees had probably already interfered with each other's growth to some extent. A range of spacings was obtained by systematic removal of alternate trees or alternate rows but there was no deliberate selection as would have been the case in a commercial thinning programme. The spacing regimes obtained were 3000, 1500, 750, 500, and 300 stems/ha. Subsequent behaviour of the trees was reported after 9 years at the new stocking regimes.

The effect of spacing on height was generally very small and the experiment gives no support to the commonly held view that crowding forces more rapid elongation — there was indeed a slight stunting at the densest spacings (Fig. 7/14). (Although height is extremely insensitive to crowding it is extremely sensitive to the conditions of a site and can often be used to evaluate sites independently of the degree of crowding of the populations.)

Annual increments in diameter increased with distance between the trees and the effects were much greater than on height. The difference between annual diameter growth at the various spacings was least at the

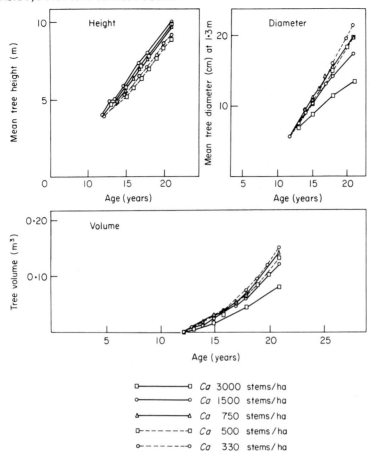

Fig. 7/14. The influence of tree spacing on the time course of increase in tree height, diameter and volume of *Picea sitchensis* (Sitka spruce). (From Jack, 1971)

start of the experiment and became greater and greater. Nine years after the spacing experiment was started the widest spaced trees were *ca* 10% shorter and 50% fatter than those closely spaced. The rather small differences in height were established quickly and maintained; the differences in diameter increased progressively with time. In this experiment involving a 10-fold variation in stems/ha there was less than a 2-fold variation in tree volume/ha. The frequency distribution of tree volumes was strongly skewed at high densities (Fig. 7/15).

A somewhat similar experiment was made with *Pseudotsuga menziesii* in the Pacific Northwest states. An unreplicated trial had been set up in

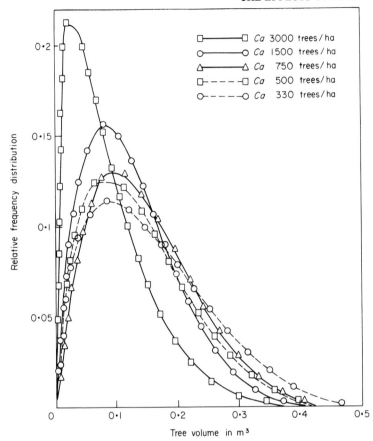

Fig. 7/15. The frequency distribution of tree volumes for *Picea sitchensis* 9 years after the establishment of the various degrees of crowding. (From Jack, 1971)

1925 on a rather uniform site with transplants placed at spacings of 1.2 x 1.2, 1.5 x 1.5, 1.8 x 1.8, 2.4 x 2.4, 3.0 x 3.0 and 3.6 x 3.6 m. Each plantation was of 1.3 ha. Plants that died in the first 5 years were replaced (Curtis and Reukema, 1970).

During the first 5 years after planting the densest plots had the highest trees but this difference had disappeared after about 10 years and subsequently the highest trees were found at the widest spacings. The population received no artificial thinning at any stage and this perhaps accounts for some of the results which seem to be at variance with common forestry experience. Some of the data are summarized in Fig. 7/16a, b and c. There was obviously density-dependent mortality and a marked

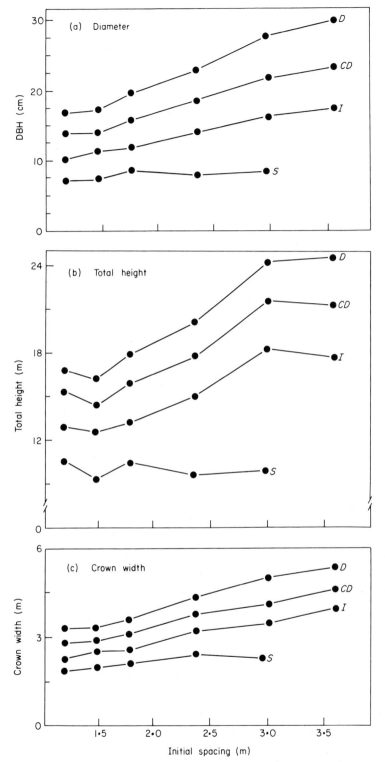

Fig. 7/16a, b, c. The influence of spacing on some components of form in *Pseudotsuga menziesii*. The populations are divided into classes according to the state of the crown D = dominant, CD = codominant, I = intermediate and S = suppressed. (From Curtis and Reukema, 1970)

effect of spacing on the frequency distribution: the higher the density the more trees fell into the suppressed class. An interesting effect of spacing appears in the relation between the volume of crown per hectare and the surface area of the crown. The more crowded trees have a crown volume that is very high relative to the surface that they expose. This may account for the higher death risk of trees at high density because it implies that the assimilatory surface is small relative to the volume of supporting tissue.

An unmanaged forest is a naturally self-thinning system with a hierarchy of dominant and suppressed trees, the death risk being concentrated within the suppressed class. In managed forests the spacing of individuals is adjusted, as they grow, by thinning to favour the maximal production of useful trunks. If a young uniform aged stand remains unthinned, a hierarchy of dominant and suppressed individuals develops in which the suppressed forms represent resources that are diverted from the mature timber producers. Intolerant species, in dense stands, tend quickly to develop a bimodal frequency distribution of volume with two clear classes of dominant and suppressed individuals, but in tolerant species a single moded distribution is commoner.

Dominance and suppression affect not only the status of whole trees in the forest hierarchy but also the status of individual branches. In the unstressed development of a tree in open parkland, branches near ground level may remain active throughout the life of the tree and the form of the whole is a massed balloon of foliage extending to ground level or to browsing height if there are livestock about. On comparable trees in a forest the lower branches will have died early in the life of the plantation and the form of the tree is a tall unbranched trunk with living canopy branches only near the top. Close spacing reduces the number of branches that are formed: in Jack's experiment with *Picea sitchensis* there were 5 branches in the first whorl above 2 m height in the most crowded trees but *ca* 7 in the least crowded. However most of the control of the extent of branching in a mature tree has occurred by the death of branches rather than by inhibition of their formation. In a dense developing stand the lower branches are shaded and die: it is very characteristic of an unthinned stand that (a) the canopy is closed, (b) the canopy forms a relatively thin layer and (c) there are abundant dead branches and their remains. A tree responds to density stress by partial mortality and such death of parts probably usually precedes and continues until the whole of a suppressed individual is dead.

The influence of crowding on the reproduction of trees

The pre-reproductive period of trees is usually long (see Chapter 22) and density stress usually makes it longer. It is common forestry experience that the reproductive activity of a tree can be promoted by releasing it from the density stress of its neighbours. The individual trees that dominate a crowded canopy produce a disproportionately large part of the seed output of a stand (Matthews, 1963). Where trees are deliberately grown for seed it is customary to plant at wide spacing or to "release" trees that are in a crowded stand. The most detailed studies of the effects of density on seed production come, however, not from forestry but from experiments with fruit trees.

A remarkable experiment was made to determine the influence of tree spacing on the behaviour of 'Jonathan' apples. The experiment was made in Hungary and was started in 1953 at the time of collectivization to discover which planting systems would be best for large cooperative or state farms. An account of the experiment is given by Verheij (1968) who visited the site when the orchard was 15 years old and analysed data from the field records. The experiment involved variations in density (100–1110 trees/ha) and various planting patterns. The patterns did not seem to have major effects and Verheij treats the experiment as a simple study of density. Fruiting tended to occur as heavy and light crops in alternate years (a characteristic of the apple when unthinned) and the fruiting data have been summarized over pairs of years.

The height of the trees was scarcely affected by density though there was a large effect on the spread of the canopy. The trees were pruned to give semi-standard, bush and spindle shapes and this specialized feature of the crop and the experiment has to be borne in mind when generalizing from the results. The cumulative yield of fruit per hectare is shown in Fig. 7/17a. In 1957–60 the yield was more or less linearly related to density and there was no evidence of interference between the trees. With the passage of time this linearity was lost and by the end of the study period there was some slight evidence that fruit production per hectare was actually reduced at the highest densities. The cross-sectional area of trunks per hectare followed a different pattern and increased both with time and with number of trees per hectare throughout the experiment Fig. 7/17b. Thus the yield of fruit per unit of trunk fell dramatically with increasing density. It would be good to know in such a situation what proportion of annual gross or net assimilation was devoted to fruit production. The comparison of fruit production with

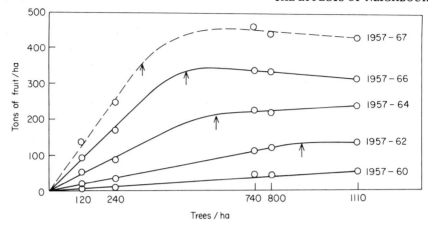

Fig. 7/17a. The cumulative yield of fruit borne by plantations of apple at various planting densities. The arrows show the density at which trees of various ages begin to affect each other's yield. (From Verheij, 1968)

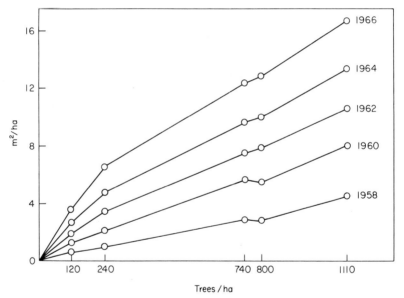

Fig. 7/17b. The relationship between planting density and basal area of trunks per hectare for apples. (From Verheij, 1968)

diameter growth is a measure of biennial reproductive activity against the accumulated dead tissue over the years and does not compare easily with measures made on non-woody plants.

Density affected the relationship between mean weight per fruit and the number of fruits per tree — two components of apple tree yield. At close spacing the fruits tended to be smaller and fewer were ripened per tree than at wide spacing. At any one spacing the more fruits borne per tree the smaller were the fruits — evidence of a correlative relationship between the fruit on a single tree.

The changes in fruit yield, growth in basal area age and spacing are brought together in Fig. 7/18. It would have been interesting to show

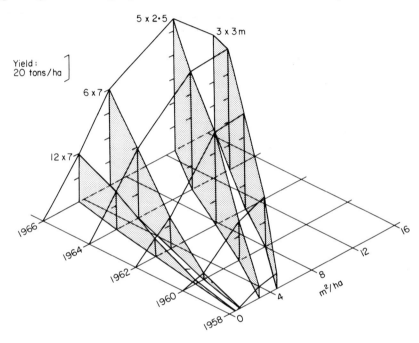

Fig. 7/18. The relationship between the yield of fruit and the growth in basal area over time in orchards of 'Jonathan' apple planted at different densities. (From Verheij, 1968)

the reproductive capacity of the plants (seeds/plant) and of the stands (seeds/ha) but seed number was not counted. A large part of the weight of the domestic apple is the enlarged receptacle — the seed is a very small fraction. If we assume seed number to be roughly proportional to fruit weight it appears that the peak of reproductive activity is reached well before there is any decline in vegetative activity. The orchards were still young in terms of the apple tree's potential age and, after a period of reproductive vigour, the trees entered a long phase of increasing reproductive impotence. The greater the stress of density the more rapidly did the reproductive activity start to decline.

The integration of individuals in a forest

In a number of forest tree species, graft unions form between the roots of different individuals. This has been studied particularly in *Pinus strobus* (Fig. 7/19) (Borman and Graham, 1959), but has been reported

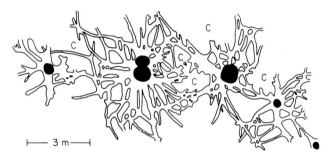

Fig. 7/19a. Diagram of a root network of *Pinus strobus* exposed on the surface of a deep partially drained peat soil. Black circles represent old stumps, and C's indicate areas where some roots have been removed by chopping.

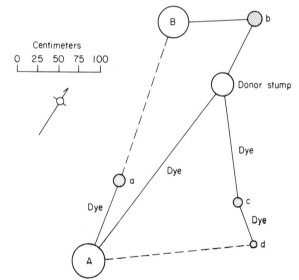

Fig. 7/19b. The pathways of a dye (acid fuchsin) injected to a donor tree and subsequently traced in surrounding trees. Note that not all trees that transmit the dye may be recorded — thus *a* received dye via A, but A was not itself stained. (Figs a and b from Bormann and Graham, 1959)

in a number of other genera (Graham, 1959, lists 56 species in 19 genera in which natural root grafting occurs including *Abies, Larix, Quercus, Tilia* and *Ulmus*). In *Pinus strobus* the grafts may sometimes be single links between pairs of trees "but often sizable clusters of trees are organically united by an extensive complex of self grafts and intra-specific grafts". In the extreme it is conceivable that an entire stand may be united in a single anastomosing network (Borman, 1960).

Graft unions are effective in the transport of water, ions and synthesized food materials, to such an extent that a graft between two trees may maintain the root system of one alive after it is felled (Bormann, 1961). The presence of root grafts may permit some degree of integration of the activities of individuals in a population. Bormann (1966) suggested that in *Pinus strobus* root grafting may hasten the death of suppressed trees though some studies with isotopic tracers suggest that suppressed individuals of *Quercus ellipsoidalis* may be supported through their connections with dominant indivudals.

In clonal plants (e.g. trees such as *Populus tremuloides*) the decay of connecting parts produces a population of physiologically and morphologically independent ramets: root grafting is the antithesis of such a process — genetically independent plants form connections which may even permit correlative control of each other's development by hormone transport. The phenomenon exists in herbaceous plants: individuals of the hemiparasite *Bartsia odontites* grown without a host form haustorial connections with each other and a few reach maturity and produce seed while strongly suppressing the remainder (Govier, 1966). It is doubtful whether any careful search has ever been made to detect the extent of root grafting in other communities of herbs.

8
Mixtures of Species.
I. Space and Proportions

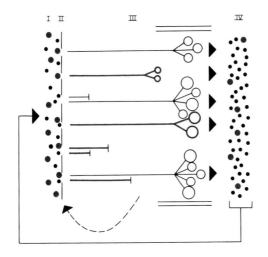

Natural plant populations are usually mixtures of species — diversity is the rule and monotony the exception. Even in the "monocultures" of agriculture and forestry continuous invasion by other species (weeds in the sense that they are plants growing in the wrong place), is the normal situation. Sometimes a single species may gain exclusive occupation of a site or a stratum and form pure stands, e.g. some reed swamps such as *Typha, Phragmites, Glyceria*, the tussocks of many alpine species (e.g.

Saxifraga) and the tussocks of mosses. However these pure stands are almost always formed by clonal spread of a very few individuals and the pure stands are then more properly regarded as the growth forms of individuals (comparable to the canopy of a single tree) rather than as populations of plants. Most of the "pure stands" of natural vegetation are the repeated parts of single individuals.

The description and quantification of community diversity and of vegetational pattern form the subjects of a whole specialized literature reviewed by Greig-Smith (1957) and Whittaker (1972). This active area of research is, however, largely couched in terms that are not particularly appropriate for the population biologist. Perhaps because much of plant sociology grew out of geography and an interest in the distribution of vegetation, diversity has been studied mainly in relation to areas of land. Conclusions from diversity studies are usually expressed as the numbers and relative frequencies of species in relation to an area of land (see Chapter 23). Population biology is concerned with the behaviour, experiences and fates of individuals; the area of land (or quadrat size) is less important than distances between neighbours.

It is interesting to put the questions of community diversity into the context of "environmental grain" (MacArthur and Levins, 1964). In a mixed vegetation containing plant type A and others, in what ways does plant type A "sample" or experience the variety of the vegetation of which it is a part? We can think of this "sampling" or experience as the extent to which the individuals of type A touch, shade, deprive of nutrients and water, or in other ways influence (or are influenced by) individuals of their own (A) and other types (B, C . . .). This is to take a "plant's eye view" of the surrounding vegetation. Taking MacArthur and Levin's concept of grain we could say that if a plant type "samples" the vegetation in the proportions in which its component species occur, it is behaving in a fine-grained manner. On the other hand if the distribution of individuals is ordered in some way, for example by some heterogeneity in the physical conditions of the environment, plants of type A might commonly have neighbours of type B and seldom of types C, D, E etc. – even though these are common, but occur in patches away from A. Then A could be said to "sample" the vegetation in a coarse-grained manner.

Some aspects of studying vegetation from the "plant's eye view" are considered in Chapter 2. It is important to emphasize in the present context that the neighbourhood concept of diversity is largely peculiar to plants and non-motile animals such as barnacles, corals, hydroids and fixed colonial animals like some social insects, e.g. ants. The spatial relationships of individuals are critical within plant populations: escape from a neighbour's influence depends on directional growth or the dispersal of progeny, and it cannot be achieved by running away.

There are three distinct categories of neighbour that may interfere with the growth of a plant in the field:

(a) Immediate neighbours may be parts of the same genet — branches or ramets of the same original seedling. A shoot on an oak tree is most likely to touch, shade or be shaded by another shoot on the same tree. The tiller of a grass is most likely to shade or be shaded by another tiller of the same (though perhaps now unconnected) genet. The roots from a tiller are most likely to grow close to and make demands on the same nutrients and water supply as other roots from the same genet. The very form of a plant as a *réiteration* of modules, maximizes this close proximity of the units of branched and clonal growth. Some forms of construction of plants may reduce the extent of intra-clonal neighbourhood relationships and long stolons or rhizomes can increase the number of interspecific contacts that a clone may make. Two grasses — *Agropyron repens*, which can place its daughter ramets at a distance of 1 m from the parent tiller in a year's growth, and the closely related *Agropyron caninum* which forms a dense tussock of tightly packed tillers — represent very different ways in which a plant samples the surrounding vegetation. *Agropyron repens* will tend to sample in a fine grained manner, but the environment for individuals of *Agropyron caninum* is coarse grained.

Although the unit modules (tillers, *réiterations*, ramets) of a single genet will be genetical identities (except for somatic change) this does not mean that they will compete with each other on equal terms. Parts of a clone or shoots on a tree will differ phenotypically, e.g. in age and size, and may often make quite different demands on environmental resources. Phenotypic differentiation between neighbouring parts of a clone is often well developed as ring patterns in which the central ramets in a dense clone are depauperate compared with those at an invading perimeter (see discussion in Kershaw, 1964).

(b) Immediate neighbours may be genets (or ramets belonging to genets) derived from different seeds and therefore usually of different genotypes. In a community of very few species, intergenotypic contacts may represent the commonest type of neighbour. Certainly, where seed dispersal is poor (and most seed is seldom dispersed far: see Chapter 2) intergenotypic contacts may well predominate in the early life of seedlings until thinning has eliminated the closest neighbours. The "struggle for existence" between neighbours of different genotypes, but of the same species, represents part of the raw material of natural selection and evolution. A struggle for existence between neighbours of different genotypes can result in the emergence of winners and losers in the preempting of resources and subsequently in the number of progeny produced and the chance of leaving descendants.

Neighbour relationships between members of the same species are particularly intriguing when the neighbours are of different sexes. This situation is found in dioecious species e.g. *Rumex acetosella, Mercurialis perennis, Asparagus officinalis* among clonal perennials, and *Cannabis sativa* among annuals. There is often some differentiation between the growth habits and demands on the environment made by the two sexes so that neighbours of different sex behave towards each other almost as if they were members of different species.

(c) Immediate neighbours may be genets (or parts of genets) of different taxa. In a community of plants that is truly diverse from the "plant's eye view", neighbours will be of a variety of species. The most likely plant to deprive a plant of type A of resources is then one of type B, C, D etc.

This discussion has been couched in terms of neighbours that deprive each other of resources, but proximity has presumably the same relevance if interference with neighbours is through toxins released into the environment or if neighbours influence each other by suppressing the activities of predators or pathogens. For example, grass leaves appear to be protected from grazing animals in some British grasslands if the leaves are close to plants of *Ranunculus bulbosus* which contains an unpleasant glycoside, ranunculin (Harper and Sagar, 1953).

The interference between neighbouring plants of different taxa is presumably often the cause of successional change within vegetation. It is also a potentially powerful selective and evolutionary force; the characteristics of neighbours B, C, D, etc. determine which genotypes among A are successful.

The various types of neighbourhood relationship can be coded as follows:

A′ . . . . . . A′   Intraclonal neighbours, no direct significance in selection.

A' . . . . . . A''   Intergenotypic neighbours, direct "struggle" for exist-
                      ence, intraspecific selection

A' . . . . . . B   Interspecific neighbours, direct role in competitive
                      exclusion, no direct role in selection.

A' · · · · · · ⌐   Interspecific neighbours, indirect role of B in deter-
A''. . . . . . ⌐ B   mining fitness of A' with respect to A''

Although neighbour relationships determine a plant's experience of community diversity, very few attempts have ever been made to study the phenomenon in detail. Sakai (1957) sowed rice in a pattern, arranging six plants to form a hexagon surrounding a central test plant. He was interested in the strength of competition of different rice genotypes and showed that the number and arrangement of these in the hexagon affected the competitive influence on the central plant. This design was modified (Harper, 1961) in a study of the interference between two annual grasses, *Bromus rigidus* and *B. madritensis*. The species were sown at a constant density and in equal proportions but the arrangements of individuals were such as to maximize contacts either within or between species. In addition four different but accurately replicated random arrangements were studied. At the end of the growing season the yield of the two species was markedly different between the arrangements (Fig. 8/1 and Table 8/I).

A much more sophisticated study of neighbours was made by Mack and Harper (1977) involving five short-lived annuals found naturally in mixed populations on sand dunes in North Wales. The species were *Vulpia membranacea*, *Mibora minima*, *Phleum arenarium*, *Cerastium atrovirens* and *Saxifraga tridactylites*. Seed of each of the five species was sown at $3700/m^2$ on flats (30 x 38 cm) of sterilized sand from the dunes. Each flat was divided into 6 squares, each 5 x 5 cm, and four of the squares received an equal density of seed of one of the four remaining species. The fifth square remained a monoculture of the "background" species and the sixth received an additional allotment of the background species. The experiment therefore included each species at single and double density and all possible combinations of the various species pairs. Immediately after germination the populations were thinned to 40 plants per square (20 of the background species and 20 of the other). The thinning was done by a randomized process in which the individuals to be removed were determined from a punched paper disc of random holes. In addition to the combinations of species in pairs a series of squares

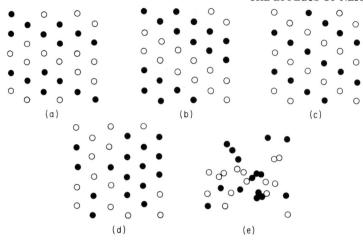

Fig. 8/1. Typical sowing grids used in experiments on interference between *Bromus rigidus* and *B. madritensis*. (a) Hexagon pattern with each individual surrounded by three of its own species and three of the associated species. (b) Hexagon pattern with each individual surrounded by four of its own and two of the associated species. (c) Hexagon pattern with each individual surrounded by two of its own and four of the associated species. (d) Hexagon pattern with the allocation of each position to one or the other species at random. (e) One of four random arrangements of seed position allocated at random to one or the other species. In each arrangement equal numbers of the two species were sown. (From Harper, 1961)

Table 8/I

The influence of the pattern of arrangement of individuals of *Bromus rigidus* and *B. madritensis* on the total yield of above-ground parts and on the proportional contribution of the species to the total yield.

Ratios of dry weights of *B. rigidus/B. madritensis*

Pattern of arrangement	Ratio of dry weights	Significant differences
Hexagon: 4:2 (b)	2.624	
3:3 (a)	1.993	
2:4 (c)	2.253	
random (d)	2.517	
Random: random H	2.776	
I	2.789	
J	2.544	
K	2.452	

* Signifies $P < 0.05$; ** $P < 0.01$; *** $P < 0.001$.

(From Harper, 1961)

was also sown with 20 seeds of each of the five species giving a combined density of 100 individuals per 5 x 5 cm square.

The position of every plant was accurately mapped and after 3, 6 and 8 months, replicate plots were harvested for dry weight and seed production. The measurements made on individual plants together with their position (as x, y coordinates) were analysed to determine (a) the varying aggressiveness of the species, (b) the distance between each plant and (c) the spatial dispersion of the most competitive neighbours around each plant.

The aggressiveness of each species was measured as the average depression it caused in the other four species compared with their growth in pure stand. The values were *Vulpia* 59%, *Phleum* 52%, *Mibora* 31% and *Cerastium* 11.25%. If we consider *Vulpia* as having an aggressiveness of 1.0, the values become *Phleum* 0.88, *Mibora* 0.52 and *Cerastium* 0.18. The aggressiveness of *Saxifraga* was so slight that it could not be measured realistically.

Each plant was considered as being at the centre of a circle of 2 cm radius divided into four equal quadrants. The number of quadrants in which a neighbour occurred around it was then counted. A plant which had many neighbours but in only two quadrants of the outer ring (1—2 cm) is represented by a map as in Fig. 8/2a. A plant with this number and arrangement of neighbours was generally larger than one with the same neighbours more evenly distributed between the quadrants (Fig. 8/2b). The neighbour effects of *Mibora* and *Cerastium* were so small that it was not necessary to correct the data for dispersion in this way. For *Vulpia* and *Phleum* the biomass in the neighbourhood of an individual was corrected for dispersion so that for the arrangement in Fig. 8/2a it was multiplied by 0.5 (since two of the four quadrants were empty) and the biomass of neighbours in Fig. 8/2b was multiplied by 1.0 because every quadrant was occupied.

The regression for individuals of *Vulpia* was calculated as:

$$\frac{\text{Biomass of}}{\textit{Vulpia}} = \alpha - \left[\log \sum_{0-0.5} (c_1 b_1 + c_2 b_2 + c_3 b_3 + c_4 b_4) + \right.$$

$$\left. 1/r \left\{ \sum_{0.51-1.0} c_1 b_1 + \ldots \right\} + 1/r \sum_{1.01-2} \left\{ \frac{c_1 b_1 + \ldots}{5} \right\} \right]$$

where $r = 1$ if all four quadrants contain *Phleum* and/or *Vulpia* neighbours

 $r = 2$ if only two quadrants contain *Phleum* and/or *Vulpia* neighbours, etc.

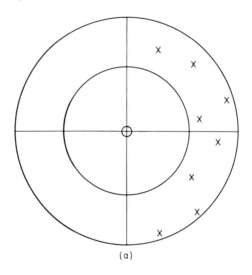

(a)

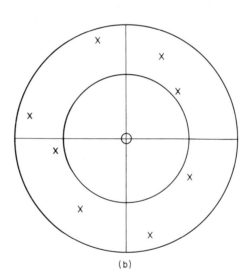

(b)

Fig. 8/2. Examples of the arrangements of neighbours x of a plant ○, described according to their distribution between quadrants of circles of various radii around the plant ○. (From Mack and Harper, 1977)

and $c_n b_n$ = the accumulated biomass (b_n) for each species times its aggressiveness as determined earlier. Up to 69% of the variation in weight of individual plants of *Vulpia* was accounted for by this regression. The remainder of the variation between plants is presumably due to variations in the time of emergence, environmental gradients in the design and genetic variability.

If two species with different aggressiveness are grown together in a mixed population, the arrangements of the individuals may be perfectly ordered with equidistant spacing and a perfectly even distribution of the two species with respect to each other. Under these circumstances the more vigorous of the two species has many "interfaces" with the weaker species and many opportunities to show its relative vigour. On the other hand, if the two species are clumped, intraspecific contacts will be frequent and the strong will tend to be pitted against the strong, the weak against the weak. Then the aggressiveness of the strong tends to be dissipated in competition with each other and the weak gain from being clumped. It is probably an important characteristic of plant communities in nature that dispersal is non-random and the persistence of the weaker competitors will always be favoured when this is the case.

The spatial arrangement of individuals relative to neighbours becomes even more important when we consider the entry of a few individuals of a new species into a population of another. While the new species is rare it will meet mainly interspecific contacts; indeed all the contacts made by the first invading individual will be interspecific. If the invader is successful and multiplies, the frequency of intraspecific contact increases and the chance of the new invaders making interspecific contacts declines. This suggests that the early phases of invasion of a community by an aggressive newcomer should be fast. This point is illustrated in Fig. 8/3a and b. The phenomenon is, of course, not restricted to mixtures of species; it will apply equally to mixtures of phenotypes and hence to the rates of spread of competitively superior phenotypes within a population of a single species. This point was made in a series of experiments by Sakai on competition between strains of rice (Sakai, 1955, 1961).

As a plant increases in size it becomes closer to its neighbours, so that time is in this way equivalent to distance (see also Chapter 6). In nature, the distance apart of individuals and the times at which they became established produce effects that are confounded and analysis is difficult or impossible. For example, in a forest, it is possible to determine the

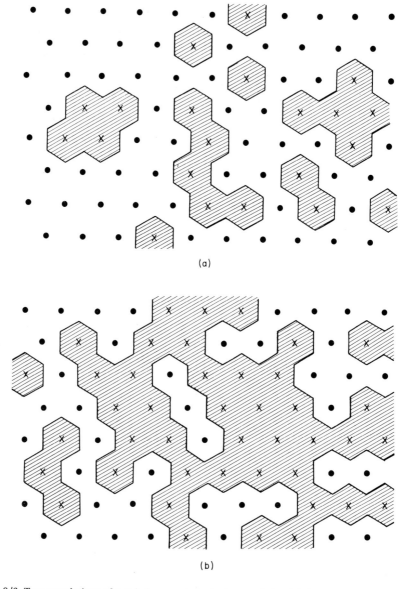

(a)

(b)

Fig. 8/3. Two populations of regularly spaced plants. In population (a) 21 of the 84 plants have been randomly assigned to species A and the remainder to species B. The number of interspecific interfaces is 86. In (b) 42 plants have been randomly assigned to species A — the number of interfaces is increased only to 110.

spatial relationships between individuals (e.g. the methods of Opie, 1968), but it is only when the trees are of the same age that the effect of distance from neighbours will account for much of the variation in the size of individuals. Similarly, it may often be possible to age the individuals in a population, but unless distance relationships are known the variation in growth between individuals may not be interpretable.

It is easy to design experiments that distinguish between the effects of time and space on competitive interactions (see e.g. experiments on single species of Ross, Fig. 6/12); either spacing or timing can be varied while the other variable is held constant. The two closely related species of annual grass *Bromus madritensis* and *B. rigidus* occur in mixed populations in the rangelands of California. *B. rigidus* has considerably larger grains and embryonic capital and when the two species are sown together in pot culture experiments *B. rigidus* rapidly achieves dominance. In equiproportioned mixtures of the two species *B. rigidus* contributed overwhelmingly to the biomass of the mixed population. However, by delaying the introduction of *B. rigidus* into the mixtures the balance was tipped decisively in favour of *B. madritensis* (Fig. 8/4). In this experiment the total final yield of the mixtures was independent of the relative sowing times of the two species. All that was altered was the relative contribution made by each to the mixture — their relative dominance.

A comparable experiment involving two very dissimilar species both characteristic of neutral permanent grassland in Britain was made (Sagar, 1959) with *Plantago lanceolata* and *Lolium perenne.* The species were sown at the same time or *Lolium perenne* was sown 21 days before or after *P. lanceolata*. Again the time of sowing of the two species relative to each other had a profound effect on the balance of the species. Whichever species was sown late was greatly reduced in dry weight and also suffered much the heavier mortality (Fig. 8/5).

It is relatively easy to arrange in an experiment that populations of two species germinate synchronously. In nature the relative times of germination are usually under precise control by the genotype and the environment. The requirements for breaking seed dormancy are often very species-specific (see Chapter 3) and the time at which potentially competing species enter a struggle for existence will be determined by the times at which dormancy is broken. In a competitive struggle for limited resources there is a great premium on early germination and growth.

The growing plant pre-empts to itself space, i.e. resources: success

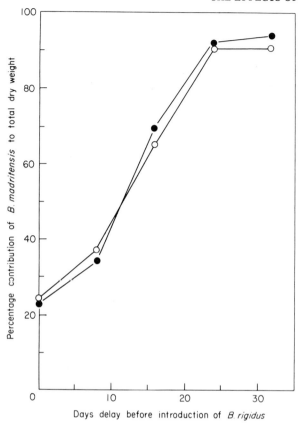

Fig. 8/4. The influence of the time of introduction of *Bromus rigidus* into mixtures of *B. rigidus* and *B. madritensis* on the proportional contribution of *B. madritensis* to total dry weight per pot. The two species were sown in equal proportions at a combined density of 30 plants per pot. Total yield per pot was unaffected by delaying the introduction of *B. rigidus*. ●, after 126 days growth; ○, after 166 days growth. (From Harper, 1961)

depends on the capture and use of resources that are then denied to a competitor. Capture of resources deprives neighbours and brings competitive advantage. Successful capture depends on (a) distance from neighbours and (b) size. Size can come from a faster growth rate, or a larger starting capital (large seed reserves) or a longer period of growth. Time, size and distance are all in a sense alternatives.

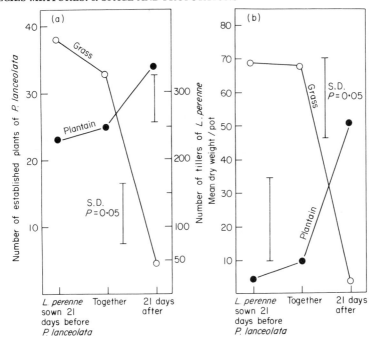

Fig. 8/5. The influence of the time of sowing *Lolium perenne* relative to that of *Plantago lanceo-lata* on the growth and establishment of the two species mixtures. (From Sagar, 1959)

Experimental models of the growth of mixed populations

In a study of the behaviour of two species grown in mixture the experimentalist is free to vary both density and proportion: it is easy to design experiments in which these two are confounded in such a way that interpretation is almost impossible.

Difficulties are particularly great in experiments of "additive" design. In these, species A is sown at a standard density and species B is sown with it at a range of densities. Such an experimental design has an obvious relevance to many field situations in which one species is invading the area occupied by another, for example, a weed invading a crop. The analysis of this experimental design is, however, peculiarly difficult because the proportional composition *and* the density of the mixture are both changed and their effects are completely confounded.

Most of the problems of additive designs are eliminated in substitutive designs. The "Replacement Series" (de Wit, 1960) involves sowing two species in varying proportions while maintaining overall density constant. There is an important artificiality in this procedure: most plant populations that change in proportions over time also change in density. However, the substitutive experiments are particularly elegant for the study of plant interactions involving two species.

Additive experiments

In additive experiments two species are grown together, the density of one is maintained constant and that of the other is varied. The first species can be regarded as an indicator and experiments of this sort can be used to compare the relative aggressiveness of a group of species to the indicator. In its simplest form an additive experiment becomes a phytometer experiment (Clements *et al.*, 1924 a, b) in which a single indicator plant is placed in a stand of a second species.

White clover (*Trifolium repens*) is commonly found in association with a number of pasture grasses and sometimes with its congener *T. fragiferum*. An additive experiment was designed (Clatworthy, 1960) to determine how an establishing seedling of *T. repens* is affected when it is grown in populations of other species. Seed of *T. repens* was sown together with varying seed rates of *Lolium perenne, Agrostis tenuis, Trifolium fragiferum* and, for comparison, with further seed of *T. repens* itself. Immediately after the seedlings had emerged they were thinned so that 10 seedlings of indicator *T. repens* grew in the presence of 0, 5, 10, 20, 40, 80 or 160 seedlings of each of the associated species. The experiment was made in pots of 22.5 cm diameter filled with garden loam and the plants were harvested after 7 months. At the time of harvest many of the plants of *T. repens* had developed stolons and, in some cases, flowers. The results of the experiment are shown in Fig. 8/6.

The dry weight of *T. repens* declined with increasing density of each of the associated species. The most damaging associate was *Lolium perenne* followed by *Agrostis tenuis. Trifolium fragiferum* was less aggressive towards *T. repens* than was *T. repens* itself.

Flowering and stolon development (numbers and length) were both reduced by the presence of associated species. Flowering was prevented by the presence of *ca* 160 plants of *T. fragiferum*, by about 160 plants of *T. repens*, by 80 plants of *Agrostis tenuis* or by 20 plants of *Lolium perenne*. Forty plants of *Lolium perenne* prevented by *T. repens* from forming stolons, whereas 160 plants of *Agrostis tenuis* were needed to

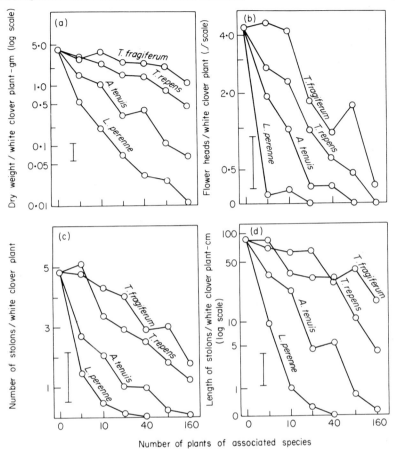

Fig. 8/6. The effect of populations of *Lolium perenne, Agrostis tenuis, Trifolium repens* and
T. fragiferum, sown at a range of densities, on the growth of *T. repens* sown at the
same time and measured after 7 months. Data have been transformed in ways appropriate to homogenize the variance.
(a) Dry weight per plant of *T. repens.*
(b) Number of flower heads produced per plant.
(c) Number of stolons per plant.
(d) Length of stolons per plant.
(From Clatworthy, 1960)

produce the same effect.

An additive experiment was made to compare the mutual aggressiveness of a group of annuals consisting of *Linim usitatissimum* in its oil
seed (linseed) and fibre (flax) cultivars and two of the weeds commonly
found with *Linum, Camelina alyssum* and *C. sativa.* Each species in turn

was used as an indicator of the aggressiveness of the other three. A single plant of the indicator was surrounded by 5, 10, 20, 40 or 80 plants of the test species. The results are shown in Fig. 8/7. Generally, the weight of each indicator declined with increasing density of the surrounding population. There was a clear law of diminishing returns: increasing the density of the test species above 20 plants per pot had an extremely small influence on the indicator.

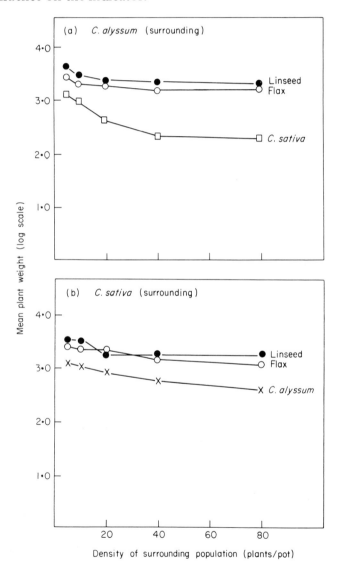

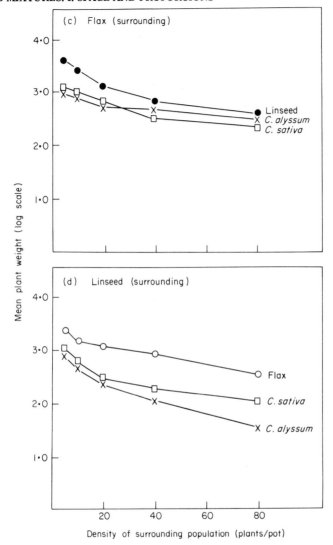

Fig. 8/7. The interactions between populations of *Camelina alyssum, C. sativa,* flax and linseed *(Linum usitatissimum)*. Each figure shows the effect of increasing the density of the background (surrounding species) on the mean weight of each of the others. (From Obeid, 1965)

This experiment established a very clear pecking order among the species. Irrespective of which species was used as the indicator, the order of aggressiveness was oil-seed *Linum* > fibre *Linum* > *C. sativa* > *C. alyssum*. This is the same order that is taken by the species if they are

arranged according to their performance in pure stands and classified
according to their height, weight, leaf area etc.

It is a question of great interest whether all groups of species can be
arranged in such a "pecking order" and whether if A > B and B > C it
necessarily follows that A > C. This question has been put to the test in

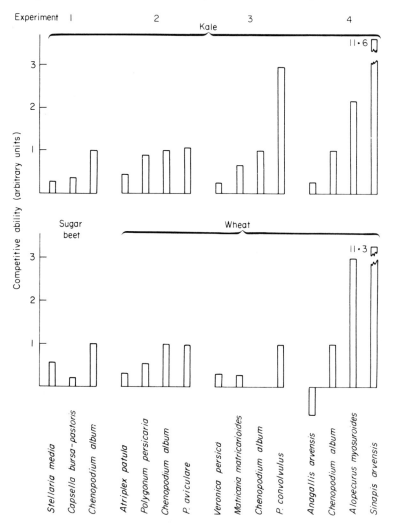

Fig. 8/8. Competitive abilities of different weed species measured by their effects on kale and
sugar beet or wheat, average of high- and low-nitrogen treatments. The unit is the
competitive ability of *Chenopodium,* the effect of other species being expressed as
a ratio to the effect of *Chenopodium* in the same experiment. (From Welbank, 1963)

an experiment of Welbank (1963) in which he grew single plants of kale, sugar beet and wheat alone and in populations of a wide variety of weeds. He measured the "competitive abilities" of the weeds by the extent to which they reduced crop growth. He included *Chenopodium album* as a control and the results from all the experiments were corrected on the basis of the effects of *C. album*. In general the order of competitive ability of the weeds was the same, irrespective of the test species, but there were exceptions. For example *Polygonum convolvulus* was less aggressive to wheat than *Alopecurus myosuroides* but the order was reversed when they were judged against sugar beet (Fig. 8/8). Many examples of carefully designed additive experiments show that the order of aggressiveness of a group of species is independent of the test species (e.g. a study of four species of *Avena* (Trenbath and Harper, 1973; Trenbath, 1974a, 1975)), but it is probably dangerous to assume this as a general rule without further experiments.

Many additive experiments have been made to measure the effect of weeds on the yield of crops. Experimental infestations of a wheat crop with *Alopecurus myosuroides* at densities of 30, 100 and 300 plants per m^2 brought about yield reductions of 12, 32 and 36% in a wheat crop (Naylor, 1972). Such an experiment permits the cost of a weed infestation to be measured.

Substitutive experiments

The main characteristic of a substitutive experiment is that the proportions of two species I and J in mixture are varied while the overall density I + J is maintained constant — a "replacement series". This experimental design was introduced by de Wit (1960), and its main use has been by de Wit and his associates at Wageningen, Holland.

The results of experiments based on the replacement series can, in theory, take any of four basic forms (Figs 8/9a—e).

Model I. The growth of two species in mixture (measured as e.g. dry weight, seed number, seed weight etc.) results in each contributing to the total yield in direct ratio to its proportion in the sown seed. This can happen in two ways. Either the density of the mixed population is so low that individuals within it do not interfere with each other (the yield of each species is then linearly related to its number of plants and competitive interactions do not occur) *or* density is great enough for plants to interfere with each other but the effect of I on J is precisely

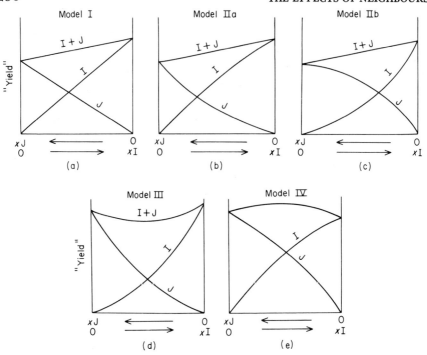

Fig. 8/9. A variety of results from "replacement series" experiments (see text)

the same as that of J on J and the effect of J on I is the same as of I on
I. In this model, in their ability to interfere with each other's growth,
the two species are exactly equivalent. In either case the yield of the
mixture is predictable from the yield of the pure stands. When the density is high enough for individuals to interfere with each other, experimental results of this form suggest that the two species make the same
demands on environmental resources. Note that the yields of the two
pure stands need not be equal; the two species may be making similar
demands on environmental resources but converting them with different
efficiencies or to different ends, e.g. bigger but fewer seeds.

   Model II. The effect of I on J is greater than of J on J and the influence of J on I is less than of I on I. Here the two species make demands
on the same environmental resources — but there is a differential
between them. The yield of the mixture is again predictable from the
yield of the pure stands. Note that the species yielding most in pure

stand need not be the most successful form in the mixture. A species that is productive in pure stand may be an ineffective competitor — the so called "Montgomery effect".

Model III. The effect of I on J is greater than that of J on J and the effect of J on I is greater than that of I on I. This is a model of mutual antagonism in which neither species contributes its expected share to the yield of a mixture. Such a situation would arise if each species damaged the environment of the other more than it damaged its own environment. One could imagine that if each species produced a specific toxin to which it was itself insensitive the results of a replacement series experiment could appear like Model III. Conceivably the same result might be obtained if I was able to make luxury consumption of a limiting nutrient thus hindering the growth of J, but J, by virtue of greater height shaded I and prevented it from exploiting its absorbed nutrients in proportionate growth.

Model IV. The effect of I on J is less than that of J on J and the effect of J on I is less than that of I on I. At first sight this looks like a model of symbiosis but it is not necessarily so. In a symbiosis each component of a mixture benefits from the presence of the other. In this model all that is required is that each species fails to suffer as much as expected from the presence of the other. The model describes situations in which, for one reason or another the species escape some measure of competition with each other. It might apply if the growth of I was limited by resource a and J by resource b. It might apply if the growth of I occurred in a different season to J so that neither interfered with the demand for water or light made by the other.

In the interpretation of the results of an experiment based on the replacement series it is important to remember the shape of a yield density curve for a single species (or a replacement series in which one species is represented by sowing dead seeds or stones). Provided that the density reaches levels high enough for plants to interfere with each other the yield curve will be something like Fig. 8/10a. Thus, if we sow a replacement series of two species neither of which responds at all to the presence of the other, we might expect to obtain a yield relationship like Fig. 8/10b.

One of the earliest experiments to be analysed as a replacement series involved mixtures of oats and barley — two species that are sometimes sown together in mixtures as a deliberate agricultural practice. Pure stands and equiproportioned mixtures were sown on land that had

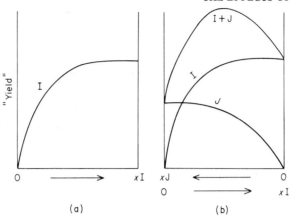

Fig. 8/10. See text

received different fertilizer treatments and that had, as a result, different pH status. The yields of the various plots were expressed as the
number of grains yielded per hectare and the results are shown in Fig.
8/11a–d. The fall in pH is associated with an increase in the aggressiveness of oats towards barley, though in the transition Fig. 8/10a to Fig.
8/10b the performance of neither species has changed in pure stand. The
transition from Fig. 8/11b to Fig. 8/11c shows a further decline in the
aggressiveness of barley but its performance in pure stand is now reduced.
In Fig. 8/10d the environment is lethal to barley and although a mixture
was sown, the result is in effect a simple density experiment with oats.

    A formal measure of the aggressiveness of one species towards another,
the "Relative Crowding Coefficient", can be derived from the results of
a replacement series experiment (de Wit, 1960).

The Relative Crowding Coefficient of I with respect to J is

$$\frac{\text{Mean yield per plant of I in mixture}}{\text{Mean yield per plant of J in mixture}} \bigg/ \frac{\text{Mean yield per plant of I in pure stand}}{\text{Mean yield per plant of J in pure stand}}$$

Or

$$k_{IJ} = \frac{\bar{x}_I}{\bar{x}_J} \Big/ \frac{\bar{y}_I}{\bar{y}_J}$$

In the mixed oat + barley populations the value of k_{IJ} (the Relative Crowding Coefficient of oats towards barley) changed from 0.9 to *ca* 20. The Relative Crowding Coefficient (RCC) is an inappropriate measure if the combined yields of the species in mixture are not predictable from the pure stands. In such circumstances it is preferable to use the concept

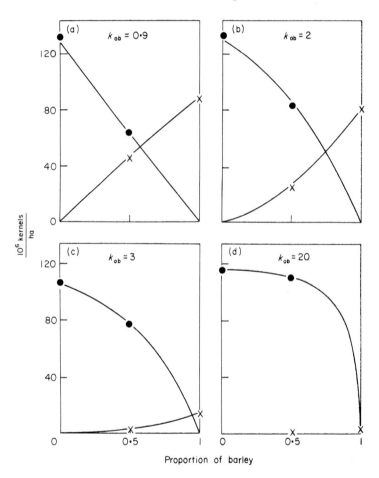

Fig. 8/11. Replacement series (substitutive) experiments with mixtures of oats (*Avena sativa*) and barley (*Hordeum sativum*) grown on soils that had received four different fertilizer treatments. (a) Calcium nitrate; pH-KCl = 4.0, (b) Nitrolime; pH-KCl = 3.7, (c) Ammonium sulphate; pH-KCl 3.2 and (d) Ammonium sulphate; pH-KCl = 3.1. (From de Wit, 1960)

of "Relative Yield".

$$\text{The Relative Yield of I is } \frac{\text{Yield of I in mixture}}{\text{Yield of I in pure stand}}$$

Or $$R_I = \frac{x_I}{y_I}$$

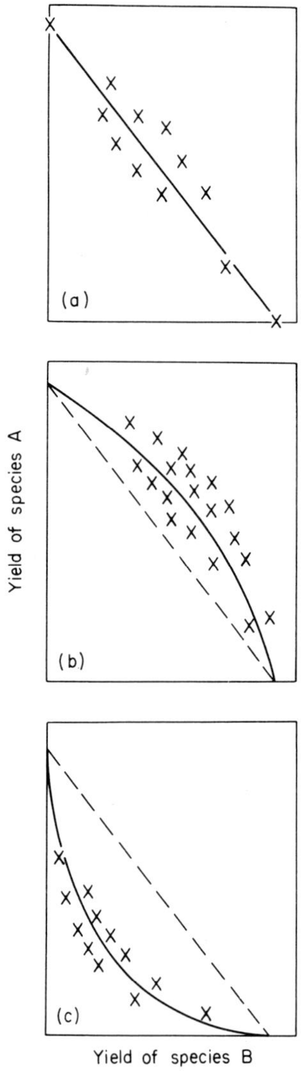

Fig. 8/12. A convenient graphical presentation of the results of a replacement series experiment (a) RYT = 1, (b) RYT > 1, (c) RYT < 1.

If the Relative Yield is calculated for both species in a mixture the sum of the two values gives the "Relative Yield Total" (RYT). The values of RYT can be used to describe the mutual relationships of pairs of species that may or may not be making demands on the same resources of the environment and may show any of the relationships in Models I–IV (Fig. 8/9). Values of RYT of *ca* 1.0 imply that the two species are making demands on the same limiting resources of the environment. Values of RYT > 1.0 suggest that the species make different demands on resources, avoid competition with each other or are showing some form of symbiotic relationship (at least with respect to combined yield). Values of RYT of <1.0 imply a mutual antagonism (Fig. 8/12a–c).

A very elegant example of the use of the Replacement Series is found in a study of mixed and pure stands of *Panicum maximum* with the legume *Glycine javanica* (de Wit *et al.*, 1966). These two species are sometimes grown together as a grass + legume association in Australia. The planting scheme was as follows:

Pot	*Panicum* plants	*Panicum* relative plant frequencies	*Glycine* plants	*Glycine* relative plant frequencies
a	0	0	8	1
b	1	0.25	6	0.75
c	2	0.5	4	0.5
d	3	0.75	2	0.25
e	4	1	0	0

This replacement design was repeated under four different treatments:

R_0N_0 Legume *not* inoculated with *Rhizobium* and *no* nitrogen fertilizer applied.

R_0N_1 Legume *not* inoculated with *Rhizobium* but nitrogen fertilizer was applied.

R_1N_0 Legume inoculated with *Rhizobium* and *no* nitrogen fertilizer applied.

R_1N_1 Legume inoculated with *Rhizobium and* nitrogen fertilizer was applied.

The plants were clipped at a height of 6 cm above the soil surface.
The dry matter yields are shown in Fig. 8/13, and they illustrate several
of the model replacement diagrams. The interpretation is made much
easier by considering the Relative Yield Totals (Fig. 8/14a, b). Without
inoculation with *Rhizobium* or the application of fertilizer nitrogen
there is some slight indication of an antagonism and when nitrogen was
applied artificially the species behaved as though they were in a simple
competitive relationship, "competing for the same space" (de Wit uses
the term "space" to imply all those resources which are required by the
plants for growth). Relative Yield Totals rise well above 1.0 after inoculation with *Rhizobium* indicating that the intensity of interference
between *Glycine* and *Panicum* is relaxed, so that they no longer make
demands on the same limiting resources and no longer compete for the
"same space". The presumption is that *Glycine* is now independent of
soil nitrogen supplies, gaining its nitrogen through the symbiotic relationship with *Rhizobium*. When *Glycine* is inoculated *and* fertilizer nitrogen
is applied, RYT values are again greater than 1.0, but the effect is less
marked than without fertilizer nitrogen.

The replacement series can also be used to examine nitrogen yields
within the various mixed cropping systems of *Panicum* and *Glycine*
(Fig. 8/14c).

Generally, when two species of grass are grown in mixtures, the yield
does not exceed that of the higher yielding pure stand (experiments of
van den Bergh (1968) and of others in the Wageningen school). Many
species have been studied in combinations under different environmental conditions and the general rule emerging is that the yield of a mixture
most commonly slightly exceeds the mean yields of the pure stands.
This is the conclusion reached by Donald (1963) reviewing earlier literature and independently by England (1968) in a very detailed study of
mixtures of *Lolium perenne* and *Dactylis glomerata*. The general conclusion is that there is no advantage to a farmer in sowing a mixture of
grass species if his aim is to maximize dry matter production under ideal
and constant conditions*. Most of these experiments can be criticized on
the grounds of artificiality, since they were mainly done in pot culture,
not in open ground. Pot culture almost certainly forces the species to
use the same limited root space and soil resources whereas in the field

---

*Snaydon (pers. comm.) has pointed out that this conclusion is only valid if it is known in advance
which pure stand will yield most. If there is uncertainty, the mixture will be on average give
higher yields than a pure stand.

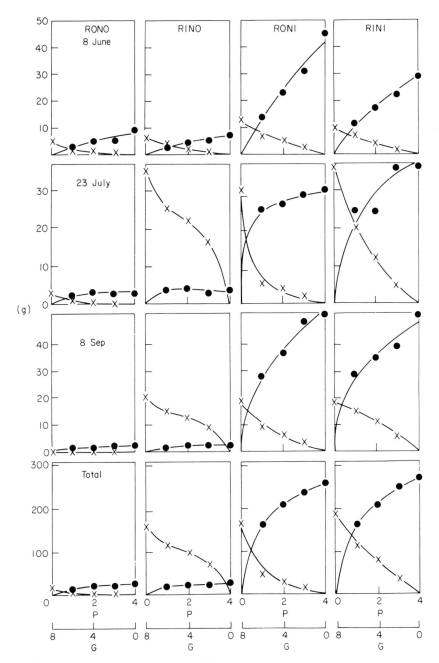

Fig. 8/13. The yield of *Panicum maximum* (P) and *Glycine javanica* (G) in replacement series experiments with and without inoculation with *Rhizobium* and application of nitrogenous fertilizer.

	No applied nitrogen	Nitrogen fertilizer applied
No *Rhizobium*	RON0	RON1
Rhizobium present	R1N0	R1N1

(From de Wit *et al.*, 1966)

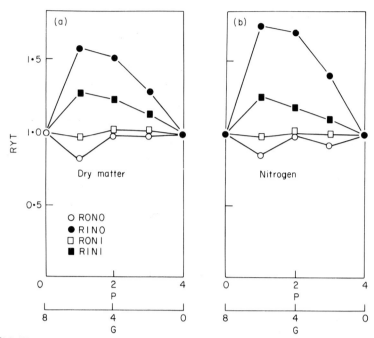

Fig. 8/14. The average relative yield totals (RYT) of *Panicum* (P) and *Glycine* (G) based on dry matter yields (a) and nitrogen yields (b). (From de Wit *et al.*, 1966)

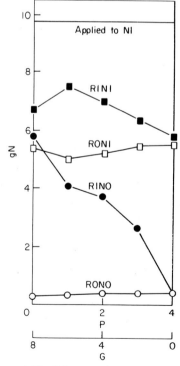

Fig. 8/14c. The total nitrogen yields of the seven harvests of *Panicum* (P) and *Glycine* (G). The horizontal line represents the nitrogen applied to the N1-treatments. (From de Wit *et al.*, 1966)

different species may penetrate different root zones and so escape some measure of interspecific stress.

The exceptions to the rule are often more interesting than the rule and there are intriguing exceptions to the rule that mixtures of grasses have RYT = 1.0. One such case is an experiment in which four species of *Avena* (*A. strigosa, A. fatua, A. sativa* and *A. ludoviciana*) were sown in all possible pure stands and mixtures (Trenbath and Harper, 1973). Values of RYT much greater than 1.0 were found when *Avena fatua* and *A. strigosa* were grown together on deep soil — but, on shallow soil RYT values approached 1.0. A subsequent analysis of rooting depth in these species (Ellern *et al.*, 1970) showed that in the deep soils, the roots of *A. strigosa* were more strongly developed in the upper layers of the soil profile but *A. fatua* contributed most to the root system deep in the soil.

Mixtures of *Festuca pratensis* and *Lolium perenne* and mixtures involving their hybrid outyield pure stands, *provided that* the populations are allowed to develop over at least a 2 year period (Whittington and O'Brien, 1968). These results have been included in a grand survey of the dry matter yields of mixtures made by Trenbath (1974b). He lists published data of the yields of 341 mixtures which have been compared with monoculture yields (Table 8/II). Clearly, situations in which mixtures behave transgressively (i.e. yield *significantly* below the lowest yielding pure stand or above the highest yielding pure stand) are very much in the minority. Perhaps it is not unexpected that any pair of species or cultivars, bred by man for their high performance in pure stands, will seldom exceed this performance when grown in mixtures; yet it is on these that experiments have been concentrated. A search for "ecological combining ability" is most likely to be successful in species or varieties that have been specifically bred for or evolved naturally towards some degree of niche separation (Harper, 1967; Chapter 24).

When two species are grown in mixture, each may modify the form of the other and when this extends to modifying the form of the seeds the effects of mixing will be carried over to the next generation. In a mixture of oats and barley, the mean weight per grain of oats was found to be greater in mixtures than in pure stands, and of barley slightly less (Fig. 8/15a). The interpretation is simple: "oats growing in a mixture are for some time before ripening surrounded by barley plants which are already ripe. These barley plants do not intercept much light and do not use minerals and water. Oats which were originally surrounded by a

Table 8/II

The distribution of biomass yields of varietal or interspecific mixtures compared with yields of their components' monocultures, based on published data of 341 mixtures. P_1 and P_2 are the yields of the higher- and lower-yielding monocultures respectively. P is the mean-monoculture yield, i.e. $\overline{P} = (P_1 + P_2)/2$. The symbol $\overline{P}+$ represents a value slightly above $\overline{P}$; it is greater than $\overline{P}$ by the quantity $0.2(P_1 - \overline{P})$. The value of $\overline{P}-$ is less than $\overline{P}$ by the same quantity. The multiplier 0.2 was chosen arbitrarily to provide categories to include mixture yields which lay "close" to $\overline{P}$. (from Trenbath, 1974b).

Crop	$<P_2$	P_2 to $\overline{P}-$	$\overline{P}-$ to $\overline{P}$	$\overline{P}$ to $\overline{P}+$	$\overline{P}+$ to P_1	$>P_1$	Author
Wheat				1	1	1	Sakai (1953)
Rice	9	7	5	2	9	4	Sakai (1955)
Either grasses		3		1		2	Aberg et al. (1943)
or legumes	1	1		2	7	3	Donald (1946)
A series of non-legumes[b]	1		1	1	9	3	Williams (1962)
Legumes	1				1	1	Williams (1963)
Flax & linseed	3	5	2	1	7	12	Harper (1964)
Grasses[a]	12	28	10	10	32	28	England (1965)
Barley	1	4		3	2		Norrington-Davies (1967)
Grasses[a]	5	5	4	2	9	15	Norrington-Davies (1968)
Grasses	2	2			1	1	Rhodes (1968)
Grasses[a]	2	5	2	1	2	4	Thomson (1969)
Grasses	1	2	2	1	4	2	Rhodes (1970)
Rye	4	4		3	12	7	Norrington-Davies and Hutto (1972)
Totals	42	66	26	28	96	83	

63.3%

39.3%   60.7%

[a]Data derived from a series of cuts.
[b]Mixtures involving a leguminous species have been omitted.

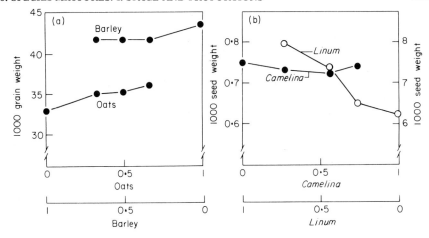

Fig. 8/15. The influence of the relative frequency of two species in mixture on the mean weight of seeds. (a) Mixtures of oats and barley; (b) mixtures of flax (*Linum usitatissimum*) and *Camelina sativa*. (From de Wit, 1960)

large fraction of barley plants are therefore able to produce during their last weeks of growth, more dry matter than oats which are surrounded by oat plants. This can only result in a higher thousand *grain* weight because the number of seeds is already fixed at that time." (de Wit, 1960). The same effect occurs in mixtures of *Linum usitatissimum* with *Camelina sativa* (Fig. 8/15b).

Ensembles of Species mixtures

A mixture of two species or forms is the simplest model of diversity for experimental study. Natural communities are usually made up of a galaxy of species and it is naturally tempting for the population biologist to try to take the complexity to pieces, like a child with a watch, "to see how it works" and then reassemble it.

One approach to understanding the relationship between the parts of a whole is to separate the elements and recombine them in all possible pairs. Quantitative genetics has developed a biometrics for an analogous problem. A comparison between two pure stands and a mixture has analogies with a comparison of two parental lines and their F_1 hybrid. For genetic analysis a diallel technique has been used in which a series of parental lines are selfed and also crossed in all possible combinations. Some measured character of the progeny is then expressed in a matrix and some form of analysis of variance and/or covariance is used to analyse relationships.

The yield of a mixture is analogous to the performance of a hybrid. If a hybrid closely resembles one parent we would say that that parent showed dominance. Similarly if the yield of a mixture closely resembles the yield of one of the contributing species in pure stand we can think of it as ecological dominance. Some mixtures of species show dominance by the higher yielding component, some by the lower yielding component (rarely); sometimes dominance is incomplete and sometimes there is transgressive behaviour in which the mixture yields less than the lowest or more than the highest yielding pure stand. All these situations are represented in Table 8/II. The results of such a diallel experiment can be analysed in a variety of ways, each of which tends to reveal some different facet of the plant interactions.

An experiment designed as a mechanical diallel that contains within it all possible combinations of n species is also $\dfrac{n(n-1)}{2}$ separate replacement series of two species, each represented by two pure stands and one equiproportioned mixture. The parameters that may be used in a biometric analysis of a mechanical diallel can be seen in Fig. 8/16 which describes them in relation to a replacement series experiment.

The main measurements made on populations in such an experiment are

(1) The yield of a pure stand of I $2\alpha_{ii}$
(2) The yield of a pure stand of J $2\alpha_{jj}$
(3) The yield of I in mixture with J α_{ij}
(4) The yield of J in mixture with I α_{ji}

(Note that the yields in (1) and (2) are twice the relevant values for comparison with (3) and (4) because the pure stands contain twice the number of plants of a given species.)

From these values a series of relationships are derived which describe the relations between the pairs of species:

(5) $\alpha_{ii} - \alpha_{jj}$ The difference between the yields of I and J in pure stand
(6) $\alpha_{ij} - \alpha_{ji}$ The difference between the yields of I and J in mixture
(7) $\alpha_{ij} - \alpha_{ii}$ The difference between the yield of I in mixture and pure stand
(8) $\alpha_{jj} - \alpha_{ji}$ The difference between the yield of J in mixture and pure stand

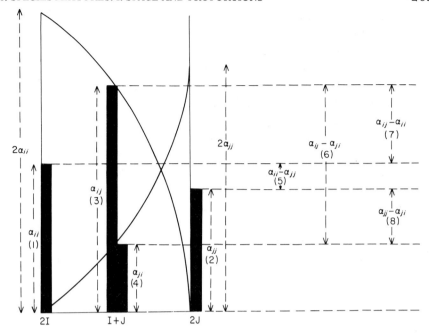

Fig. 8/16. Diagram to illustrate the various measures of growth that are used in analyzing the interaction between two species I and II grown together in a replacement series. The graph on the left is an idealized result from a replacement series experiment.

The competitive advantage of species I over species J is then given by (7) – (8), i.e $\mathcal{Y}_{ij} = \frac{1}{2}(\alpha_{ij} - \alpha_{ii}) + \frac{1}{2}(\alpha_{jj} - \alpha_{ji})$

The yield of the mixture *taken as a whole* may fall below that predicted from the pure stands (or exceed it). This element of the interaction between species is given by:

$$\delta_{ij} = \underset{(8)}{\tfrac{1}{2}(\alpha_{jj} - \alpha_{ji})} - \underset{(7)}{\tfrac{1}{2}(\alpha_{ij} - \alpha_{ii})}$$

$$= \underset{(1)+(2)}{\tfrac{1}{2}(\alpha_{ii} + \alpha_{jj})} + \underset{(3)+(4)}{\tfrac{1}{2}(\alpha_{ij} + \alpha_{ji})}$$

Various authors have obviously found it difficult to discover appropriately descriptive words for $\mathcal{Y}_{ij}$ and δ_{ij}. The effects that are described by these equations are perhaps best described as "relational" and "summational".

Individual species may have characteristic "relational" and/or "summational" characteristics that they show in most mixtures of which they form a part; these are general properties of the species. Other relational and summational characteristics may be peculiar to a particular combination of two species — these are specific properties of a defined species pair. There is a clear analogy with "general" and "specific" combining ability in the jargon of the geneticist.

From the matrix of data resulting from a mechanical diallel can be derived matrices of differences and matrices of sums and an analysis of variance can then be used to determine the magnitude of general and specific effects (main effects and interactions) that arise from growing species in mixtures.

The biometrical analysis of genetic diallels has become a sophisticated tool in the hands of plant breeders and agronomists. The statistical forms of handling the data depend in essence on the calculations described above but there are many different forms of analysis of variance and covariance that have been applied.

The direct use of the analysis of variance is found in studies of Sakai (1955), Williams (1962) and in successively more sophisticated forms (McGilchrist, 1965; McGilchrist and Trenbath, 1971). The latter paper modifies the treatment of the data to bring it into line with the concept of Relative Yield Total by considering proportional changes due to mixing of species rather than absolute changes.

The results of an analysis of variance may be coupled with a covariance analysis in which the behaviour of species in mixtures is tested for its regression on the performance of the species in pure stands. This again has an origin in genetics in the biometrical analysis of genetic diallels by Hayman and Jinks (see Dickinson and Jinks, 1956). It has been used in the analysis of competition between cultivars of *Linum* (Harper, 1964a), and for mixtures of grasses and of cereals (Fig. 8/17) (Norrington Davies, 1967). It may be useful sometimes to distinguish between "beta" competition (differences between two species grown together that are a function of their growth separately) and "alpha" competition in which the individual species are uniformly increased or decreased when grown with the other species. This distinction can be drawn (though not perfectly) from covariance analysis (Norrington-Davies, 1967; Durrant, 1965). An example of the application of the Norrington-Davies method to two different mechanical diallels is shown in Fig. 8/16.

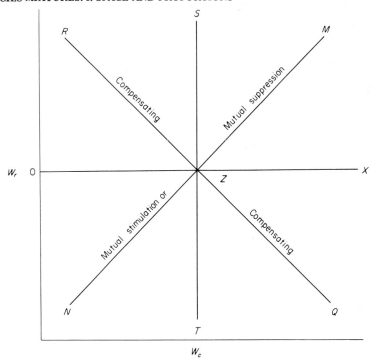

Fig. 8/17. Generalized interpretation of the W_r/W_c graph of plant interactions in a mechanical diallel; r and c are the two components within a cell of the diallel matrix. Each is represented repeatedly within a column or a row of the matrix and W_r and W_c are the covariances of the rth species onto all species grown alone and of the cth species onto all species grown alone.

"If species influence one another in mixtures such that in each mixture both species change by the same amount in the same directions, although not necessarily the same amounts or directions for all mixtures, the parents would be spread out on a line of unit slope, NM passing through Z. Then in general if the species are increased (mutual stimulation) the small species will be above the X axis on ZM, and the large species below on NZ; if the species are decreased (mutual suppression) the large species will be on ZM and the small species on NZ. Species would be expected to deviate from RQ and MN and to help in the interpretation of the deviations, another line, ST, is drawn perpendicular to the X axis through Z. Species lying to the left of ST decrease the weights of larger species with which they are grown in mixtures and increase the weights of smaller species with which they are grown. Species lying to the right of ST increase the weights of larger species with which they are grown and decrease the weights of the smaller species with which they are grown. Species above the X axis are increased in weight when grown with larger species or decreased in weight when grown with smaller species. Species below the X axis are decreased in weight when grown with larger species and increased in weight when grown with smaller species." (From Norrington-Davies, 1967)

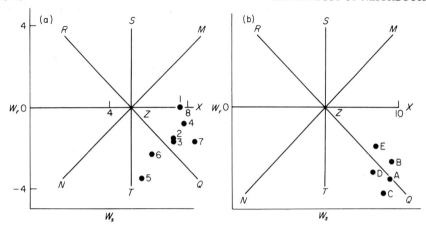

Fig. 8/18. Graphical analysis of the competitive relationships between species in two mechanical
diallels.
(a) Data from a 7-species diallel made by R. M. Moore and J. D. Williams (quoted in
Williams, 1962). 1 = *Lolium rigidum,* 2 = *Trifolium subterraneum,* 3 = *Brassica
tournefortii,* 4 = *Rapistrum rugosum,* 5 = *Centaurea calcitrapa,* 6 = *Marrubium
vulgare* and 7 = *Echium plantagineum.*
(b) Data from a 5-species mechanical diallel made by Norrington-Davies (1967).
A = *Hordeum agriocrithon,* B = *H. spontaneum,* C = *H. hexastichon,* D = *H.
distichon* and E = *H. intermedium.*

In both W_r/W_s graphs the species are in the area ZXTQ indicating that on average
species increase the weight of larger species with which they are grown and are themselves decreased when grown with larger species.

In (a) the species are spread out to a much greater extent in the direction of a line
of unit slope indicating greater variation between them; at one extreme, species 5 is
close to the line ZT which indicates that it is itself strongly influenced by the other
species without itself consistently influencing others, whereas at the other extreme
species 1, on the X axis, is influenced little by the other species but has a pronounced
influence on the growth of others. In (b) the points are close to ZQ indicating competitive effects which are mainly compensating. The $(W_r + W_s)/P$ correlation is positive
in (a), so that the mixture means tend to the values of the smaller species, but in (b)
it is negative and the means tend to the larger species. (From Norrington-Davies,
1967)

The effects of neighbours as environments

To a plant breeder it is very important whether the characteristics of his
species are stable in the face of environmental variation. In the study of
plant populations it is also interesting to know whether a particular
pattern of behaviour is stable or responsive to changes in the environment. To a plant, its neighbours are vital components of its environ-

ment. It is appropriate to ask whether plants or populations have stable behaviour when the surrounding species are changed. A procedure for studying "homeostasis" in face of the physical environment was developed by Finlay and Wilkinson (1963) who studied the yield of varieties of barley in three contrasted regions of Australia in different years. The procedure described here is developed directly from Finlay and Wilkinson's method but instead of the different environments being places and years they are represented by the presence in a mixture of different associated species. Such an analysis is illustrated for nine grassland species in Fig. 8/19 (from Jacquard and Caputa, 1970). The horizontal

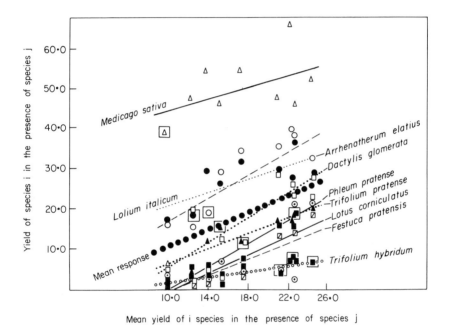

Fig. 8/19. Linear regressions between the yield of a species i in the presence of a species j (Y_{ij}) and the mean yield of species i in the presence of each of the other species (Y_j). Pure stand □. (From Jacquard and Caputa, 1970)

axis represents the mean performance (yield) of all the species against each associated species. The vertical axis is the performance of *each* species against each associate. Thus an idealized species that behaved in a perfectly average fashion would give values on the graph with a slope of 1.0. A species whose yield was very sensitive to the nature of its associates would give a slope of >1.0 and a species with very stable behaviour a slope of <1.0. Species with generally high yields will have regression lines lying towards the top of the figure, and vice versa.

Two critical values are now extracted for the description of each species: (a) the mean yield in the face of all associates and (b) the slope of the regression. These values are presented graphically in Fig. 8/20. The procedure picks out *Lotus corniculatus* as a low-yielding species with high responsiveness to associated species: given a weak associate it

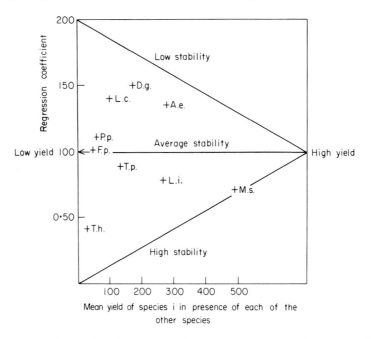

Fig. 8/20. The stability of yield of nine grassland species when grown in all possible combinations of species pairs. The regression coefficient is the slope of the lines in Fig. 8/19. The list of species is given in Fig. 8/19. (From Jacquard and Caputa, 1970)

is capable of greatly increased growth but is very sensitive to vigorous neighbours. *Dactylis glomerata*, though on average higher yielding, is even more sensitive to the nature of its neighbours. In marked contrast *Trifolium hybridum* is very insensitive to competitive pressures but regularly low yielding. *Medicago sativa* combines near-average stability with a very high average yield. There are serious problems in the statistics of this procedure that still need to be sorted out but these do not destroy the essence of the idea.

A whole diallel can be compared in contrasted environments to determine whether the relationships between pairs of species are altered by, for example, physical conditions or grazing. Only one attempt seems to have been made to use the diallel in this way. Rousvoal and Gallais (1973) made a 5-species diallel from *Lolium perenne, Dactylis glomerata, Holcus lanatus, Agrostis stolonifera* and *Trifolium repens* and clipped the swards at 3- or 6-weekly intervals (Fig. 8/21). *Trifolium repens*

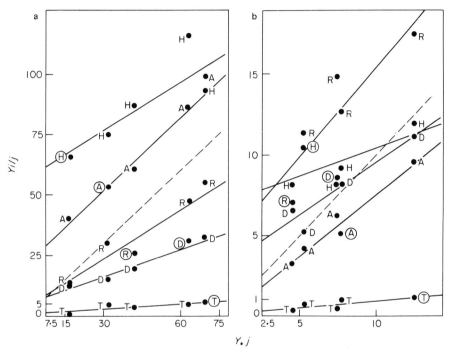

Fig. 8/21. The relationships between the yield of a species i in the presence of a species j (Y_{ij}) and the mean yield of species i in the presence of each of the other species ($Y_{.j}$) for five species of permanent grassland: *Dactylis glomerata* D, *Lolium perenne* R, *Agrostis stolonifera* A, *Holcus lanatus* H, and *Trifolium repens* T.
(a) "Pasture" regime — clipped every 3 weeks;
(b) "hay" regime — clipped every 6 weeks.
(From Rousvoal and Gallais, 1973)

retained its characteristic low yield and high stability under both regimes. *Dactylis glomerata* had high stability in the "hay" regime but was much more reactive to the nature of its associated species in the frequently cut situations ("pasture"). *Lolium perenne* also had reduced stability in the frequently cut plots.

Jacquard and Caputa (1970) have made comparisons of different bio-metrical procedures for handling diallel data — other examples are by Gallais (1970) and Chalbi (1967). In a series of recent papers (Jacquard, 1968; Rousvoal and Gallais, 1973) the biometrics have been used to unravel highly complex sets of biological interactions. Interest in ecology has centred for so long on the descriptions of ensembles of species that the time is ripe for understanding the processes within these ensembles. Pastures are complex but easily manipulated systems suitable for the early stages of such a programme but there is no fundamental reason why a large variety of natural vegetation types should not be analysed in this way. It must, however, always be borne in mind that experimental models are simplifications — often gross simplifications — and that even the subtleties of a multispecies diallel analysis in different regimes still omit the timing and spacing elements that can over-ride species differences in a struggle for existence.

9

Mixtures of Species.
II. Changes with Time

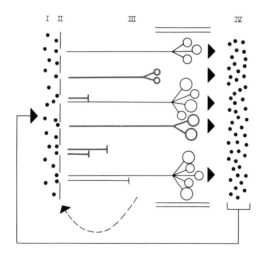

The previous chapter was concerned with the growth of species in mixture assessed by the amount of growth made at a point in time. It is often more interesting and important to know about the time sequence of the events in an interaction. The responses to interference between neighbours can take the forms of (i) a reduced growth rate of individuals, (ii) an increased death rate of the parts of individuals, and (iii) an increased death rate of whole individuals (Chapter 6). In most studies,

(i) and (ii) have not been separated and the character of the density-stressed plant, if it has survived, is then the resultant of the two forces, slower gain and more rapid loss, which result in its being smaller than a comparable plant with unrestricted growth.

The relationship between neighbours of different species is usually studied as a function of density and plant size. In monocultures this function seems to have a very constant form — the 3/2 thinning law (Chapter 6). Two experiments have been made to study the changes in numbers (density) and plant weight in mixtures of two species.

Populations of *Sinapis alba* and *Lepidium sativum* were grown in an equiproportioned mixture in a pot experiment (Bazzaz and Harper, 1976), in which 400 seeds of each species were sown together in 30 x 40 x 10 cm flats at two fertility levels. The low fertility soil was John Innes Compost No. 1, a medium normally used to supply only the needs of seedlings in horticultural practice. The high fertility medium was John Innes No. 3, a compost used for the growth of healthy mature plants. Thirty flats were sown to give five harvests x two fertility levels x three replications. At each harvest the plants were cut at ground level and the number of plants present and their individual dry weights were recorded. Figure 9/1 shows the relationship between mean plant weight and the number of survivors taking both species together as a single population. The regression has a slope of -1.594, conforming closely with the normal situation in monocultures. However, when the species are analysed separately it can be seen (Fig. 9/2) that almost all the mortality has been suffered by *Lepidium* and almost all the growth made by *Sinapis*.

In comparable experiments made with monocultures most of the mortality occurred among the smaller and weaker members of the population. In a mixed population this discrimination falls between the two species. The survivorship curves of the two species in mixture are strikingly linear (Fig. 9/3a, b) and the slope is steeper at the higher fertility — the mortality risk is constant with time but increased by fertility. Indeed by the last harvest *Lepidium* had been completely eliminated in the high fertility plots though at low fertility a few weak and straggly individuals were still present. It is likely that when the last *Lepidium* plants had been eliminated from the mixtures the whole of the mortality risk would have transferred to *Sinapis* but the experiment did not last for sufficient time for this to be tested.

A second experiment involved mixtures of *Raphanus sativus* and

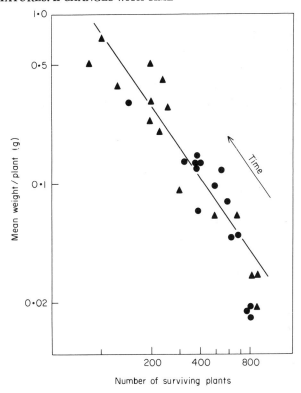

Fig. 9/1. The relationship between the number of survivors and mean plant weight in a
  mixture of *Lepidium sativum* and *Sinapis alba*. The plants of the two species are
  treated as members of a single population in the graph.
  • = low fertility ▲ = high fertility.
  (From Bazzaz and Harper, 1976)

Brassica napus grown at three fertility levels and the experiment included
pure stands of both species and differently proportioned mixtures.
Again the mixtures like the monocultures, grew and thinned along a 3/2
thinning line, the survivorship curves were close to linear and the rate of
mortality was increased at the higher fertility levels. The two species in
this experiment had closely similar growth forms and there was only
very slight difference between them in the sharing of the mortality risks
and the growth made. Each species reacted to the other much as it
reacted to members of its own species (White and Harper, 1970).

   In most experiments made to study interference between species,
mortality has not been followed over time. The study of time sequences

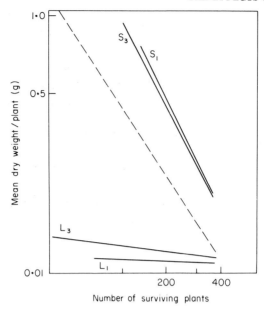

Fig. 9/2. The relationship between the number of survivors and mean plant weight in a mixture of *Lepidium sativum* and *Sinapis alba*. S = *Sinapis*, L = *Lepidium*, 3 = low fertility, 1 = high fertility. (From Bazzaz and Harper, 1976)

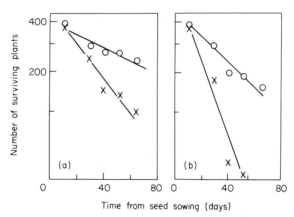

Fig. 9/3. Survivorship curves for *Lepidium sativum* (x) and *Sinapis alba* (○) grown in mixture: (a) low fertility; (b) high fertility soil. (From Bazzaz and Harper, 1976)

is very time-consuming because it is rarely possible to count survivors accurately without greatly disturbing them and the structure of their canopy. Hence it is usually necessary to use destructive sampling at each harvest and the size of the experiment is then multiplied by the number of harvests to be taken. Even when the great labour of a time study has been made it is rarely that mortality has been followed, usually because the species chosen grow clonally and the identity of genets can not be easily determined (e.g. experiments with grasses and clovers pose this problem acutely because tiller and stolon connections are rapidly lost). Most detailed studies of time sequences have therefore been confined to recording changes in weight per plot (or per ramet) and while having much interest in terms of production have lost much of their potential genetic interest.

The development of mixed populations of two grasses or a grass and a clover have been followed in "replacement series" experiments. Mixtures are either sown or planted from tillers and stolon units. The relationship between the components of a mixture over time can be described by the "Relative Replacement Rate" (de Wit, 1960; van den Bergh, 1968).

If the yield of species i in the presence of equal numbers of itself is α_{ii} and its yield in the presence of equal numbers of j is α_{ij} then the

Relative Yield of i in mixture is $\dfrac{\alpha_{ij}}{\alpha_{ii}} = r_i$. Over a time interval $m-n$ we

can then express the change in the two species in a mixture as

$$nm\rho_{ij} = \frac{r_i^n \big/ r_j^n}{r_i^m \big/ r_j^m}$$

A gain by species i in the interval between the mth and the nth harvests is then represented by $\rho j > 1$.

The Relative Replacement Rate (RRR) is a parameter of the behaviour of two species in mixture over time and can be plotted against time to give "course lines". The validity of this analytical technique depends on two important caveats:

(1) changes in density during the experiment. During the growth of a mixed population of grasses or clovers the density of tillers or stolons is likely to change. The values of RRR will then follow changes in proportions against an unknown change in density stress.

(2) changes in proportion during the experiment. The relative advantage of one species over another may be frequency-dependent and there are certainly combinations of grasses which show frequency-dependent interactions. Graphs of RRR against time will then show relative changes in proportions against unknown changes in absolute proportions. Both de Wit and van den Bergh are aware of these problems and in a study

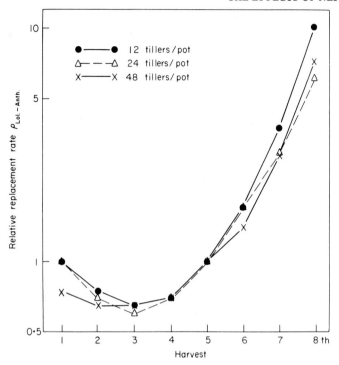

Fig. 9/4. The Relative Replacement Rates of *Lolium perenne* with respect to *Anthoxanthum odoratum* over eight successive harvests. All the replacement rates are expressed relative to the situation at the fifth harvest. (From van der Bergh, 1968)

of RRR of *Lolium perenne* against *Anthoxanthum odoratum* an experiment was started at three different densities. The course lines are shown in Fig. 9/4, and in this case the results are not from wholly destructive sampling but from successive clippings every third week. At low densities *Anthoxanthum* gained as a proportion of the mixture and then after the third or fourth harvests began a steady decline in favour of *Lolium*. At high densities *Lolium* had the advantage almost from the beginning. In this case RRR was not independent of density and this may be generally true during periods when the total yield is itself increased with density. De Wit (1960) and van den Bergh (1968) found in this case that "experimental results were not at variance with the supposition that ρ is independent of the relative yields (i.e. RRR is frequency-independent) when RYT = 1. This means that a species should gain at any frequency when it gains at one frequency, but this has not been tested for extreme relative frequencies" (de Wit *et al.*, 1966).

In brief, the caveats for using RRR require that yield of a mixture should have reached a ceiling yield and that RYT = 1. Many of the course lines that have been drawn for pasture species do not wholly satisfy these restrictions but are nevertheless extremely interesting descriptions of the processes of change in mixtures of species.

Figure 9/5 shows the course lines for a mixture of *Dactylis glomerata*

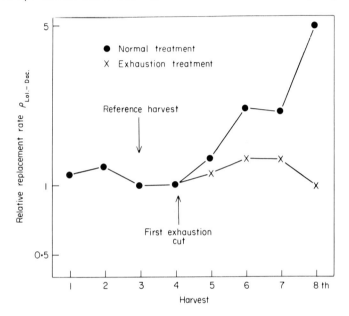

Fig. 9/5. The Relative Replacement Rates of *Lolium perenne* with respect to *Dactylis glomerata* grown together in mixture with and without an "exhaustion" treatment — see text. (From van den Bergh, 1968)

and *Lolium perenne* planted as 12 tillers of each species per pot (24 tillers in pure stands). The plants were clipped at 3-week intervals and after the fourth harvest an exhaustion treatment was applied, in which about half the pots were given a further clipping 2 days after the main harvest clip. Under the exhaustion treatment both species remained in relative balance but *Lolium perenne* rapidly gained over *Dactylis glomerata* under the normal 3-weekly regime of clipping. The interpretation given for this result is in terms of the different amounts of photosynthetic tissue left to the two species after clipping — *Lolium perenne* bears only a brown stubble but *Dactylis glomerata* after cutting has a

stubble of green tillers. The effect of the exhaustion treatment is therefore to delay the recovery of *Lolium* with respect to *Dactylis*. It is
interesting that in the exhaustion treatments the RYT of the mixture
increased steadily from 1.03 to 1.11 at the eighth cut though under the
"normal" treatment RYT remained at *ca* 1.0. Thus the mixture had a
yield advantage over the pure stands under the more penal clipping
regime but not in the more productive normal system (van den Bergh,
1968).

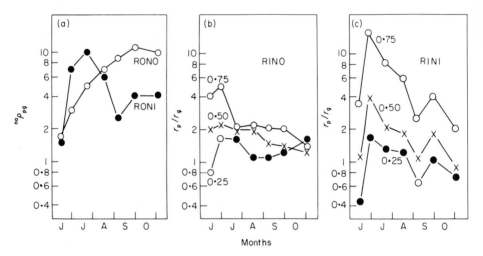

Fig. 9/6. Changes in the Relative Replacement Rate with time of mixtures of *Panicum maximum*
        and *Glycine javanica*.
        R_0N_0 — not inoculated with *Rhizobium*, no fertilizer nitrogen.
        R_0N_1 — not inoculated with *Rhizobium*, fertilizer nitrogen given.
        R_1N_0 — inoculated with *Rhizobium*, no fertilizer nitrogen.
        R_1N_1 — inoculated with *Rhizobium*, fertilizer nitrogen given.
        The course lines for (b) and (c) show populations developing from three different
        frequencies of *Panicum* in the mixtures. (From de Wit *et al.*, 1966)

Alopecurus pratensis has been found to have a lower sodium content
than other grasses in experiments at Wageningen. An experiment was
therefore designed to study interference between *Alopecurus* and *Lolium
perenne* at different K/Na ratios. When the K/Na ratio was high *Lolium
perenne* steadily increased its dominance of the mixture but *Alopecurus*
had the advantage when the potassium was replaced by sodium.

  A further example of the use of course lines comes from the experiment on the growth of *Panicum maximum* and *Glycine javanica* discussed in Chapter 8 (Fig. 8/3). The course lines reveal the dynamics

of the species interaction. In the absence of *Rhizobium* the mixtures moved steadily towards dominance by *Panicum* when no fertilizer nitrogen was applied but with applied nitrogen an initial surge by *Panicum* was followed by its giving way to vigorous growth by *Glycine* (Fig. 9/6a). In the presence of *Rhizobium* the situation was quite different. With *Rhizobium* present and no applied nitrogen the mixtures (with very different starting proportions) converged on a balance strongly favouring *Glycine* (Fig. 9/6b). This was a frequency-dependent response; a mixture with a low starting proportion of *Glycine* increased this proportion and a mixture with a high starting proportion of *Glycine* decreased the proportion. With *Rhizobium* and applied nitrogen there was very little frequency-dependence and the population with various starting proportions of the two species tended to march in parallel as the species changed their relative dominance (Fig. 9/6c).

Ratio diagrams

In the experiments described above a mixture is followed through time and the changes in its proportions are recorded as course lines. An alternative procedure is to start an experiment with mixtures sown at a variety of proportions and to detect the change in each after a lapse of time. The proportions at the end of a period are then plotted against the proportions at the beginning. This can be a sensitive and revealing procedure for detecting frequency-dependent interactions. Five basic types of interaction can be envisaged (Models I–V).

Model I (Fig. 9/7a). The proportion of the two species remains unaltered after a period of growth together. This is an idealized situation of two perfectly identical components of a mixture (at least in relationship to each other) — the balance of species is subject only to random variation and it is doubtful whether such a real situation could be found.

Model II (Fig. 9/7b). Species I gains an advantage in mixtures at all proportions. The advantage is measured by the vertical distance apart of the actual ratio line and the theoretical line of "no advantage". If the lines are parallel the advantage is independent of frequency.

Mixtures might have been made by sowing seeds and harvesting seeds or by planting ramets and harvesting ramets. In either case we can follow the fate of one particular mixture IJ from planting to harvesting and imagine it then being replanted at its new ratio and the process repeated from generation to generation. The predicted changes in such

mixtures are shown by arrow lines. The model predicts the relentless
progress towards extinction of species J.

Mixture III (Fig. 9/7c) is the converse of Model II. Species I moves to
extinction and at a speed dependent on the distance apart of the two
parallel lines (the selective or successional advantage).

Model IV (Fig. 9/7d). This is a frequency-dependent situation. The
minority component is always at an advantage and it is not possible to
specify which species will be at a successional or selective advantage
without specifying the frequency. If a high proportion of I is sown, J
gains in the mixture and if a high proportion of J is sown, I gains. Mixtures will always tend to change towards an equilibrium frequency.

Model V (Fig. 9/7e). This is also a frequency-dependent situation.

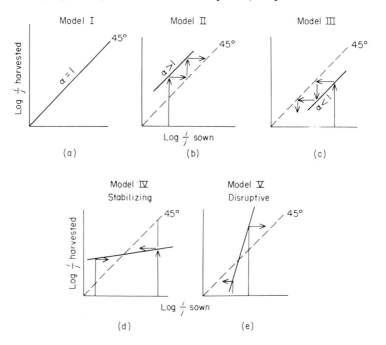

Fig. 9/7. A series of "ratio diagrams" (see text)

The majority component of a mixture is an advantage. If a high proportion of I is sown, J goes to extinction and vice versa. The behaviour
of a mixture cannot be predicted without a knowledge of the starting
frequency of the species and there is no equilibrium mixture. The mixture is disruptive, in contrast with Model IV which is stabilizing.

Note that pure stands do not appear in these models at all; they are all concerned with real mixtures.

The frequency-independent models are illustrated in experiments with cereals and with grasses. A mixture of oats and peas was sown in a replacement series and the competitive advantage of the oats proved to be quite independent of the proportions in the mixture (de Wit, 1960): the oat was always the winner (Fig. 9/8). This experiment involved both

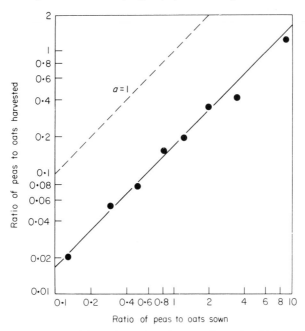

Fig. 9/8. A ratio diagram plotted from the results of an experiment in which peas and oats were grown in mixture. The x axis shows the proportions of peas to oats sown and the y axis the ratio harvested. (One pea seed is taken to be equivalent to 4.5 units for calculating sowing and harvesting ratios). (From de Wit, 1960)

sowing seeds and harvesting seeds. A similar result was obtained from mixtures of *Linum usitatissimum* with *Camelina sativa* which behaved in a wholly frequency-independent fashion. The interaction of *Linum* and *Camelina* is particularly interesting because *Camelina* has been claimed to exert its depressant action on *Linum* by liberating toxic substances.

A series of mixed populations of *Anthoxanthum odoratum* and *Phleum pratense* grown in a controlled environment showed amazingly close balance and frequency-independence (Fig. 9/9). On average *Anthoxanthum* maintained a very slight advantage. These two species

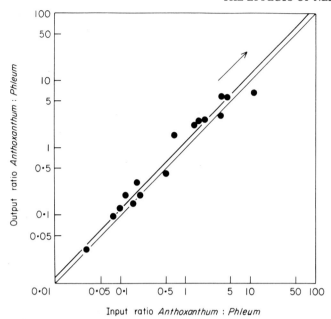

Fig. 9/9. A ratio diagram showing the relationship between the ratio of tillers of *Anthoxanthum odoratum* to those of *Phleum pratense* over a period of recovery from cutting. (From van den Bergh and de Wit, 1966)

behaved quite differently when sown in an experiment out of doors (Fig. 9/10). There was now clear frequency-dependence. Mixtures with low proportions of *Anthoxanthum* gave increases in this species, but mixtures with high proportions of *Anthoxanthum* favoured *Phleum*. The equilibrium mixture was one with a ratio of about four tillers of *Anthoxanthum* to one tiller of *Phleum*.

Trifolium repens (clover) and *Lolium perenne* (ryegrass) are another pair of grassland species that show frequency-dependent behaviour. When grown at low light intensity, mixtures with a high proportion of ryegrass or of clover showed a trend towards a more equitable mixture. At high light intensities (under this management regime) the clover moved to dominance. Note (Fig. 9/11) that there was still some frequency-dependence of the mixture at high light intensities — the advantage of clover was greater at low than at high densities. Indeed the results of this experiment may imply that there is a stable equilibrium between the species even at high light intensities but that it balances at a very, very low ryegrass content.

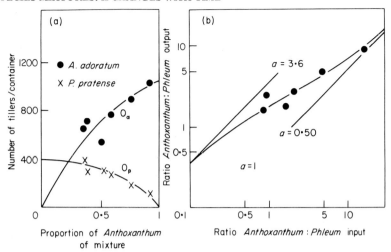

Fig. 9/10. The results of an experiment in which *Anthoxanthum odoratum* and *Phleum pratense* were grown together out of doors. Figure 9/12a shows the results as a replacement diagram and 9/12b shows the ratio diagram. (From van den Bergh and de Wit, 1960)

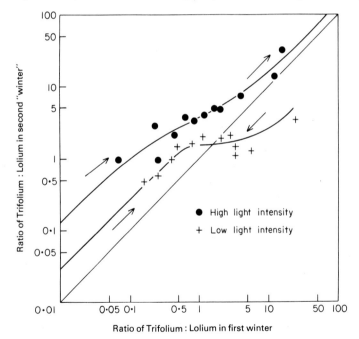

Fig. 9/11. A ratio diagram showing the changes in proportion of a mixture of *Trifolium repens* and *Lolium perenne* during a season's growth. The experiment was made at low and high light intensities. (From Ennik, 1960)

290　　　　　　　　　　　　　　　　　THE EFFECTS OF NEIGHBOURS

A curious example of a frequency-dependent response was found by Putwain and Harper (1972) for the two sexes of *Rumex acetosella*. This species is dioecious and both sexes spread clonally by means of root buds. Establishment from seed yields populations with males and females in equal proportions. However, establishment from seed is an extremely rare phenomenon in natural pastures and probably occurs only after some severe disturbance (Putwain *et al.*, 1968). The sexes

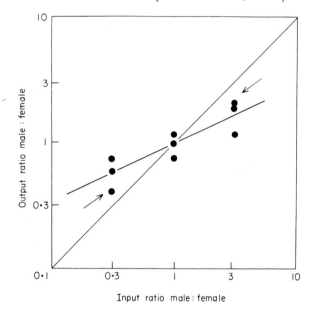

Fig. 9/12. Ratio diagram relating the sex ratio of *Rumex acetosella* at the beginning of the season to the ratio at the end. The biased sex ratios were deliberately made by planting ramets in desired proportions. (From Putwain and Harper, 1972)

may therefore spend many years developing as clones by root budding, and it might be expected that the sex ratio would then change. The males have the relatively light responsibility of pollen production and tend to be smaller plants than the females which bear the full burden of setting and ripening seeds. Yet, the sex ratios remain balanced, even in very old pastures. An experiment was made in which ramets of the two sexes were planted to give populations with grossly unbalanced sex ratios 3:1 and 1:3 as well as balanced 2:2 populations. The results after a season of growth show that the sexes responded to each other in a frequency-dependent manner (Fig. 9/12) and the equilibrium mixture was close to 1:1.

Frequency-dependent situations arise when each species (sex) suffers more from its own density than from that of its neighbours and implies some form of ecological differentiation such that the two forms "make different demands", "compete for different resources", "compete at different times". The two sexes of *R. acetosella* do in fact have a different seasonal rhythm of growth. The males are precocious and their canopy overtops that of the females in May but by June the females have overtopped the males and the latter are often starting to senesce (Fig. 9/13). A computer model (GRODEV) has been used by Trenbath

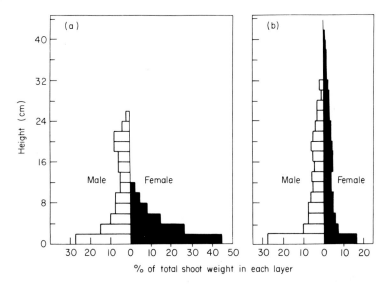

Fig. 9/13. Vertical distribution of shoot weight of male and female plants of *Rumex acetosella* on (a) 21 May and (b) 12 June. (From Putwain and Harper, 1972)

(1976) to study this situation and, provided that he gives a high light saturation value to each sex during the time of its height dominance, the model is self-stabilizing.

Combinations of density and frequency

In Chapter 8 additive experiments were criticized because they tend to confound the effects of density and of proportion. Experiments designed as Replacement Series escape this problem by maintaining density constant (sometimes by defining conditions when it seems not to matter).

Several experimenters have, however, managed to design experiments with mixtures of species that effectively combine variations in density and proportion.

Such an experimental design was applied to study the behaviour of poppies (*Papaver* spp.) in mixed populations. Five species of *Papaver* are found as weeds in agricultural crops in Britain usually in mixtures of 2, 3, 4 and sometimes 5 species within 1 m² (Harper and McNaughton, 1962). *Papaver rhoeas, P. dubium* and *P. lecoqii* bear smooth capsules and have a more erect growth habit than *P. argemone* and *P. hybridum* which have spiny capsules and a more recumbent growth habit. The experiments were broadened by the inclusion of *Papaver apulum* a Mediterranean species but *P. hybridum* was omitted because it was difficult to obtain sufficient seed.

The species were sown in equiproportioned mixtures at "effective" densities of 50 seeds (25 of I and 25 of J) and 400 seeds (200 of I and 200 of J) per 30 x 30 cm plot. This gave a comparison of densities at constant proportion. In addition plots were sown at a density of 225 seeds (25 of I + 200 of J and 200 of I + 25 of J) − different proportions but at constant density. The plants were harvested after seed had ripened and the number of surviving plants was counted and the dry weight determined. The experiment included all combinations of the five species in pairs.

The results of the experiment have been expressed as the chance of a seed producing a mature plant in the various combinations (Fig. 9/14). Each quadrilateral is represented at its four corners by the four combinations of a single species pair. The far left point represents survivorship of species A in a low-density equiproportioned mixture; the far right corner is the survivorship of A in an equiproportioned mixture at high density. The point marked • is survivorship of A in an unequal mixture in which it is in the minority and the remaining corner is for the survivorship when it is the majority component of the mixture. In virtually all cases there was clear density-dependent mortality − survivorship at density 50 was much greater than at density 400. In 16 out of the 20 cases the survivorship of a species was poorer when it preponderated in a mixture. This effect may be summarized by saying that density-dependent mortality was largely species specific − each species reacted primarily to and suffered most from its own density in a mixture. The four cases where this was not true involved *P. apulum* (three cases) and *P. dubium* in the presence of *P. lecoqii*. *P. apulum* was the odd man out in not being a

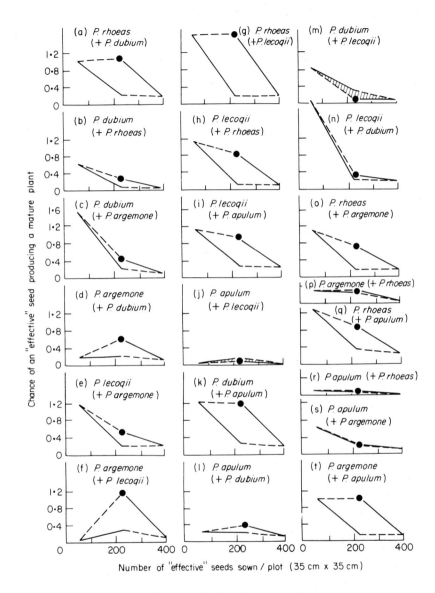

Fig. 9/14. The influence of "self-"——— and "alien-" – – – – thinning in two-species mixtures of *Papaver* spp. (see text for description of experimental design). Cross-hatched figures are those in which alien-thinning exceeded self-thinning. (From Harper and McNaughton, 1962)

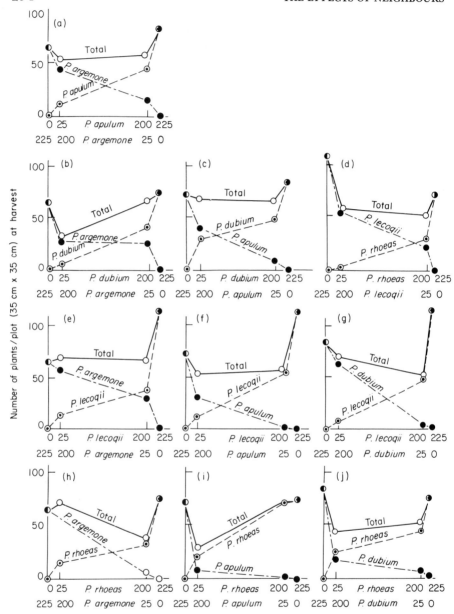

Fig. 9/15. The numbers of plants surviving to maturity from mixed sowings of poppy species at constant density and varied proportions. The figures on the horizontal scales indicate the number of seeds of each species sown on the plot. (From Harper and McNaughton, 1962)

member of the communities in which the others are found. *P. lecoqii* is taxonomically and morphologically very close to *P. dubium* and each apparently reacts to the other much as it reacts to the density of its own sort.

The experimental design included pure stands sown at density 225 seeds per plot. It is therefore possible to plot the results as a replacement series to include these pure stands (Fig. 9/15). The number of plants in the mixtures tended to fall below, sometimes far below, the numbers in the pure stands. Apparently the pure populations suffered quite disproportionately from the introduction of a minority of a second species. Such a minority makes a quite disproportionate number of interspecific contacts (Fig. 8/3) and the effect is perhaps not so surprising as at first sight. This is one of the very few experiments in which mixtures have been studied at extreme proportions and perhaps these warrant more study. The total yield per plot tended to be intermediate between the yields of the two pure stands so the mixed populations must have recovered in the growth of individuals what they lost in numbers (Harper and McNaughton, 1962).

Two legumes, *Medicago sativa* and *Trifolium pratense* behave in mixtures very much like the poppies. An experiment by Black (1960c) included mixtures with species ratios of 1:5 and 5:1, both at an intermediate density, and equiproportioned mixtures at high and low density. The chance of a seed producing a plant was again strongly density-dependent and each species suffered more from its own density in a mixture than from the density of its associate (Fig. 9/16). One can distinguish "self-thinning", the consequences of a species own density in a mixture, from "alien-thinning" the consequences of the density of the associate. Alien-thinning is less powerful than self-thinning, and (over the time period of the experiment) these two species formed a potentially stable mixture. A gain in abundance by either tended to favour the survival of the other: each species was its own worst enemy.

The experimental models of *Papaver* spp. and *Medicago/Trifolium* are both self-stabilizing mixtures but other species show a clear advantage to one component of a pair. *Rumex crispus* and *R. obtusifolius* were grown in mixtures (Cavers, 1963) (see Fig. 9/17) and in this case *R. crispus* suffered more from the presence of the alien species, *R. obtusifolius*, than from self-thinning. *R. obtusifolius* suffered more severe self-thinning than alien-thinning. Clearly the survivorship of these species did not lead to any stable mixture. This model is unusual in that density-

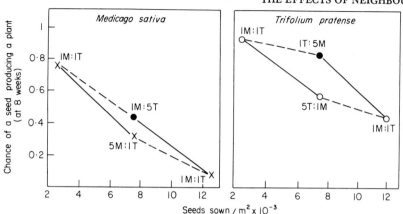

Fig. 9/16. The influence of density on "self-thinning" and "alien-thinning" in mixed populations of *Trifolium pratense* (T) and *Medicago sativa* (M). The extreme densities, 2500 and 12 500 seeds/m², were composed of equal numbers of the two species. The intermediate density, 7000 seeds/m², was either with a preponderance (5:1) of one species or the other.

The difference between the chance of a seed producing an 8-week-old plant at the two differently constituted intermediate densities reflects the different degrees of thinning in predominantly "self" and predominantly "alien" populations. In each diagram the broken line indicates "alien-thinning" and the continuous line "self-thinning" (i.e. inter- and intra-specific effects respectively). Calculated from data in Black (1960). (From Harper, 1967)

dependent mortality was not occurring — rather the reverse. The chance of a seed producing a mature plant was *increased* by density and the whole model contrasted very strongly with that of the *Papaver* spp. and *Trifolium/Medicago*. A situation in which density increases seedling survival has been reported before for *Rumex* (Harper and Chancellor, 1959) and introduces a further complexity into the study of interspecific interactions. The benefits of density in the establishment of *Rumex crispus* and *R. obtusifolius* are certainly not extended into the phase of seed production (Fig. 9/18) when the chance of a seed producing a seed is strongly density-dependent both to self and alien densities.

The various experimental models described in this chapter define with more or less precision the relationships between pairs of species in terms of growth or reproduction. They describe formal relationships between parts of a whole. For the most part they pose but go no distance towards answering questions about mechanisms. The models make the first and critical move from the study of simple systems (monocultures) towards complexity. There have been few attempts to model

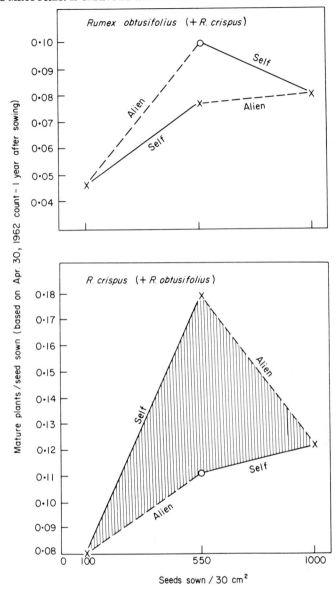

Fig. 9/17. The influence of density and proportions on the establishment and survival of mixtures of *Rumex crispus* and *R. obtusifolius*. In each figure the populations at low and high density are composed of equal proportions of the two speices. At the intermediate density the populations are predominantly (500/550) of one species. The survivorship of the predominant species is shown as ○. (From Cavers, 1963)

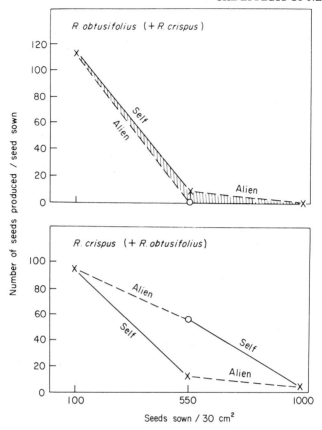

Fig. 9/18. As Fig. 9/17 but expressed as number of seeds produced per seed sown. (From Cavers, 1963)

plant mixtures of more than two species, and both the logistics of the experiments and the language needed to describe the complex results pose great problems at present. There are, however, at least two attempts that have been made to study mixtures of three species. Baeumer and de Wit (1968) made an experimental study of a mixture of oats with long- and short-stemmed peas and Haizel and Harper (1973) studied mixtures of a cereal with two associated weeds. The latter experiments were designed to answer a number of questions about interactions between species in complex situations. Like any experiment involving two species, a three-species experiment must include each species in pure stand for comparison but should also include the relevant combinations of the species in pairs. The design might be made as a replacement series in which the total density in the 1, 2 and 3 species populations is always constant or may contain additive elements. The number of variants in the experiment is limited mainly by the time and labour involved in doing the work. The species chosen for the study were barley (*Hordeum sativum*), mustard (*Sinapis alba*) and wild oats (*Avena fatua*). Each species was sown at an "index" density to give 12 plants at harvest: in fact 24 seeds were sown and thinned to 12 after emergence. Populations contained 12, 24 or 36 plants depending on the number of the index density sown per pot. Thus BMO implies that a pot contained 12 barley plants, 12 mustard and 12 wild oat plants. B implies that the pot contained 12 barley plants, BB 24 and BBB 36 barley plants. BM implies a pot containing 12 barley plants and 12 of mustard.

One of the aims of the experiment was to test the effect of removing a species from a mixture at different stages in growth and so there was superimposed on the design an additional feature in which an index unit (12 plants) or in some cases two index units (24 plants) were cut at ground level 3 weeks after sowing. Thus a population that started growth as BMO might suffer removal of the 12 oat plants or the 12 mustard plants or both. It is then symbolized by BMŎ or BM̊O or B̊M̊Ŏ. Such loss might be the equivalent of a selective grazer choosing to eat one of the species from the mixture, a disease that attacked and killed one of the species — or, in agricultural practice, the selective removal of one component by a herbicide. Similarly, a mixture BM- can be regarded as a three-species mixture in which the third species was killed before germination, the equivalent of selective predation on the seed, differential seedling mortality or a pre-emergence herbicide.

For each species the design include two densities of pure stands (three in the case of barley) so that intraspecific density effects could be compared with interspecific influences. For example B in the presence of O (BO) can be compared with B in the presence of B (BB) and B in the presence of OO (BOO) can be compared with B in the presence of BB (BBB).

The effects of the various species and treatments are shown (Fig. 9/19a, b, c) as the depressions caused to the growth of an index population, and all comparisons are relative to the performance of a pure stand at 12 plants per pot. Barley and mustard proved very sensitive to self-density effects. At 12 plants per pot these species were near ceiling yield and any increases in density were absorbed by reduced individual plant growth. Wild oats proved much less sensitive to self-density. The order of aggressiveness of the three species depended on the chosen indicator. Thus against barley: barley > oats > mustard, against wild oats: barley > mustard > oats and against mustard: mustard > barley > oats. This finding contradicts the view that there is some intrinsic competitive measure that can be used to order the aggressiveness of a group of plants irrespective of the test species (see discussion of Fig. 8/7).

The effects on a test species of increasing the density of an associate are not additive — there is usually a law of diminishing returns. Similarly the effects of two species J and K acting together on I are usually less

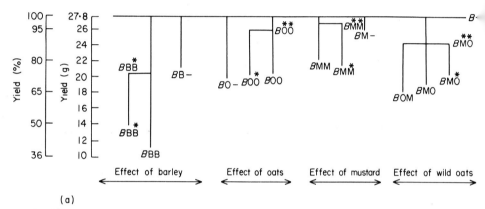

(a)

Fig. 9/19a. Reduction in the yield of barley (index) caused by the presence of associated species. See text. (From Haizel and Harper, 1973)

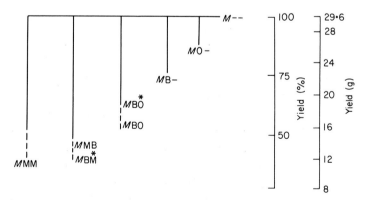

Fig. 9/19b. Reduction in the yield of white mustard (index) caused by the presence of associated species. See text. (From Haizel and Harper, 1973)

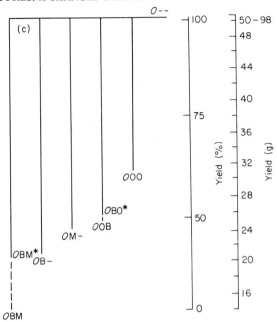

Fig. 9/19c. Reduction in the yield of wild oats (index) caused by the presence of associated
species. See text. (From Haizel and Harper, 1973)

than the sum of the effects of J and K acting separately, but the combined effect of barley + wild oats on the yield of mustard was greater
than the sum of their separate effects. Certainly the effects of mixtures
of species are not additive.

A great part of the harm done to a plant by the presence of another
apparently occurred in the first 3 weeks of growth. This fact is widely
recognized in agricultural practice and in horticulture where early weed-
ing of a crop may be very important.

Models of the sort described in this chapter can describe relationships
between species with precision in terms of changes or rates of change in
weight, numbers or reproductive activity. Sometimes the models can
destroy previous generalizations: it is clearly not true that yielding
ability is correlated with aggressiveness, *or* that mixtures always outyield pure stands, *or* that mixtures have the mean yield of pure stands,
or that competitive abilities are additive, *or* that "competition" is most
intense between members of the same species. Each of these statements
can be true for specific cases but can be clearly shown to be untrue in

others. The models have destroyed the generalizations. Equally important, the models can show the possibilities of types of interaction that
may help to understand happenings in the more complex real world:
species may interact to form self-stabilizing mixtures; mixtures may
sometimes outyield the component species in pure stand; a species with
high yield may fail in mixture with one of low yield; species may interact with each other so that their effects in complex mixtures are not
additive.

Experimental models have a further role. Like all models and
analogues, population models have to be designed. The act of designing
the model focuses the mind; it crystallizes the nature of the problems in
the real world of nature in the attempt to mimic parts of it in simple
analogues.

General theory

Questions about the behaviour of pairs of species growing together in
the same environment have lain at the core of population biology since
Darwin wrote the third chapter of "The Origin of Species". Because
population biology developed almost entirely in the hands of zoologists
and, because higher animals have determinate growth, competition
theory has been concerned with changing numbers. Models of the growth
of simple populations of single species have been concerned with dN/dt
(e.g. Verhulst's model of the logistic growth of populations (see Chapter
1)). Models of mixtures of two species have again been described in terms
of the numbers of individuals (see Chapter 1). A study of competition
between populations of two species of animal is usually made by releasing known numbers of the animals into a limited environment and
following the changes in birth rates and death rates or the composition
of the population over a number of generations (e.g. Park, 1955). The
experiments on competition between plants are usually made over a
single generation, sown at a range of densities, and the observer studies
changes in weight or size of the developing plants and differential mortality; but scarcely ever is a mixed population followed over generations.
There are, however, the most striking analogies between the models of
interaction between animal populations and models of plant interaction,
as might be expected from the fact that the animal has the potential for
exponential increase in numbers and the plant has the potential for
exponential increase in size (weight or number of modules of construc-

tion). Figure 9/20 illustrates a variety of the models of two-species interaction used for plants and for animals:

(a) Replacement series diagrams, showing the variety of theoretical
and real relationships between two species of plant sown at constant
density but at a range of proportions (Fig. 8/9a–e).

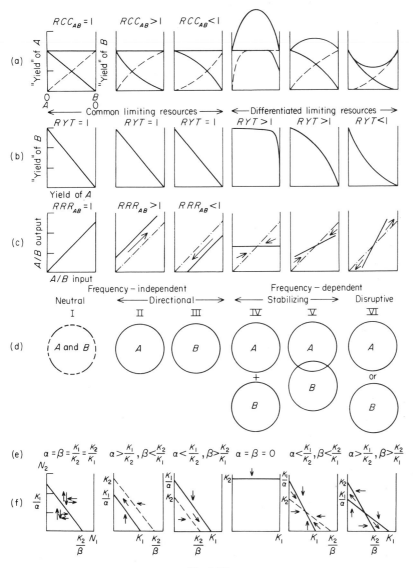

Fig. 9/20.

(b) Relative yield diagrams for two plant species grown in a replacement series as (a) but with the yield of species A plotted against B (Fig. 8/12a–c).

(c) Ratio diagrams obtained from a replacement series experiment and in which the ratio of the species present after a period of time is plotted against the ratio of the species sown or planted (Fig. 9/7a–e).

(d) The niche relationships of the two species corresponding to the models above and below expressed as Venn diagrams. A circle defines the fundamental niche of a species in two dimensions (it is of course really a multidimensional space). The area of a circle can be taken as defining the carrying capacity of the environment for a particular species. Models I, II and III imply that the two species make demands on the same resources of the environment and are mutually exclusive of each other. Models IV and V describe non-overlapping or partially overlapping niches and Model VI describes a situation of unique alternatives of which one or the other is possible but not both.

(e) The algebraic description of the relationships between two animal species changing in number in the presence of each other: these are the formal solutions to the Lotka-Volterra equation in the form used by Hutchinson (1965).

(f) Graphical description of the changes in numbers of two species developing in each other's presence, corresponding to the algebraic form above (from Slobodkin, 1964a).

10

The Limiting Resources of the Environment

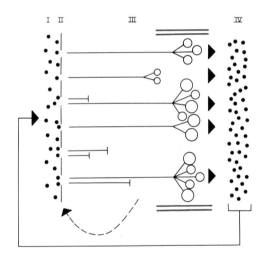

As a young plant develops from zygote to embryo or as a ramet develops as a vegetative off-shoot, its requirements are at first drawn from reserves supplied by the parent. These have been laid down in the embryo before it ripened and became separated from the parent, or are supplied during the actual process of establishment through the physical links that are usually maintained, for a time at least, in a clonal system. Sooner or later the young plant either exhausts or becomes independent of these parental supplies and depends for its further growth on its own ability to extract energy, nutrients, gases and water from the environment. These resources are limited and may on occasions be insufficient to allow unrestricted growth. Furthermore, individual plants may develop in such proximity that each lowers the availability of resources to the other; neighbours may then exaggerate an insufficiency of resources or create a deficiency where there was ample to support the growth of an isolated individual.

Light, water, nutrients, oxygen and carbon dioxide are consumable "supply factors" in contrast to the "quality" or "condition" factors of the environment, such as temperature, which may be unfavourable and so restrict growth but are not in any sense consumed.

Most of the development of population biology has been concerned with animals and in both experimental and theoretical studies there has been an appropriate emphasis on the interactions between animal populations and their food. All animals derive their food from other living organisms or their products — the food supply is itself a potentially exponential growth system or its product. The food of an animal is self-renewing. The food of green plants does not reproduce and is not the result of a reproductive process. The population behaviour of plants may therefore be expected to differ in important ways from that of animals.

Light as a resource factor

The supply of light to an area of land is the most reliable of the environmental resources for plant growth. Its reliability lies in the regularity of diurnal and annual cycles of light intensity. The periodicity of radiation at the surface of the earth's atmosphere is very constant because it depends simply on the repeated movements of the earth in relation to the sun. The proportion of incoming radiation which reaches the earth's surface is altered by the thickness of the layer of atmosphere through which it has passed. Variations in cloudiness bring the only serious unpredictability into the light climate at any point on the surface of the earth. Even the uncertainty of cloud cover is insufficient to prevent prediction of intensity of radiation from month to month with considerable accuracy (Fig. 10/1), and illumination can be calculated from the radiation intensity with some confidence (Black, 1960a).

Light occurs as a stream of quanta which, if they are to be used in plant growth, must be trapped by a photosynthetic structure and converted immediately into chemically bound energy. In complete contrast with water and nutrients, light cannot accumulate in the environment

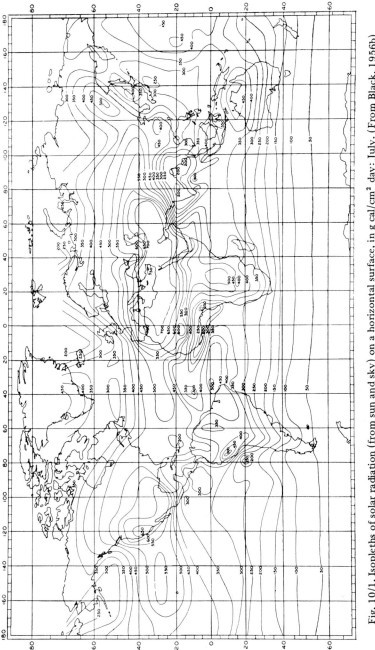

Fig. 10/1. Isopleths of solar radiation (from sun and sky) on a horizontal surface, in g cal/cm² day: July. (From Black, 1956b)

for later use. Neighbouring plants may interfere with and reduce each other's supplies of light by direct interception. The first leaf that intercepts an incoming ray of light may reflect it (in which case it may hit another leaf), absorb it and convert it directly into photosynthetic product or into heat (in which case it may provide latent heat of evaporation of water), or transmit it, filtered, so that it reaches the lower leaves dimmer and spectrally altered.

Evidence that light may become a resource in short supply comes in part from measurements of the amount of light penetrating through leaf canopies. In a series of plant communities in Japan the percentage of full daylight reaching the ground ranged between 0.2 and 37% (Monsi and Saeki, 1953). The highest values occurred in forests of *Pinus densiflora* (28—37%). *Larix leptolepis* (13—25%) and *Castanea crenata* (13—22%). Most values, however, fell between 0.2 and 15% (see Table 10/I). This type of analysis has now been made on many natural and cultivated forms of vegetation; situations in which only negligible light penetrates through a canopy to reach the ground level are legion (Donald, 1963). Most of the very low values are found under herbaceous communities of grassland or weedy crops. The intensity of light under vegetation commonly falls well below the compensation point at which photosynthesis is balanced by respiration. This implies that the lower leaves in such canopies are light-starved even though the upper leaves may be fully illuminated. The self-shading, of leaf by leaf, of leaf by branch, and mutual shading by the parts of neighbouring plants produces a profile of light intensity down through vegetation. This can bring the light resources of a population to a level at which parts of or whole individuals are light-starved.

There are many serious technical difficulties in the detailed analysis of the distribution of light within vegetation. It is difficult to design equipment which senses both direct and diffuse radiation in the way that a leaf would do. The changes in quality of light that accompany changes in intensity down through a canopy are also hard to measure but may be important as it is the photosynthetically less effective radiation that is transmitted through leaves (Kasanaga and Monsi, 1954). Moreover, the light passing down through a canopy of vegetation is not a continuous gradient of reducing intensity but a moving dappled pattern of direct light added to a background of diffuse light (Evans, 1956). The technical limitations in the measurement of light within vegetation have been reviewed by Anderson (1964, 1966) and after reading her papers it is

Table 10/I
The penetration of light through canopies of various
woodland and herbaceous communities in Japan
(from Monsi and Saeki, 1953)

Plant community	Relative Light Intensity (% of full daylight)
WOODLANDS	
Pinus densiflora	28—37
Larix leptolepis	13—25
L. leptolepis under a field layer of *Sasa nipponica*	1.5—2
Castanea crenata	13—22
Quercus crispula	7—14
Q. crispula under a field layer of vegetation	1.4—4.6
Chamaecyparis obtusa	5—15
Shiia cuspidata	2.5—5
Phyllostachys nigra	
in summer	0.2—0.5
in winter	2.5—4

MOUNTAIN GRASSLANDS	Leaf Area Index	Minimal Light Intensity (% of full daylight)
Sasa nipponica	4.9	0.5
Miscanthus sinensis		1.3

LOWLAND MEADOWS		
Miscanthus sacchariflorus		
+ *Thalictrum simplex*	5.4	2.6
+*Sanguisorba tenuifolia*	4.3	1.4
+*Humulus japonicus*	3.6	0.1
+*Glycine soja*	5.7	1.4
Phragmites communis		
+*Thalictrum simplex*	4.8	1.8
+*Sanguisorba tenuifolia*	5.1	0.8

the brave man who researches in this area. In the following parts of this chapter and also Chapter 11, these limitations must be borne in mind when interpreting the results of experiments.

The individual leaf, isolated from the plant and from neighbouring leaves and exposed to controlled light intensities, characteristically increases its rate of photosynthesis linearly with increasing light intensity, before a maximum value is reached beyond which further increases in intensity produce no change in the photosynthetic rate (a group of species largely from the tropics are exceptions to this rule -- see below). However, whole plants, because they create a light profile, bear different leaves in different light intensities. The growth of the whole plant integrates the activity of its separate parts. Thus when layers of leaves shade each other in a canopy, the topmost leaves may be light saturated; an increase in the intensity of light supplied will then not affect their rate of photosynthesis, but lower leaves in shade in the same canopy may still respond to the increase in light that penetrates through the light-saturated upper layer. The whole canopy may then continue to react to increasing light intensity at values far higher than will saturate an isolated leaf.

In the field, light varies in intensity, duration, quality, direction and angle of incidence both in daily and annual cycles. Leaves vary between species and on the same plant in the angle at which they are borne and consequently in the time of day at which they cast the greatest shadow. If we are to account for the use of light as a resource by a population of plants it becomes necessary either to integrate all these variables as has been attempted very successfully in computer integrations (e.g. de Wit, 1965) or alternatively to study the physiology of whole plants or whole populations rather than that of parts. A number of studies have been made of the whole plant physiology of reactions to light. Many of these have involved growing plants or populations under shades which cut down incident radiation by a constant fraction. An ideal shade would ensure that the plants receive different light intensities, but the same day-lengths and daily rhythms of change in intensity and duration as unshaded plants. In practice most shades bias the intensity according to the direction of incoming light. For the analysis of this type of experiment different parameters are used from those in much of laboratory physiology.

(i) *The Net Assimilation Rate (NAR or E)* This is a measure of the efficiency of a whole plant or of a population as an assimilating system, expressed as the increase in dry matter per unit leaf area. The NAR is commonly expressed as:

$$\frac{W_2 - W_1}{t_2 - t_1} \times \frac{\log_e L_2 - \log_e L_1}{L_2 - L_1}$$

where W_1 and W_2 are plant dry weights at times t_1 and t_2 when the photosynthetic areas are L_1 and L_2. Values of the NAR may then be expressed in terms such as g cm^{-2} day^{-1}. Note that this is a measure of net, not gross assimilation — it is a crude expression of photosynthesis minus respiration.

(ii) *The Leaf Area Ratio (LAR)*

$$\frac{L_2 - L_1}{W_2 - W_1} \times \frac{\log_e W_2 - \log_e W_1}{\log_e L_2 - \log_e L_1}$$

This is a measure of the area of leaf exposed by a plant expressed as a fraction of the plant — usually cm^2 g^{-1}.

(iii) *The Relative Growth Rate (RGR or R)*

$$\frac{\log_e W_2 - \log_e W_1}{t_2 - t_1}$$

This is a measure of the growth rate of the whole plant, a compound interest function, and may be expressed in such terms as g g^{-1} day^{-1}. When a plant is growing freely and increasing in weight exponentially, unrestricted by the presence of neighbours or its own self-limitation (e.g. self-shading), this expression is formally equivalent to the intrinsic rate of natural increase of a population (r) (see Chapter 1).

It is an important property of the Relative Growth Rate that RGR = NAR x LAR. All three of these parameters may be used to study a whole plant growing in isolation from its neighbours, a plant in a population, a whole population, or an area of vegetation. In addition three further parameters are used only for the analysis of populations or vegetation:

(iv) *The Leaf Area Index (LAI or L)* This is a measure of the area of photosynthetic surface expanded over a given area of ground (e.g. cm^2 cm^{-2}, ha ha^{-1}) or as a simple unitless integer.

(v) *The Leaf Area Duration (LAD or D)* This is an integrated value of the Leaf Area Index over a period of time, usually a growing season.

$$LAD = \tfrac{1}{2}(LAI_n + LAI_{n+1})(t_{n+1} - t_n)$$

It can therefore be used to describe the extent and duration of the light-trapping apparatus of a population of plants from the early stages of seed establishment, in which only expanded cotyledons are present and much light falls on bare ground, through the period of maximum LAI, when lower leaves in dense populations may be in near darkness, and ending in leaf fall or senescence. It is expressed as Leaf Area Index— Time units.

*(vi) *The Crop Growth Rate (C)* This is a measure of the growth rate of vegetation per unit area of land, expressed e.g. as g m^{-2} day^{-1}.

When individual whole plants of temperate species are grown under a series of shades it is generally found that the Net Assimilation Rate increases linearly with the logarithm of the relative light intensity (Blackman and Wilson, 1951a, b), though there may be a departure from strict linearity at very high radiation intensities of 700–750 cal cm^{-2} day^{-1}. This relationship is illustrated in Fig. 10/2 for four contrasted species. The Net Assimilation Rate takes a value of zero at a relative light intensity at which photosynthesis just compensates for the respiration of the whole plant.

The Leaf Area Ratio also is linearly related to the logarithm of relative light intensity but in this case declines with increasing light. If this trend were extrapolated LAR would fall to a value of zero, the extinction point, at some value of light intensity 3–30 times that of full daylight — this extinction point is, however, a purely hypothetical value. The Relative Growth Rate, being the product of LAR and NAR, has values of zero at the compensation point (NAR = 0) and, in theory, at the extinction point (LAR = 0) and rises to a peak value which may occur at more or less than full daylight (Fig. 10/2). In Britain *Geum urbanum*, a perennial woodland species, has its maximal relative growth rate at a little over 50% daylight, *Helianthus annuus* (the cultivated annual sunflower) at 71% daylight, *Fagopyrum esculentum* (buckwheat) at 95% daylight and *Trifolium subterraneum* (subterranean clover) at intensities far higher than full British daylight (Blackman and Wilson, 1951b).

In experiments in which whole plants or populations are grown under shades, all parts of the plant are of course not at the same light intensity; a light profile still develops within the canopy as upper leaves shade those further down. One major effect of shade is to slow down the rate of photosynthesis relative to respiration and the concept of *Net* Assimilation Rate focuses attention on the resultant of these two activities, rather than on either process alone. Growth in dry weight depends on the photosynthetic activity of the whole plant exceeding its total respiratory load. The respiratory burden may become so great that even an efficiently photosynthesizing plant may make no growth.

*Slight and usually fairly insignificant variations in the form of calculation of many of these parameters can be found in the work of different authors and a useful review of the forms, mathematical derivations and underlying assumptions involved was given by Radford (1967).

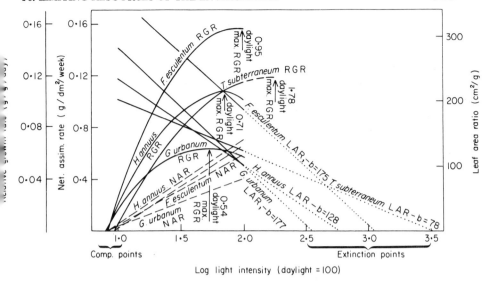

Fig. 10/2. The relationship between Net Assimilation Rate, Leaf Area Ratio and Relative Growth Rate for *Fagopyrum esculentum*, *Trifolium subterraneum*, *Helianthus annuus* and *Geum urbanum* over a range of light intensities. Arrows indicate the light intensity at which the relative growth rate will be greatest. (From Blackman and Wilson, 1951b). Reproduced with permission from *Annals of Botany* 15, 373–408

The vertical distribution of photosynthetic and dependent tissues above ground has been measured for a number of species. Each leaf layer exists in its own particular light regime and the ways in which the activities of the leaves combine to determine the yielding properties of a whole population are idealized in diagrams by Donald (1961). Figure 10/3, for example, models three stages in the development of a grass or cereal. In the first two stages the plants are shown bearing an increasing number of leaves and the contribution of each leaf layer to the assimilation of the whole population is assigned a theoretical value. Leaves are assumed to maintain a constant respiration rate until they begin to senesce. The photosynthetic activity of each leaf is reduced according to its depth in the profile. In parts (a) and (b) of the diagram only the net assimilation rate of the foliage is considered but in (c) the non-photosynthetic tissues are added to give the net assimilation of the population as a whole.

Donald's concept of the way in which growth rate alters with time is shown in Fig. 10/4. In the early phases of development of a canopy the *rate* of leaf production is shown increasing, together with an increase in

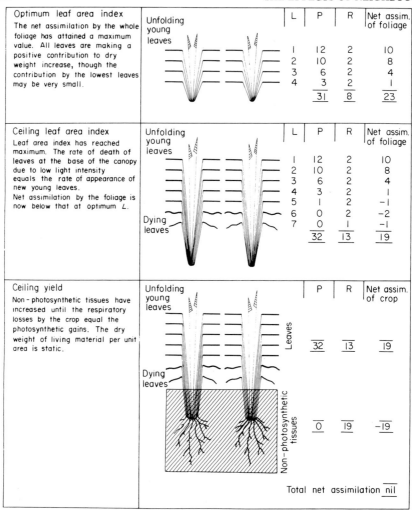

Fig. 10/3. Model to illustrate the relationships between the photosynthetic activity (P) and respiration (R) of different leaf layers (L) in a grass or cereal population and the effects of these and the dependent root system on the net assimilation rate of the crop. (From Donald, 1961)

non-photosynthetic parts, until the canopy reaches an "Optimum Leaf Area Index". This is the state shown in Fig. 10/3a in which every leaf is making a positive contribution to dry weight increase, though that from the lowest leaves may be very small. Leaf area per plant is shown as increasing beyond this optimum and the Leaf Area Index becomes supra-

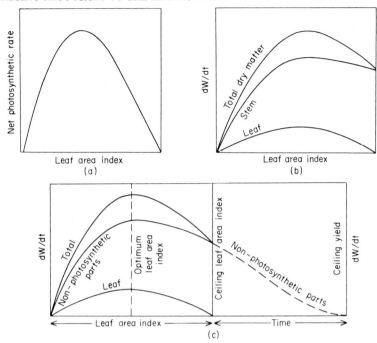

Fig. 10/4. Suggested general relationships between Leaf Area Index and the rates of net
   photosynthesis and production of dry matter. (From Donald, 1961)

optimal; the crop growth rate now starts to decline. The rate of increase
of foliage also continues to decline to zero. At this stage, although new
leaves may be produced, the loss of others low in the canopy balances
the gain: this is the "Ceiling Leaf Area Index" (Fig. 10/3b). A canopy
at this stage is still producing excess assimilates and some of these are
used in the growth of more non-photosynthetic structures. This respiratory burden increases until it matches the assimilatory activity of the
canopy. The Net Assimilation Rate of the plants and the population
then becomes zero: this is "Ceiling Yield" and the dry matter present
per unit area is static.

   Donald's version of the relationship between the Leaf Area Index, crop
growth rate and the balance between respiration and photosynthesis is
not universally accepted and may be too simplified. However, many
experiments with a variety of crop plants give broad support to the
essential features of Donald's interpretation. There is, for example,
evidence that the density of foliage may develop to a supra-optimal
level. Watson (1958) sowed high densities of kale and after a dense

canopy had developed he thinned the crop to leave values of LAI of 5.3, 4.2, 2.7 and 1.6. The subsequent crop growth rates were 78, 130, 101 and 88 g m^{-2} week^{-1}. In an experiment with *Trifolium subterraneum*, Davidson and Donald (1958) found a maximal growth rate at LAI = 4.5 whereas at LAI = 9 the rate of increase of leaf area fell to zero and dry matter production of the crop declined by 30%. An analysis of the growth of a woodland of *Pinus sylvestris* by Ovington (1957) showed that the rate of production of leaf weight increased over the first 20 years to a ceiling of 4000 kg ha^{-1} year^{-1}. The rate of production of total dry matter reached a ceiling of 22 000 kg ha^{-1} year^{-1} after 25 years. The respiratory load continued to increase after this time so that the crop growth rate declined to only 8000 kg ha^{-1} year^{-1} after 55 years. There are, however, other studies which show crop growth rate increasing not to a peak but to a plateau as LAI is increased (e.g. Shibles and Weber, 1965, with soyabeans). Their data suggests that there is no burden involved in bearing excess LAI. There remains much room for argument and experimentation about the behaviour of leaves low in the canopy. Do they die as soon as they reach compensation point (Saeki, 1960) or does the structure of a developing canopy alter in such a way that the penetration of light to the lower leaves is improved (Verhagen *et al.*, 1963)?

The concept of an optimal Leaf Area Index has emerged repeatedly in studies with a variety of species by Japanese workers. Figure 10/5

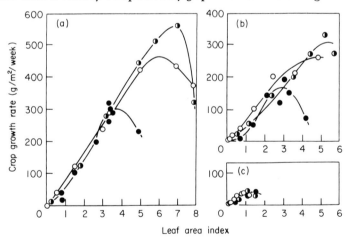

Fig. 10/5. The relationship between leaf area index and crop growth rate in populations of *Helianthus annuus* (sunflower) grown at three light intensities A = 100% light, B = 60% light and C = 23% light. (From Hiroi and Monsi, 1966)

shows the relationship between Crop Growth Rate and Leaf Area Index in stands of *Helianthus annuus*. The optimal LAI varied, as would be expected, with the light intensity when this was artifically regulated by shading. Very sophisticated treatments of the relationship between the crop growth rate, the Leaf Area Index and the radiation intensity were made by Black (1963) with subterranean clover (*Trifolium subterraneum* in Australia (Fig. 10/6) and by Takeda (1961) with rice in Japan. In

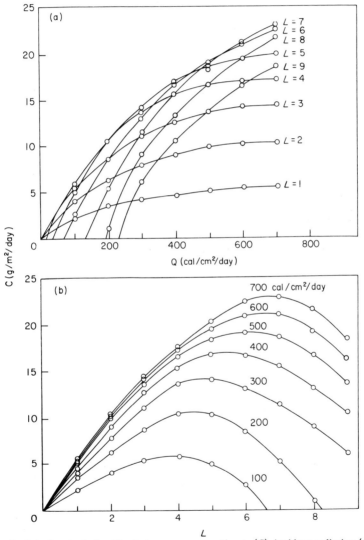

Fig. 10/6. Calculated relationships between crop growth rate (C), incident radiation (Q) and Leaf Area Index (L) for *Trifolium subterraneum*. (From Black, 1963)

both of these studies the crop growth rate was related directly to the real rather than the relative intensity of incident radiation. Takeda's data show that at 100 cal cm^{-2} day^{-1} a rice population loses weight if it carries an LAI of more than 7 and that towards the end of the growing season an LAI of more than 4 produces a negative Crop Growth Rate. The greater the intensity of radiation the greater was the optimal LAI. Thus an LAI of 7 which leads to zero crop growth rate at radiation intensity of 100 cal cm^{-2} day^{-1} is the optimal LAI at about 200 cal cm^{-2} day^{-1} and LAI in excess of 10 may be required for maximal crop growth rate at 500 cal cm^{-2} day^{-1}. The optimal Leaf Area Index for subterranean clover shifts in the same way with increasing radiation but with this species the optimal LAI does not rise much above 7, even at 700 cal cm^{-2} day^{-1}. Black was able to estimate the percentage utilization of light energy at different intensities and different values of LAI. The important point emerged that maximal utilization of incident light does not occur at the LAI that is optimal for growth of the crop but usually at values far below this.

Studies like those of Black and Takeda take the emphasis away from individual plants and lay it instead on the behaviour of areas of land covered with leaves — a further stage in the progression from the physiology of leaves through the physiology of whole plants to the physiology of whole populations. The population acquires a holistic physiology within which the individual plants are subordinated in the physiology of the whole. Such subordination of the individuals within the whole is easily shown by sowing populations at a range of densities. Figure 10/7 shows the results of growing *Trifolium subterraneum* at 4, 16 and 36 plants dm^{-2} and following the progress of growth of the canopy and the distribution of light within it. The dark zone (less than 5% of incident radiation) appears first in the high density population and last in the low density population — but after 156 days' growth the three populations have become virtually identical (Stern, 1965).

Populations of *Helianthus annuus* sown at different densities gave essentially similar results (Fig. 10/8) and Hiroi and Monsi (1966) showed that the number of dead old leaves, deep in the canopy, was much greater at the higher densities. There was a degree of over-compensation for increasing density — the most dense populations had a lower yield at the final harvest than those at medium and low density. Hiroi and Monsi varied not only plant density but also the light intensity by shading their plots. The Crop Growth Rates of all populations declined with time but

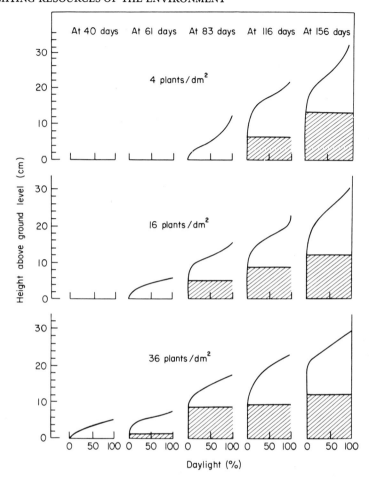

Fig. 10/7. The influence of plant density on the light profiles within populations of *Trifolium subterraneum*. The cross-hatched zone is the region within the canopy to which less than 5% of incident light penetrates. (From Stern, 1965)

the low densities maintained a high growth rate for longer. With the passage of time the yields of plant material per unit area tended to converge, irrespective of the starting density (see Chapter 8). The yield at which the populations converged was dependent on the light intensity but the density of plants was relatively unimportant: their performance had been integrated within the limits that the environmental resources allowed.

A plant population, growing within an environment of limited light resources, adjusts its structure and its growth rate to the supplies. Perfect

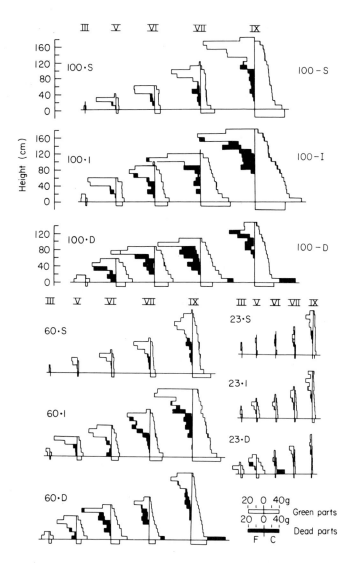

Fig. 10/8. The distribution of tissues in stratified clips down the profile of populations of
Helianthus annuus (sunflower). The values are shown per plant with photosynthetic
structure to the left and support structure to the right. 100, 60 and 23 are the values
for % of full daylight and S, I and D are increasing densities. (From Hiroi and Monsi,
1966)

adjustment is impossible because the environment changes: the LAI that is optimal for noon on a clear day will be supra-optimal in the morning and afternoon of the same day or at noon on a cloudy day. The Leaf Area Index that is optimal at noon on a clear summer day will be supra-optimal on a clear day at noon in the spring or autumn. Some seasonal adjustment of LAI does occur: populations of *Trifolium repens* change both LAI and the height of the canopy with the changing seasons in ways that tend to optimize crop growth (Brougham, 1962) but the production and loss of leaves are not fast enough to adjust to changes on shorter time scales. Canopies are therefore compromises — balances between respiratory costs and photosynthetic advantages — in environments that are partly created by the canopies themselves and which vary in time.

Most of the evidence that shortage of light restricts plant growth in the field is circumstantial, consisting of demonstrations either that almost all incident light is intercepted by a canopy, or that artificially reducing the light intensity produces results like those observed in crowded populations. It would be interesting to make experiments in which the light intensity in the field is increased locally and to determine whether there is a growth response. If extra illumination were supplied to the vegetation on a forest floor does a great flush of growth occur in the ground flora? Areas of canopy can of course be removed and the response of the ground floor observed, but such experiments are inconclusive proofs of limiting light because the treatment automatically affects the demands made on nutrients and water as well.

Johnston *et al.* (1969) made one of the rare experiments to increase the level of light falling in a population of leaves in the field. They inserted wide-spectrum fluorescent tubes between the rows of a close canopied crop of soyabeans, placed at the top, middle or bottom of the canopy. They also laid strips of white polythene on the ground between the rows to increase reflection. The light added was equivalent to 250 and 500 watts m^{-2} of ground area in crops planted at 100 and 50 cm row distances respectively. The seed yield of the crop was partitioned into the bottom, middle and top zones. The additional light boosted seed production in the two lower shaded zones (Fig. 10/9). The assimilation rate of the lower leaves increased by 73% when the lights were turned on. In the same way, opening the canopy to allow sunlight to penetrate to the lower leaves increased their assimilation rate by 258%. In this experiment two soyabean varieties were used: 'Wayne', having a

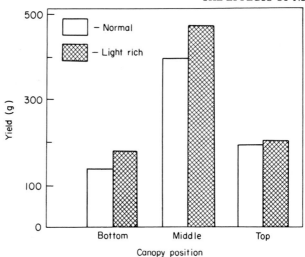

Fig. 10/9. The effect of increasing the intensity of radiation at various positions within a
canopy of *Glycine soja* (soybean) by inserting wide spectrum fluorescent tubes
between the rows of the crop. (From Johnston *et al.*, 1969). Reproduced from
Crop Science 9, 579, 580 by permission of the Crop Science Society of America.

dense leaf canopy and 'Ansoy' which had a less dense canopy. When the
canopies were opened the bottom leaves of the two varieties assimilated
at the same rate, but in closed canopies the bottom leaves of 'Ansoy'
assimilated at 69% and of 'Wayne' at 19% of their activity in full sunlight.

Whereas it is difficult to augment the light intensity of a naturally
illuminated plant population, it is very easy to add nutrients or water.
Conversely it is very difficult in an experiment to reduce the levels of
water or nutrient supply to an established population of plants but it is
rather easy to do so with the light. These differences have a profound
effect on the type of information we have about limiting resource
factors in nature.

One further point needs to be made about light as a resource and it
emphasizes a major difference between a plant's reaction to a shortage
of supply factors and the reaction of most animals. Although we may
crudely regard whole plants as interfering with each other's light supplies,
in reality this interference occurs between individual leaves. A leaf is
successful in pre-empting light supplies if it overlaps another, be it only
by a millimeter. Thus of two neighbouring plants A and B, the topmost
leaf of A may overlap and shade the topmost leaf of B, but another leaf
of A may be overlapped and shaded by a leaf of B. There is no compar-

able situation in higher animals, though somewhat similar effects may occur between individual members of colonial animals, e.g. corals. The curiously population-like structure of an individual higher plant (Chapter 1) permits different parts of the same organism to behave to some extent as discrete units, even interfering directly with each other's activities.

Carbon dioxide as a limiting resource factor

When the intensity of light falling on an isolated leaf is increased, the assimilation rate commonly increases linearly to a plateau value at which some other factor becomes limiting. This plateau can usually be raised by increasing the external concentration of carbon dioxide and the implication is that the rate of supply of this resource can therefore control the rate of photosynthesis. It seems reasonable to enquire whether such a process has population consequences — in a rapidly assimilating canopy does the density of leaves lower the carbon dioxide supply to a level at which the individuals in a population suffer but an isolated individual would not? An answer to this question must depend on (a) evidence that the levels of CO_2 fall in photosynthesizing canopies and (b) evidence that the rate of photosynthesis is reduced at such lowered concentrations. It is also helpful if (c) it can be shown that supplementation of the CO_2 in a canopy increases its assimilation rate more than is the case for an isolated leaf or plant in the same environment.

Carbon dioxide flux within canopies

The concentration of CO_2 in the atmosphere away from a photosynthetic surface is about 300 parts per million and shows only slight variation from place to place. Movement of CO_2 to a leaf surface occurs both by gaseous diffusion and by turbulent transfer. Turbulence falls off rapidly down through a canopy but is a function of wind speed, even deep in a corn crop. Wright and Lemon (1966a) were able to show that turbulent transfer is several orders of magnitude greater than molecular diffusion, even deep within the crop. Nevertheless measurable zones of CO_2 depletion occur within a stand of vegetation (Fig. 10/10a). The assimilation rate of the crop increased during a period of wind, suggesting that the rate of supply of CO_2 was indeed limiting (Fig. 10/10b). In these studies the depletion zone, was as expected, greatest and extended

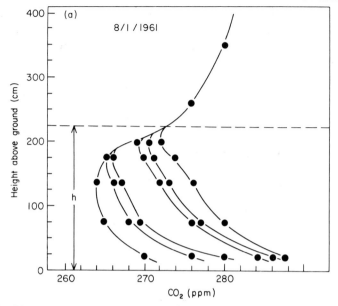

Fig. 10/10a. CO_2 profiles within and above a maize crop for the period 1200–1700 hours (h is height of the crop).

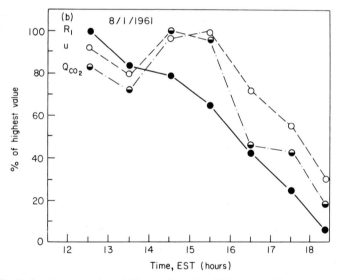

Fig. 10/10b. An hourly comparison of the windspeed at 400 cm, total fixation of CO_2, and incident radiation — all expressed as percentages of the corresponding highest intensity for the afternoon hours. Data from a crop of maize. (From Wright and Lemon, 1966b). Reproduced from *Agronomy Journal* 58 (1966), 266–267 by permission of the American Society of Agronomy deepest into the canopy at noon and became less marked and higher in the canopy in the afternoon. In the evening carbon dioxide levels rose rapidly to values of 300–380 p.p.m. as respiration exceeded photosynthesis and the direction of the CO_2 flux was reversed (Fig. 10/11). The lowest concentrations observed by Wright and Lemon in the corn crop were *ca* 264 p.p.m. There is no reason to suppose that the CO_2 levels around a single isolated plant would depart significantly from that of the surrounding air. The experiments therefore provide prima facie evidence that CO_2 depletion may be a real population effect.

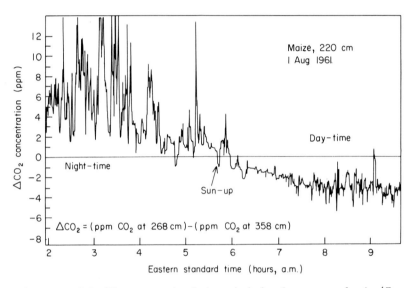

Fig. 10/11. Trace of the CO_2 concentration during a single day above a crop of maize (*Zea mays*) showing the nature of the variations, the effects of night-time stability and daytime CO_2 fixation. (From Wright and Lemon, 1966b). Reproduced from *Agronomy Journal* **58** (1966), 266–267 by permission of the American Society of Agronomy

Effect of CO_2 concentration on assimilation

A carbon dioxide compensation point has been recognized in many species (this is a level of CO_2 below which assimilation does not occur), a level below which the plant cannot reduce the concentration further. The CO_2 compensation point lies at 30–70 p.p.m. for most temperate plants (Black *et al.*, 1969). In marked contrast, a group of tropical and subtropical species (and *Spartina* spp. among temperate genera) can reduce the CO_2 level in the atmosphere to 5 p.p.m. or even lower. These two classes of plants, which have been called C3 and C4 plants, appear to invoke partly different pathways of carbon assimilation which may involve enzymes with different affinities for CO_2. Representative species from the C4 plants are listed in Table 10/II. Of course neither group of plant is likely to reduce the CO_2 content of its surrounding atmosphere to values near the CO_2 compensation point unless the leaves are in enclosed chambers: more important is that by maintaining exceedingly

Table 10/II
Plant species with C4 photosynthetic systems
(from Black *et al.*, 1969, 1971 and others)

Amaranthus albus	*Zea mays*
A. palmeri	*Cynodon dactylon*
A. retroflexus	*Paspalum notatum*
A. edulis	*P. distichum*
Atriplex spongiosa	*Echinochloa crusgalli*
A. rosea	*E. stagnina*
A. semibaccata	*Digitaria sanguinalis*
Portulaca oleracea	*Setaria italica*
P. grandiflora	*S. glauca*
Kochia scoparia	*Chloris gayana*
Salsola kali	
Panicum miliaceum	*Sorghum vulgare*
P. bulbosum	*Sorghum halepense*
Andropogon gayanus	*Eleusine caracana*
Eragrostis chloromelas	*Saccharum officinale*
E. pilosa	*Cyperus rotundus*
E. brownei	*C. bowmanii*
	C. eragrostis

Some genera contain both C3 and C4 species, e.g.

Euphorbia maculata (C4)	*Euphorbia corollata* (C3)
Panicum miliaceum (C4)	*Panicum lindheimevi* (C3)
Cyperus rotundus (C4)	*Cyperus papyrus* (C3)
Atriplex rosea (C4)	*Atriplex hastata* (C3)

Some orders include C-A-M, C3 and C4 plants, e.g. Caryophyllales, Euphorbiales and Asterales (Evans, 1971).

low CO_2 levels within intercellular spaces, C4 plants are able to maintain a steeper diffusion gradient and so speed up the movement of CO_2 to the sites of photosynthesis.

There are other important differences between C4 and C3 plants. The

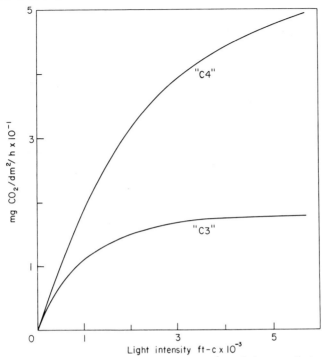

Fig. 10/12. Typical forms for the relationship between the rate of photosynthesis and light intensity for C3 and C4 plants. (From Black, Chen and Brown, 1969). Reproduced with
permission from *Weed Science* 17, 338—344

leaves of C4 plants continue to increase the rate of photosynthesis up to
and beyond full light intensity, whereas that of C3 plants tends to reach
a plateau (Fig. 10/12). C4 plants also have more efficient water usage,
transpiring only about half as much water per unit of dry weight gain as
C3 species. The temperature optima of photosynthesis are also higher in
C4 species and members of this group are almost exclusively tropical;
many inhabit arid or saline areas where an efficient water economy may
be of crucial value.

If neighbouring plants lower the CO_2 level in a canopy it is difficult
to escape the conclusion that density will affect assimilation rates. Moreover, if two members of a population maintain different gradients, one
may gain CO_2 at the expense of another. A neat experiment demonstrating this point was made by Mansfield (1968) who enclosed leaves
of maize (C4) and *Pelargonium* (C3) together in a chamber in bright
light. The CO_2 compensation point of *Pelargonium* alone was 65—89
p.p.m. but in the presence of the maize leaf values of 9—19 p.p.m. were

obtained. "Since the compensation point established for the two leaves together was below that of *Pelargonium* alone, the latter must inevitably have suffered a net loss of CO_2 which was balanced by a net gain in the maize."

A third group of plants exists with a different CO_2 assimilating system — Crassulacean acid metabolism (C-A-M). In this group CO_2 is largely assimilated in darkness and photoconverted to photosynthate in daytime when the stomata may remain closed, protecting the plant against water loss. Such a mechanism could theoretically allow Crassulacean species to draw on CO_2 resources at different times from C3 and C4 species with which they were growing. There is no evidence that such an association has any significance in nature.

Effects of CO_2 supplementation

An experiment was made by Egli *et al*. (1970) to determine the effects of supplementing CO_2 levels within a canopy of soybeans (*Glycine soja*). They enclosed small populations of soyabean plants growing in the field. The enclosure was constructed of transparent mylar film and the ground was covered with plastic to prevent the leakage of CO_2 from the soil. They monitored the levels of CO_2 inside the enclosure by means of an infrared gas analyser and this was connected to a CO_2 controller which fed pure CO_2 into the canopy as soon as the levels fell below a pre-set value. Increasing the CO_2 levels from 300 to 600 p.p.m. gave a linear increase in the net assimilation rate. At 600 p.p.m. the daily net assimilation of three different varieties of soybean was increased by 53, 75 and 84%. There was also a marked reduction in transpiration, presumably as a result of the partial closure of stomata which usually occurs when supra-normal concentrations of CO_2 are applied to leaves.

It seems very likely that both the disadvantages of a lowering of the CO_2 level by the photosynthesis of a plant population and also the advantages of photosynthetically efficient plants are confounded with the water relations of the plant. At lowered CO_2 levels, stomata must remain open wider or for longer if the assimilation rate is not to fall. Extra water loss will then occur. The genus *Atriplex* includes both C4 and C3 species. *A. spongiosa* is a C4 species and *A. hastata* a C3 species. During early growth, *A. spongiosa* maintains higher net assimilation rates than *A. hastata* but *A. spongiosa* devotes a lower fraction of its assimilates to producing new leaves and a much higher fraction to root, stem and fruits. The greater efficiency of the C4 system is exploited not in making much bigger plants, but in making a more effective water-conserving organism with greater reproductive efficiency (Slatyer, 1970).

Water as a consumable resource

The supply of water to an area of land is often the least reliable (in time) of all the resources needed for plant growth. Even in areas of high annual rainfall, rainy days are difficult to predict and within a rainy season there is much day-to-day variation. Where there are marked wet and dry seasons, the date of onset of the rainy season usually has a large element of uncertainty. In arid regions the irregular timing of rainfall is as important to plant growth as the low annual average. Light supplies are not stored. In contrast rainfall is stored in the soil or in lakes or ponds; plants draw water primarily from these stored supplies which act as a buffer that increases the certainty of an uncertain resource. The water regime beneath an arid grassland community in central Australia illustrates this point (Fig. 10/13). Between October 1954 and November

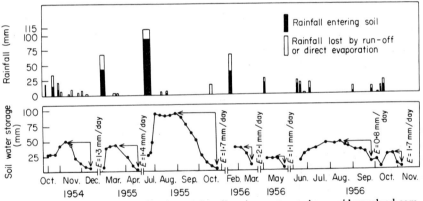

Fig. 10/13. Rainfall and progressive changes in soil moisture storage in an arid grassland community in central Australia from Oct. 1954 to Nov. 1956. (From Winkworth, 1970) Reproduced with permission from *Agric. Meteorol.* 7, 387–399

1956 there were 27 occasions on which rain fell. Light rains did not contribute to the store in the soil because the water was lost almost immediately by evaporation. Some of the water arriving in storms was lost in runoff. The rainfall that entered the soil contributed to a volume of stored water which was then drawn upon through succeeding rainless periods. One heavy rainfall in July 1955 followed a period of extreme water deficit; this charged the soil reservoir with a store of water that was gradually reduced over the next three rainless months to the starting condition of aridity.

A terrestrial plant serves as a wick connecting the water reservoir of the soil with the atmosphere. As a plant develops and expands both its root system and its leaf area, it increases the pathway of water loss from the soil. As the soil dries, both the tension by which water is held in the soil and the resistance to its movement through the soil, increase: the "plant wick" reflects these changes by reduced water content and increasing resistance to water flow; stomata may close, the assimilation rate of the plant declines, growth ceases and the plant may die.

The latent heat of evaporation of the water lost in transpiration is provided directly or indirectly by solar radiation. Thus light intensity is concerned in both photosynthesis and transpiration and the reactions of plants to light and water are entangled. One leaf overtopping another will shade it and so reduce both its rate of assimilation *and* water loss. Direct sunlight, absorbed by a leaf and converted to heat provides part of the latent heat of evaporation, the remainder is drawn from advective energy, carried in air currents from somewhere else that had been heated by the sun. The rate of loss of water from a mass of vegetation is governed, like that from an open lake of water, by the amount of latent heat available. The potential transpiration of an area of vegetation that (a) completely shades the ground, is (b) short and of uniform height, (c) green and (d) never short of water, is determined by meteorological factors (radiation and air movement) rather than by any special biological properties such as the nature of the species, the plant density, or the fertilizer regime (Penman, 1956). The limitations on this rule are interesting in the context of population biology.

(a) *Complete coverage of the ground.* If a significant fraction of incident radiation strikes bare ground, the immediate surface of the soil may become dry and cease to lose water freely. If there is no plant, acting as a wick through the exposed soil, it becomes to some extent self-sealing against water loss. It follows that at very low densities of plant populations, or in early development before complete leaf cover has been achieved, water is conserved in the soil reservoir that would otherwise have been lost. Under these conditions the temperature of the soil surface may rise sufficiently to be lethal to young seedlings.

(b) *Even height.* Uneven canopies of vegetation cause wind turbulence and a consequent increase in the local expenditure of advective energy. This may result in a greater rate of water loss than might be expected from a smooth canopy such as an extensive, even stand of wheat. This advective effect becomes particularly marked at the margins of vegetation in arid regions and in vegetation with a mixed profile such as scrub and savanna.

(c) *Green*. Variations in the greenness of different plant species is relatively unimportant, but a deciduous forest has a markedly different capacity for absorbing radiation in its leafy and leafless stages. Furthermore the development of glaucous or hairy leaves may increase the amount of light reflected.

(d) *Freely supplied with water*. If the supply of water in and from the soil is inadequate to maintain the potential evaporation rate of the canopy, the radiant energy is dissipated as heat. In very bright sunlight the consequent heating of the leaves may be lethal.

The reservoir of soil water available as a resource for plant growth is that which is held tightly enough by the soil's absorptive forces to withstand the pull of gravity but loosely enough to be extracted along the pressure gradients created by the absorptive forces of the root. The volume of this reservoir is of course dependent on soil properties, particularly the size distribution of the pores (there are some differences in the forces exerted by different species which blur the lower boundary of the available soil reservoir but the amount of water that can be extracted at this limit is usually very small and probably not very important). What fraction of the reservoir is available to each individual plant is largely determined by the volume of soil that is intimately explored by its root system. Dense populations tend to suffer water shortage earlier in their life than sparse populations. This may happen because the sparser population is later in covering the ground with leaves and so some water is conserved longer in the soil. Sparser populations suffer less stress from their neighbours by shading, and nutrient depletion. These more vigorous plants have time to develop a more extensive root system than those grown densely and hence tap a larger reservoir of water in the soil (Fig. 10/14).

It is rare in crop monoculture for all individuals in a population to be at the same size and developmental stage at the same time. In natural mixed populations it is even rarer. The variance in size above ground is reflected in a variation in the depth and extent of root systems. When water stress occurs it is therefore likely to be unevenly sensed by a population and some individuals may react before others. Where two or more species are growing together there may be sharper distinctions

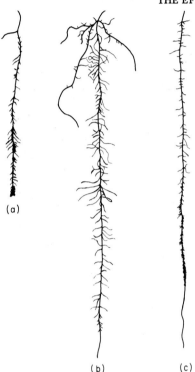

(a)

(b)   (c)

Fig. 10/14. The effects of age and plant density on the development of the root system of
Bromus mollis. (A) Plant grown for 30 days at 1.2 cm spacing; (B) plant grown
for 45 days at 1.2 cm spacing; (C) plant grown for 45 days at 0.6 cm spacing.
(From Neilson, 1964)

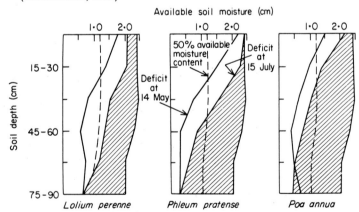

Fig. 10/15. Water deficits at different depths in the soil in an orchard of young apple trees
planted with *Lolium perenne* (perennial ryegrass), *Phleum pratense* (timothy) or
Poa annua (annual meadow grass). (From Milthorpe, 1961 after Goode)

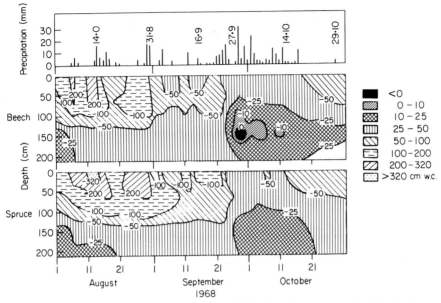

Fig. 10/16. Changes in soil moisture tension beneath beech and spruce woodland during the period August–October 1968. Values for rainfall are shown above. (From Benecke and Mayr, 1971). Reproduced with permission from "Integrated Experimental Ecology" (H. Ellenberg, ed.). Chapman and Hall, London

between types of root system and also the transpiring canopies. The water resources may then be used at different times and/or drawn from different zones in the soil (Fig. 10/15 and 10/16). A precociously developing species may use up water resources and deprive a later grower in a struggle for existence that can lead to a winner and a loser. Alternatively, species may exploit different zones of the soil profile and so avoid a direct zone of conflict.

Each of the groups of supply factors required by higher plants presents its own special problems to the investigator and the nature of these is revealing of the ways in which a resource behaves in nature. In studies of water the empirical difficulties are two-fold. It is very difficult or impossible to reduce the amount of water in the soil reservoir while a population of plants is growing in that soil and it is almost equally difficult to increase the water available to a population except by bringing part or all of the soil to field capacity. When water is applied to the surface of a dry soil it does not distribute evenly — instead the surface layer is brought to field capacity. Further water applied to the system now drains through this layer and extends the depth of the zone that is at field capacity. A quite sharp line may separate the wetted and the underlying dry zones — the "wetting front". Thus an experimenter cannot at all readily maintain a soil at any steady state that is drier than field capacity. This is a problem often forgotten by experimenters who think that by continually watering soil volumes to constant weight they are assuring an evenly distributed water supply; in fact they are usually supplying surface roots with water at field capacity and leaving deeper roots in drought. Physical techniques for controlling water tension by using, for example, types of pressure membranes, cannot control the water tension in volumes of soil large enough to support single small plants, and certainly not populations.

This experimental difficulty emphasizes the nature of water behaviour in nature. Different intensities of rainfall fill the soil reservoir to different depths; the water is not rapidly redistributed and is therefore differentially available to roots placed at different levels in the soil. The strategy of root placement is then particularly important. The occasional rain in an arid zone may only wet the uppermost layers. The development of a wide spreading superficial root system maximizes the chance of tapping this reserve. In contrast, in areas in which the profile is wetted to depth during a rainy season that is regularly followed by a long rain-free period, a more appropriate strategy may be to concentrate root development on a perennial exploration at depth (e.g. *Medicago sativa*), particularly if short-lived and rapidly growing annuals are efficient at exploiting the surface water.

There appears to be very little direct evidence that, during their extension through the soil, roots interfere with each other's direction of growth or, by pre-empting soil volumes to themselves, exclude other roots which develop nearby. There are, however, reports that the root systems of neighbouring plants sometimes appear to remain more discrete than might be expected (Russell, 1966).

Two species with innately different root formations will of course tend to exploit soil volumes differently and two species of *Avena*, *A. fatua* and *A. strigosa*, illustrate such a difference between quite closely related species. When growing in a mixture *A. fatua* exploits deeper layers of soil more thoroughly than does *A. strigosa* which is a species characteristically cultivated on thin soils (Table 10/III).

One experimental test of a hypothesis of mutual root exclusion was made by growing paired plants of oat and of oats and peas in pot culture (Litav and Harper, 1967). One plant in each pot was fed with $^{14}CO_2$

Table 10/III

Estimated percentage of total root weight at each
depth of *Avena fatua* and *A. strigosa* grown in
mixture (M) at two harvest dates

Depth (cm)	H1		H2	
	A. fatua	*A. strigosa*	*A. fatua*	*A. strigosa*
0–10	64.0	36.0	51.2	48.8
10–20	66.6	33.4	57.5	42.5
20–30	75.9	24.1	56.0	44.0
30–40	82.3	17.7	70.3	29.7

From Ellern *et al.*, 1970.

which enabled its root system to be recognized autoradiographically.
The pot of soil was sliced vertically at right angles to the plane joining
the centres of the two plants and random samples of root, 1 cm long,
were removed from the exposed surface. The nearest root to each
random sample was also removed. Pairs of "nearest neighbours" were
then autoradiographed and the frequency of association of labelled and
unlabelled fragments did not depart from that expected on random distribution of the root systems of the two species. If one of the plants in
a pot was treated with urea (applied to the leaves) there was some
evidence of an associative relationship between the roots of the two
species. In no treatment was there the slightest evidence of mutual
exclusion. This rather surprising result needs support from studies with
other species before it can be accepted as at all general.

Much mystery still surrounds the mechanisms by which the direction
of root growth is determined. Geotropisms, chemotropisms and hydro-
tropisms may all play a part and the presence of a neighbouring plant
might be expected to influence local hydration and chemical concentration and so be potentially capable of affecting where a root grows. There
are obvious differences between the innate root patterns of different
species which can be interpreted as adaptively significant. However the
behaviour of whole populations of roots is dreadfully obscure. Until
more detailed evidence is available it seems that root growth in populations is adequately described by the following generalizations.

(i) Root growth is dependent on shoot growth and although the ratios
of root to shoot may be varied, population effects which depress the
rate of shoot growth are reflected in reduced root development. Indeed
defoliation of a shoot leads directly to partial root death and consequent
readjustment of the root:shoot ratio (see Chapter 12).

(ii) Directions of root growth are partly determined by innate characteristics of the branching pattern, but there is a highly localized branching response of individual rootlets to the condition in which they find themselves. This may lead to extensive very local proliferation (Weaver, 1926) of one part of one root.

(iii) On present evidence there is no indication that roots of different plants interfere with each other to the extent of a mutual exclusion from specific soil zones, though innate differences in root form of different species growing together may sometimes give the impression that a proximate process of exclusion has occurred.

Mineral nutrients as resource factors

Plants obtain their mineral resources from the substrate and many of the conditions that govern their availability as resources for plant growth are similar to those affecting water. Both are withdrawn from reservoirs, which have local characteristic forces of supply and retention. Most of the soil minerals are retained in the soil by some physical or chemical linkage with insoluble soil constituents and are in a rapid dynamic equilibrium with ions in the soil solution. When a nutrient is removed by a root there is a local lowering of the concentration and a diffusion gradient is created; nutrients then tend to diffuse along this gradient. Only ions in the soil solution can diffuse and most are fixed on the soil colloids for much of the time, but the equilibrium between the fixed and the mobile fraction ensures that the ionic concentrations in the soil solution are buffered. The nitrate ion is the great exception to this rule, not being held to soil colloids and being wholly mobile in the soil solution. Potassium and nitrate ions exemplify two classes of nutrient behaviour. The potassium ion is strongly buffered and local depletion in the soil by root absorption is regularly made up by the release of bound potassium. The local concentration is maintained by buffering. When the local concentration of nitrate ions is lowered there is no local buffering and the concentration in the soil solution is made up by diffusion. "The zone of depletion for an ion like potassium spreads outwards more slowly than it does for nitrate; though the amount crossing the boundary for a given lowering of the solution concentration is in fact greater" (Nye, 1968; Fig. 10/17).

As a plant withdraws nutrients from the environs of its root surface the supply will fail if the rate of local replenishment is inadequate. The relative rates of withdrawal and replenishment determine the size of

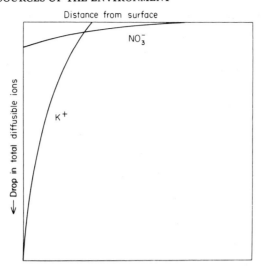

Fig. 10/17. Depletion zones for K^+ and NO_3^-. The lowering of the soil solution concentration is the same, but the buffer power of K^+ is ten times greater. (From Nye, 1968)

" depletion contours" that surround an absorbing rootlet. A root in the soil will begin to interfere with the nutrient resources available to another when their depletion zones overlap. The mobility of phosphate ions is extremely low and steep depletion contours will appear near to an absorbing root surface. During a season of plant growth a phosphate ion is unlikely to move a greater distance than the diameter of a root hair. This extreme localization of depletion zones minimizes the chance that a rootlet of one plant will interfere with the availability of phosphate to another. In contrast, the mobility of nitrates is high, depletion contours are less steep and the proximity of the root to the ions is less important; depletion zones are likely to be wide and so roots are more likely to affect each other's levels of supply. It is perhaps not surprising that density stress in plant populations can sometimes be relieved by adding nitrate fertilizers, but rarely by adding phosphates.

This simplified picture of nutrient behaviour in soil is made slightly more complicated by the existence of processes of mass flow of the soil solution towards the root surfaces, brought about by the transpiration stream. This is most likely to be important in the case of NO_3, SO_4, Ca and Mg (Barber et al., 1962). The argument about the relevance of mass flow to nutrient uptake is complex (Tinker, 1968) particularly because in populations of plants the flow paths of water may be unevenly distributed among the individuals. In a general sense, however, both mass flow and nutrient diffusion are likely to maximize nutrient flow towards the plants that have the greatest growth, because it is these that have tapped the largest volume of soil, tend to transpire most water and thus pull the greatest mass flow of nutrients to their root systems. Transpiration increases with the development of leaf area until there is complete leaf coverage, and it also increases with the potential transpiration. These two processes result in mass flow being maximal after midsummer; by this time the greater part of nutrient uptake has often taken place.

Luxury consumption of nutrients is possible in plants — an excess over immediate needs may be absorbed by the plant in its young stages and subsequently be redistributed in the plant as it grows. This phenomenon is used agriculturally, for example in the cultivation of sisal, where the effective application of nutrients as fertilizers cannot be made to reach the root system with certainty during the dry season. Excess is therefore applied during the early stages of growth of the crop. It is not clear how far such luxury consumption plays a part in the aggressiveness of one species over another in nature but there is evidence in cereals (which exemplify annual strategies) that 90% of the total nitrogen and phosphorus content of the mature plants has been absorbed before the plant has achieved 25% of its final dry weight (Williams, 1955).

A process akin to luxury consumption is, however, undoubtedly of great significance in some plant populations. This is the maintenance of absorbed minerals within the tissues of perennial plants. In environments of low nutrient status and particularly where there is great risk that nutrients will be lost from the soil by leaching there is a benefit to the individual in having long life and progressively accumulating an individual store of nutrients. The deciduous habit in a perennial tree releases some nutrients to the soil and these are now at risk, both to leaching and to uptake by another individual. The evergreen habit and efficient systems for internal circulation permit the individual plant to retain control of a share of nutrients that may be in short supply.

Just as a plant shoot is a population of units, a single root system is a population of repetitive units each with root cap, meristem, absorbing zone etc. These units are not randomly arranged in the soil and the branching pattern (frequency, angle and length) has elements of order though there is apparently much more variation, or noise, in the development of a root pattern than that of a shoot. Patterns of rooting appear to go a long way to ensure that the absorbing zones of different roots on the same plant do not interfere with each other. However, as pointed out earlier, there is little evidence to suggest that roots from different plants affect each other's rooting pattern. If it is true that the roots of an individual plant are "spaced" but those of neighbouring plants are random with respect to each other, we might expect greater interference between root systems than between parts of one system. This is an area of study that is important for understanding the physiology of populations and the manner of interference between individual plants below ground desperately needs further study.

Nutrients are rarely, if ever, evenly distributed with depth in the soil (Fig. 10/18). Thus, as with water resources, there is the possibility of a

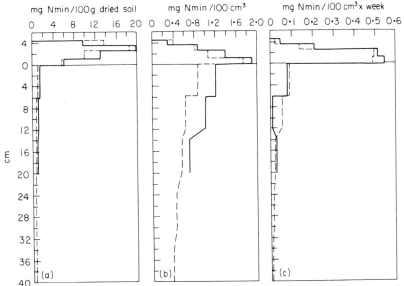

Fig. 10/18. Vertical distribution of field Nmin content and of net mineralization in the soil of a beech stand (a) Yearly average of the field Nmin content, based on dry weight (b) Yearly average of the field Nmin content, based on soil volume. (c) Yearly average of the net mineralization, based on soil volume.————— = 1967; — — — — — = 1968. Re. (b) and (c): The length of the horizontal columns shows the field Nmin content or the mineralization in the different horizons under an area of 100 cm² for a depth of 1 cm. (From Runge, 1971). Reproduced with permission from "Integrated Experimental Ecology" (H. Ellenberg, ed.). Chapman and Hall, London.

division of territory between species in the zones that are exploited — a spatial avoidance of areas of conflicting soil exploitation (see Fig. 10/19). Alternatively members of a single species may adopt a "Jack of all trades" strategy and the role of water absorption and nutrient uptake be divided between surface and deep roots on the same individual (ring-

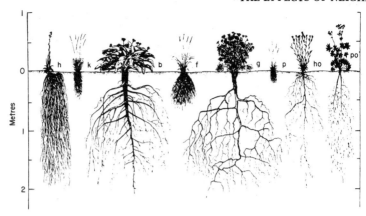

Fig. 10/19. Schematic bisect showing the root and stem relations of important prairie plants, drawn from photographs and data obtained by the excavation and examination of 325 root systems of these 8 species: (h) *Hieracium scouleri*; (k) *Koeleria cristata*; (b) *Balsamorhiza sagittata*; (f) *Festuca ovina ingrata*; (g) *Geranium viscosissimum*; (p) *Poa sandbergii*; (ho) *Hoorebekia racemosa*; (po) *Potentilla blaschkeana.* (From Weaver, 1919)

culture techniques for the cultivation of tomatoes make use of this ability, providing nutrients through a small volume of soil at the surface and water to the deeper parts of the root system).

It is quite as difficult to identify the effects of nutrients as a resource in populations of plants as it is to disentangle the effects of light and water. A major effect of supplying nutrients to vegetation may simply be to speed up the time at which light becomes limiting. Similarly the application of nutrients may make possible a greater growth of roots which are then capable of tapping greater water resources. This interaction may itself bring forward the time at which light becomes limiting. The existence of these interactions makes it difficult to interpret fertilizer experiments. It is perfectly easy to imagine a situation, e.g. a crop in a nutrient-deficient soil, in which isolated plants grow in a stunted fashion: the growth may be so poor that individual plants do not grow large enough to interfere with each other. The root systems are stunted and do not draw on the same supplies of the limited nutrients. The addition of nutrients to such a population can easily be envisaged as causing an increased growth so that the individuals shade each other. At no stage need the plants have reduced the nutrients available to each other yet an apparently density-dependent nutrient response would be obtained.

A clear example of a truly density-dependent plant response to nitrogen comes from an experiment of Lang *et al.* (1956) in which the optimal density for yield of grain was higher with increasing nitrogen application (Fig. 10/20).

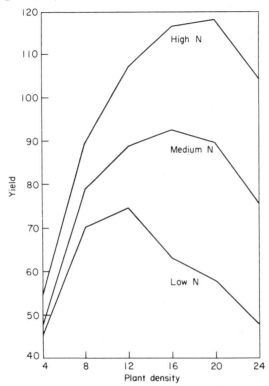

Fig. 10/20. The relationship between density, nitrogen supply and yield in maize (*Zea mays*). (Densities were given in 10^3 plants per acre and yields in bushels per acre). (From Lang et al., 1956). Reproduced from *Agronomy Journal* 48 (1956), 284–289 by permission of the American Society of Agronomy.

The technical problems in the way of understanding nutrient demands by plant populations are greatly complicated by the existence of mycorrhiza:

> Harley considered that in some cases the root was so unimportant as to be an artifact. This is because practically all plants in the natural condition are mycorrhizic and it is through mycorrhiza that nutrient uptake from soil occurs (Gunary, 1968).

Given also that roots are below ground and their study immensely time-consuming, it is not surprising that the greater part of the study of interference between plants has concentrated on leaves and light.

Oxygen as a limiting resource

It is difficult to imagine situations in which the supply of oxygen to the above-ground part of plants ever becomes deficient as a result of the mutual demands made by members of the population. The concentration is usually high (*ca* 20%) and the mobility in the gaseous phase is also very high. This is not necessarily the case in soil. The diffusion rate of oxygen in water is 10 000 times slower than in air and the presence of water films in soil, tortuous diffusion pathways amongst soil particles and water films on root surfaces all hinder oxygen movement. However, the affinity of terminal oxidase systems for oxygen is enormously high and the rate of aerobic respiration in the root is unlikely to be hindered unless the oxygen concentration somewhere in the root approaches zero. Indeed it seems that the metabolic activities of roots are unlikely to be reduced unless the oxygen concentration at the root surface itself reaches very low values. (Greenwood, 1969).

Aquatic plants and also terrestrial species permit oxygen movement through the tissues from the shoots to the roots. In aquatic species, e.g. rice, the flux of oxygen may indeed be outwards from the root and support an aerobic rhizosphere (Greenwood, 1969; Luxmoore *et al.*, 1970a—d). Any interference that roots may exert on the oxygen supplies of each other in the soil can only be important if the transport of oxygen down through the shoots is insufficient to meet root demands. Luxmoore *et al.* (1970) defined the Percentage Plant Aeration (the proportion of root oxygen derived from the shoot) as

$$\frac{\text{Flux into top of root} - \text{Rate of radial loss}}{\text{Total respiration rate}} \times \frac{100}{1}$$

and they have developed models of root aeration which suggest that as the root radius approaches zero, PPA approaches zero but as root length tends to zero, PPA approaches 100%. They argue that the dependence of roots on soil oxygen is only likely to become important in the case of long, thin roots. Even when the root has morphological and anatomical characters that make internal oxygen transport slow, it seems that the supply of oxygen from the soil is rarely likely to be limiting. Soil air usually differs only slightly in oxygen concentration from that of the atmosphere but local zones of oxygen depletion (pockets of anaerobic conditions) may arise in water-saturated soils where the depletion contours of oxygen are steep because of the low diffusibility of oxygen in water. The oxygen concentration in water-saturated aggregates was found to fall from the value of air-saturated water to zero over a distance of 0.1 cm if respiration rates were very high. Even short periods (e.g. 1 day) of anaerobiosis may be seriously damaging to root systems. Greenwood (1969) concludes, however, that such anaerobic zones are unlikely to persist in soil except in waterlogged conditions and then only for the occasional day after heavy storms. It is of course possible that neighbouring roots affect each other's oxygen supply by exaggerating the speed with which these zones in the soil become anaerobic. On present evidence this seems likely to be a rare occurrence because (a) the natural inhabitants of waterlogged soils generally have high values of PPA, (b) on well drained soils, oxygen deficiency is unlikely to occur, and (c) if the soils are very wet the depletion zones of oxygen around a root are likely to be very narrow and the contours steep.

Oxygen is a resource consumed in large quantities by the roots of growing plants — about nine times their volume of oxygen gas each day at 29°C (Lemon and Wiegand, 1962) but of all the resources needed for plant growth, oxygen seems the least likely to be reduced to a limiting level by the growth of neighbouring plants.

Much of this chapter has been concerned to analyse those resources of plant growth that are likely to be involved in interference between neighbours. Of the many consumed resources, light and water are obviously strong candidates, though in subtly different ways. Among the nutrients, nitrate is probably important and among the gases, CO_2. The critical issues isolating these resources from others, which may be in short supply but are not a source of conflict between plants, is whether the depletion zone created by a part of one plant extends far enough to include part of another.

It proves very difficult in practice to discover, when groups of plants are clearly interfering with each other's growth, just what is the mechanism involved. Even a demonstration that a density stress is relieved by the addition of a particular resource may be insufficeint evidence that that resource was itself the source of the density stress. The addition of water to a population of plants may for example increase nitrate availability and relieve a shortage, and there are many similar interactions which can mislead the investigator. Bleasdale (1966) suggested that resource shortages responsible for density stress might be detected by examining the changes in the form of a plant (its allocation of assimilates to various types of organ) and comparing the influence of density on these parameters with the effect of adding various resources. He analysed data of Hozumi and Ueno (1944) from an experiment in which a population of turnips (*Brassica napus*) had been grown at three densities and supplied with all combinations of two levels of water supply, two levels of light intensity and two levels of fertilizer application. Figure 10/21

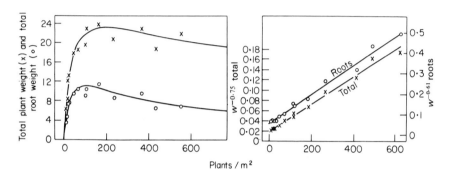

Fig. 10/21. The relationship between plant density and the fresh weights per unit area of tops + tap-root (total weight) and tap-root alone (root weight) for red beet (*Beta vulgaris*). Note that both total and root yield decline at the higher densities. w = weight per plant. (From Bleasdale, 1966)

illustrates the relationship between the fresh weight per plant and the fresh weight of tops (leaves plus petioles). These values have a linear, allometric relationship, with a slope of 1.2. Variations in density produce changes in the allocation of weight to tops which are of the same order as changes in light and water supply. However, although lowering the fertility level reduced total plant weight, it reduced the weight of tops far more than did a similar reduction of plant weight by density. Bleasdale concluded that these plants were not interfering with each other's nutrient supplies.

Many of the experiments aimed at discovering the source of density stress in populations have been made with mixtures of two species and are discussed in Chapter 11. The evidence from experiments on interference between pairs of species has almost always emphasized the interactions between the supply of different resources rather than enabled an investigator to pick out one particular resource as being critical in a struggle for existence. It seems probable that the same sorts of interactions occur within intraspecific effects. The fact that changes in the behaviour of leaves feed back onto the behaviour of root systems and the behaviour of root systems in turn reacts back onto the development of shoots makes interactions between above-ground and below-ground resources almost inevitable.

11

Mechanisms of Interaction between Species

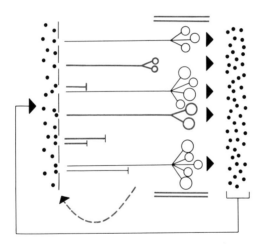

Experimental models, such as the replacement series, diallel designs and additive experiments are effective means for describing the behaviour of species in mixtures, but contribute nothing directly to understanding mechanisms. It would be satisfying to be able to take two species and, by analysis of their behaviour when grown together, to ascribe the success of one over another to a particular morphological feature, a particular pattern of life cycle or a simple physiological trait.

The presence of a plant growing in the region in which a seed falls can

affect in a great variety of ways the chance that the seed will produce a
plant. A still incomplete list includes the following:

  (a) reducing light intensity
  (b) changing light quality
  (c) transpiring limited water
  (d) changing the humidity profile
  (e) absorbing limiting nutrients
  (f) providing limited nitrogen
  (g) sheltering or excluding predators (or sheltering the predators
       of predators)
  (h) favouring or reducing pathogenic activity
  (i) encouraging defecation or urination in the neighbourhood
  (j) providing rubbing posts or "play" objects and so encouraging
       local trampling
  (k) raising the soil level (accumulation of organic matter)
  (l) liberating selective toxins
  (m) changing soil reaction

The analysis of which particular factors act in any one effect of
neighbour on neighbour is bound to be extremely difficult and has rarely
been achieved in experimental populations; it is infinitely harder in
nature.

In practice, the establishment of causation is usually barred to the
scientist, who has to be content with correlation. The problem that then
arises is what sorts of correlation are acceptable in lieu of a causative
proof. In pathology, another science in which interactions between
organisms are involved, a similar question has been: "how do we determine that a particular organism is the causal agent of a particular
disease?" The famous series of "postulates" put forward by Koch (1890;
Stapp, 1961) required the following tests:

  (1) The infective agent must be cultivated from the plant, and its
purity tested.

  (2) After inoculation with the pure culture of the isolated micro-
organism, a plant previously in perfect condition must show the same
pathogenic symptoms as the diseased plant.

  (3) The bacterial agent must be re-isolated from the experimentally
infected plant.

  (4) The re-isolated micro-organism and that originally inoculated
must be tested for identity.

Such rigorous proof has rarely been applied to any ecological interaction; it may indeed be too much to hope for such rigour in field studies but it is nevertheless vital to be aware of the limited value of correlation as a proof of causation. Nowhere is this more important than in establishing the nature of ecological interaction between species.

Any comparison between two species must involve differences, some slight, some large, that may affect their success in a struggle for existence. The problem then becomes that of deciding which among the many differences will be chosen for study. It is usually possible, by an appropriately designed experiment, to show relevance for any chosen biological difference. If relevance cannot be found it is arguable that the experimenter chose the wrong experiment! There are no open minds in science and each experimenter has predispositions and limited expertise. If experiments are designed to discover whether plants compete for nutrients, it is highly likely that nutrients will turn out to be important and the same is true of light, water, toxins, predators, pathogens, etc. Many experimenters have been concerned with the role of light in plant interactions: experiments are then designed with optimal supplies of water and nutrients and, not surprisingly, light proves to have a vital role. There is room for much scepticism in the interpretation of competition experiments, particularly when they are applied to the field. This chapter is written with the deliberate intent of introducing such an element of scepticism!

Some of the methodological problems involved in unravelling an interaction between species are illustrated in the seemingly simple task of deciding whether such interaction takes place above or below ground.

Interactions above and below ground

An experiment was designed by Donald (1958) in which the root systems of two species could exploit the same soil volume while their canopies were prevented from intermingling and shading each other *or* the canopies could intermingle freely but the root systems were prevented from meeting. *Phalaris tuberosa* and *Lolium perenne* were chosen for the study and grown in pots with partitions between them above and/or below ground (Fig. 11/1a). Two levels of nitrogen supply were included in the experiment which had sufficient replicates for two destructive harvests to be taken. The two species represented an aggressor (*Lolium*

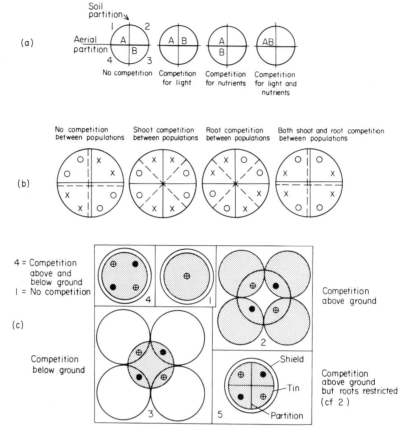

Fig. 11/1. Experimental designs used by various authors to separate root and shoot 'competi-
    tion'.
    (a) Design used by Donald (1958). A and B are the positions of two competing plants
        in relation to an aerial and a soil partition.
    (b) Design used by Snaydon (1971). ○ and x are the positions of two species — the con-
        tinuous line is an aerial partition and the dashed line is a partition in the soil.
    (c) Design used by Aspinall (1960). ⊕ = one subordinate plant, ● = eight dominant
        plants. Large white circle = radiation shield; stippled circle = pot with sand.

perenne) and a suppressed species (*Phalaris tuberosa*) and their behaviour
in the variously partitioned pots is summarized in Table 11/I.

The arrangement of plants in this experiment was unusual because of
the constraints imposed by the partitions. Complete mixing of plants of
the two species was not possible and so a test plant of one species was
grown with a population of the others. For this reason the values in

Table 11/I

Yield of species in mixture	Shoots and roots separated	Roots separated	Shoots separated	Shoots and roots intermingled
Lolium	4.71	4.19	4.31	4.72
Phalaris	4.67	3.19	1.17	0.32

Table 11/I for the two species cannot be read to give yields per pot.

Phalaris was depressed 32% when its shoots were intermingled with *Lolium*, 75% when its roots were intermingled and 93% when both roots and shoots intermingled. Clearly the effects above ground and below were not additive: there was interaction.

Donald summarized the nature of the interactions (Table 11/II), and

Table 11/II

The effects of competition by an aggressor species (A) on a suppressed species (B)

Effects	Competition for Light only	Competition for Nutrients only	Competition for both Light and Nutrients
Direct	(*a*) Intrusion of A into light environment of B	(*c*) Intrusion of A into nutrient supply of B	B suffers: (*a*) *Reduced light supply*
	Reduced light supply for B	*Reduced nutrient supply for B*	(*c*) *Reduced nutrient supply*
Indirect	(*b*) As a result of reduced light supply	(*d*) As a result of reduced nutrient supply	(*b*) *Reduced capacity to exploit the nutrient supply*
	B has reduced capacity to exploit its own nutrient supply	*B has reduced capacity to exploit its own light supply*	(*d*) *Reduced capacity to exploit the light supply*
Interactions	Interaction of (*a*) and (*b*)	Interaction of (*c*) and (*d*)	Interactions *ab, ac, ad, bc, bd,* and *cd* plus any higher order interactions

From Donald, 1958.

much earlier Clements *et al.* (1929) had described the similar interaction
that they envisaged in "competition" for light and water:

> The beginning of competition is due to reaction when the plants are so spaced
> that the reaction of one affects the response of the other by limiting it, the
> initial advantage thus gained is increased by accumulation, since even a slight
> increase in the amount of energy or raw material is followed by corresponding
> growth, and this by a further gain in response and reaction. A larger, deeper or
> more active root system enables one plant to secure a larger amount of the
> chresard [water available for growth] and the immediate reaction is to reduce the
> amount obtained by the other. The stem and leaves of the former grow in size
> and number and thus require more water; the roots respond by augmenting the
> absorbing surface to supply the demand and automatically reduce the water con-
> tent still further. At the same time the correlated growth of stems and leaves is
> producing a reaction on light by absorption, leaving less energy available for the
> leaves of the competitor beneath it, while increasing the amount of food for the
> further growth of absorbing roots, taller stems and overshadowing leaves.

The effect of a single set-back to one component of a mixture of species
has repercussions backwards and forwards between the members of the
population which potentially affect most of the vital activities of both
species.

The use of partitions in an experiment introduces serious complica-
tions. A partition above ground will intercept light. A partition below
ground that prevents two root systems from intermingling also divides
the soil resources and so may deprive a vigorous plant of some of its
potential rooting space, irrespective of the presence or absence of the
second species.

A subtle modification of Donald's experiment was made by Aspinall
(1960) who attempted to separate above and below ground effects in
the interaction between *Polygonum persicaria* and barley. In this experiment a plant of *Polygonum* was grown with eight plants of barley or a
plant of barley with eight plants of *Polygonum*. As in Donald's experiment, the treatments included opportunities for plants to interfere with
each other through the root, shoot, root + shoot or all forms of interference were denied. The volume of soil available to each species was
kept constant and all plants were grown with shades though they were
in different positions (see Fig. 11/1c). The results of both Donald's and
Aspinall's experiments are summarized in Fig. 11/2.

Generally the effect of barley on *Polygonum* was greater than that
of *Polygonum* on barley. *Polygonum* suffered from the presence of
barley below ground before significant effects were apparent above

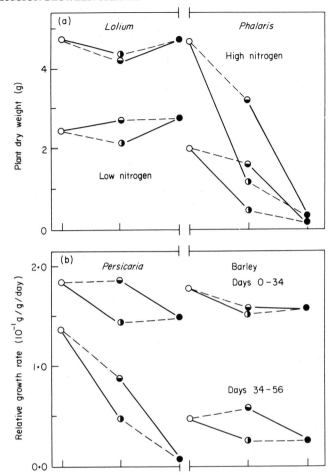

Fig. 11/2. (a) Interaction diagrams showing dry weight of *Lolium* and *Phalaris* after growth
      for 72 days in soils of initially low and high levels of nitrogen.
   (b) Similar diagrams showing the relative growth rates of *Polygonum* and barley
      over days 0–34 and 34–56 from emergence.
   ● roots and shoots intermingled   ◕ roots separated
   ◑ shoots separated   ○ roots and shoots separated.
   (a) From Donald, 1958.
   (b) From Aspinall, 1960.

ground. In the second half of the growth period there were clear effects
both above and below ground but the below-ground effects were still
the stronger. *Polygonum* was capable of making significant growth after
barley ripened and some of its success as a weed is probably due to this
ability to use a period when resources are unavailable to barley — a

temporal escape of *Polygonum* from the effects of its early-maturing neighbour. It remains an open question how far the emphasis on root effects that emerges from both of these experiments and those of Snaydon (1971), reflects simply the fertility level at which the experiments were made. Probably, if the experiments had been done at yet higher levels of nutrient supply, interference between the shoots would have appeared stronger than that between roots.

The nature of interactions

A plant may influence its neighbours by changing their environment. The changes may be by addition or subtraction and there is much controversy about which is more important. There may also be indirect effects, not acting through resources or toxins but affecting conditions such as temperature or wind velocity, encouraging or discouraging animals and so affecting predation, trampling, etc. It is often extremely difficult to separate these effects in the field though it is not so hard to imitate them in the laboratory. In practice the laboratory provides the means for demonstrating the possibilities of different sorts of effects making it more likely that they will then be recognized in the field.

Deprivation of Resources

It is reasonable to expect that neighbouring plants may shade one another and there may result a mutual or a one-sided depression of growth due to deprivation of light. Similarly the roots of neighbouring plants may draw on limited resources of nutrients and also, by drawing on restricted water supplies, may desiccate each other's root systems. The simplest way to identify a mechanism of interaction is to grow together two forms of a single species that differ in some well defined character. If it can be assumed that the two forms do not differ in other respects the interaction can be defined with some precision. The experiment of Black (1958) in which he grew together populations of *Trifolium subterraneum* with large and small seeds (Chapter 6 and Fig. 6/26) enabled him to isolate mutual shading as the interaction predominantly involved. Plants from large seeds established a canopy quickly and in mixed populations quickly excluded most of the incident light from the neighbours derived from small seeds. In that experiment Black measured light intensities below the canopy. It is conceivable that if he had measured potassium levels (or some other parameter of soil resources) he might have found equally strong evidence that plants from large seeds monopolized the soil volume and its resources.

A somewhat similar experiment by Black (1960b) compared the growth of three varieties of *Trifolium subterraneum* that differed in petiole length. The varieties were 'Tallarook' (short), 'Bacchus Marsh' (intermediate) and 'Yarloop' (long petioled). In each comparison of a pair of varieties in mixture the longer-petioled form was successful in dominating the canopy and the productivity of the mixture (Figs 11/3, 11/4). There may, of course, have been equally important differences between the varieties below ground. The experiment is, however, per-fectly clear in demonstrating that long petioles may bring an advantage to a variety in a mixture. There would seem to be no advantage in the possession of long petioles except in a struggle for existence with shorter-petioled forms. This is a splendid example of the way selective forces

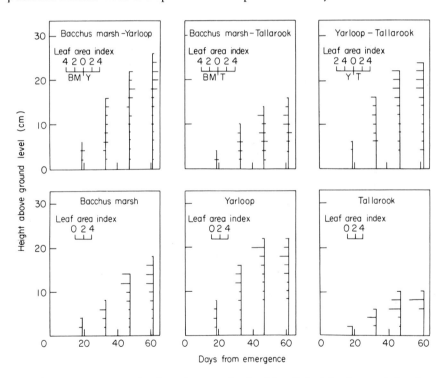

Fig. 11/3. The distribution of leaf area (expressed as Leaf Area Index) for three varieties of *Trifolium subterraneum* (subterranean clover) in mixtures and in pure stands. (From Black, 1960)

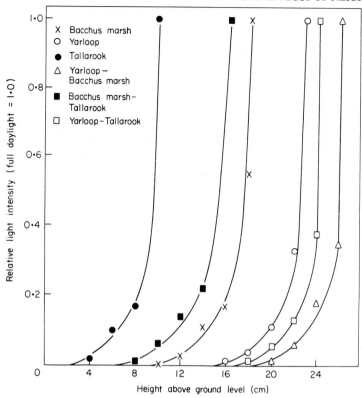

Fig. 11/4. The profiles of light intensity within swards of *Trifolium subterraneum* (subterranean clover) of three varieties in pure stands and mixtures (cf. Fig. 11/3). (From Black, 1960)

may operate within a species to favour an intrinsically less efficient growth form (see Chapter 24).

When species, rather than forms of a species, grow together it is much less easy to identify the differences between them that are relevant in determining success or failure. The more properties that are held in common by the two species the more likely is the chance of picking out the relevant differences. Comparisons between closely related species are therefore particularly helpful:

> As species of the same genus have usually, though by no means invariably, some similarity in habits and constitution, and always in structure, the struggle will generally be more severe* between species of the same genus, when they come into competition with each other, than between species of distinct genera.
>
> (Darwin, 1859)

*One might add "and protracted" to emphasize that an intense struggle between well balanced forces is not likely to be resolved quickly.

In one such interspecific comparison, populations of *Trifolium repens* and *T. fragiferum* were grown in mixtures. These two species have very similar vegetative form and are sometimes found cohabiting in old neutral grasslands (Harper and Clatworthy, 1963). The species differ in seed size, *T. fragiferum* having the larger seeds and the larger cotyledons after germination. *T. repens* is, however, the first to germinate and hence the first to start photosynthesis. After 9 weeks of growth the canopy of *T. repens* overtopped that of *T. fragiferum* but the latter was capable of developing longer petioles and by the 15th and 21st weeks *T. fragiferum* predominated in the upper canopy (Fig. 11/5). The experiment was ended after 21 weeks and the long-term outcome cannot be predicted; probably much would depend on the cutting or grazing regime which would be likely to favour *T. fragiferum* if the sward remained tall or *T. repens* if it was frequently defoliated. Clearly at least three major differences between these closely related species (seed size, germination time, petiole length) all played a part in a complex interaction that determines the relative efficiencies of the two species in the capture of light.

A similar experiment was made with three species of clover by Williams (1963) who grew *Trifolium subterraneum, T. hirtum* and *T. incarnatum* in pure stands and mixtures. All three species are winter annuals. At the time of seedling emergence *T. hirtum* and *T. incarnatum* exposed larger areas of cotyledon than *T. subterraneum* and exposed larger and higher unifoliate and first trifoliate leaves. In mixtures of *T. subterraneum* and *T. incarnatum* the former maintained the higher rate of dry matter production although the latter species exposed the greater area of leaf. Apparently the critical factor was that a few leaves of *T. subterraneum* overtopped and thus shaded the abundant leaf area of *T. incarnatum*. As the mixed populations of these species developed, a pattern of changing relationships appeared in which subterranean clover "which appeared to have the least photosynthetic capability in the seedling stage, was able to become dominant as it attained full canopy development". The results of this experiment were formally analysed to show changes in Relative Growth Rate, and Net Assimilation Rate (Table 11/III).

Density stress may produce changes in plant form (Chapter 7) and such changes appear in mixtures of species as well as in pure stands. In Williams' experiments with *Trifolium* spp. one of the effects of vigorous associates on the weakly competitive *T. hirtum* was a rapid increase in

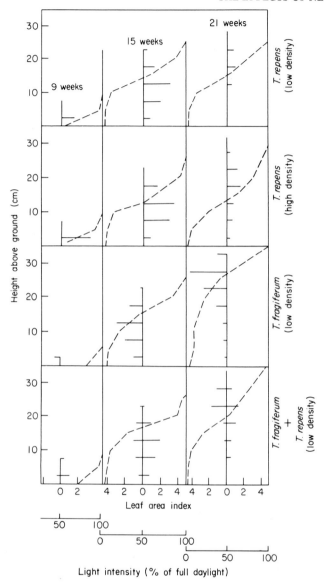

Fig. 11/5. The distribution of foliage and light intensity within stands of *Trifolium repens,*
T. fragiferum and a mixture of equal proportions of the two species at a combined
density equal to that of the pure stands. Horizontal unbroken lines indicate the area
of leaf borne within 5 cm intervals above ground level. The area of leaf of *T. repens*
is plotted to the right of the vertical and *T. fragiferum* to the left. The broken line
in each section of the figure represents the light intensity as a percentage of full daylight. (From Harper and Clatworthy, 1963)

Table 11/III

Relative growth rate and net assimilation rate on a leaf-area and a
leaf-weight basis in simple and mixed communities of clover

Community component	Days	Relative growth rate (mg/g day)[a]			Net assimilation rate					
					Leaf-area basis (mg/dm² day)[b]			Leaf-weight basis (mg/g day)[c]		
		0–58	58–79	79–99	0–58	58–79	79–99	0–58	58–79	79–99
Subterranean clover alone		43	48	28	40	30	19	91	98	71
Rose clover alone		44	39	27	49	32	20	92	72	63
Crimson clover alone		44	46	29	39	28	18	88	90	66
Subterranean (with rose)		41	53	35	40	36	25	88	111	91
Rose (with subterranean)		42	44	10	51	41	8	88	87	23
Crimson (with rose)		45	48	39	42	32	27	90	94	92
Rose (with crimson)		43	40	12	49	35	9	88	79	29
Crimson (with subterranean)		45	47	24	42	30	15	91	91	56
Subterranean (with crimson)		42	52	34	44	35	25	89	108	88

From Williams, 1963.

[a] Expressed as mg shoot-weight increase per g total shoot weight per day.
[b] Expressed as mg shoot-weight increase per dm² leaf area per day.
[c] Expressed as mg shoot-weight increase per g leaf weight per day.

the ratio of leaf area to leaf weight — a well known reaction to lowered light intensity.

The unequal capture of light by grasses and clovers in mixture seems to account for part of the differential between these species in mixtures. An experiment of Stern and Donald (1962) followed the development of the canopy over time under four regimes of nitrogen application (Fig. 11/6). The effect of the nitrogen levels was to change the balance

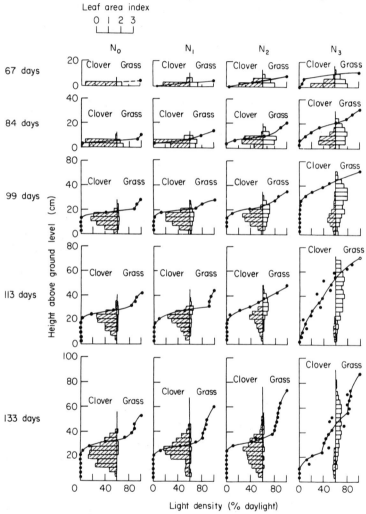

Fig. 11/6. Vertical distribution of the leaf area index of grass and clover under four nitrogen treatments together with the profile of light density relative to daylight. (From Stern and Donald, 1962)

of species profoundly so that after 133 days the sward at low nitrogen level was wholly dominated by clover but at high nitrogen the grass was dominant. Nitrogen supply affects the clover and grass differentially and although the application of nitrogen increases the early leaf production by clover in the mixtures it markedly increases the height of the grass canopy. The grass (adequately supplied with nitrogen) overtops the clover and the advantage is progressive, leading to the almost total suppression of the clover. At first sight such an experimental result might have been interpreted as purely a problem in nitrogen nutrition. With no applied nitrogen the nodule-bearing and nitrogen-fixing legume was at an advantage — it evaded a struggle for existence for limiting nitrogen supplies. However, given adequate nitrogen the grass became the winner. Yet it is clearly unreal to separate the partitioning of nitrogen resources from the partitioning of incident radiation. The experiment starts as a single factor experiment but quickly turns itself into a study in interactions between factors.

Jelinowska (1967) studied the behaviour of *Medicago sativa* sown together with barley. Agricultural practice sometimes permits the barley crop to be harvested and *Medicago* continues to grow subsequently. Unfortunately the influence of barley is commonly to prevent effective establishment of *Medicago*. The hypothesis that barley affected *Medicago* by shading was tested by creating an artificial barley canopy from green celluloid strips. The growth of plants of *Medicago* under the celluloid canopy was depressed by about 20% compared with their performance in full light. However, a barley crop depressed *Medicago* by 70% — moreover the yield depression of *Medicago* was apparent very early in the growth cycle before shading by barley seemed likely to have been effective. Subsequent experiments showed that the performance of *Medicago* in the first and second years became independent of whether barley had been sown or not, provided that high doses of potassium fertilizers were applied. The dependence of *Medicago* for survival on abundant potassium was even more marked if the barley had received fertilizer nitrogen. Again an enquiry that started to examine a simple "competition" for a single resource led inescapably to the study of an interaction.

Many of the very simplest field experiments on the role of light in a struggle for existence between plants seem never to have been attempted. It would be of immense interest to use standard plant units (phytometers), either pots freely supplied with water and nutrients or small water culture vessels with a test plant in each, and to place these units in transects across vegetation through open land, grassland, scrub and woodland. The growth made in a unit time might not be due solely to the radiation received but at least the soil variables would have been excluded. The usually preferred alternative is to place physical measuring devices for light intensity and quality in the habitats to be compared (see Chapter 10). The measures then obtained leave little doubt that light intensity is dangerously reduced for most species by an overtopping canopy.

The demand for water by a pair of species provides an opportunity for a resource in short supply to be shared unevenly. However, the curious nature of water as a resource makes it yet more likely that when it is in short supply interactions will be forced with other resource factors. The effects of water shortage are most apparent in a lowered rate of assimilation so that differences in mutual shading are likely to be less important than when water is abundant. The uptake of some nutrients, particularly nitrogen as nitrate ions, may depend on water flow through the soil to the roots. A shortage of water will then manifest itself in a reduced nitrogen supply. Plants grown at high density or in shade tend to adjust their root/shoot ratio in favour of shoot. Thus an experience of a neighbour's shade may result in a feebler or shallower root system and a lowered soil volume that can be tapped in periods of water shortage. Milthorpe (1961) generalizes to one general principle: "the greater the amount of leaf growth made before plants come into contact with each other, then the more extensive is the root system and the less likely is the plant to suffer from drought."

The fact that different species may root at different levels in the soil may mean that they avoid tapping the same resources of water. This will be particularly relevant as the available water supply nears exhaustion because the movement of water through the soil is then greatly hindered and the plant depends on zones of soil intimately associated with the root hairs. Evidence that root extraction zones really are different comes from comparisons of available soil moisture in apple orchards sown with different species of grass (Fig. 10/15). *Poa annua* reduced the soil moisture level very much less in the deeper parts of the profile than *Lolium perenne*. The apples grew more slowly when the orchards were sown with *Lolium* than with *Poa*, but of course the uptake of minerals by the apple trees might also have been affected by the grasses and it would, as in most cases, be foolish to ascribe the apple—grass interaction to "competition for water".

Where water supplies are short and unreliable the natural density of
plants may be so low that mutual demands between neighbours may
appear unlikely. "When we reach arctic regions or snow capped mountains, or absolute deserts, the struggle for life is almost exclusively with
the elements" (Darwin, 1859). This generalization has seldom been put
to the test. Friedman (1971) studied the fate of seedlings of *Artemisia
herba-alba* at various distances from adult plants of *Zygophyllum
dumosum* in the Negev Desert of Israel. The adult shrubs were at least
10 m apart and all other perennials present within a radius of 8 m were
removed. Some seedlings (29) arose naturally at various distances from
Zygophyllum plants and in addition 4–5-week-old seedlings of *Artemisia*
were deliberately planted at distances of 50, 100 and 200 cm from
Zygophyllum plants. The transplanted seedlings received a watering at
the time of planting and after 3 and 6 days. The growth of survivors
was measured as the total length of branches (Fig. 11/7) and the percentage mortality is shown in Table 11/IV. Control plants were followed
which established naturally or were introduced at a distance of at least

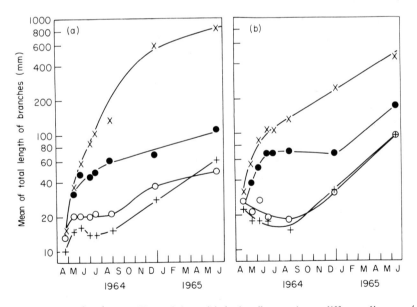

Fig. 11/7. Length (mm) of seedlings of *Artemisia herba-alba* growing at different distances (cm)
from adult *Zygophyllum dumosum* shrubs. (a) Transplanted: x, control; •, 200 cm;
o, 100 cm; +, 50 cm. (b) Naturally growing: x, control; •, 136–170 cm; o, 90–106 cm;
+, 47–65 cm. (From Friedman, 1971)

Table 11/IV
Percentage mortality of seedlings of *Artemisia herba-alba*
transplanted at different distances from adult *Zygophyllum*
dumosum shrubs. (From Friedman, 1971)

Distance of seedlings from *Z. dumosum* (cm)	Dates of measurements							
			1964				1965	
	24 Apr	16 May	30 June	16 Nov	28 Dec	17 Jan	29 May	4 Nov
50	45.0	47.0	47.6	47.6	47.6	47.6	47.6	55.4
100	27.1	37.0	37.0	41.0	41.0	41.0	41.0	53.0
200	17.7	28.0	28.3	33.0	33.0	33.0	33.0	40.7
Control	8.5	8.5	8.5	8.5	8.5	8.5	8.5	17.0

10 m from any living perennial. Clearly proximity to a plant of *Zygophyllum* reduced the survivorship and growth rate of *Artemisia*. The effect on the growth rate extended to severe suppression even within the zone 100–200 cm from a shrub — the zone of extension of the roots of *Zygophyllum*. The experiments tempt one to discuss limitations of water but conceivably nutrition depletion zones may also be involved and the possibility of toxic substances liberated could also be explored. Without doubt, in this extremely arid environment a density effect on growth rate and mortality is clear.

The interactions between neighbouring plants are so complex, tight and apparently impossible to disentangle that it may be wise to cease to look for simple effects. De Wit (1960) avoided the questions very deliberately and went so far as to say of a search for limiting resource factors in plant interactions that to subdivide the complex into particular components is ". . . not necessary, always inaccurate and therefore inadvisable." Instead, he wrote about "competition for space", a composite of all growth resources and the interactions between them. Hall (1974a) joined issue with de Wit by arguing that ". . . assuming there is one limiting factor, regardless of the degree to which this affects all other aspects of growth, identification of this factor may enable the situation to be rectified completely; in this case, the question of whether or not interactions might occur might even become irrelevant."

Hall examined the nitrogen relations of *Chloris gayana* (Rhodes grass) and *Stylosanthes humilis* (Townsville stylo) a legume. He distinguished two types of interference between the species: "competitive interference" (whereby one species directly affects the growth of the other by competing for a resource or resources potentially available equally to both) *and* "non-competitive interference" (in this case symbiotically fixed nitrogen which at least in the early stages of the experiment would be available to the legume but not to the grass). In the experiments (Vallis *et al.*, 1967) small amounts of ^{15}N, a stable isotope, had been applied to the soil so that the nitrogen available to both species in the soil could be distinguished from the rhizobially fixed nitrogen which was unlabelled. The experiment with *Chloris* and *Stylosanthes* was designed as a Replacement Series (Chapter 8) and the relative yields and the Relative Yield Totals (RYT) were determined after 13 weeks of growth. *Stylosanthes* was severely depressed by *Chloris* and the yields of *Chloris* in mixture were considerably higher than the yields in monoculture. The plants were assayed for nitrogen which could be separated, by means of the isotopic marking, into that derived from the soil and that derived from rhizobial activity. The results are shown in Fig. 11/8 as replacement diagrams. Clearly the RYT for dry matter is well above 1.0 as also is the RYT for total nitrogen. However RYT for soil nitrogen has a value of *ca* 1.0. This result would seem to indicate quite clearly that the two species are making demands on the same limiting pool of soil nitrogen and that rhizobially fixed nitrogen is an addition to this pool that is not equally available to both species. In Hall's terms the soil nitrogen is a source of competitive interference between the species and the rhizobial nitrogen a source of non-competitive interference.

In a second experiment *Setaria anceps* (Nandi setaria) and *Desmodium intortum* (greenleaf desmodium) were sown as a replacement series in pots and after 21 days half of the pots received potassium fertilizer as $KHCO_3$ (112 kg ha^{-1}). The plants were cut to give a first harvest after 63 days, the potassium treatment was repeated and a further harvest was taken 31 days later. The harvested material was analysed for nitrogen, phosphorus, potassium, calcium and magnesium. The results are illustrated in Fig. 11/9a, b. The presence of *Setaria* severely reduced the growth of *Desmodium*, which showed marked foliar symptoms of potassium deficiency. However, the application of potassium strongly stimulated the growth of *Desmodium* so that its growth in mixtures approached the growth in pure cultures. In the absence of applied potassium (Fig. 11/9a) the RYT value for potassium taken up by the mixtures was *ca* 1.0, i.e. the species were apparently drawing on

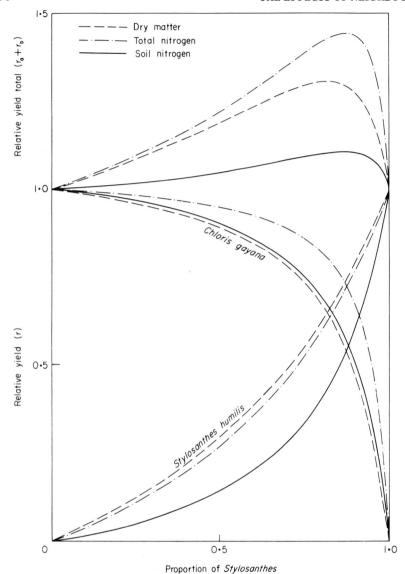

Fig. 11/8. The relative yield total and the relative yields of *Chloris gayana* and *Stylosanthes humilis* grown in a replacement series. (From Hall, 1974a)

a common source. This was also true for phosphorus and total dry matter. Only for nitrogen was the RYT markedly in excess of 1.0. In the presence of potassium the situation was completely different (Fig. 11/9b) and RYT values for total yield, phosphorus, potassium and nitrogen all

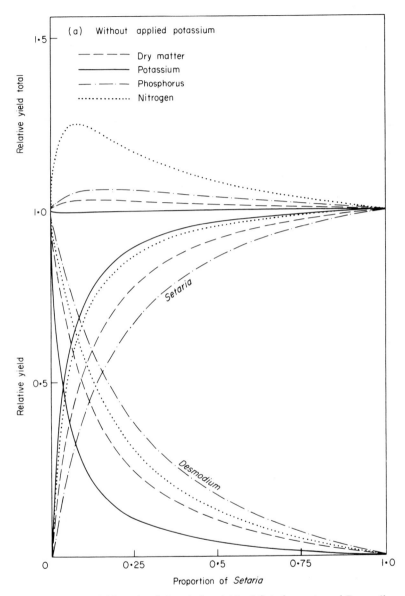

Fig. 11/9a. The relative yield total and the relative yields of *Setaria anceps* and *Desmodium intortum* grown in a replacement series without added potassium. (From Hall, 1974b)

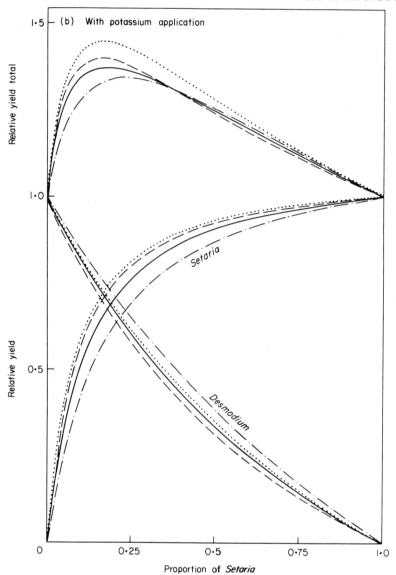

Fig. 11/9b. As Fig. 11/9a but with added potassium.

greatly exceeded 1.0. Hall's interpretation is that there was competitive interference between *Setaria* and *Desmodium* with respect to potassium. Only when potassium was applied did *Setaria* cease to have its marked depressing action on *Desmodium*. The fact that RYT for both potassium and for dry matter was about 1.0 implies that both species were limited by a common pool of limiting resource. RYT exceeded 1.0 when potassium was applied indicating that there was then no longer a common limit to the growth of both species. The nitrogen supply to *Desmodium* via *Rhizobium* was not immediately available to *Setaria* and represented a resource limit not held in common by the two species.

This approach has considerable potential value and differs sharply from most of the studies of interference between plants in three critical respects: (1) the object of competitive interaction was tentatively identified by visual symptoms — the characteristics of particular deficiencies are of course often recognizable; (2) observations on dry matter yield were coupled with tissue analysis and (3) the use of the Replacement Series made rigid comparisons of species behaviour possible between mixtures and pure stands.

There is a very extensive literature in which it is demonstrated repeatedly that the balance between a pair of species in mixture is changed by the addition of a particular nutrient, alteration of the pH, change in the level of a water table, application of water stress or of shading. These experiments had significant historical importance in emphasizing that the interaction between a pair of species was a function of the environment in which the interaction occurred and an anecdotal value in defining, for a specialized condition of environment and species, the effects of a particular change. It is very doubtful whether such experiments have contributed significantly either to understanding the mechanism of "competition" or to generalizing about its effects.

The role of toxic substances

Some of the depressive effects of a plant upon its neighbours are so striking that an interpretation based on the monopolization of resources has often seemed inadequate. An alternative is obviously that some plants may release into their environment toxic materials that harm or even kill neighbours. This hypothesis might seem easy to test but in fact there are very great difficulties in demonstrating that toxicity effects are other than laboratory artefacts.

Almost all species can, by appropriate digestion, extraction and concentration, be persuaded to yield a product that is toxic to one species or another. The extracted cell sap of most plants will plasmolyse and kill tissues of the same species. It is no proof of a toxic interaction between plants to demonstrate that toxins can be isolated from the aggressor. The history of research into toxic interactions has included many *causes célèbres*: startling claims that have subsequently been abandoned in favour of some other interpretation (Pickering and Bedford, 1919). The literature has recently been surveyed by Rice (1974) who is convinced of the reality of toxic interactions in nature. The nature of the evidence differs from one situation to another and is best described as a set of examples.

(a) *Helianthus annuus*. This plant is an important weed in early stages of old field successions and aqueous extracts of its tissues inhibit the growth *in vitro* of some of the other species found in the same situation. In the field, definite zones of reduced growth of the associated species could be detected, and the dry weights of plants of *Erigeron canadensis* and *Rudbeckia hirta* were significantly lower near to *Helianthus* plants. Wilson and Rice (1968) tested the soil 0.25 m and 1.0 m from the *Helianthus* plants so as to detect any difference in fertility. They found no significant differences. Strangely they compared the soils for pH, organic carbon, total phosphorus and total nitrogen. This seems a very odd choice: the tests would more appropriately have been for available phosphate, potassium, nitrate and ammonium nitrogen, the minerals most usually found in short supply after a period of plant growth. Their conclusion that the reduced mineral content of the soil was not responsible for the *Helianthus* effects is therefore not justified. Plants of various species were then grown in soil from beneath sunflower plants and 1 m away and the soil from beneath the sunflowers supported smaller plants than the control soil. The conclusion that "the soil experiments indicated that substances do get into soil from sunflower plants, and have a differential effect on the growth of associated species" is not justified from this evidence.

The next comparison was between the growth of a series of test species in 10 cm pots of soil to which had been added either 1 g of air-dried sunflower leaf powder or 1 g of dried peat moss. The growth of seedlings of *Aristida oligantha* was stimulated and all other species tested showed reduced growth on sunflower-augmented soil compared with the growth on peat-augmented soil. This experiment is susceptible to different interpretations. Addition of readily decomposable organic matter to soil normally results in a sudden microbial demand for nitrogen — nitrogen deficiency is a common immediate consequence of ploughing straw (or even adding glucose) to soil. Peat is highly resistant to microbial decomposition and does not have the same effect. It would have been interesting to have included in this experiment additions to the soil of leaf powder from a series of other species not suspected of containing a toxin.

A different approach was to use an experimental design due to Martin and Rademacher (1960) in which leachates were led from pots of *Helianthus* growing in quartz sand onto pots of quartz sand containing test plants. The pots of *Helianthus* and the test pots were arranged alternately down a staircase and a complete nutrient solution was pumped from a reservoir at the bottom to the top of the staircase. The solution then dripped from pot to pot until it reached the reservoir and was recirculated for 4 h daily. A control ladder contained only test plants. All species (including *Helianthus*) suffered from inclusion in a *Helianthus* circuit of leachate. The experimental results are compatible with a hypothesis of toxin production or of nutrient depletion. The critical question is whether the mineral composition of the circulated solution was changed by passage along the ladder. Without doubt the composition of the circulated solution was changed because it supplied the needs of all the growing plants on the staircase, and although the test species were seedlings the plants of *Helianthus* were 30–45 cm high when transplanted from the field into 10 cm pots and might be expected to make heavy nutrient demands.

In a further experiment the leaves of sunflower were sprayed with a fine mist of water and the leachate collected and applied to test pots of soil sown with test species. The leachate was not phytotoxic when the test species were grown in quartz sand but marked reductions were obtained in seedling dry weight in soil. It is important to know the composition of the leachate. Many experiments, particularly those of Tukey (1971), have shown that inorganic and organic materials in variety leach readily from leaves. If sufficient sugars had been present in the leachate to stimulate microbial activity the observed results might have resulted from nitrogen depletion as suggested earlier for additions of leaf powder.

Wilson and Rice reported a number of effects of leachates on the germination of test species and these are more difficult to explain without a toxin hypothesis. Rather consistently all test species seemed to show some depression of germination, including *Helianthus* itself. The concentration of the leachates is an important unknown in many such experiments and osmotic effects are conceivable though not very likely.

This series of experiments gives suggestive, but not conclusive, evidence of a toxic interaction in the specialized laboratory conditions — they provide no evidence for a toxicity phenomenon in the field. Many of the experiments described by Rice in his textbook on "Allelopathy" are comparable to the above with sunflower. It is an extraordinarily difficult task to design an experiment that conclusively tests the toxin hypothesis of plant interaction (see e.g. comments by Kershaw, 1964).

(b) *Prunus persica* — the peach. Interference between individuals of the same species has sometimes been interpreted as being due to toxic interaction and indeed it is suggested in the results of Rice's experiments with sunflower (see above). Perhaps the most successful demonstration that there is an unusual interaction between plants, sufficiently unusual to support a toxin hypothesis, comes from studies with peaches by Hirano and Kira (1965). There had been a long history of reports that it was difficult to establish peach orchards on areas that had previously carried this crop (Proebsting and Gilmore, 1941). Hirano and Morioka (1964) studied the susceptibility of seedlings of eight different species of fruit tree to leachates from sand cultures of these same species. They found that seedlings of the peach, fig and grape suffered severely from leachates from their own species, while the persimmon and the mandarin suffered scarcely at all. Hirano and Kira (1965) set up a series of density experiments involving pure stands of these two species. Kira had already been involved in a large number of research studies on a wide range of plants and with his colleagues had developed "laws" of constant yield and of thinning (Chapter 6). He was particularly well equipped to examine aberrant plant behaviour.

The effect of varying density on the growth of the persimmon seedlings was quite normal: yield per unit area rapidly became independent of density. Self-thinning was also "normal" with the biomass of the crowded populations continuing to increase with the progress of the self-thinning. The peach behaved quite differently: the higher biomass was obtained from the lower densities and self-thinning continued even during periods when biomass was steadily declining (cf. "normal" self-thinning, Chapter 6). This most curious behaviour of the peach populations occurred only in glasshouse experiments and nothing like it was found when the experiments were repeated outdoors. The results of the glasshouse experiments are compatible with a toxin interpretation, incompatible with a resource limit interpretation. Other interpretations, such as density-dependent pathogenic activity, are not ruled out.

(c) *Agropyron repens*. The difficulties in the way of establishing a toxin hypothesis of plant interaction are well demonstrated in the example of *Agropyron repens*. This perennial grass seriously depresses yield in cereal crops and extracts from *Agropyron* are toxic to a number of test species (e.g. see Osvald, 1947). Elegant field experiments (Buchholtz, 1971; Bandeen and Buchholtz, 1967) show that a crop of maize infested with *Agropyron* is reduced by 2.74 kg for every kg of weed present compared with reductions of 0.27–2.2 kg for infestations of annual weeds. Maize crops infested with *Agropyron repens* commonly show symptoms resembling those of nitrogen and potassium deficiency and the leaves of maize grown with *Agropyron* had strikingly reduced levels of N and K. Application of heavy rates of N and K fertilizer improved but did not nearly restore the corn yield. Buchholtz (1971) suggested that in some way *Agropyron* made nutrients unavailable to the maize plant or prevented the maize plant from taking up nutrients. The reduced nutrient uptake by the maize was certainly not accounted for by the amounts absorbed by *Agropyron*. Maize and *Agropyron* were grown together in pots and a few of the roots of maize were allowed to grow into vessels containing a complete nutrient solution. Maize plants treated in this way showed spectacularly improved growth even though the greater part of the root system was in *Agropyron*-infested soil.

The influence of *Agropyron repens* on crop plants and *Impatiens parviflora* was tested in a series of experiments by Welbank (1961, 1962, 1964). Comparisons of growth of the test species + *Agropyron repens* showed that the depression could be almost entirely overcome by watering and/or nitrogen application. The influence of more frequent watering was to increase nitrogen uptake by the test species. These results contrast with Buchholtz and Bandeen's findings with maize + *Agropyron*. Welbank's experiments included the introduction of *Agropyron* residues into the soil (Welbank, 1960, 1963) and there was no indication of any toxic interaction unless the residues decomposed anaerobically, in which case persistent toxic residues were found. Residues at least as toxic are produced by *Medicago sativa*, *Lolium multiflorum* and *Agrostis tenuis* under the same conditions. Any direct liberation of toxins from living roots of *Agropyron* seems unlikely because Welbank obtained good germination and seedling growth of test species on a substrate composed of a living root mass of *Agropyron*!

(d) *Camelina*. Toxin production has been invoked to explain the depression in yield of *Linum usitatissimum* by *Camelina* spp. Grümmer

and Beyer (1960; see also Grümmer, 1955, 1961) found that there was
no great depression of the growth of *Linum* when its roots were intermingled with *Camelina, provided that* the plants were watered from
below: strong depression of *Linum* occurred if watering was from above.
Linum in pure culture was depressed by watering through a sieve containing *Camelina* shoots. Grümmer claimed that toxins were liberated
from the shoots of *Camelina.* However, mixed cultures of *Camelina
sativa* and *Linum* grown in replacement series by de Wit (1960) showed
no interaction that was not compatible with resource limitation. Indeed
Linum appears to be the stronger aggressor in mixtures with *Camelina
sativa* or *C. alyssum* (Obeid, 1965). Failure of one research worker to
repeat the findings of another may simply imply that the species contains
ecotypes that do not behave in the same way. Some strains of *Camelina*
may perhaps produce toxins and others not.

(e) *Grevillea robusta.* Species that produce toxic effects on others in
the laboratory may be ineffective in the field. The leaf litter of forest
species in south Queensland, Australia, inhibited seed germination in
the laboratory but appeared to have no ecological significance in the
field (Webb *et al.*, 1961; Cannon *et al.*, 1962). However, the distribution of *Grevillea robusta* suggested that its distribution might be affected
by an interference other than through limited resources (Webb *et al.*,
1967). This tree grows poorly in pure stands and fails to regenerate
effectively except among other species, e.g. *Araucaria cunninghamii.* In
plantations of *Grevillea*, seedlings of this species develop a characteristic
blackening at the tips of the leaves and the seedlings die. The same
symptoms could be produced even when attempts were made to prevent
shortages of light, water and nutrients. The blackening symptoms
followed contact of the seedling roots with actively growing roots of
older plants of *Grevillea.* Pathogens could not be found in the necrotic
leaves but root pathogens can produce leaf symptoms and could conceivably cause the observed effects. Leachates from *Grevillea* foliage
also caused some depressions in the growth of seedlings but neither the
black necrosis nor death.

(f) *Desert Shrubs.* Many of the suggested feedback processes affecting
recruitment and establishment of seedlings have concerned desert plants.
In arid regions associations and exclusions of species are often very
striking. Many species of annual plants characteristic of Californian
deserts are confined to sites beneath shrubs (Went, 1942). Species such
as *Malacothrix californica, Rafinesquia neomexicana, Phacelia distans*

and *Emmenanthe pendulifera* are largely restricted to sites under shrubs of *Franseria dumosa, Larrea tridentata, Opuntia echinocarpena* and other woody species. The shrub *Encelia farinosa* is remarkable in seldom supporting such a flora. The leaves of *Encelia* contain a growth-inhibiting substance which in aqueous or ether extracts kills tomato seedlings in water culture in 1–3 days. The growth of tomato seedlings is depressed even by the application of *Encelia* leaves to the top of a sand culture, and irrigation through the mulch. The toxin has been identified as 3-acetyl, 6-methoxy benzaldehyde, synthesized and shown to possess an inhibitory action closely similar to the natural compound (Gray and Bonner, 1948a, b).

Encelia farinosa is very widely distributed in the Colorado desert together with two other shrubby species, *Franseria dumosa* and *Larrea tridentata*. *Franseria* contrasts so remarkably with *Encelia* in its ability to support an association of herbs that the two were obvious species to compare in relation to potential toxin production. *Franseria* was found to produce a growth inhibitor even more potent than that of *Encelia*, yet *Franseria* grows in association with abundant annual herbs. *Franseria*, *Encelia* and *Thamnosma* all produce water-soluble toxic materials which not only inhibited the growth of the tomato (a curiously irrelevant test species) but also *Malacothrix, Chaenactis* and *Cryptantha* seedlings in water culture. The first two of these species are typically shrub-dependent annuals whereas the third is typically shrub-independent.

Encelia farinosa has a characteristic habit of branching above ground in contrast with, for example, *Franseria* which bears rhizomatous shoots and forms an intricate system of branching at ground level; a consequence of this different growth form is that shrubs of *Franseria* accumulate wind-blown soil and debris. One interpretation of the characteristic shrub-dependent flora of *Franseria* and many of the other shrubs is that the accumulation of this wind-borne debris is necessary for the formation of an appropriate seed-bed; such a seed-bed is absent under *Encelia* (Muller and Muller, 1956). It is significant that open-ground species, i.e. those not normally dependent on shrubs, are found growing freely under *Encelia*.

... the abundant growth of shrub-dependent herbs beneath dense shrubs of *Franseria dumosa* in spite of its production of a growth inhibitor, even more potent than that of *Encelia*, is in itself ample proof that the presence of a growth inhibitor does not automatically constitute an advantage to the shrub that produces it. But not only does the inhibitor fail to operate in the case of *Franseria*, it also fails to operate in the case of *Encelia*, beneath which open ground species thrive, and beneath which shrub dependent herbs are absent only because *Encelia* has failed to change the character of the soil sufficiently. . . . It is extremely doubtful that any advantage could accrue to a desert shrub in which a truly effective growth inhibitor is produced. The critical period for such a shrub would embrace its seedling and sapling stages during which its root system would still be sufficiently shallow to suffer competition from shallow rooted annual herbs. During this stage of development the shrub is far too young to produce a leafy crown of effective size and toxic effect. By the time this stage had developed such a crown, its root system has long since outgrown serious competition from the shallow rooted annual herbs . . . A natural habitat, even in a relatively simple community of the desert, is far too intricate a system of influences and factors, physical and biological to hope that there may be found a single factor controlling the life of a perennial species. An explanation when it is arrived at will be at least as intricate as the situation it seeks to describe (Muller, 1953).

Muller's critical attitude towards the toxin feedback hypothesis in the case of *Encelia* is all the more interesting because he became convinced later of the reality of toxic interactions between species in nature, for example in the role of *Salvia leucophylla* in annual Californian grasslands (Muller, C. H., 1965; Muller, W. H., 1965; Muller *et al.*, 1964a; Muller and del Moral, 1966). It is characteristic of this species to produce or be associated with conspicuous surrounding zones free from grasses and other herbs. Annual grasses and herbs are usually absent from the interiors of shrub areas dominated by *Salvia leucophylla* and also by *Artemisia californica*. A halo-zone of bare soil commonly extends 60–90 cm beyond the canopy of the shrub branches. Terpenes occur in *Salvia* species and are toxic to annual grasses and herbs. Muller hypothesized that inhibitory terpenes are volatilized and moved in the vapour state, inhibiting particularly the germination of seeds and the growth of seedlings that they reach. Vapour transport was considered likely because the distance between a shrub and the edge of the inhibited zone in the field extends well beyond the crown of the shrub or the lateral spread of the root system. Assays of the toxicity of *Salvia* material to many of the local species were made by germinating seed on moist filter paper within a sealed culture dish that held small beakers containing plant material. Many of the species found naturally adjacent to thickets of *Salvia* were extremely sensitive to inhibition of seedling growth in these tests, including *Bromus rigidus, B. mollis, B. rubens, Avena fatua, Stipa pulchra* and *Festuca megalura*. The terpenes were isolated and characterized chromatographically, and samples of the atmosphere within shrubs of *Salvia* contained detectable quantities of the most toxic

terpenes. Terpenes are highly soluble in lipids and Muller suggested that
they were absorbed through the cuticle of young seedlings. Terpenes are
also adsorbed on soil and the surface crust of soil adjacent to *Salvia*
thickets contains detectable quantities of two terpenes, cineole and
camphor; moreover, seedlings grown in surface soil taken from bare
areas around *Salvia* were significantly inhibited.

This group of studies by Muller and his coworkers goes far in establishing correlations between toxin production by *Salvia* and the bare halo
areas in surrounding vegetation. A more complete causal explanation
could perhaps have been approached by deliberately placing a potted
Salvia outside a halo area looking for a response in the surrounding
vegetation, or by deliberately blowing air from inside a *Salvia* shrub onto uninhibited grassland. Criticism came, however, from a different
quarter and is an outstanding demonstration of the dangers of using
purely correlative information in determining ecological causation.

It had been suggested that the bare zones around *Salvia* shrubs were
created by the grazing or trampling activities of animals, or the seed
feeding activities of birds which might forage more actively near shrubs.
This hypothesis was tested by Bartholomew (1970) in a variety of ways.
He counted the number of rabbit faeces per unit area along transects
out from *Salvia* shrubs. Faeces were densest close to the *Salvia* shrubs
and their number fell off steeply across the bare zone into the grassland.
A similar pattern occurred around shrubs of *Artemisia*. Around shrubs
of *Baccharis* the highest density of faeces occurred at about 1 m from
the shrubs. Of course the distribution of faeces does not necessarily
reflect the feeding activity of the rabbits but does emphasize that the
presence of a shrub affects the behaviour pattern of the rabbit and that
different species of shrub affect behaviour in different ways.

Bartholomew also placed feeding stations (millet seeds on pieces of
sandpaper) in the bare zones around shrubs of *Salvia* and in the grassland around the halo area. In a 24 h period 86% of the seed was removed
from the halo area and only 12% from the grassland. Live traps were set
for small mammals which were suspected to be at least in part responsible for taking seed. 75 traps in the halo area trapped 28 mice but a
similar number of traps placed 2 m outside the halo zone trapped only
one mouse.

The most critical test of the role of small mammals in the *Salvia* halo
effect was made by erecting small exclosures 30 x 30 cm and 15 cm
high composed of 0.6 cm mesh. A set of these exclosures was also pre-

pared with one side left open so that small mammals might enter but which would offer closely similar microclimatic conditions of shade and effect on precipitation to the complete exclosures. Exclosures were placed in the halo area around *Salvia* shrubs and also near shrubs of *Baccharis pilosa* and left in position from April 1969 to April 1970. The dry weight of plant material was then measured within each exclosure (Table 11/V).

Table 11/V

The effect of exclosure on the growth of herbage in the halo zone around shrubs of *Baccharis pilularis* and *Salvia leucophylla* (Bartholomew, 1970)

		Dry weight of herbage	
		Exclosure with one open side	Complete exclosure
Halo zone	*Baccharis pilularis*	0.79 ± 0.17 g	15.04 ± 1.92 g
	Salvia leucophylla	0.50 g	11.62 g
Beyond halo zone		18.21 g 13.28 g	—

The fact that vigorous plant growth occurred within the halos when exclosures were erected suggests that the toxin hypothesis is unnecessary to account for the observed pattern of vegetation. Bartholomew's experiments do not of course throw any doubt on Muller's careful demonstration that *Salvia* produces toxins and that these are distributed in the environment around the shrub; the toxins may, however, be irrelevant in the suppression of neighbouring plants.

The distribution of individuals in populations of the creosote bush (*Larrea divaricata*) strongly suggests influences on seedling establishment by mature plants and at distances beyond the range of the canopy:

> The creosote bush is spread with amazingly even spacing over the desert, and this is especially obvious from an aeroplane. The spacing apparently is due to the fact that the roots of the bush excrete toxic substances which kill any seedlings which start near it. The distance of spacing is correlated with rainfall; the less rainfall the wider the spacing. This probably means that rain leaches the poison

from the soil so that they do not contaminate so wide an area. We commonly find young creosote bushes along roads in the desert where the road builders have torn up the old bushes.

(Went, 1955)

A general relationship between the density of *Larrea* and rainfall has also been shown by Woodell *et al.* (1969), who found that in regions of relatively higher rainfall seedlings and young plants tended to occur in a clumped distribution, to be nearly randomly distributed in intermediate climates and to show an over-dispersed or ordered pattern in the lower rainfall regions. They suggested that both the differences in density and in pattern could most readily be explained in terms of water shortage and the relative stress placed by neighbouring plants on each other during periods of water shortage. The toxin interpretation is made even less likely because although *Larrea* extracts are toxic to some test species they are inactive against *Larrea* itself (Knipe and Herbel, 1966)!

Two recent discoveries complicate the interpretation of population feedback in *Larrea*. In common with some other species, *Larrea* affects the water relations of the soil surrounding the plants, reducing soil wettability. A consequence of this is that water droplets falling on the soil surface remain on the surface and evaporate, whereas droplets falling further away from the plants quickly penetrate the soil and escape direct evaporation. This process which is not completely understood, results in drier soil surfaces near *Larrea* plants (Adams *et al.*, 1970). It has also recently been shown for a related species, *Larrea tridentata*, that successful germination depends on burial of the seed by small mammals (Chew and Chew, 1970).

There are perhaps some reasons for expecting that toxic interactions will not be common in higher plants:

(i) higher plants have been shown rapidly to evolve tolerance to environmental toxins such as lead, nickel, copper and zinc (Turner, 1969). Similarly, in the relatively short history of the use of herbicides there are already reports of the development of herbicide-tolerant forms among previously intolerant species (Harper, 1956).

(ii) Many highly complex organic molecules are quickly broken down in soil, particularly after the microbial population has had time to adapt to the presence of these potential metabolites. It would be surprising if compounds regularly released from plants were not dealt with quickly in the same way.

The literature concerning interactions between plants is unsatisfactory and yet stimulating. There are few examples of mechanisms of inter-action that have been clearly established as ecologically relevant or are rigidly proven — it is rather like the literature of pathology before Koch. It is certain that one cannot establish a cause for a field phenomenon by a process of exclusion — the number of possible factors to be excluded is too great. Hence it is not possible to argue that an interfer-ence phenomenon in the field that is not explicable in terms of light, water or nutrients must be due to toxins. It is not safe to assume that interference where no toxins can be found is necessarily due to shortages of water, light or nutrients. There are too many other ways in which a plant may influence its neighbours, for example sheltering pests, carry-ing pathogens, encouraging foraging birds and mammals, sheltering slugs and snails. Sandfaer (1968, 1970a, b) carried out a long series of experi-ments to examine the performance of barley varieties in mixtures. He obtained low RYT values in a number of cases; this might suggest inter-ference by some means other than resource shortage. Ultimately it appeared that one of the varieties was a symptomless carrier of barley stripe mosaic virus which it transmitted to other varieties with which it grew in mixture.

Koch's postulates give some guidance for future work, particularly in the emphasis he places on "symptoms". A large part of the study of mechanisms of plant interaction has concentrated on gross measure-ments such as life or death, or weight. The more successful attempts to understand interference have involved studying symptoms. Often the effects of a toxin or a nutrient deficiency are very specific: leaf tip blackening in *Grevillea*, potassium deficiency symptoms in *Desmodium*, sudden wilt (effect of walnut on seedlings; Massey, 1925; Bode, 1958), change in leaf area/weight ratio in *Trifolium hirtum* (Williams, 1963). Sometimes a more sensitive procedure may involve tissue analysis, parti-cularly if a nutrient effect is posulated. Leaf analysis was used, for example, by Welbank (1962, 1964) to analyse interference between *Agropyron* and beet or *Impatiens*, and by van den Bergh (1968) to explain the very curious effects of K/Na balance on the outcome of interference between *Lolium* and *Alopecurus* (see Chapter 9). A symptom, the more specific the better, may be recognized in the field, repeated in the laboratory and reproduced deliberately in the field. The experimenter can therefore hope to approach some sort of procedure equivalent to satisfying Koch's postulates. So long as attention is concentrated on qualities as vague as "growth" or "germination" it is unlikely that much progress will be made in understanding interference: "If we were to try in imagination to give to one species an advantage over another in the struggle for existence probably in no one case would we know what to do" (Darwin, 1859). Subsequent research has greatly widened our imagination but not contributed much to our understanding.

The Effects of Predators

12

Defoliation

When one animal eats another the prey is usually killed, whether it be a shrimp eaten by a whale, a mouse by an owl or a fly by a frog. When a plant is the prey of an animal the prey may continue to live. A great many animals use only the parts of plants as food, leaving systems that are capable of regeneration. There are few comparable activities among animals feeding on animals, for example the predators of corals and colonial hydroids. An animal grazing on a plant is comparable to a bird that feeds on earthworms but which leaves a part of each worm to regenerate a meal for another day!

Some animals behave essentially as parasites on plants, not consuming tissues but tapping a supply of resources within the plant (e.g. aphids, some eelworms, scale insects). These organisms parallel very closely the feeding style of some animals on animals (e.g. fleas, ticks, tapeworms, trypanosomes). Some animals do kill the plant on which they feed but usually only when the prey are seeds or seedlings (e.g. slugs and snails) or when a grazing activity is continued so long that the plant is weakened and dies, probably from other causes to which it is now more sensitive.

By far the greatest part of the literature concerning animals as predators on plants is in zoology, agriculture and forestry. Unfortunately there is no satisfactory terminology for discussing plant-animal interactions; herbivore, herbivory, phytophagous organisms, phytophagy, grazers, grazing, all have rather specialized meanings in the context of

385

animal behaviour. I have chosen to use the general word "predator" to describe an organism that eats another organism and to include in this general term those animals that eat the parts of a plant but do not necessarily kill (cf. Janzen, 1970), as well as those that kill.

A zoological treatment of the animals that eat plants emphasizes the way in which the eating is done (e.g. the shapes of mouth-parts) or the dietary value of what is eaten. To the plant it may matter little whether part of a leaf is removed by the scissors of a leaf-cutting ant, the file of a snail, the bite of a cow or the peck of a bird. What matters is the amount of organ removed and the importance of this organ in the integrated growth of the whole plant, and ultimately the likelihood that the bitten plant will leave descendants. Similarly it may matter little to a leaf if its current assimilates are diverted into the stylet of an aphid or to the growth of a new root or leaf — though there may be very different consequences to the fitness of the whole plant. The present discussion of animal predators on plants has therefore been concentrated on the reaction of the plant to damage whether it be loss of or injury to leaf, root, seed or seedling.

Defoliation

Many of the effects of a defoliating animal on a plant population are illustrated by a series of simple experiments in which the slug *Agriolimax reticulatus* was allowed to graze experimental populations of the grass *Lolium perenne* (Hatto and Harper, 1969).

Surface-feeding slugs like *Agriolimax reticulatus* damage establishing populations of grasses by chewing through the young shoots at ground level, eating the meristematic region at the base and leaving the felled leaves uneaten on the soil surface. They do much more damage than is represented by what they actually consume.

Large seed trays (38 x 22.5 cm) were sown with seeds of *Lolium perenne* and marked out into 5 x 5 cm squares. Two slugs were allowed to graze one half of the tray (= *ca* 47 per m^2) and were excluded from the other half. Each sward was grazed for 1 week beginning 3 days after the seedlings had emerged. The slugs feed mainly at night, and each morning the damaged shoots in each square were counted and ringed and the damaged pieces of leaf were removed. The remaining above-ground parts of the grass were harvested at the end of the week. The results were as follows:

Yield of half tray before slugs were introduced
   (mean of 15 replicates)   0.17 g
Yield of half tray after 1 week without slugs   1.27 g
Yield of half tray after 1 week with slugs   0.93 g
Loss in yield   0.34 g (26.8%)

During the course of the experiment the sward was growing exponentially at a rate dependent mainly on the area of leaf exposed by each plant.
Leaves that are "felled" are a loss to the crop but are also a loss to the
photosynthetic capital of the system and its potential for further growth.
A further experiment was made to study the effect of grazing over a
longer time period. Trays were sown as before but divided into three
parts to permit a split plot experimental design with the following treatments:

(a) No slugs present
(b) Two slugs present for 5 days before harvest ($= ca$ 70 slugs/m^2)
(c) Two slugs present from 10 until 5 days before harvest
    ($= ca$ 70 slugs/m^2)

Harvests were made at 5 day intervals on replicate trays to allow
estimates to be made of the effects of grazing.
   The results are shown in Fig. 12/1 and Table 12/I.

Table 12/I
The influence of the timing of slug grazing on the
yield (g) of experimental populations of *Lolium perenne*

Harvest (8 replicates)	Time after seedling emergence			
	5 days	10 days	15 days	20 days
Yield of control (a)	0.10	0.37	0.65	0.96
Yield of treatment (b)	0.03	0.24	0.55	0.90
Percentage loss $\dfrac{a-b}{a} \times 100$	70	35	15	6
Yield of treatment (c)	—	0.10	0.30	0.86
Percentage loss $\dfrac{a-c}{a} \times 100$	—	73	54	10
No. of days retarded	17	10	2	1

The younger the seedlings the more severe and long lasting were the effects of damage. The effect on the productivity of the grass can be expressed as a time delay: grazing is equivalent to setting back the growth of the sward by x days (see Table 12/I).

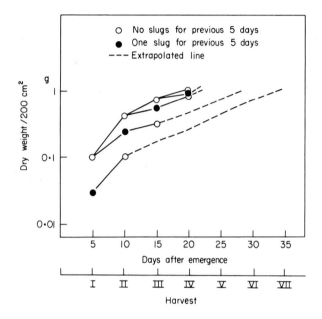

Fig. 12/1. The dry weight of populations of seedlings of *Lolium perenne* (perennial ryegrass) after various periods of grazing by the slug *Agriolimax reticulatus*. (From Hatto and Harper, 1969)

Grazing effects on a plant population may be a function of the numbers of plants present as well as the number of animals, and both these variables were introduced into a further experiment. Undivided flats (333 cm²) were sown with grass seeds at densities corresponding with 6.7, 13.5, 33.6 and 100.9 kg of seed per ha. Four densities of slugs (0, 1, 2 or 3 per flat) corresponded to 0, 30, 60 and 90 slugs per m². The plots were harvested after 5 or 15 days grazing. The results are shown in Table 12/II. The growth made by the grass was determined by the balance between the amount of leaf available and the density of slugs. The lowest density of grass seedlings could not make significant growth in the presence of even one slug and even the highest density of grass failed to increase in yield in the presence of three slugs.

Table 12/II

The influence of the density of slugs grazing on *Lolium perenne* sown at four densities. Mean dry matter yield (g) of above-ground parts of *Lolium perenne* (6 replicates)

Grass density (equivalent in kg seed sown per ha)	Slug density/flat			
	0	1	2	3
Harvest 1				
(after 5 days) 6.7	0.009^e	0.002^e	0.003^e	0.002^e
13.3	0.013^{de}	0.009^e	0.005^e	0.004^e
33.6	0.038^{cde}	0.006^e	0.010^e	0.002^e
100.9	0.086^{bc}	0.050^{bcde}	0.024^{de}	0.012^{de}
Harvest II				
(after 15 days) 6.7	0.035^{cde}	0.008^e	0.005^e	0.004^e
13.3	0.053^{bcde}	0.008^e	0.014^{de}	0.006^e
33.6	0.147^a	0.069^{bcd}	0.005^e	0.011^e
100.9	0.361	0.272	0.099^{ab}	0.008^e

(Means that carry the same superscript do not differ significantly at $P = 0.05$)

The pattern of damage caused by the slugs was not random. If feeding was randomly distributed in the swards, the number of seedlings damaged per square (5 x 5 cm) might be expected to conform to the Poisson distribution, i.e. that $\frac{\text{variance}}{\text{mean}} = 1$. This ratio was calculated over 15 replicate trays for the damage done per square each night and also for the final pattern of damage. In the final count, and in all but one of the eight individual nights of observation, the variance/mean ratio was very significantly ($P < 0.001$) greater than 1.

The effect of slug grazing is to damage some plants in a population and, unless predation is very heavy, to leave others undamaged; predation is differential between plants. To demonstrate this effect plants that had been damaged by slugs were marked with plastic rings so that their regrowth could be followed. Flats of *Lolium perenne* were sown and divided into three sections. Each part received a different treatment: (a) no slugs = control; (b) one slug present until harvest (equivalent to 35 slugs per m²); (c) one slug present until 5 days before harvest when it was removed; see Table 12/III.

Many of the plants recovered after grazing but their contribution to the yield of the sward was minute compared to their numbers. A high

Table 12/III

The contribution to yield by slug damaged plants. Days are counted from the time of seedling emergence = day 0. (See Fig. 12/1)

		Total yield of above-ground parts (g)	Yield of damaged plants (g)	% weight of damaged plants	% numbers of damaged plants	% death of damaged plants
(Mean of 10 replicates)						
Slug present from day 5 to day 10. Plants ringed on day 10 and harvested on day 15.	Slug present on days 10–15	0.488	0.007	1.4	17.5	28.2
	Slug absent on days 10–15	0.378	0.008	2.2	29.1	47.2
Slug present from day 5 to day 15. Plants ringed on day 10 and harvested on day 20.	Slug present on days 15–20 (i)	1.036	0.013	1.3	23.7	37.6
	Slug absent on days 15–20 (ii)	1.137	0.009	0.8	13.6	34.3
Slug present from day 5 to day 15. Plants ringed on day 15 and harvested on day 20.	Slug present on days 15–20 (i)	1.036	0.001	0.1	1.8	43.2
	Slug absent on days 15–20 (ii)	1.137	0.002	0.2	4.4	56.7

(Mean of 20 replicates)

Slug present from day 5 to day 10 and present or absent from day 10 to day 15.	Plants ringed on day 10 and harvested on day 15.	0.433	0.008	1.7	22.8	39.4
Slug present from day 5 to day 15 and present or absent from day 15 to day 20	Plants ringed on day 10 and harvested on day 20.	1.087	0.011	1.1	18.5	36.3
	Plants ringed on day 15 and harvested on day 20.		0.001	0.1	3.1	53.0

proportion of the damaged plants died, particularly when the remaining sward had made further growth. The experiment illustrates the important feature that defoliation can be disastrous for some individuals within a sward, but need have no serious effect on its yield as a whole.

If the loss of a leaf from a plant simply reduced the photosynthetic area, the effects of defoliation would be relatively easy to assess. Loss of a leaf would then be equivalent to the loss of the time necessary to replace the lost photosynthetic area. It would be a setback equivalent to reducing the plant to an earlier stage of growth. The effect on the plant of defoliation would depend on the proportion of its total leaf area that was lost, the age of the lost leaf and its position in the canopy. In practice the effects of defoliation may be much less or much more than are simply accountable in this way. If a leaf is removed, the defoliated individual may find itself in a subordinate position within the hierarchy of a dense population — the effects of defoliation on the individual are then much greater than simply a loss of photosynthetic surface. A contrasting effect is that if part of a plant's foliage is removed, the remaining leaves may function at greater photosynthetic efficiency. It appears that leaves usually function at levels of photosynthetic efficiency well below that of which they are capable and may be stimulated, for example by defoliation, to assimilate more rapidly (Maggs, 1964; Sweet and Wareing, 1966 (Fig. 12/2); Khan and Sagar, 1969; Neales and Incoll, 1968). "The effects of environmental hazards can thus be buffered by the plant, the subsequent reductions in yield being far less than would be expected" (Marshall and Sagar, 1968).

Although the effects of partial defoliation may be compensated for, at least in part, the effects of a severe defoliation can have repercussions right through the integrated activities of the whole plant. In many pasture grasses there is no major store of reserves to be drawn upon for regrowth after defoliation and current assimilates cannot meet the demand from newly developing tissues. The demand is met by remobilization of other materials, for example proteins, and there is a quite sudden reaction in the rate of root growth. After defoliation the roots are thinner, shorter and the rate of initiation of new roots is slowed down. In *Dactylis glomerata* the rate of nutrient uptake declines within 24 h of defoliation (Davidson and Milthorpe, 1966a, b). Severe defoliation results in "root pruning", thus reducing the aggressiveness of the plant towards others both above and below ground (Troughton, 1960). The effect is shown diagrammatically in Fig. 12/3.

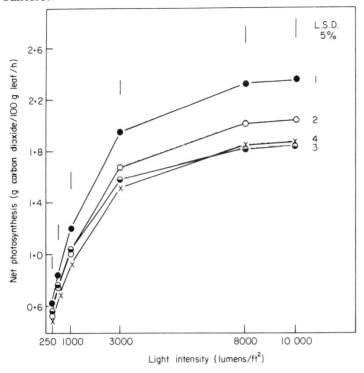

Fig. 12/2. The influence of defoliation on the rates of net photosynthesis of 11-week-old
seedlings of *Pinus radiata:*
1. all fully expanded leaves removed 6 days before measurement of photosynthesis
2. one third of the leaves removed before measurement
3. all fully expanded leaves removed immediately before measurement of photosynthesis
4. no leaves removed.
(From Sweet and Wareing, 1966)

The defoliation of trees, at least in the seedling stage, can increase the
efficiency of the remaining leaves (see Fig. 12/2), but severe defoliation
can have disastrous effects, particularly when it is repeated over several
years. Forest trees are particularly sensitive to defoliation by epidemic
insect attack and the effects on growth are dramatically recorded in the
annual growth rings (Fig. 12/4) (Varley and Gradwell, 1962, 1967;
Williams, 1967). After trees have suffered severe attacks from defoliating insects they often develop a flush of new foliage (Kulman, 1971).
The new growth may be photosynthetically more efficient, but in
deciduous trees it is present for a relatively short time and the tree has

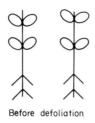

Before defoliation

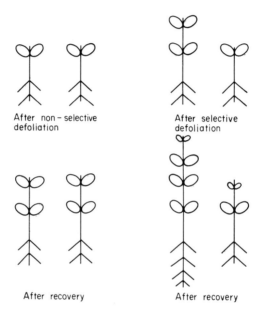

Fig. 12/3. Diagrammatic description of the effects of selective and non-selective defoliation within a density stressed population of plants.

had to draw on resources and mobilize a new supply of nutrients to replace those taken by the predator. The loss of the mineral nutrients in a leaf may indeed be a much more serious matter to the plant than the loss of carbon products.

The modular structure of plants gives them an age structure (see Chapter 1); a leaf passes through phases of juvenility to maturity and senescence. The value of the leaf to the rest of the plant changes during this period. The leaf starts as a dependent structure drawing on resources from other parts of the plant — it matures and becomes itself a net

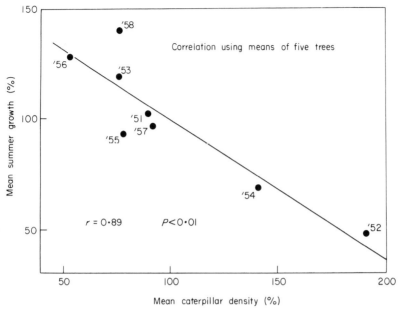

Fig. 12/4. The relationship between the annual radial increment of summer wood and the mean density of caterpillars feeding on *Quercus robur*. (From Varley and Gradwell, 1962)

exporter of organic materials, reaching a peak of activity and then de-clining (Milthorpe and Moorby, 1974). The leaf senesces and becomes an exporter of minerals before it is fully expanded (see Fig. 1/9) and the effects of defoliation can be expected to reflect these changes in leaf activity. A defoliator that takes only young leaves can be expected to harm the plant more than one that takes only old and senescing leaves that are already almost entering the category of detritus. The· young leaf is generally the more nutritious, both as a source of protein and of minerals, and is less packed with indigestible structural com-pounds such as cellulose and lignins. Older leaves also accumulate secon-dary compounds that may make them distasteful. It appears that almost all predators of plants take young tissue preferentially. Almost all the common species of Lepidoptera that feed on *Q. robur* eat only young leaves and the older leaves are not capable of supporting rapid larval growth. As oak leaves age the protein content falls, the content of tan-nins increases and the tannins combine with proteins to form indigestible complexes (Feeney, 1968, 1969, 1970; Feeney and Bostock, 1968). The adjustment of the life cycles of Lepidoptera to the tannin content

of the oaks is shown in Fig. 12/5. *Diurnea fagella* is an exceptional
species that does feed on older leaves but has a very low rate of larval
growth. Some insects that feed on *Papaver* spp. prefer the young leaves
which are relatively poor in alkaloids, but insects that feed on Umbelliferae prefer the old leaves which appear to be less odorous than the
young ones (Ehrlich and Raven, 1965).

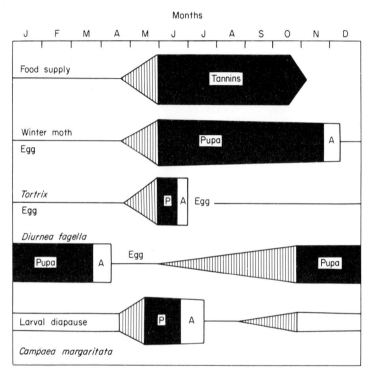

Fig. 12/5. Diagram of seasonal change in quantity and quality of oak leaves and the life-cycles
of representative Lepidoptera.
Larval growth periods are shown by vertical shading. Note the slow growth of species
feeding on mature leaves which contain much tannin. These species are relatively uncommon. Almost all common species feed only on young leaves before condensed
tannins are laid down. (From Varley, 1967)

The removal of leaves, and particularly whole shoots, tends to stimulate the development of dormant buds and increase branching. This
effect is shown particularly well in the reaction of pines to tent caterpillars which often defoliate the leading shoot and, by killing it, alter
the growth form of the tree. Grazing changes the whole pattern of mor-

phological development. *Calluna vulgaris* has a natural cycle of growth
when free from damage by grazing or fire. "During the period of establishment plants are small, rather pyramidal in shape, and cover is incomplete. Later, after about six years, plants pass into the building
phase in which production per unit area becomes maximal (Barclay-
Estrup, 1970) and the characteristic branching pattern produces a dense
canopy. From an age of about 15 years there is a decline in the rate of
growth of peripheral shoots and a tendency for the canopy to thin out
near the centre of the dwarf shrub. This is followed by the degener-
ate phase in which the central branches tend to collapse and die forming an expanding gap in the centre of the plant" (Gimingham, 1971). If
the plants are clipped (simulating grazing) by removing about 80% of
the current shoot growth, the plants are maintained with a compact
habit, like young shoots, and if the clipping is done every year the plants
form a dense cushion (Grant and Hunter, 1966; Mohamed and Gimingham, 1970). The regular clipping continually stimulates the outgrowth
of dormant buds and the plants continually renew their shoot systems
instead of accumulating a much larger mass of ageing tissue. Grazing or
clipping reduces the intensity of flowering and presumably of seed production in *Calluna*. As in so many other instances, a process that pro-
longs the vegetative life of an individual plant has repercussions on its
seed production and hence on its ability to disperse and place descendants elsewhere.

By analogy with the behaviour of trees and shrubs it might be expected that the partial defoliation of grasses would lead to changes in the
dominance structure amongst the tillers. In fact the defoliation of some
tillers of a whole genet of *Lolium multiflorum* increases the export of
carbohydrates from undamaged tillers and directs this export preferentially to the defoliated tillers, with the result that the defoliated tillers
are quite rapidly restored to their position in the tiller hierarchy. Tillers
apparently retain a potential interdependence that can be called into
action to redress an imbalance due to differential defoliation (Marshall
and Sagar, 1965, 1968).

Damage by herbivores that includes buds as well as leaves is clearly
likely to be especially damaging — not only are photosynthetic organs
removed but also the sites for their regeneration. Many browsing animals,
e.g. elk and deer, take young shoots and have an effect on the geometry
of the browsed tree. Perhaps the most damaging herbivore for the life of
a tree is one that takes cambial tissue and destroys the ability of the

plant to maintain a regenerating link between the root and shoot system.
The goat (like the rabbit and grey squirrel) is a master in the art of debarking trees and when the cambial ring is destroyed the tree is killed or
forced to regenerate from stump sprouts. This is one of the rare examples
of a predator that directly kills mature plants.

In Colorado, the porcupine (*Erithezon epixanthum*) feeds during
winter months almost exclusively on the phloem layers of pinyon pine
(*Pinus edulis*) and ponderosa pine (*Pinus ponderosa*). The stripping of
phloem may kill whole branches but an incomplete ring of damage may
permit the tree to regenerate compression wood. A branch may be
damaged repeatedly, year after year. One branch of pinyon pine 12.5
cm diameter and 3.6 m long, bore 43 feeding scars (*ca* 6000 cm^2) and
had scars representing every year but three from 1926 to 1946. One tree
of 45 cm diameter bore scars that could be dated to 1841, 1886, 1901
and 1942 — four peak periods of porcupine abundance in Colorado
(Spencer, 1964) (Fig. 12/6). The effects of porcupines on pines, like
most biological interactions, is more complex than it appears at first
sight. The porcupine preferentially selects trees that have been attacked
by the root fungus *Leptographium* and can detect infected trees before
the disease is apparent to the forester. Infected trees retain unusually
high concentrations of sugars and starches in the upper branches and
this apparently makes them more palatable to the porcupine. If the tree
dies, death might be ascribed to *Leptographium*, to the activities of the
porcupine, to the interaction between them or to the competitive suppression of the damaged trees by the healthy neighbours. Presumably
the damage done by porcupines also increases the likelihood of damage
from other sources, particularly wound parasites. The example illustrates
the difficulty of ascribing ultimate causes to ecological phenomena.

Insect defoliation of the aspen (*Populus tremuloides*) by tent caterpillars illustrates the problem of assigning the role of defoliators in a
forest (Churchill *et al.*, 1964). The aspen suffers severe defoliation by
tent caterpillars (*Malacosoma disstria*) in the Eastern and Great Lake
States of U.S.A. Defoliation is often complete and is followed by re-
foliation. As with many defoliating insects, the populations of forest
tent caterpillars develop over years to epidemics which then subside. A
heavy outbreak in Minnesota started in 1948, reached a peak in 1951–3
and subsided in 1959. During a defoliation period, growth may be
retarded by 90% with a further 15% in the following year (Churchill *et
al.*, 1964; Duncan and Hodson, 1958). The main understorey species,

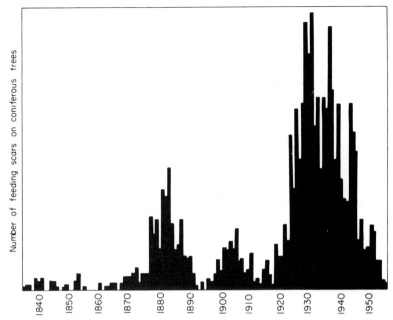

Fig. 12/6. The relative abundance of feeding scars caused by porcupines (*Erithezon epixanthum*) in Colorado where the animal feeds mainly on *Pinus edulis* and *P. ponderosa*. (From Spencer, 1964)

Abies balsamea increased its growth rate during periods when the aspen was defoliated. Three primary causes of death could be recognized among aspen populations: the fungi *Hypoxylon* and *Nectria* and insect borers. All three causes of death were most abundant among trees that had a past history of severe defoliation. Mortality of trees in the forest was generally related to past defoliations except, surprisingly, in the suppressed class of the aspen population. Perhaps other causes of death were of paramount importance among those aspens that were already weak.

Many phytophagous insects do not defoliate but probe within the plant tissues for food. The Heteroptera search for areas that are particularly rich in proteins and other nutrients, and cause damage far in excess of the amount of tissue that they actually consume by choosing meristems, flower buds and young or ripe fruits and seeds as a food source (Southwood, 1973, who gives a valuable comparison of the feeding habits of the various orders of Insecta). There are very few analyses of the effects of non-defoliating insects on the biology of the plant host. An exception is a study of the effects of aphid infestations on the growth of sycamore (*Acer pseudo-platanus*) and lime (*Tilia* x *vulgaris*) by Dixon

(1971a, b). A lime tree 14 m in height with 58 000 leaves may carry
1 070 000 aphids (*Eucalipterus tiliae*) at a time. Lime saplings that were
allowed to develop aphid infestations achieved only 7—8% of the weight
increase of uninfested plants. The infested plants bore the same number
and size of leaves and developed the same area of annual ring increment;
the infested leaves were 22% heavier and contained 10% more nitrogen
per g than uninfested plants at the time of leaf fall. The main effect of
infestation by aphids was to put a stop to root growth. In the year
following an infestation, saplings that had previously been infested bore
only half the leaf area of controls though the leaves were greener and
had a higher assimilation rate.

The effects of infestations of *Drepanosiphon platanoides* on young
sycamores were quite different. Leaf efficiency apparently increased
but the aphids reduced the growth rate of all parts of the plant. Clearly
the interaction between aphids and their plant hosts is a much more
complex affair than simple defoliation.

A number of animals eat roots or damage root systems by forming
galls. Populations of *Tipula* larvae in grassland commonly reach levels
of over 1 500 000 per acre and 20—30% increases in the yield of grass
can be obtained if these populations are destroyed (White and French,
1968). After the larvae have fed during the winter—spring period the
yield of a pasture usually recovers rapidly. Despite the rapid recovery of
the yielding capacity of the population in a grass sward, the balance of
species within it may be changed by the predator's activities — particularly if the *Tipula* larvae kill local patches which may then be recolonized
by different species. Like so many cases of predator damage to plants
the recovery rate of the population is often rapid provided that predation is evenly distributed. It is localized patches of damage that do most
harm to production because the survivors are not sufficiently close together to fill in the gaps. *Tipula* is a selective feeder and is particularly
damaging to *Trifolium repens*.

There appear to be situations in which plants may have too many
roots and derive real benefit from root pruning. Wheat plants, forced to
rely on only a single seminal root, used much less water but ripened
much more grain than individuals with a fully developed root system
(Passioura, 1972). The plants were grown in soil and allowed a fixed
amount of soil water; those with a reduced root system developed extremely high water tensions within the xylem and they transpired less
water but the water supply in the soil lasted longer and the plants were

able to fill and ripen grain. In contrast, the plants that possessed a full root system tapped the volume of soil water more rapidly and it was quickly used up so that they failed to set seed. Presumably the plants with the reduced root system, although potentially fitter, would have failed in a mixture with normally rooted individuals. Only in a population in which *all* plants had been deprived of their main roots would the advantage of a reduced root system be apparent.

There are a number of reports of root feeding activities that produce no decline in plant yield (Harris, 1974). Most of these are for crop plants where the neighbours to a damaged individual expand to fill the available space. Damage is most likely to be seen where predation affects the competitive balance between two or more forms or species. An experimental demonstration of the effects of a predator on the balance between species was made with mixtures of oats and barley sown in a replacement series (Sibma *et al.*, 1964). The plants were grown in containers of soil infested or uninfested with the cereal root eelworm (*Heterodera avenae*). The barley was resistant to the eelworm and the oats susceptible. In the absence of the eelworm oats were strongly aggressive towards barley in the mixtures (relative crowding coefficient $k_{ob} = 6$) but in the presence of *Heterodera* the situation was completely changed (Fig. 12/7) and the oats and barley were almost equally balanced

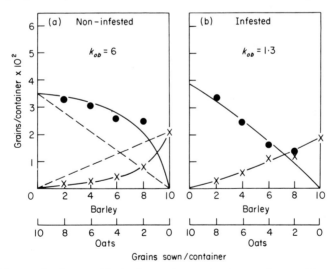

Fig. 12/7. The growth of oats and barley in a replacement series experiment with and without an infestation of the eelworm *Heterodera avenae*.
x——x = barley ●——● = oats. (From Sibma *et al.*, 1964)

(relative crowding coefficient k_{ob} = 1.3). The yield of oats in pure stand was *not* reduced by the presence of the eelworm; the predator affected the balance of species within a mixture without affecting either the total yield of the mixture or the yield of the susceptible species in pure stand.

Effects of defoliation on reproduction

Any description of a predator's activities that is appropriate for a study of the population biology of plants would describe effects on births, deaths, immigration and emigration. There appears to be no study that contains anything approaching the necessary detail. Most of the effects of predators have been detected in effects on some useful economic product such as timber volume or grain yield (Jameson, 1963).

One deliberate experiment was made to determine the effects on fruiting caused by defoliating forest trees in a tropical deciduous forest in Costa Rica (Rockwood, 1973). Six species of tree were selected: *Acacia farnesiana, Bauhinia ungulata, Cochlospermum vitifolium, Crescentia alata, Gliricidia sepium* and *Spondias purpurea*. Plants were selected and matched in pairs to serve as experimental and control trees. Most species produced second crops of foliage after defoliation, though *Cochlospermum* was an exception. The second crops of foliage, which were also removed from the experimental trees, were usually poorer than the first. (The regrowth of *Crescentia* was removed by a chrysomelid beetle which did not eat the leaves of the control tree, apparently a case of a palatability preference for young growth.)

The result of defoliations was almost to eliminate seed production in the year of the treatment (see Table 12/IV).

The time of defoliation is critical in determining the type of plant response. If the leaves are removed after the inflorescence has been formed, the effect is usually to increase seed abortion or for the plant to produce smaller seeds. If defoliation precedes the time of formation of an inflorescence, one may not be formed or it will be smaller. In this case, however, seed size is unlikely to be affected. Maun and Cavers (1971) removed the cauline leaves of *Rumex crispus* in an attempt to simulate the activities of defoliating caterpillars. The defoliation was done at the stage of anthesis and there was scarcely any effect on the number of seeds produced, but there was a major effect in reducing seed size. Similar results had been obtained with *Trifolium incarnatum*

Table 12/IV
The seed yield of defoliated and control trees
(From Rockwood, 1973)

Species		Number of fruit		Weight of fruit	
		Control	Experimental	Control	Experimental
Acacia farnesiana		1554	7	1999	9
Bauhinia ungulata		249	16	413	21
Cochlospermum vitifolium		38	18	1020	485
Gliricidia sepium		4195	860	31 505	5453
Spondias purpurea		10 635	0	45	0
Crescentia alata	(a)	626	641	241	217
	(b)	288	37	93	18
	(c)	363	75	135	24

(a) before, (b) 6 months after and (c) 1 year after defoliation.

(Knight and Hollowell, 1962) and with wheat and oats (Womack and Thurman, 1962). The mean seed weight per panicle of *Rumex crispus* was reduced from 2.6 to 1.3 and the weight of individual seeds from 1.44 to 0.86 mg. Defoliation had no effect on the percentage germination of seeds though the requirements for dormancy-breaking were less stringent. The smaller seeds from the defoliated plants contained less food reserve, produced smaller seedlings, and when germinated in the dark the seedlings from defoliated plants had a maximum hypocotyl extension of 3 cm less than the controls.

A consequence of defoliation that may sometimes be even more important than a reduced seed crop is delayed reproduction. The removal of the fully expanded leaves of *Trifolium subterraneum* had the effect of delaying flowering by as much as 30 days (Collins and Aitken, 1970) and it is not difficult to imagine how this might throw completely out of joint the evolved synchronization of phenology and climate.

13

Seasonality, Search and Choice

Vegetative growth has a more or less clear seasonal cycle (even in the humid tropics); flowering and the setting of fruit and seed are usually short episodes set within the rhythm of vegetative growth. Nectar, pollen fruits and seeds are the most seasonal food resources.

When an environment is predictably cyclic it usually pays an organism to anticipate events rather than to respond to them (Levins, 1968). It is therefore not surprising that the life cycles of most plant eaters are triggered in essential phases like dormancy, diapause and the oestrus cycle by external clocks (usually photoperiod) so that the demand for food coincides in general with its availability. Only if the seasons lack a predictable rhythm, e.g. rains in a desert, has it paid to respond to favourable conditions as they appear rather than to predict them with an external or physiological clock.

A food supply that is only available for a short part of a season can satisfy the demands of a predator with a similarly short period of demand. Thus a predator that takes only the seeds of a particular species may have a life cycle in which the egg is deposited in the seed, the larvae grow at the expense of the developing embryo and a period of diapause then carries the life cycle round to the next time at which seeds are being formed. Alternatively, the predator may take a variety of foods and subsist on other diets in seasons when there is no seed about. A further alternative is to store seed during the glut period and stretch out the products of a period of glut over the intervening periods of famine.

405

A food supply that is available for long and reliable periods can support predators with very different life cycles — either long-lived herbivores that include many seasons within a life cycle or the completely contrasting short, rapidly repeating generations of aphids. In the latter case, continuous parthenogenetic reproduction allows populations (of a single genet) to grow rapidly and to "track" seasonal changes in abundance of the food plant. Obviously the shorter the interval from birth to giving birth, the more accurately can the population of predators adjust to an increasing availability of food. "Aphids . . . have well-developed homoeostatic mechanisms which can enable a population to adjust quickly to the highly variable and rapidly changing quality and quantity of their food supply . . ." (Way and Cammell, 1970).

A very few plant foods are almost completely non-seasonal and the predator can then be a specialist *and* long lived. Bark and timber beetles illustrate such a solution, but in general, in an environment of seasonal food supply, predators must be specialists with a short feeding life, *or* must store food, *or* accept a variety of diets.

The ways in which man adjusts the food needs of farm livestock to the seasonal rhythm of crop production correspond in many ways to the evolutionary solutions found in nature. The annual cycle of grass production by *Lolium perenne* on farms in New Zealand is shown in Fig. 13/1. The cycle is highly predictable with a characteristic decline in

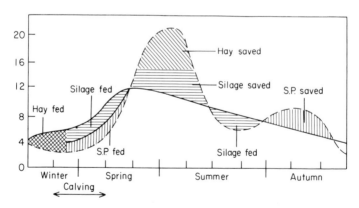

Fig. 13/1. The annual cycle of production of forage by *Lolium perenne* and the annual cycle of food demand by livestock. Grass production is shown as a broken line and food demand as a continuous line. Production in glut periods is conserved (as hay or silage) and fed during periods of "famine". S.P. is excess forage left in the field in the autumn and fed with supplements in the spring. (From Hamilton, 1956)

the winter months when the insolation and temperatures are low and in midsummer when there is a characteristic "midsummer gap" associated with the slow vegetative growth during flowering. Cattle or sheep grazing a pasture of this species will experience periods of famine and glut (unless the density of stocking is set so low that the food needs are met during the periods of lowest productivity). During periods of food shortage the animals may choose to eat old leaves or flowering stems of *Lolium* that they ignore during a glut period; they may lose weight and starve. During periods of food abundance the animal may be choosy in the parts that it eats. The characteristics of the pasture may change under such varying pressures — unpalatable species may establish in the sward and more palatable forms of *Lolium perenne* may decline (see Chapter 14).

The farmer deals with this problem in a variety of ways: by controlling the breeding period so that the augmentation of stock is seasonally planned, by storing excess grass during periods of glut (as silage or hay) and feeding it during famine seasons, by slaughter of excess livestock before famine seasons, by exporting the animals to a region where food is not in short supply (transhumance) or by growing other species of crop (e.g. cereals, brassicas) to be eaten during the famine periods. All of these and other solutions to the problems of seasonal food supply can be found among herbivores in nature:

(a) seasonal oestrus cycle;
(b) storage from glut periods, cf. caching of seeds by small mammals;
(c) slaughter — autumn death of workers in colonial Hymenoptera;
(d) migration to alternative food sites, cf. geese with winter and summer feeding sites (Yelverton and Quay, 1959);
(e) change of diet (polyphagy), perhaps the commonest solution in nature to the problem of seasonal food availability (Fig. 13/2).

In nature solutions exist that are not available to a farmer:

(f) dormancy or diapause — the animal may enter a seasonal period of inactivity or dormancy when it opts out of a need for food.

The longer the life of a predator, the more continuous its need for food, the more likely it is to experience periods of glut and famine. The population densities of long-lived predators are likely to be determined by what is available during the seasons of shortage, and this limits the damage that they may do during periods of glut. If it were not for this

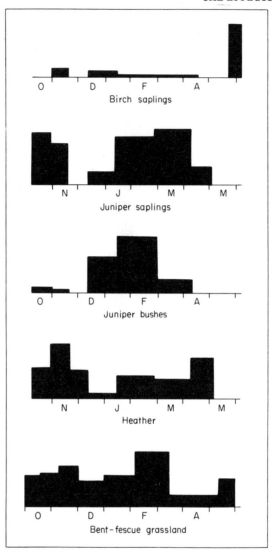

Fig. 13/2. The species chosen by red deer for grazing and browsing in Glenfieshire, Scotland. (From Miller *et al.,* 1970)

limitation it is doubtful whether man could harvest food crops. In practice crops are grown as extensive areas and ripen over short periods and few seed predators can multiply quickly enough to cope with this temporary superabundance of riches. If crops ripened continuously through the year the density of predators could be maintained at higher levels, causing damage of a quite different order of magnitude. In agricultural systems in which rice is grown to yield three crops a year the pest problem from leafhoppers (Delphacidae) is enormously greater than in areas of once-a-year cropping (MacQuillan, 1974).

Predator satiation is a function of the distribution of the food supply in space as well as in time. Darwin (1859) pointed out that it is almost impossible to harvest grain from a small plot of cereals (it is all taken by birds) whereas a field of grain will suffer proportionately much less damage. The space and time elements in predation come together to make predators an acute problem for plant breeders. The cereal breeder usually grows his trial varieties in small plots and includes varieties that ripen at different dates so the food supply for birds and small mammals is nicely spaced out over a much longer period than on a farm; the risks of crop loss are correspondingly higher.

A predator on seeds can cope with a limited season of glut if it has a long resting phase between crops and it is not surprising that many of the specialist seed predators are insects (with diapause) and eelworms (with resting stages).

Polyphagy and monophagy

The simplest solution to the problem of seasonal food supplies is to be able to accept alternatives. This is the advice given by Marie Antoinette to the bread-starved people of Paris: "Let them eat cake". However, cake costs more than bread and for any predator there are costs to be set against the otherwise obvious advantages of polyphagy. The main disadvantage of polyphagy is that efficient collection and digestion of food requires specialization. A beak adapted to dehusk seeds is likely to be inefficient for grazing leaves. A gizzard appropriate for grinding hard seeds is inappropriate for a diet of soft fruit. A linked disadvantage of polyphagy is that it appears to be relatively easy for a plant to evolve defence against a generalist predator, by a change that demands specialization to overcome it.

The evolutionary problems in the development of generalist and specialist feeding habits have been discussed at length by MacArthur and Pianka (1966), Emlen (1966, 1968) and Schoener (1969a, b). MacArthur and Pianka interpreted feeding habits as strategies of behaviour that optimize the time or energy spent by the organism: "an activity should be enlarged as long as the resulting gain in time spent per unit food exceeds the loss". There are some dangers in this view, in particular it is by no means clear that the allocation of an organism's time and energy budget determines its fitness, i.e. its chance of leaving descendants. A food-collecting habit may be energetically inefficient (for example collection and storage of seeds greatly in excess of requirement) but contribute to fitness if the habit deprives competitors. The selfish feeder would seem to have the advantage in natural selection and more attention should perhaps be focused on a search for components of fitness rather than components of efficiency when interpreting natural systems; efficiency does not necessarily contribute to fitness.

The concept of optimization is helpful in focusing attention on the way food is distributed in time and space and the time and energy spent in search. Potential food may be distributed in a fine-grained fashion, the predator meeting it in the proportions in which it occurs, *or* in a coarse-grained fashion when the predator meets it in patches. The "grain" of the resource can be a feature of both space and time: a seasonal food supply is coarse-grained in *time* to a long-lived predator. In general the specialist is favoured in the fine-grained system and the generalist in the coarse: see the discussion in MacArthur and Pianka (1966) and in Pianka (1974).

Climatically the "ideal" tropical rain forest is fine-grained, the yearly variations in mean daily temperature may be less than the daily range. In such a relatively constant and predictable climate the tree canopy is most often a mixture of a great number of intermingled tree species (Ashton, 1969; Richards, 1952). To a bird, ranging through such a habitat, food is presented in a fine-grained fashion — fine-grained in both space and time. This is the ideal condition for the evolution of specialists, and the contrast between the feeding behaviour of the birds in such a community and in a temperate woodland has been elegantly described by Cain (1969):

> It occurred to me while studying birds in British Guiana that in the course of the year the common thrush in England passes through ecological niches appropriate to many different families of South American birds. When it is pulling up worms it is an ibis, when it is eating snails it is an Everglades kite, when it is boring into large fruit it is, I suppose an icterid, when it is gulping down small fruit it is a cotingid or manikin (piprid), when it is taking caterpillars off leaves it is probably some form of tyrannid or formicariid. Now all these activities can be stable modes of life throughout the year in the non-seasonal wet tropics. In England a bird that tried to live by any one of them alone would be extinct in a year. Specialisation of niche therefore can go on to an extent which is almost impossible

in the temperate zone, except, of course, to those migrants who specialise in England in the summer (when England makes its nearest approach to tropical luxuriance) and then go to the tropics to maintain the same mode of life.

The grain of the environment may of course be quite different for a short-lived species of predator with dormancy or diapause, that selects the phase of the seasons in which to feed, or in space, for an animal with a very short range of search, such as a snail: what is a uniform mixture of species and habitats to a mobile bird will appear a coarse patchwork to a caterpillar. The question of whether a generalist habit, a "Jack of all trades" life, is more efficient than the life of the specialist depends essentially on the way the individuals sample the environment as well as on the pattern of environmental variation in space and time.

A very short-lived predator can be successfully monophagous in a seasonally and spatially patchy environment provided that it selects its patches. Some of the most extreme examples of food specialization are found among the insect predators of seeds and fruits whose food supply is strictly limited both in space and time. Members of the genus *Rhagoletis* are flies that feed on fruit, cherries, blueberries, apples, currants, rose hips, walnuts and tomatoes. The species are not completely monophagous but restrict their attacks to plant hosts within a genus or closely related genera. The precise interlocking of the life histories of the predator and its plant prey depend on precision in seasonality, search and choice. Adults may travel more than a mile during their 20—30 day life and they recognize and home onto suitable host trees from a distance. They recognize tree shape (and are attracted to appropriate models) and they are sensitive to colour (they distinguish between differently coloured models). On the tree the fruits serve as highly specific rendezvous for courtship (they will mate only on appropriately shaped fruits or models) and there appear to be very specific olfactory stimuli involved as well (Prokopy *et al.*, 1973). Such precise selection for food and courtship offers an opportunity for very precise evolutionary radiation. Slight specific changes in host or predator can drastically upset the precision of the relationship and it is not surprising that this genus contains many groups of sibling species and has rapid sympatric evolution of predator and prey (Boller and Bush, 1974): e.g. the apple-infesting form of *Rhagoletis, R. pomonella*, first appeared about 100 years ago as a variant of a form infesting *Crataegus*. Although the morphology of the two forms scarcely differs, the phenology of the flies has already diverged to give precise coadaptation to the different hosts (Fig. 13/3) (Bush, 1969a, b).

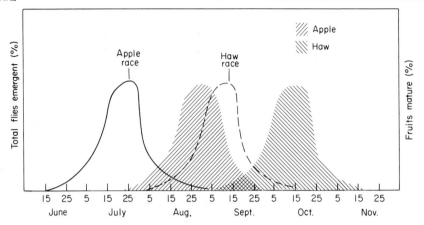

Fig. 13/3. Emergence period of the apple and hawthorn races of *Rhagoletis pomonella*. The cross
hatched curve represents the approximate period of fruit maturation when larvae are
leaving the fruit and pupating in the soil. (From Bush, 1969b)

The ability of predators and parasites to form specialized races means
that it may be useless to search for classical predator–prey cycles. The
population dynamics is continually confounded in the population
genetics. Just as in simple competition experiments between a pair of
species (e.g. the flour beetles *Tribolium confusum* and *T. castaneum*)
the genetic composition of the populations changes during the course
of an experiment (Lerner and Ho, 1961; Lerner and Dempster, 1962),
so in a predator–prey interaction both may suffer profound genetic
change. The activity of a predator applies a selective force to the prey
which declines until a resistant form of the prey takes over. The new
form of prey selects a more virulent predator or parasite and in turn a
new form of prey or host is favoured. These internal shifts in the genetic
composition of predator and prey, parasite and host are wholly characteristic of monophagous feeders: ". . . specialization is a deepening rut
in evolution" (Huffaker, 1964). The term monophagy is used to include,
in an extreme form, the *obligate* alternation of two species as host or
prey as with *Puccinia graminis* alternating between a grass and *Berberis*
as hosts, *Cronartium ribicola* alternating between *Ribes* and *Pinus* and
the *Heliconius* butterflies alternating between *Anguria* and *Gurania*
which provide pollen and nectar for adult butterflies and *Passiflora* spp.
which are the larval food plants (Ehrlich and Gilbert, 1973).

Once a highly specific prey–predator or host–parasite relationship is
established, the mutual evolution of both organisms seems to lead

remorselessly to finer and finer degrees of specificity. The population dynamics of both predator and prey then involves partners — or opponents — in a game played on a more or less rapid microevolutionary scale (see general discussion in Pimentel, 1964; Pimentel et al., 1963, 1965; and the discussion by Ehrlich and Raven, 1965, of the coevolution of butterflies and their host plants).

Monophagy and polyphagy are not clearly defined categories. No animal is wholly polyphagous and monophagy may in its most extreme form mean that a predator will accept only a specialized race of a prey species or that it will accept several species in a closely related group. All feeding activities involve elements of choice and search and both repellents and attractants may be involved in determining the specificity. The caterpillars of Tyria jacobaeae L. feed on many species of Senecio but not on Senecio viscosus which bears dense glandular hairs. If the glandular material is dissolved away in methyl alcohol E. jacobaeae will accept S. viscosus. When the glandular material is painted onto normally acceptable species the larvae refuse them. The glandular material is clearly repellent (Merz, 1959).

The larvae of the cabbage white butterflies Pieris brassicae and P. rapae feed on host plants in families that contain mustard oil glycosides (Cruciferae, Capparidaceae, Resedaceae, Tropaeolaceae) but will eat flour, starch or even filter paper if it is smeared with the juice of Bunias (Cruciferae) (Verschaeffelt, 1910) and will eat the leaves of some other species of plant if they have been treated with the mustard oil glycosides sinigrin or sinalbin (Thorsteinson, 1960). In this case the mustard oils are acting as attractants.

The evolutionary sequences involving attractants and repellents are presumably: (i) a palatable host is selected by the activities of a predator and forms of prey that are relatively unpalatable are favoured; (ii) the new relatively unpalatable host is now a selective force favouring a form of the predator that can break through the palatability barrier; (iii) the specialized predator has access to a food that is denied to other, less specialized predators; (iv) the predator evolves dependence on the feature that guarantees its specialized resource — what had evolved as a repellent has become the attractant (Ehrlich and Raven, 1965). There is a clear parallel in the evolution of some pathogenic bacteria from sulphonamide susceptible through sulphonamide resistant to sulphonamide-dependent forms.

One microevolutionary consequence of the evolution of a predator—

prey system is the development of polymorphisms in respect of palatability. Three species of *Lupinus, L. bakeri, L. caudatus* and *L. floribundus*, are common in sagebush areas in the basin of the Gunnison River, Colorado. The larvae of the small butterfly *Glaucopsyche lygdamus* are the major predators of lupin inflorescences in the area and cause seed losses approaching 100% in *L. floribundus* and 0–10% in the other species (Dollinger *et al.*, 1973). The lupins contain a variety of alkaloids. In areas where the activity of *Glaucopsyche* is restricted by other factors, the plants contain small amounts of single bicyclic alkaloids in their inflorescences. Where lupins are readily available as food for *Glaucopsyche* the alkaloids are in high concentration. The individuals that suffer heavy predation contain many alkaloids but each plant contains exactly the same mixture. In contrast, plants that are only lightly predated contain three or four isomers of lupanine and closely related tetracyclic compounds in individually different mixtures. The authors argue that individual variability is in itself an antispecialist defence mechanism. These observations agree well with the view expressed earlier in this chapter that a monophagous predator–prey system has peculiar properties as a unit of population behaviour: the lupin–*Glaucopsyche* interaction would appear to be a continuous evolutionary flux. Any model of interactions between predators and prey that assumes genetically invariant species in classic oscillations must be quite unreal in such situations.

Any mechanism of defence against predators is liable to be broken by an evolutionary change in the predator. Some forms of defence are more likely to be broken than others. Chemical defences may be rather easily broken and when the defence has been broken by one species of predator (usually a specialist) the most likely next line of defence may well be some subtle molecular modification of the first defence. This would explain how members of a homologous series of chemicals are often found within the same species and the same individual: e.g. the alkaloids of *Lupinus* (Dollinger *et al.*, 1973), the alkaloids of *Senecio* (Barger and Blackie, 1937) and the polymorphism of terpenes both phenolic and non-phenolic in *Thymus vulgaris* (Passet, 1971).

Morphological barriers may be less readily broken. Spines, tough leaves, long hairs and prickles confer longer-term protection against polyphagous as well as some monophagous animals. The leaves of holly (*Ilex aquifolium*) can be eaten by the larvae of *Lasiocampa quercus* provided that the spiny margin has been removed (Merz, 1959). In such cases there is no reason to expect polymorphism – the predator is already taking a variety of foods and the potential prey has achieved long-term protection by one unique morphological feature.

An odd anecdotal observation suggests that the genetics of feeding habits may have direct effects on the nature of vegetation. In the New Forest, southern England, ponies have traditionally wandered freely and maintained open grassland areas tightly grazed. The prickly *Rubus fruticosus* has been excluded from the grassy areas by the ponies which have eaten it. Genetic improvement of the New Forest pony by cross breeding with Arab stock is reputed to have changed the ponies' feeding habits, and they no longer eat *Rubus* which now spreads rapidly in the grazed areas.

Simply to taste nasty is a poor defence against predators unless the predator can learn to avoid the plants that taste bad. If a plant has to be bitten every time before the animal rejects it the defence has been largely a failure. Some defences seem to work in this inefficient fashion: the cyanogenic glycosides protect plants of *Trifolium repens* and *Lotus corniculatus* against slugs and snails. The predator nibbles and then leaves cyanogenic forms but nibbles and then consumes the forms that lack cyanogens (see Bishop and Korn, 1969; Jones 1962, 1966, 1972, 1973; Crawford-Sidebotham, 1971). The slug or snail does very little damage to the plants that it nibbles and discards. More commonly noxious plants, in addition to containing repellent chemicals, possess distinguishing characteristics of appearance or smell that enable the animal to recognize without tasting. The importance of visual symbols is illustrated by the reaction of cattle to *Senecio jacobaea*. This plant contains a group of alkaloids, is clearly unpleasant to livestock and is usually avoided. On rare occasions animals may eat it and the effects are fatal; sub-lethal symptoms are apparently not known. Animals that eat toxic quantities are usually those that have been reared in the absence of the plant and suddenly been moved to an area where it is abundant. Animals that have grown up with the plant avoid it: they will, however, eat it quite readily after it has been dried as hay *or* after it has been sprayed with a hormonal weed killer and has assumed a twisted shape *or* after it has been cut and allowed to wilt in the field. The animal recognizes and avoids only the normal growth form of the plant and it is this that determines whether the animal will eat it (and die) or avoid it (and live) (Harper, 1958; Harper and Wood, 1957.)

The whole area of study of search, recognition and avoidance has been deeply studied in the feeding habits of animals on animals and most of what we know about crypsis, warning coloration and pattern, mimicry and camouflage comes from predator—prey interactions where the prey are animals. Botanists have been reluctant to accept precisions of adaptation that are commonplace to zoologists and often seem reluctant to see the animal as a powerful selective force in plant evolution except in the curiously acceptable realm of adaptations to pollination! It may be that much of the fantastic variation in leaf form, variegation, dissection and marking that is known in the plant kingdom is accounted for by the selective advantage to the plant of associating unpalatability with a visual symbol.

A rare example of a study in depth of co-adaptation between specifically monophagous predators and their plant prey is Gilbert's study (Gilbert, 1975) of the butterfly *Heliconius* and its food plant *Passiflora*. The *ca* 350 spp. of *Passiflora* are tropical vines of the New World with some of the most striking intra- and interspecific variation in leaf and stipule shape known in any genus of the plant kingdom. Locally, populations of *Passiflora* are difficult to find and are most easily discovered by using egg-laying butterflies as clues! Herbivore damage is often very extensive and plants are rarely in flower or fruit. There are apparently limits to the number of *Passiflora* species to be found in an area — there are never more than 5% and usually 2—3% of the total 350 species present together in the same habitat. Distasteful chemicals, cyanogenic glycosides with and without alkaloids are present in the host species and there is high specificity in the species of *Heliconius* that feed on particular species of *Passiflora*. Almost every species of *Passiflora* is unique from the others in the species of *Heliconius* or combination of species that prefers it for oviposition in a particular area. A few species of *Passiflora* bear hooked trichomes on the leaves; these immobilize *Heliconius* larvae and apparently allow these species of *Passiflora* to escape predation (Gilbert, 1971).

The butterfly is visually sophisticated and learns the position of particular vines, returning to them on regular visits. There are 45 species of *Heliconius*, each specializing on its own sub-set of *Passiflora* species. The leaf shape of *Passiflora* varies enormously between species in a habitat, although the leaves of tropical trees are in general more noted for the monotony of their shape. The "gestalt" of sympatric species of *Passiflora* is entirely different (Fig. 13/4). There are convergences of leaf shape among *Passiflora* species onto other common (but to *Heliconius* inedible) species. So close are the convergences that taxonomists

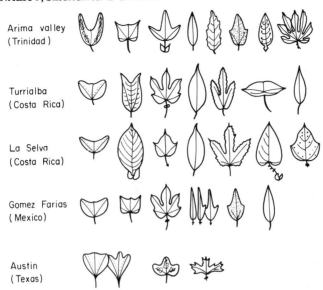

Arima valley (Trinidad)

Turrialba (Costa Rica)

La Selva (Costa Rica)

Gomez Farias (Mexico)

Austin (Texas)

Fig. 13/4. Variation in leaf shape among groups of sympatric species of *Passiflora*. (From Gilbert, 1975). Reproduced with permission from Gilbert and Raven "Coevolution of Animals and Plants". Copyright © University of Texas Press, Austin, Texas.

have named the *Passiflora* species after other genera that they resembled: *P. dioscoreaefolia, P. morifolia, P. capparidifolia, P. laurifolia* etc. The situation strongly suggests that sophisticated recognition patterns among the butterflies have fed back as evolutionary influences on plant form. The suggestion is made even stronger by the discovery of egg mimics in *Passiflora cyanea* and *P. auriculata*, glandular outgrowths from the leaf that mimic to an incredible extent not only the size and shape but also the golden colour of *Heliconius* eggs at the point of hatch. *Heliconius* detects and rejects shoots of *Passiflora* that carry eggs or young larvae. It is difficult to interpret the egg structures as other than egg mimics.

There are other cases in which the search by an animal for food may have forced mimicry in the plant kingdom. A remarkable parallel series has been shown in leaf shape, form and texture between pairs of species in the genus *Cliffortia* (Rosaceae) and *Aspalanthus* (Fabaceae) (Dahlgren, 1971) which strongly suggests parallel evolution under predation and it has been reported that the Australian mistletoes which are palatable converge in leaf form on the leaf shape of their relatively inedible host trees, mainly *Eucalyptus* and *Casuarina* spp. (B. A. Barlow and D. Wiens, pers. comm.). Any author writing a book on the population biology of

plants in 1985 may well have to devote a whole chapter to the significance of mimicry in the plant kingdom.

Palatability, preference and choice in polyphagous species

Probably all polyphagous animals demonstrate an order of preference
when offered a range of potential food plants. In a classic early experiment Linnaeus (1762) showed that sheep were choosy about the species
they ate from a number on offer. Since that time a large number of
records has appeared of the apparent preferences of animals feeding on
mixed vegetation. Some of the literature refers to field observations
such as a report on elk browsing in Norway during the winter (Lykke,
1965). During the summer months the elk are widely dispersed and do
little damage to the vegetation but they congregate in the winter. The
preference order among winter browse species was *Juniperus* > *Salix* >
Sorbus > *Pinus* > *Picea. Alnus,* although it was abundant, was scarcely
touched; *Pinus*, although well down the preference order, was severely
damaged by the elk and 50% of regenerating pines suffered. Field studies
of this sort are complicated to interpret because of the variations in
densities and patchy distributions of the browse species. Formal experimental designs are needed for a rigid ordination of preferences. A plantation of three species of *Pinus, P. strobus, P. resinosa, P. banksiana* and
Picea glauca, was established in Ontario, Canada in 1955. 500 specimens
of each species were planted in rows approximately parallel to the contours of a 140 ft ridge with north- and south-facing slopes. Species were
assigned to rows randomly in series of four. The experiment had not
been designed to study animal behaviour but when deer accidentally
broke into the plantation they had a free choice among the four species
within a few steps of any point. Table 13/I shows the incidence of
browsing and Table 13/II summarizes the growth in height made by
damaged and undamaged trees. The preferred species was jack pine
(*P. banksiana*) followed by white pine (*P. strobus*). Red pine (*P. resinosa*)
was only lightly browsed and white spruce (*Picea glauca*) was ignored.
Most of the browse damage was to the terminal leading shoots and *P.
banksiana*, which has an initial superiority in growth rate over the other
species, suffered severe stem deformation and often formed multiple
leaders as a result of damage by the deer. Most browsing damage occurred
on the ridge top and the south slope and the author (Horton, 1964)

Table 13/I
Browsing incidence on planted trees

Location	Winter 1956—1957				Winter 1958—1959				Winter 1960—1961			
	wP	rP	jP	wS	wP	rP	jP	wS	wP	rP	jP	wS
	Per cent											
North slope	16	4	78	0	1	0	23	0	2	0	51	0
Ridge top	41	–	92	–	12	0	64	0	5	0	79	0
South slope	40	28	86	0	13	2	67	0	29	0	81	0
All	31	19	84	0	9	1	48	0	17	0	70	0

Table 13/II
1956—1959 height growth of damaged and undamaged trees

Location	White pine		Red pine		Jack pine		White spruce
	U	D	U	D	U	D	U
	Inches						
North slope	29	28	28	25	57*	44	19
Ridge top	23	18	22	20	43*	31	21
South slope	27	22	28*	22	44*	32	19

*Significant difference at 0.05 level by t-test.
Legend for Tables 13/I and 13/II: wP – white pine, rP – red pine, jP – jack pine, wS – white spruce,
U – undamaged, D – damaged by browsing.

From Horton, 1964.

attributed this to the relative freedom from snow cover of these areas
compared with the north slope. This is a good example of a difference
in animal behaviour that, over a period, could easily lead to a major
difference in the dominant plant species in different areas. The damage
correlates with a climatic factor but the cause lies in the activity of the
deer.

A similar study, but for a grazing animal, was made by Archer (1973).
Horses and ponies were allowed to graze on randomized plots of 30
species and varieties of grass and clover. She timed the periods spent by
the animals on the various plots in two separate grazing periods. The
results are shown as an order of preference in Fig. 13/5. The effect of
plant species was significant at $P < 0.001$. There was a significant interaction ($P < 0.05$) between horses and species showing that individual
horses differed in the time that they spent on the different species. The

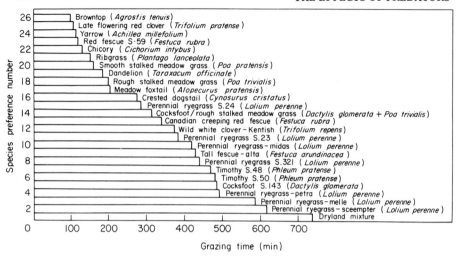

Fig. 13/5. The time spent by horses grazing on plots of different species and mixtures of grasses and clovers. (From Archer, 1973)

favourite species was clearly *Lolium perenne* in three of its cultivars but other strains of the same species (S23 and S24) were well down the list of preference. The most popular plots were those sown with "drylands mixture" a mixture of five species: *Lolium perenne, Festuca pratensis, Dactylis glomerata, Phleum pratense* and *Trifolium repens*. The horse is well known as a choosy feeder and the results of this experiment suggest that sometimes variety itself may be sought. Experiments of this sort allow an animal free choice among randomly placed but clearly defined options. In practice, both in nature and in agricultural systems, grazing animals meet a complex pattern of intermingled species in which the pattern itself changes from place to place.

The white fronted goose (*Anser a. albifrons*) grazes maritime pastures in winter and its behaviour was studied (Owen, 1971) on a salt marsh with more or less clearly defined zones of vegetation. The geese chose to graze specific areas (as shown both by mapping the distribution of faeces and studying the distribution of defoliation) in the order of preference *Agrostis stolonifera* > *Lolium perenne* > *Festuca rubra* > *Hordeum secalinum* > *Juncus gerardii*. *Agrostis stolonifera* was clearly the preferred species. Over the 3 years of the study, populations of wigeon (*Anas penelope*) increased greatly and they too preferred *Agrostis stolonifera* — in fact the wigeon deprived the geese of much of their first choice in diet, an example of competitive exclusion from the preferred food niche.

Studies of grazing preference are usually made with a single species of animal on a variety of foods. A few comparisons have been made between animal species under comparable conditions. Cattle and sheep make different choices from amongst rows of *Lolium perenne, Dactylis glomerata, Festuca pratensis, F. elatior, F. rubra, Phleum pratense* and *Phalaris tuberosa* (Cowlishaw and Alder, 1960). Most studies of palatability are made with plants in the height of the growing season but this comparison between cattle and sheep was made in the depth of winter (February and March in England) and the diet available to the animals was the overwintered old herbage. The preferred species were those that were still green (*Phleum, Trifolium, Poa trivialis* and *Lolium*); cattle were generally more fond of *Festuca pratensis* than were sheep, and sheep were more prepared to eat *Dactylis* than were the cattle. The experiments included some mixed populations of plants and there were indications (as in the experiments with horses) that the animals preferred the mixtures.

The ability of animals to select a diet depends on their mouth parts. A sheep has a precise and small bite and is capable of extreme precision in its choice of plants, or parts of plants, in a pasture — it appears to discriminate in a pasture between leaves of *Trifolium repens* bearing different leaf mark polymorphisms (Cahn and Harper, 1976b; Charles, 1968) — but the cow which rolls its tongue round a tuft of herbage and pulls can scarcely have any fine scale of control over its diet in an intimately mixed sward of many species.

Indirect measures of preference, such as the time spent by an animal on different plots or measures of the amount of plant material present before and after grazing, are not as satisfactory as direct measurements of what the animal has actually eaten. The most direct way to measure food choice is by analysis of faeces, the contents of the stomach or of the oesophagus of the grazing animal. Oesophageal samples are probably the ideal as the sample has not then been digested or mixed by the churning action of the rumen. Faecal analysis is effective when the diet is mostly composed of grasses; their silicified epidermal walls are very distinctive so that identification is relatively easy and they are not badly damaged in a rumen (Martin, 1964), but the leaves of more tender dicotyledonous plants break down rapidly and are difficult to identify. Rumen contents can be extracted from the living animal after insertion of a rumen fistula but small animals must usually be killed to determine the stomach contents.

The two common small rodents of British woodlands, the wood mouse (*Apodemus sylvaticus*) and the bank vole (*Clethrionomys glareolus*) were trapped and killed in Wytham Wood, near Oxford, England. The trapping was done by using cheese as bait — clearly this was the preferred diet! The contents of the caecum (sometimes the small intestine as well) were analysed microscopically (Watts, 1968). *Apodemus* was mainly a seed eater but during periods of seed scarcity it turned to arthropods as an alternative diet. As a third preference, *Apodemus* took the buds and shoots of green plants. *Clethrionomys* scarcely touched animal food but ate fruits and seeds, though primarily those with a soft seed coat or pericarp. This species also ate the leaves of woody plants (preferring them to the leaves of herbs) and during the winter ate large amounts of dead leaves. The two species tended to make different choices among foods when food was abundant and to rely on different fall-back choices when food was scarce. *Clethrionomys* ate some roots though they were not an important part of the diet. It is an interesting question how much predation of root systems is done by small mammals particularly those of open grassland, but unfortunately it is extremely difficult to identify and quantify root material in stomach contents.

In western Queensland, Australia, kangaroos graze on sheep pastures and a study of their diets was made by examining stomach contents. Two species of kangaroo, *Macropus giganteus* (the grey kangaroo or scrubber) and *Megaleia rufa* (the red kangaroo), are regarded as pests by some sheep farmers. The climate of the area is strongly seasonal, and the flora is rich in grasses and dicotyledons, particularly Chenopodiaceae and Leguminosae. A large number of identifications could be made to species level by microscopic examination of powdered dry stomach contents; 13 grass species could be distinguished although members of seven genera including the important *Aristida, Chloris* and *Digitaria* could not be separated; 25 species of dicotyledon could be identified but six genera were difficult. It was not possible to distinguish genera and species in members of the Malvaceae and Sterculiaceae although they could easily be separated from members of other groups.

The paddock that the authors studied in detail contained three plant associations, a dis-climax *Astrebla* association, a *Eucalyptus populnea—Acacia aneura* association and a *Eucalyptus melanophloia—Triodia mitchellii* association (Holland and Moore, 1962).

The animals selected different components of diet from the paddock and this was apparent even from the gross appearance of stomach contents. Kangaroo stomachs always contained green herbage, whereas the sheep took yellow, dried-off herbage particularly during the hottest weather. The grey kangaroos are mainly grass eaters, taking 65% grass and 36% dicotyledons in January when the pastures were "lush with a germination of annuals and with good growth of perennials". At the same period of the year the sheep and red kangaroos ate a mixture of about 46% grass and 54% dicotyledons. Grasses contributed increasingly to the diet of all three species as the pastures became drier and the weather colder, but the differences between grey kangaroos on the one hand and sheep and red kangaroos on the other remained. There were, however, major differences in the species chosen by the sheep and red kangaroos (Fig. 13/6). In January the sheep diet was largely Malvaceae with smaller amounts of *Portulaca oleracea*, *Kochia tomentosa* and *Dactyloctenium radulans*; they did not eat *Astrebla pectinata*. At the same time the red kangaroos were eating *Chenopodium*, *Marsilea drummondii*, *Goodenia glabra*, *Eragrostis* spp., *Astrebla pectinata* and *Triodia mitchellii*. Rather surprisingly the sheep, not the kangaroos, were common browsers (Griffiths and Barker, 1966).

A division of food resources between a group of cohabiting herbivores is to be expected — a partitioning of the food niche. Differentiation between the grey and red kangaroo in feeding habits may imply an evolutionary history of competitive exclusion — it is less expected that the sheep as an introduced mammal should interniche (annidate) so neatly with the two resident marsupials. It would be extremely interesting to determine whether the dietary habits of herbivores in mixed populations remained the same when they were maintained as single species populations. Are the differences in dietary habit due to ultimate factors (terminology of Lack, 1954), the fixed evolutionary consequences of past mutual selection, or to proximate factors — the immediate effects of food shortages created by one species on what remains to be eaten by others?

Detailed analysis of dietary choice has usually been made with domestic animals and although the findings are not necessarily transferable directly to wild animals they are highly relevant, particularly in exposing the hazards both technical and philosophical in understanding dietary selection. Sheep show a clear order of preference between the parts of a grass when they are offered it in pure stand. The order is young leaf > old leaf > green stem > dry stem (Arnold *et al.*, 1966). Captive grouse

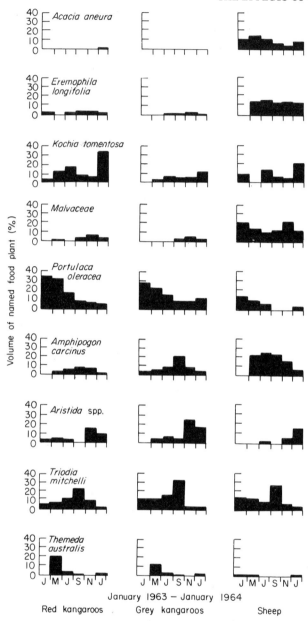

Fig. 13/6. The contribution of nine major food plants to the diet of grey kangaroos (*Macropus giganteus*), red kangaroos (*Megaleia rufa*) and sheep within some Australian plant associations. The figures represent contributions by volume to stomach contents). (From Griffiths and Barker, 1966)

(*Lagopus lagopus*) strongly prefer 3—4 year old heather for grazing
(Moss *et al.*, 1972). Species differ in their growth cycles and so some of
the selectivity shown by grazing animals between plant species may
really be the selection of young leaves from the spectrum of growth
phenologies in a mixed plant community (Arnold, 1964). The grazing
animal wanders in the course of its grazing activities and is a patchy defoliator. During the period when a grass plant is not eaten it may pass
from a palatable to an unpalatable stage in its phenology and be ignored
when the animal returns to it. Such a plant may then be at an advantage
over its neighbours and develop to a vigorous tussock, a well known
characteristic of *Dactylis glomerata*. The patchy grazing habit of the
animal, coupled with differences in acceptability with plant age, leads
to heterogeneities in a pasture, tussocks and lightly grazed areas, in a
previously homogeneous vegetation. Where there are intrinsic differences
in palatability between species in a mixed population such heterogeneity
is exaggerated still further.

Grazing animals defoliate differentially with respect to height in a
sward and when one species overtops another, the one lowest in the
canopy may escape grazing. In the early spring the growth of *Trifolium
repens* is commonly overtopped by grasses and sheep may then choose
to remove the grass and leave the clover (Arnold, 1964). Selection of
diet sometimes depends on the age of the animal as well as of the plant.
Clethrionomys and *Apodemus* (bank vole and wood mouse) took a
smaller proportion of seeds in their diet when young but accepted a
wider variety of foods than adults (Watts, 1968). Studies made on sheep
suggest that there is a learning process and differences between the feeding habits of different flocks make generalization very difficult. Flocks
of sheep, like many ungulates, tend to form into sub-flocks which divide
up a territory between them. The sub-flocks tend to be genetic families
and the plant species available as diet may differ markedly between
territories (Hunter, 1964).

The wood pigeon (*Columba palumbus*) has marked food preferences
among seeds, preferring wheat, maple peas, peanuts and green peas to
hemp and maize and preferring all these to millet, rice, rape, mustard,
linseed and sunflower. This order of preference has been established
even before the birds leave the nest and if naive birds are reared on
poultry pellets they still show the same preference order between offered
seeds when they are 3 months old. Nevertheless the woodpigeon will
cease to feed on its preferred food when the search brings uneconomic

diminishing returns. Mixed batches of wheat and peas were offered in the field to wood pigeons and 5 birds captured on the first day contained 1592 grains of wheat and 0 peas. Three birds taken on the fifth day contained 85 wheat grains and 55 peas and 7 birds examined after the eighth day had taken 0 wheat and 121 peas (Murton *et al.*, 1963). A pigeon takes about 50 g of cereal per day and if food is relatively abundant can collect this amount in 1—2 h. When the density of grains falls below *ca* 1 g per 1.8 dm² the birds tend to switch to other foods.

One of the most irritating features to emerge from acceptability studies is the frequent apparently absurd behaviour of the animals. It is well known that grazing animals graze most intensely that part of a pasture that they can just reach under or over a fence, e.g. a temporary electric fence across an otherwise uniform pasture. There seem to be elements of novelty seeking or exploration that play a part in dietary choice.

The very concept of palatability is difficult to define when it involves not only differences between plant species, but effects due to plant age and spatial distribution, animal age, social structure and possibly learning effects superimposed (Ivins, 1952, 1955).

> Preference can be adequately described only in terms of the relationship between the absolute and relative abundances of all potential foods, the 'risk' of each food and the proportion of each food in the diet. It will probably not remain constant either in time or space. The explanation of observed preferences entails many further factors. Observers should thus be extremely cautious and thorough when they speak of food preferences.
>
> (Emlen, 1966)

It seems a reasonable guess that the choices made by an animal among a variety of foods on offer are adaptive; that in some way the choice of diet maximizes fitness; that the most nutritious material is taken in relation to the energy spent in search collection and digestion. The protein-rich and digestible C3 species of plant seem to be commoner food sources for animals than the dietetically less useful C4 plants (see Chapter 10) Caswell *et al.*, 1973).

> Inasmuch as food preferences may be at least partially controlled by genetic factors, one would assume that natural selection has favoured those genotypes which predispose their owners to favour the 'right' foods, that is, those that yield most in net energy and nutrients per time to their predators.
>
> (Emlen, 1968)

Although the principle of optimal behaviour is very attractive (it is a facet of the philosophy of evolutionary optimism which holds that

nature ensures the best in the best of all possible worlds) there is at least
one alternative way of looking at food choice. In the event of a struggle
for existence between two herbivore populations with limited but various
food resources it is likely that either one species will go to extinction *or*
that the species will diverge in feeding habits. In the latter case it may
subsequently be of selective advantage, not only for these different feeding habits to become fixed genetically but for the animals to develop
different behaviours and physiologies that fit them better to use the
types of food that they have been forced to take as a result of competitive exclusion. It is absurd to expect that a red kangaroo and a grey
kangaroo both choose from a pasture what is dietetically "best". It is
more sensible to look for the ways in which they cope with the different
diets that competition, with its threat of exclusion, forces them to adopt.

A great deal of energy has been put into trying to discover whether
animals do indeed make optimal dietary choices. In the case of sheep
there is clear evidence that the diet selected is of greater nutritional
value than the average of the plant material on offer, but this is probably only true when there is plenty of good food available (Meyer *et al.*,
1957; Arnold, 1962). The rumen or oesophageal fistula allows the actual
choice made by the animal to be compared with the herbage on offer.
Hamilton *et al.* (1973) found that when the best herbage was in short
supply the diet selected by the sheep was poorer than they could have

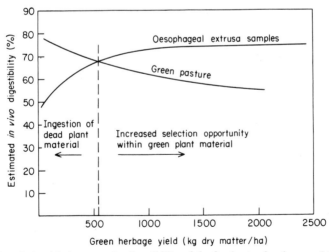

Fig. 13/7. The relationship between the digestibility of the food eaten by sheep and that of
the herbage on offer plotted against the yield of green herbage. (From Hamilton
et al., 1973)

obtained, though when herbage was abundant they selected the most digestible fractions (Fig. 13/7).

Birds select between seeds that are offered as food. The shape of the beak determines to a large extent the sorts of seeds that can be collected and eaten as well as the ability to dehusk hard-coated seeds. Birds with large bills tend to take larger food items than their smaller-billed relatives (Lack, 1947). Willson (1971) captured eight species of finch in the field and fed them a mixture of the seeds of eight plant species in the laboratory for a week. Subsequently she measured the choices made by each bird species in a series of tests. The seeds differed greatly in size, from sunflower (*Helianthus annuus*) 8.3 per g to thistle (*Guizotia abyssinica?*) 343.5 per g. The calorific value per kernel varied from 440 for sunflower to 8 for millet. The number of calories per gram of kernel ranged from 7325 for sunflower to 4387 for canary grass (*Phalaris canariensis*). There was no indication that the birds maximized their calorific intake per seed "captured". All the birds chose mostly small seeds though the small-billed species ate *only* small seeds and birds with bigger bills took a wider variety of sizes. The different seed types took different times to dehusk but, at least for the smaller seeds, seed preference was quite independent of husking speed. For all birds that were tested, the most frequently eaten seeds included those that yielded the most calories per minute but there was no correlation in rank orders and commonly eaten seeds included types with a low caloric yield per minute. Again, as with larger mammals, it may be quite wrong to expect to find optimal behaviour on the part of a single species when its evolution will have been partly, perhaps largely, determined by the competitive pressures from other species to take different foods rather than the best.

The way in which competition can cause exclusion from a preferred niche is clearly shown in studies of flatworms (triclads) which feed on other aquatic animals (there is no example quite so clear for animals that feed on plants). When the two species *Dugesia polychroa* and *Polycelis tenuis* were both offered damaged specimens of oligochaets, gastropods and arthropods, both preferred oligochaets. When undamaged prey were offered separately the oligochaets were again eaten in greater numbers by both triclads. In Britain, where both species of triclad are found together *Dugesia* is forced to take mainly gastropods because of the greater success of *Polycelis* in competing for the preferred food, but in parts of Canada where *Polycelis* is absent, *Dugesia* takes predominantly oligochaets. In experimental populations, the persistence of *Dugesia* in the presence of *Polycelis* is only possible if gastropods are available to provide a "food refuge" for *Dugesia* (Reynoldson and Young, 1965; Reynoldson and Davies, 1970; Boddington and Metterick, 1974; Reynoldson and Bellamy, 1973).

When an animal has a range of potential foods on offer and is polyphagous it may choose to concentrate upon the commonest and only switch to an alternative when the previously common food has been depleted (apostatic behaviour). Such behaviour is interpreted as due to the formation of search images. A polyphagous feeder behaves for periods as a strict monophage (e.g. the feeding of wood pigeons on mixed seed). Such behaviour is perhaps of great importance to the plant population biologist because the point of switch (the density of prey at which the search image changes) determines the density of prey that are left, e.g. seeds that are left to germinate. The formation of search images may increase the efficiency of search — it is probably always easier to find something if you know *precisely* what you are looking for. Most demonstrations of visual selection that appear to involve feeding images have been done with birds (de Ruiter, 1952; Tinbergen, 1960; Beukema, 1968). One of the effects of apostatic selection is to penalize a common prey and so give a relative advantage to rarer forms. It is a frequency-dependent process and so tends to maintain diversity if it is applied by a predator to a mixed population of prey. It may, similarly, stabilize a polymorphism within a prey species, penalizing each morph that becomes abundant in a population (Allen and Clarke, 1968; Clarke, 1962). Apostatic selection has been shown to operate in the choice of food by small mammals (Soane and Clarke, 1973) and olfactory clues may apparently be used as search "images" as well as the better known visual clues.

The plant population biologist is perhaps more interested in what is left after predation than in what is actually eaten — the ability of a plant to recover after defoliation depends on the residues that remain to regenerate. The ability of a population to reestablish from seed depends on the seeds that are left after predation. The greater part of the study of animal predation on plants has been conducted from the animal's "point of view". Rather differently designed experiments are needed before the animal's role can be properly assessed from the plant's "point of view": in particular, experiments are required in which marked, counted seeds are placed at risk to predation and attention is concentrated on what is left. A few generalizations do seem to emerge about the interactions between predators and seeds:

(i) Plant embryos, like eggs, are generally palatable and do not carry the load of toxic compounds, alkaloids, glycosides and terpenes that are in the leaves and fruits of so many species. There are exceptions to this generalization but the contrast between the toxicity of seeds and leaves remains very striking (Orians and Janzen, 1974).

(ii) Thick seed coats and pericarps seem to provide some protection against predation, if only by wasting the predator's time (3.1—70.1 s per seed in studies of desert rodents (Rosensweig and Sterner, 1970).

(iii) Small seeds escape predation better than large seeds, partly because they are more quickly buried or fall into cracks in the soil and escape detection, and partly because there is a high ratio of search effort per unit reward.

(iv) Well dispersed seed escapes predation more effectively than seed dropped in patches.

(v) Seeds bearing awns or spines tend to escape heavy predation — especially if they speed up the process of burial as in *Avena fatua*. Awns may also protect seed from pre-dispersal predation. M. F. Horne (unpublished) showed that birds heavily predate the ears of 'Proctor' barley which has ears that droop at maturity and long parallel awns that do not get in the way of the birds when feeding. Primitive barleys with awns that are carried at right-angles to the ear escape predation and in a mixture of 'Proctor' with the primitive forms, 'Proctor' was protected from birds by the awns of the intermingled primitive varieties.

Search

The seed or plants that are left after a predator has taken his toll are what is available to grow and form a new generation of plants. What the predator leaves depends on his demands and the efficiency of his search. A predator's search pattern may ultimately determine both the abundance and distribution of a plant. Feeding involves a search for materials that conform to an image or range of images (not necessarily visual).

A monophagous feeder will starve when its search activities fail to discover enough food to support its activities, whereas a polyphagous feeder changes diet. The critical issue for the population biology of plants is the point at which the monophage starves or migrates and the polyphage decides to eat something else. Search (and capture, which for an animal feeding on a plant includes an activity like dehusking the seed) requires time and energy. "Search and find" is a game played by two teams — seekers and sought. The seekers gain by discovery, the sought gain by escape. Any adaptation in the seeker which increases his chance of making discoveries brings counter-selection on the sought to favour more effective escape. Dispersal is one ploy in a strategy of escape. It increases the energy spent in search and therefore increases the chance of not being found.

A good example of the way in which the dispersal of plants protects against predation is found in the behaviour of the butterfly *Melitaea harrisii* whose larvae feed on *Aster umbellatus*. In an area studied by Dethier (1959) in Maine, U.S.A., *Aster* is the only acceptable food plant for the larvae. The butterfly lays its eggs in batches of 20—400 on the leaves of *Aster*, the larvae feed gregariously until the third instar when they disperse, over-winter and then start feeding again as solitary feeders. The young larvae eat their way down from the top of a plant to the base and as a single plant is never enough to satisfy the needs of a colony they are forced to migrate. A colony from a small batch of eggs will be able to grow to quite large caterpillars before migration but if the batch of eggs is large or the plant is small the larvae quickly defoliate the plant and migrate when very small. Dethier observed a number of populations of *Aster* and the migratory behaviour of *Melitaea* larvae. Not all larvae succeeded in finding new plants. Successful search varied with the distance to be travelled and the size of the larvae; the period of migration exposed the caterpillars to the risk of starvation and put them at the mercy of predators. A typical study is illustrated in Fig. 13/8: 30 larvae from each of four neighbouring clusters were colour marked and the four plants were then defoliated. After 24 h the larvae had distributed themselves among the surrounding plants. Isolated plants clearly escaped discovery in this time and the chance of escape was the greater because *Melitaea* larvae wander at random with respect to the position of host plants. Larvae never succeeded in reaching plants farther away than 120 cm in the course of a migration. The adult butterfly, of course, has a much greater range of search than the larvae and there is some evidence that the chance of populations of *Aster* being attacked by *Melitaea* is again a function of the search range of ovipositing females.

A classic model of the interaction of dispersal and search range is described by Huffaker (1958). He examined populations of a herbivorous mite which fed on oranges. Huffaker "dispersed" oranges among rubber balls of the same size and also coated the surfaces of oranges in such a

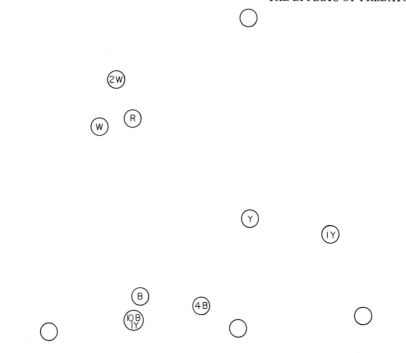

Fig. 13/8. Map of the positions of plants of *Aster umbellatus* of which four, W, R, Y and B, bore egg clutches of *Melitaea*. The first instar larvae were marked White, Red, Yellow or Blue. The plants were defoliated and the migrating larvae 24 h afterwards were distributed as shown. (From Dethier, 1959)

way as to leave varied areas of skin exposed to the mite. In this way the amount of food (habitable site) could be clumped or evenly distributed and many small or few large feeding areas could be left uncoated on the oranges. When a 2-orange feeding area was distributed over a 4-orange range, the total density of mites reached *ca* 8000 but when the same feeding area was distributed over a 40-orange range the population density of mites was only *ca* 5000.

The search range of a predator may be involved in the regulation of population size of *either* predator *or* prey *or* both. If the predator is a monophage it suffers when the prey are widely dispersed (even though they may be abundant). A monophage is, however, very unlikely to destroy a population of its food plant, if only because some individuals are likely to lie outside its search range and the predator is likely to suffer catastrophically before the food plant is completely killed. In contrast it is quite easy to see how the persistent activities of a polyphagous feeder may lead to extinction of food plants because a relentless pressure can be maintained on a specially attractive food species while the predator's populations are maintained on alternative food sources. When we look for plant populations that may be regulated by predators we might expect that in the case of monophagous predators search range will often be critical in determining plant density, but in the case of polyphagous feeders the level of rarity at which the animal gives up the search and changes its feeding image is more likely to be important.

Although botanists have rarely looked for patterns of animal behaviour to explain the abundances of species the importance of predation and particularly the search range of predators was clearly recognized by Ridley (1930):

> In almost every plant the greatest number of its seeds fall too near the mother plant to be successful, and soon perish. Only the seeds which are removed to a distance are those that reproduce the species. Where too many plants of one species are grown together, they are apt to be attacked by some pest, insect, or fungus. It is largely due to this . . . that one-plant associations are prevented and nullified by better means for dispersal of the seeds. When plants are too close together, disease can spread from one to the other, and can become fatal to all. Where plants of one kind are separated by those of other kinds, the pest, even if present, cannot spread, and itself will die out, or at least become negligible.

Much of the stress of this chapter is laid on specific behavioural patterns of animals and the ways in which these interact — often in fast evolutionary fluxes — with aspects of the morphology, chemistry and life cycle pattern of the prey. Virtually all of the work reported is that of zoologists and the research is zoocentric. If an animal plays a critical role in determining the distribution and abundance of a plant the activity is not likely to be obvious. A quite new type of experimental approach to plant ecology is required if the role of animals is to be exposed. The most likely experimental procedures for displaying the role of predators are (i) exclosure experiments and (ii) the deliberate increase in the local density of the plant that is studied. This latter procedure which has scarcely ever been followed outside agriculture and horticulture is the most likely to reveal what forces bring the density back to the normal level.

14

The Role of the Grazing Animal

Much of our knowledge of the role of grazing animals in the population biology of plants comes from agriculture where deliberately introduced mammals are managed on natural or artificial grasslands. There is a vast literature in agronomy which is gradually being absorbed into the sciences of nature conservation and reserve management as it is realized that the botanical characteristics of a grassland community reflect its history of grazing pressures (Duffey *et al.*, 1974). An introduction to the agronomic literature has been set in an ecological context by Spedding (1971).

One outstanding study of grazing effects was made early in the science of ecology by Tansley and Adamson and stands apart from the main body of literature on grazing. They attempted to study the influence of the rabbit (*Lepus cuniculus*) on the composition of British chalk grasslands. The rabbit is believed to have been introduced to Britain by the Normans but probably did not develop large feral populations until 1750 (Sheail, 1971). Tansley and Adamson (1925) placed exclosures on the floristically very rich grassland of chalk. The effect over the first 6 years was the rapid "degeneration" of the community to a monotonous vegetation dominated by the grass *Zerna erecta* (Fig. 14/1). Over longer periods exclusion of the rabbits allowed natural establishment of shrubby species in a succession towards scrub (Hope-Simpson, 1940). This experiment contributed a predictive element into the ecology of grassland and when in the years following 1954 myxomatosis rapidly destroy-

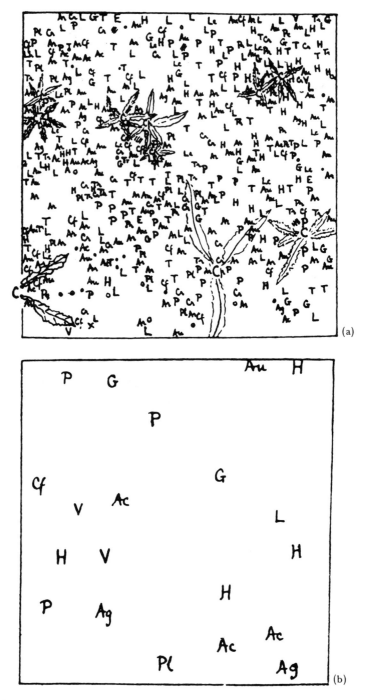

Fig. 14/1. The influence of rabbit grazing on the floristic composition of chalk grassland (a) chart of an area of 25 cm square under moderately heavy rabbit grazing (b) the same area after exclusion of rabbits for 6 years. The symbols represent different species and the background in (b) is composed of the grass *Zerna erecta (Bromus erectus)*. (From Tansley and Adamson, 1925)

ed much of the rabbit population in Britain the exclosure effect was
displayed on a very large scale. It is clear that the great floristic diversity
of the chalk grasslands owes its existence to the continued activity of
the rabbit, and in particular to the selective grazing of the potentially
dominant grasses.

A second example of the control of the composition of a plant community by the activities of defoliators is a comparison of two contrasting
areas of grassland near Oxford in the Thames valley. One area, Port
Meadow, has been grazed by cattle, horses and geese (no sheep) for nearly 900 years except for the years 1643—45 of the Civil War when it was
cut for hay. Two meadows close by have been cut regularly for hay, and
livestock only allowed to graze the "aftermath". Cutting for hay on
these meadows is traditionally done rather late in the summer. The
grazed and mown meadows are so close and on the same soil type that
they represent an unreplicated comparison of management practices.
The flora of the meadows was described in 1937 (Baker, 1937) and 95
species of higher plant were present on the area as a whole; 56 species
were in the grazed pasture and 69 in the hay meadows; 26 species were
restricted to Port Meadow and 39 to the hay meadows. The flora of
Port Meadow is composed of perennial grasses (e.g. *Lolium perenne,
Cynosurus cristatus, Phleum pratense, Poa trivialis* etc.) plus laterally
spreading clonal dicots (*Prunella vulgaris, Bellis perennis, Ranunculus
repens*) and some rosette-forming species without stolons or rhizomes
(*Taraxacum officinale, Ranunculus bulbosus*). The hay meadows contain perennial grasses but also a high proportion of annual grasses (e.g.
Bromus spp.) and short-lived perennials (*Anthoxanthum odoratum*)
while the dicotyledonous flora is predominantly of plants that bear
erect shoots and cauline leaves (*Silene floscuculi, Rhinanthus crista-galli*).

Since 1937 two species have appeared in Port Meadow which were
not in the 1937 lists: *Senecio jacobaea* and *Veronica filiformis*. The
former has a growth form alien to the grazed pasture, forming tall
flowering stems in its second and later seasons of growth. It escapes
grazing because of its extreme unpalatability. *Veronica filiformis* has
entered the community apparently as an escape from Oxford gardens
and forms a low tangle of thin shoots in the base of the grazed sward.

The contrast between the two grasslands is essentially between a
situation where to be tall is disastrous and one where success in a struggle
for existence depends on being tall. The mown system prevents dominance ever being gained by plants with perennial woody stems but every

year the vegetation is allowed to grow tall and rosette or creeping growth forms have a poor chance of survival under a high canopy. The system of management allows early-flowering species to complete a large part of their potential growing season and to set seed before hay is cut. The grazed meadow is under continual grazing pressure and the successful growth forms depend on meristems being borne at or below ground level, extensive clonal growth, little dependence on establishment from seed, leaf areas that are borne close to the ground and escape grazing or the possession of chemical or physical protection against the grazer (alkaloids, glycosides, spines etc.).

A classic series of experiments in the literature of agronomy are the studies by Martin Jones (Jones, 1933a, b, c, d) of the effects of controlled grazing by sheep on the composition of pastures. In one experiment a very old permanent pasture was taken for a grazing experiment. The sward consisted chiefly of *Agrostis* species together with *Holcus lanatus*, *Festuca ovina*, *Festuca rubra* and some other grasses together with *Trifolium repens*. Plots were grazed by sheep which were managed in different ways: (a) plots were grazed very closely, particularly in the spring; (b) grazing was delayed until the beginning of May; (c) plots were rotationally grazed, i.e. the herbage was eaten off fairly closely at each grazing and the sward was then rested from grazing until the regrowth had approached 12.5 cm in the summer months, 10 cm in April and October and 5 cm during the winter; (d) swards were overgrazed in winter and spring and undergrazed in summer and autumn. This latter treatment approximated to continuous grazing with a constant number of stock which were therefore partly starved during periods of slow crop growth yet could not consume all the growth made during the high season of production.

Imposed on the series (a), (b), (c) and (d) of sheep management was a difference in fertilizers applied which contrasted applications of phosphate plus nitrogen with phosphate alone. Unfortunately Martin Jones did not make detailed botanical records but he did group plant species into four categories: (i) perennial ryegrass (*Lolium perenne*), (ii) useful grasses, (iii) white clover (*Trifolium repens*) and (iv) useless grasses plus other weeds. These broad categories were sufficient to show clearly the magnitude and direction of changes within the plant populations (Fig. 14/2). Intense grazing in the spring produced a spectacular increase in the proportion of *Trifolium repens* in the sward and a lesser but still striking increase in clover was obtained from balanced rotational grazing.

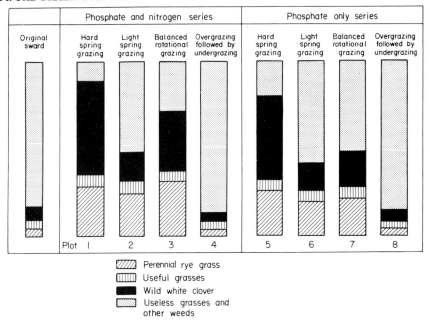

Fig. 14/2. The influence of four different grazing regimes on the botanical composition of an old permanent grassland. (From Jones, M. G., 1933e)

After treatments (b), light spring grazing, and (d), overgrazing followed by undergrazing, the proportion of clover present was not much altered from that in the original sward. The proportion of *Lolium perenne* present was increased by every management treatment except (d).

The changes in composition of the swards in Martin Jones' experiments were very rapid although they had apparently retained a stable composition for many years previously. He interpreted the results of this experiment in terms both of the competitive interaction between the species and of the selectivity of the grazing animal. The species originally present are known to vary in "palatability" to sheep, *Lolium perenne* and *Trifolium repens* being particularly sought after. These species make strong growth when grazing is restricted. The time at which grazing occurred affected the balance of species because the grasses normally start producing new leaves earlier than *Trifolium repens*, so that if grazing is allowed early in the season it penalizes the grasses more than the clovers which are then favoured because they develop in a less dense sward. If grazing is delayed the grasses form a closed canopy, putting *Trifolium repens* at a disadvantage.

The supreme influence of the biotic factor, however, is two-fold first, there is the selective grazing: this becomes manifest when plenty of fodder is available, the palatable plants being sought after; or again when the habit of growth becomes so prostrate that it becomes almost inaccessible to the animal, in which case the erect growing plants are taken. Secondly the biotic influence is exerted through differential weakening of the various species owing to variations in the intensity of grazing at different times in the year, and to inherent differences between the capacity of different species and strains of plant to withstand defoliation.

The relative abundances of species in grassland may be changed by altering the nutrient status of the soil or by specific grazing practices. The two factors interact but the effects of the grazing animal can override quite major nutritional effects. An area of native hill pasture at Llety-ifan-Hên at 270 m in Cardiganshire, Wales was laid out in an experimental design involving the addition of nitrogen, potassium and phosphatic fertilizer with and without liming (Milton, 1940, 1947). The plots were either exposed to free grazing by the hill sheep or were fenced. The fenced plots were grazed but on a cycle of graze—rest—graze—rest. During the grazing phase of the cycle the sheep grazed most of the species present and during the rest periods all the species made regrowth from a more or less similar degree of defoliation. On the open plots the sheep were always free to come and go as they pleased and to select whether to graze a plot or not. Under these conditions grazing was very selective. Profound changes in the species composition of the plots followed the treatments and these were followed over a period of 16 years. The response to fertilizer treatment was very marked and occurred rapidly on the fenced plots but in marked contrast the same fertilizer treatments on the open plots produced very slow change. The uncontrolled access of sheep to the unfenced plots meant that all changes to more palatable species were hindered by the selective grazing.

Major changes in the composition of plant populations can be obtained by grazing as opposed to cutting and Fig. 14/3 shows the results of only 2 years of different managements on the composition of a hill pasture.

Grazing is a complicated process varying in subtle ways between animals and producing different reactions among plants. The cow rolls her tongue round a bunch of grass and pulls. If the plant is well rooted the leaves break but some plants may be physically uprooted by such a tearing action. In a grassland containing *Agrostis stolonifera* cattle often remove whole groups of weakly rooted tillers and leave them scattered over the ground. Sheep bite leaves between incisors of the lower jaw and

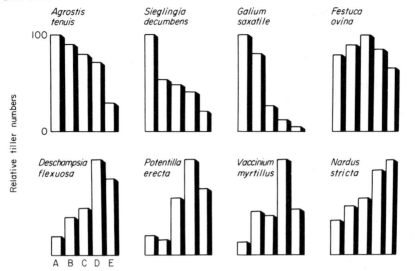

Fig. 14/3. The influence of two years of grazing and mowing treatments on the *relative* numbers of tillers of various species in *Agrostis*-Fescue upland pasture. (Redrawn from Jones, Ll. I., 1967)

A = frequent close grazing
B = moderate grazing
C = light grazing
D = mown ungrazed
E = unmown—ungrazed.

a pad on the upper jaw whereas the rabbit cuts leaves with teeth on both jaws. It is probable that the effects of different grazing animals on a sward could be assessed by comparing the numbers of leaves bearing different types of scar left by these very different techniques of defoliation. Many of the grazers present on pastures have distinctive methods for removing leaves: the slug pares away a part of a leaf with its radula, whereas geese bite sideways. Some of the grazing techniques permit very precise choice between individual leaves on a pasture (e.g. the sheep may distinguish between individual leaves of clover and select large leaves and specific leaf marks (Cahn and Harper, in press), but the cow and the horse are bound to be far less precise. The rabbit and the goat are probably as precise as the sheep.

    The imitation of grazing by deliberate clipping cannot possibly mimic many of the subtleties of real grazing. A clip is non-selective except in so far as it affects different layers in the canopy. A clipping or mowing gives an even defoliation whereas the activity of the animal is patchy.

Sheep tend to graze small patches *ca* 16 x 16 cm and then move away to another patch (Morris, 1969), giving a mosaic of heavily and lightly grazed areas. In a heavily stocked pasture of *Lolium perenne* sheep returned to graze marked tillers every 7—8 days and, when the stocking level was lower, at intervals of 11—14 days. The animal removed *ca* 40% of the green leaf length of the tiller (Hodgson, 1966) and the younger tillers were much more likely to be taken than the old. In such a situation reproductive tillers may often escape damage and this is particularly the case in grasses like *Cynosurus cristatus* and *Lolium perenne* in which the interval between one visit by sheep and the next is sufficient for the flowering tillers to elongate, harden and become unpalatable.

Grazing may kill plants if it happens when they are in the seedling stage. The most sensitive species are those with epigeal germination — one bite will usually remove all the leaves and with them the whole of the shoot meristems of the plant. Epigeal species carry all their reserve meristems above the hypocotyl (e.g. many dicotyledonous weeds) so that defoliation is almost always lethal. The growth forms that withstand grazing are usually hypogeal (the grasses *par excellence*) and grow clonally as tussocks or by lateral growth.

Damage done to a plant growing in isolation may set it back in the course of its growth. It may delay the time taken before a perennial flowers and sets seed, and it may reduce the number of flowers and seeds that are formed. The effect on an isolated plant will then be to slow down the rate at which its population expands and the speed with which its descendants occupy a habitable site. For a plant growing in a crowded population the effects may be much more severe, particularly if the damage is not done uniformly to neighbours in a population (see Fig. 12/3). The presence of the animal affects the balance of other density-dependent processes. Thus if a struggle for existence in a sward concerns the partitioning of light as a limited resource, a defoliator may change the proportions of the light captured by two species or forms and hence their rate of regrowth. The effects can be seen most strikingly in the alterations of the balance of species in a pasture but there are other intraspecific changes which are of short and long term importance.

The speed of the genetic changes that occur in different grazing regimes is quite extraordinary. Brougham and Harris (1967) sowed pastures with two forms of *Lolium perenne* 'Grasslands Ruanui' and 'Grasslands Manawa' in an equiproportioned mixture together with white clover. The plants could be classified into types according to (i) a fluorescent property of seedlings, (ii) the development of awns, (iii) the time of after-math heading and (iv) a 1−4 scale varying from 1 = narrow leaves with prostrate habit and dense tillering to 4 = wide leaves, an erect habit and loose tillering. The experimental swards received different grazing treatments: (i) continuous and close, (ii) moderate and (iii) lax grazing. Trends in the representation of the types in the pastures are shown in Fig. 14/14.

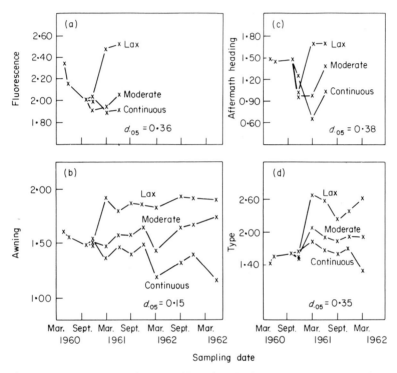

Fig. 14/4. Changes in the genotypic composition of a mixed population of forms of *Lolium perenne* under lax, moderate and continuous grazing treatments. (From Brougham and Harris, 1967)

After $2\frac{1}{2}$ years *ca* 90% of the plants remaining in the lax grazing treatment had morphological characters like the Manawa ryegrass whereas in the continuous grazing treatment 85% of the plants were of Ruanui character. What is even more remarkable is the speed of the changes: within a month from sowing the mean fluorescence score changed from 1.85 to 2.34. The major changes in representation of the two forms of ryegrass had occurred within 4 months of the start of the different grazing treatments. Most ecological change probably involves evolutionary flux — this example (and others, e.g. Charles, 1961) point to the speed and the extent of this process which is usually ignored. As in so many other examples in this book the focusing of attention on the population biology of a species immediately exposes processes in the realm of population genetics.

Modelling the population biology of a grassland system

The effects of grazing animals on a habitat containing several species of plant involve complex interactions, more suited to analysis by computer than the mind. Goodall (1967, 1969) pioneered this approach with a relatively simple model which takes account of the following variables:

(a) *Meteorological factors*. He used real data for Kalgoorlie in Western Australia: mean monthly temperatures, random samples from rainfall records supplemented with samples from empirically determined relationships between rainfall sequences in different months of the year.

(b) *Soil moisture*. He calculated this as a water budget similar to the procedure used for calculating irrigation need.

(c) *Plant growth*. He assumed that each plant responds independently to its inanimate environment, i.e. he did not take into account interference between neighbouring plants. Growth rates were calculated as relative growth rates, a function of temperature and available soil moisture.

(d) *Movement of animals*. Grazing animals tend to spend more time near fence lines and watering points than elsewhere and their activities depend on the availability and palatability of the plants available. The number of sheep minutes of grazing time was taken to be proportional to the product of two factors: (i) preference of the sheep for the area (a function of the distance from the fence line and water) and of the water stress of the vegetation (measured as evaporation rate), and (ii) the effective quantity of forage, defined as the sum of the quantities of various species weighted with respect to their palatabilities. The model therefore includes elements of choice and search on the part of the predator.

(e) *The amount of forage consumed by the animal*. This was taken to be a function of that on the paddock as a whole and was distributed over the paddock in proportion to the preference factors described in (d). Within a unit area the amount of each species consumed was regarded as a function of the quantity of that species present as a ratio to that of all less palatable species.

(f) The animals were considered to increase in weight as a function of the food they eat and their weight at the time.

(g) *Livestock mortality*. This was taken to be mainly dependent on the weight of the animals.

The characteristics of the five principal species are shown in Table 14/I and the hypothetical paddock is shown as a map in Fig. 14/5.

The elements of the model are a series of differential equations (Fig. 14/6) and Goodall used these to determine changes in floristic composition of the paddock, and the numbers and weight of the animals in

Table 14/I

Characteristics of the five principal species composing a hypothetical pasture (from Goodall, 1969)

Species	Growth rate	Sensitivity to temperature	Sensitivity to soil moisture	Palatability
1	High	Moderate	Moderate	High
2	High	Moderate	High	Low
3	Low	Moderate	High	High
4	Moderate	High	Low	Low
5	Moderate	Low	High	Moderate

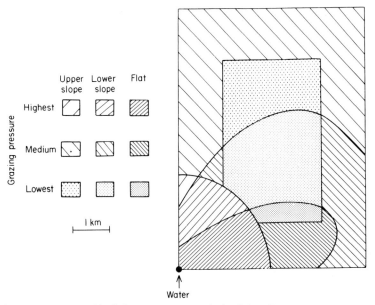

Fig. 14/5. Hypothetical paddock for a computer analysis of the effects of grazing pressures on species composition. (From Goodall, 1967)

four combinations of conditions — heavy and light stocking rates and in seasons of high and low rainfall. The results are shown after 1 and 4 months of simulant grazing in Table 14/II. The model is of course fictitious, but the situations that it describes are readily imagined. The results

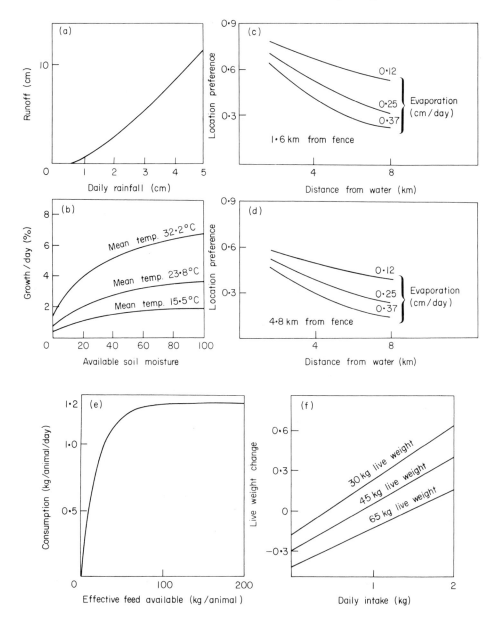

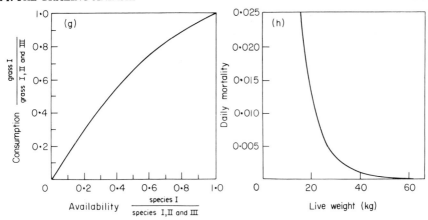

Fig. 14/6. Relationships between variables involved in a simulant grazing programme involving
sheep on a paddock of mixed species compositon. (a) run-off and daily rainfall in
paddock unit I, (b) forage growth rate of Species I, temperature and soil moisture,
(c) and (d) sheep preferences for locality and distance from fence and water and
evaporation rate, (e) consumption of forage and amount available, (f) consumption
of most palatable species and the proportion of these among the whole, (g) live weight
change of sheep in relation to forage intake and (h) mortality of sheep in relation to
live weight. (From Goodall, 1967)

are necessary consequences of the equations fed into the model but these
are not far from representing real common situations. The conclusions
are those that are known from the real world of grazing situations — extinction of some species, relative increases by unpalatable species, mortality of predators during periods of food shortage, species compositions
in the various subpaddocks that reflect an interaction between climatic
conditions and grazing pressures. To introduce competitive interactions
between the plant species would have increased the verisimilitude of the
model, but would have involved a 5 x 5 matrix of interactions and so
made the whole incredibly complex (though yet more realistic). To have
introduced genetic variation within the species (both of plant and animal)
would have been a further step towards realism, but at the same time
would have completely lost the value of the simplified model.

Most models serve their main role in ordering an intangible number of
interacting variables. This model does this admirably and could be applied without very much change to other predator—vegetation systems
such as the feeding activities of small mammals on a woodland floor where
forest regeneration rather than animal yield is the relevant interest. Other
models of predator—prey interactions where the predator is a herbivore

Table 14/II

Changes in the species composition of a hypothetical pasture and changes in livestock weight and numbers in a computer study of grazing effects (from Goodall, 1969)

Rainfall	Stocking rate	Months	Weight of forage					Stock numbers	Stock mean live-weight (kg)
			Sp. 1	Sp. 2	Sp. 3	Sp. 4	Sp. 5		
	Low	0	01	42	29	179	24	2000	50.0
	High							4000	
Average[a]	High	1	50	51	11	214	26	3964	49.9
		4	4	36	0	504	26	3834	40.7
Average[a]	Low	1	82	55	18	227	28	1984	51.0
		4	72	102	6	706	36	1941	55.0
Low[b]	High	1	37	40	10	167	22	3964	49.6
		4	1	11	0	127	15	3766	38.8
Low[b]	Low	1	62	43	17	197	23	1984	31.8
		4	8	33	2	327	21	1937	51.7

[a] 1.41 inches in first month, 2.71 inches in 4 months.
[b] 0.19 inches in first month, 1.04 inches in 4 months.

and the prey are plants are of great interest (for example seed predation in a demographic model of annual plants in a desert (Wilcott, 1973), predation on the seeds of trees (Vandermeer, 1975) and the effects of defecation in a grassland system (Spedding, 1971)).

Secondary consequences of grazing

It would be quite wrong to suppose that the effects of grazing animals in the population biology of plants are dominated by defoliation. During the course of an animal's life it may affect plant growth in a great variety of ways. "Grazing animals frequently sit, lie, scratch and paw on the pasture in addition to walking, running and jumping on it" (Spedding, 1971). Moreover they deposit dung and urine. The localized effect of each of these activities is to produce mosaic effects, heterogeneities within a grazed system that add to the heterogeneities caused by patchy grazing. The activities of a grazing animal disrupt any trend to monotony within vegetation. Once heterogeneity exists the activities of animals tend to exaggerate the variations within a community.

The deposition of faeces is one of the more conspicuous effects of grazing animals. A 350 kg cow voids *ca* 34 kg of faeces (5–6 kg dry weight) and covers *ca* 0.75 m² of ground each day (MacLusky, 1960). A large faecal dropping represents a disaster for the plants beneath and around it. The effects are at least 4-fold: (i) smothering and exclusion of light from the plants; (ii) a local disturbance of the nutrient relations of the pasture which may extend beyond the faecal patch; (iii) changes in the pattern of grazing around the patch which animals tend to avoid; and (iv) the creation of an island for colonization by new species or invasion by the more vigorous lateral-spreading species in the neighbourhood.

The smothering effect of dung has species-specific effects. Few grasses can grow through a solid dung patch, particularly if it has dried in the sun, but a few dicots (e.g. *Cirsium arvense*) can emerge through a dung patch or even raise it. The growth habits of some species (e.g. the stoloniferous habit of *Ranunculus repens*) allow the plant to spread quickly over a dung patch even if it does not root in it.

The decay of sun-baked dung may take several years as it cracks and decomposes and the presence or absence of appropriate agents of decomposition can control the speed with which a patch disappears. The absence from Australia of dung beetles capable of dealing with mammalian dung (marsupial dung presents no problem) has meant that the long persistence of dung patches is a serious problem in Australian grasslands; sufficiently serious to have stimulated the deliberate introduction of dung beetles from other countries (Hughes, 1975).

When a dung patch does break up and disappear it will often leave a bare patch suitable for colonization. The colonists may already be present as seeds that have survived passage through the animal's gut and subsequent entombment in the dung patch, e.g. *Poa annua*.

In the immediate circle around a patch of dung there is a commonly a zone of increased plant growth which probably is usually due to local stimulation from the nitrogen released from the dung. The stimulation may also result from an unwillingness of animals to graze close to their own dung and subsequently their unwillingness to eat plants that have grown rank because they have been ignored. Within this zone of taller, ungrazed herbage the balance of competitive interactions is changed in favour of species that have high canopies or that can sprawl upon the support of the ungrazed sward, e.g. *Agrostis stolonifera*. The effects of supplementing the dung deposited by sheep in a permanent grassland are shown in Fig. 14/7.

The fouling of grassland by dung warrants much more detailed study in natural as well as agricultural systems. There is a frequently made suggestion that the fouling effect is species-specific, that an animal is unwilling to take herbage fouled by a member of its own species but willing to take it if the fouling was by a different species. If this effect is real and general it would mean that more effective utilization of a pasture would be gained from grazing mixed populations of herbivores and that in nature there may be species-specific, density-dependent control of the availability of food.

The effects of local deposits of urine are less obvious. A cow produces 1–2.5 kg of dry matter per day as urine, returning most of the daily intake of potassium and some of the nitrogen in a readily available form. In potassium-deficient communities the urine effect can be seen as local patches of dark green foliage against a background of paler sward in which the leaves show the characteristic tip burn of potassium deficiency. Both faeces and urine are nutrient rich, representing mineral elements collected over a large area and deposited in a small area. The grazing animal is therefore a diversifier of the fertility regimes in its environment.

It is not only the larger mammals that create islands of disturbance within plant communities. The rabbit (*Lepus cuniculus*) uses latrines in

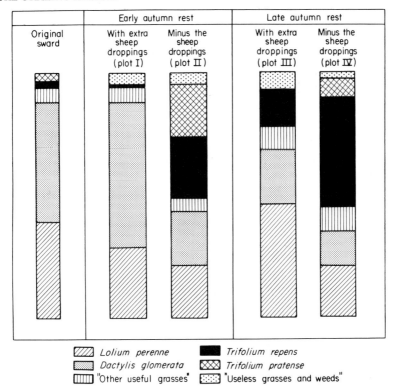

Fig. 14/7. The effects on composition of a permanent grassland of adding additional sheep faeces and of resting the pasture from grazing in early and late autumn. The effects are shown after one year of treatment. (From Jones, M. G., 1933e)

which a whole colony may deposit nutrients that have been harvested over a wide area. In addition the rabbit burrows and scrapes soil surfaces, creating local devegetated zones. The population density of *Iberis amara* (an annual of chalk grassland) seems to be largely dependent on the frequency of rabbit scars in a continuous grass sward. The mole is another local disturber of vegetation, creating local islands of soil with a beautiful seed bed consistency, burying existing plants and providing local conditions for seedling establishment. In a 3-year study of *Trifolium repens* in a permanent grassland in North Wales, M. A. Cahn (unpublished) found only two seedlings that survived to form established plants, and both occurred on molehills. The mounds have charactcristic patterns of colonization, constituting miniature successional sequences set within a rather stable community.

A finer pattern of heterogeneity is created by hoof marks. A cow walks 2–3 miles a day and a sheep 3–8 miles (Hafez and Schein, 1962). Hoof pressures are surprisingly great (740–920 g/cm^2 for sheep and 1280–1600 g/cm^2 for cattle: Spedding, 1971) and when animals are running, the downwards and sideways thrust may be severely damaging both to the plant and to the structure of the soil. The effects are extremely localized, except near gateways and watering points (Thomas, 1960), where extensive areas may be dominated by treading-resistant species such as *Polygonum aviculare* and *Poa annua*. However the localized hoof marks may be of critical importance in the long term population dynamics of a community. They provide local disturbances in which seedlings may establish within a community in which otherwise new genets seldom appear. Figure 5/8 illustrates one carefully mapped population of seedlings arising after a local episode of trampling by cattle on the soil when it was wet. In general it is difficult to separate the consequences of trampling on a pasture from the other effects that a grazing animal has on the ecosystem, though the effects of trampling divorced from grazing and defecation are well known from the activities of tourist man (Streeter, 1970).

There is an enormous anecdotal literature about the non-grazing effects of grazing animals but is difficult to generalize and to put forward a coherent thesis; the effects (like those of grazing (Ellison, 1960)) are often very specific to the vegetation and to the species of animal involved. Elephants can convert luxurious woodland grassland to a treeless condition by consumption, debarking and trampling (Buecher and Dawkins, 1961). The hippopotamus can impose a characteristic vegetation on a river margin (Lock, 1972).

Most of what we know about the effects of grazing animals on vegetation comes from studies of the larger mammals and marsupials. It may be a serious mistake to underestimate the role of smaller grazers present in the same communities.

The woodpigeon (*Columba palumbus*) illustrates the ways in which food preference of a predator may feed back onto the vegetation from which it selects. The woodpigeon moves through the year from one food source to another, taking the flowers and fruits of trees in their season, sown grain from arable fields and ripe grain from stubble, weeds, leaves and seeds (Fig. 14/8). The bird apparently forms accurate food images and concentrates its feeding on specific items of diet that change through the year. Not all seasons offer adequate diets and, following a post-breed-

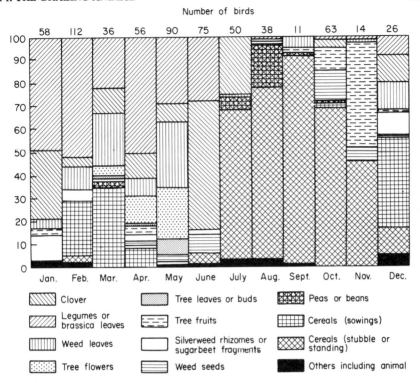

Fig. 14/8. The diet of wood pigeons (*Columba palumbus*) throughout the year in Cambridge-shire, England. (From Murton *et al.*, 1964)

ing season peak in the size of pigeon populations, the availability of grain left in stubble determines survivorship of the pigeons in early December. At this stage some birds die of starvation and others emigrate. In February and March the pigeons studied by Murton (1965, 1966) depended on the availability of clover (*Trifolium* spp.) leaves for food; there was no limitation of food supply through the rest of the season. This observation makes the point (stressed by Lack, 1954) that there may be only one point in the year, or one food in the annual cycle of diet, where the populations of the predator are limited by a shortage of food. This is the critical period — this is the critical food.

Murton concentrated a study on the relationship between the pigeon and clover. He studied an area of 2647 acres near Newmarket where 95% of the pigeons' diet from December to March is clover leaves. He used exclosure cages to measure clover growth in the absence of pigeon graz-

ing. The birds feed gregariously with a stable flock structure that is re-
established when a flock moves to a fresh field. The nature of the movement of a flock determines that those birds in the middle and rear get
more food than those in front. The birds walk and peck as they walk.
As the density of clover leaves falls, the number of pecks per minute
remains constant but the birds increase the number of paces per minute
until at 100 leaves per 30 cm square the peck rate starts to decline. The
front birds are affected first and are most deprived of food. They lose
weight and suffer mortality or migrate and leave the flock. In this way
the population of a flock adjusts to the availability of food supply and
most birds remain in good condition: the hardship is not equally shared.
When the clover supply "runs out" the birds leave it and switch to
another food, usually brassicas in the studied area. This change of diet
may also happen if there is snow and the clover is covered. On the forced
diet of brassicas the mean body weight of the pigeons falls and the
juveniles suffer most.

During January the pigeons avoid fields with a clover leaf density of
< 100 leaves per 30 cm square but visit such fields later in February. The
feeding habit and social organization of the flocks has the effect that the
predation of clover leaves is continued until the leaf density falls below
a critical level. At this point the bird changes diet. The density of the
clover leaf populations is then a direct consequence of the predators'
behaviour. Murton compared the rate of clover regrowth after defoliation
with that of his exclosure plots — defoliated plants developed new leaves
much more rapidly than the controls and by May the grazed clover plots
had the same clover density as the controls.

The short-tailed vole (*Microtus agrestis*) requires about 2000 cm^2 of
grassland per individual per day and when populations are high there
may be direct competition between voles and sheep for food (Chitty
et al., 1968). As part of a vole's feeding is done below ground on root
systems the effects on the growth of above ground parts may be greater
than the grazing of leaves by sheep. An estimate of the numbers and
activity of the slug *Agriolimax reticulatus* in an old pasture in North
Wales suggests that the natural populations of this species damaged or
ate as much grass on 1 ha of land as one quarter of a bullock (J. Hatto
and J. L. Harper, unpublished). The grazing fauna of a grassland habitat
is extremely diverse, including both polyphagous feeders with palatability
preferences and monophages such as the larvae of several Lepidoptera.
There is very little information about the role that these small herbivores

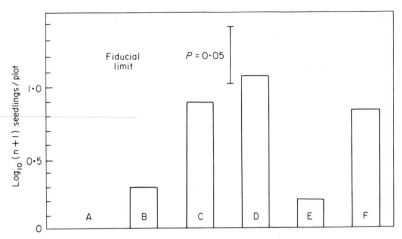

Fig. 14/9. The effect of defoliation of permanent grassland on the establishment of daisies
(*Bellis perennis*) from seed.
(A) No defoliation
(B) One defoliation in early summer
(C) Regular defoliation every time the sward reached 15 cm in height
(D) As C but the sward maintained short in May
(E) Frequent *cutting as D but maintained short* through the year
(F) Vegetation *killed with paraquat before seed* of *Bellis* was sown.
(From Foster, 1964)

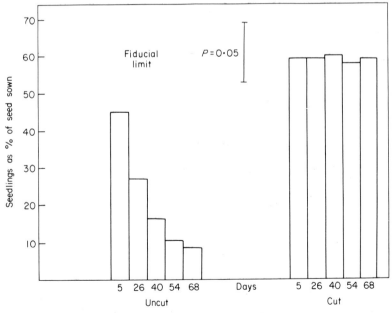

Fig. 14/10. The establishment of daisies (*Bellis perennis*) from seed sown at different times
into a sward of *Lolium perenne* with different regimes of defoliation. The swards
were defoliated by cutting to within 1.0—2.5 cm of the soil surface at 6—14 day
intervals. The dates are the times at which seed of *Bellis* was sown after that of
Lolium perenne. (From Foster, 1964)

may play in the population dynamics (and productivity) of grassland.

The densities of individual species within a complex plant and animal population are likely to be influenced by many interactions. It remains an act of faith that a few critical factors play an overwhelming role in regulation of populations and that these can be discovered by experiment, particularly by perturbing the system and observing the reaction. The densities of the daisy seem to be controlled largely by the vigour of the associated populations of grasses and manipulations of these give a quick response in the population dynamics of the daisy (see Figs 5/15, 5/16). In these experiments grazing was simulated by clipping the sward but it is easy to see how the frequency and intensity of defoliation of grasses can feed back onto the germination and establishment of other species: the complexity of such a grassland community is not so great as to defy the unravelling of causes and effects.

15

The Predation of Seeds and Fruits

The crop of seeds produced by plants may have four roles in their population biology:

(a) *The replacement of individuals in a population that die.* In an annual plant the seed is the critical link between generations; for a perennial and particularly a clonal plant, death may be a rare event and the responsibility of the seed in maintaining populations may be very minor.

(b) *The increase in population size locally,* e.g. during phases when the carrying capacity of the habitat has not been reached during colonization phases or recovery after catastrophes.

(c) *The colonization of new areas at a distance from the parent population,* i.e. dispersal.

(d) *The display of genetic variation,* because it is usually through seed production that the variety of recombinants is hazarded to the environment (though seed production may not always have been preceded by a sexual process — e.g. apomicts). The predation of seeds may, in theory, be important in any of these four roles. In practice it is exceedingly difficult to determine whether the predation of seeds matters in nature. A great many studies have been made of the magnitude of predation but very few help to determine whether predation is relevant either to the evolution or the population biology of plants. This chapter brings together some of the evidence, some of the hypotheses, but the questions will remain at the end largely unanswered.

Seeds are almost always seasonal products of plants' growth, the

examples *par excellence* of high quality glut foods. The predators fall
into two classes, those that are short lived and synchronized with the
phenology of the plant (mainly insects that complete their larval stages
within the developing or ripe seed, pod or inflorescence), and longer-
lived, polyphagous feeders that take seeds during the season of abundance (with or without storing a surplus) and eat other foods in other
seasons or places.

Seed predators may take seeds before or after dispersal. For the most
part the specialized short-lived seed eaters take seeds before dispersal, or
at least the predator's eggs are laid in the seed before it leaves the parent
plant. The longer-lived mammals and birds may take seeds before or after
dispersal and be specialized to one or the other feeding style. A pre-
dispersal predator reduces the size of the seed crop but leaves the dispersal process and pattern unchanged; its effect is the same as if the plant
produced fewer seeds. A post-dispersal predator finds seeds in a pattern
determined by the dispersal system and leaves a residue with a pattern
determined by his behavioural characteristics of choice and search. A few
species of predator take seed before *and* after dispersal, e.g. the goldfinch
(*Carduelis carduelis*) feeds in the autumn on seeds in the heads of thistles
(*Cirsium* spp.) and after seed dispersal continues the search for seeds
scattered on the ground.

Predation on seeds before dispersal has some obvious advantages to
the predator; he has jumped a queue of seed feeders most of which wait
for seed to fall to the ground. However, predispersal predation demands
a degree of specialization, in particular the ability to fly or to climb.
One of the many advantages of height among plants is that it places the
seed crop out of reach of earth-bound predators.

Predispersal predation

If seed is eaten before dispersal the effect on the population biology of
the plants depends on the subsequent pattern of seed dispersal. The
shapes of some seed dispersal curves have been described in Chapter 2
and some of these are redrawn in Fig. 15/1 on a semilogarithmic scale
to show the change in range of dispersal if the seed crop is halved (e.g.
by predation). A reduction in the seed crop of 50% will halve the number
of seeds reaching any point in the dispersal range. It will also alter the
distance from the seed crop at which any particular density of seeds will
land. The figure shows the effect in terms of distance and although pre-

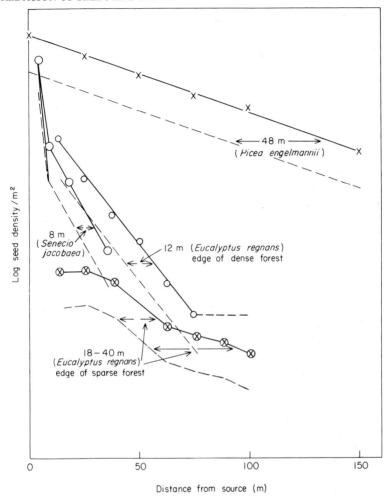

Fig. 15/1. The relationship between distance from seed source and the logarithm of seed density for *Senecio jacobaea, Picea engelmannii* and *Eucalyptus regnans*. For each curve the effect of halving the seed density is shown as a broken line to illustrate the effects that pre-dispersal predation might have on the distances of seed transport.

dispersal predation reduces the distance at which successful colonizations may be expected, the effect is not large.

Wind-dispersed seeds (and probably most seeds) fall close to the parent plant. Where seed is shed densely other causes of mortality such as the many effects of overcrowding (Chapter 6) bring plant density down to the supportable population size. If the density of seeds is such that density-dependent processes will thin the population, predation may simply have removed seeds that had no future, that were doomed to die from the stresses of density. Predation is relevant in the control of population size if it carries the seed density below that to which the plant population will be reduced by later density-dependent processes. This generalization is crudely represented in Fig. 15/2.

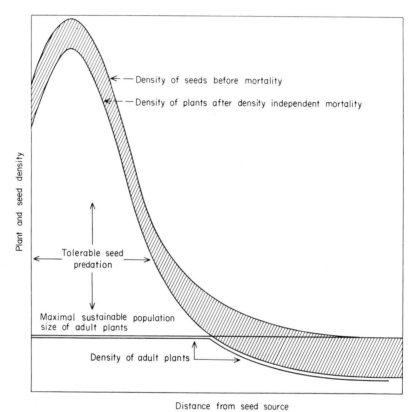

Fig. 15/2. An idealized relationship between seed dispersal with distance and the density of adult plants showing the level of predation that is tolerable without affecting the size of populations of adult plants. In effect seed predation is tolerable if it takes what is doomed to die through density dependent causes.

In expanding populations the role of the predator may be much more important; it then reduces the rate of increase of the population. It is perhaps here, in the speed of local colonization, that a seed predator is most likely to have profound effects on a plant population. It is, however, very difficult to find convincing examples in the literature to illustrate this argument.

Two examples of predispersal predation have been studied in the course of biological control of weeds. The bruchid beetle, *Apion ulicis,* was introduced to New Zealand in an attempt to control the density and spread of gorse (*Ulex europaeus*). The beetle has successfully reduced the seed crop of *Ulex* by 98% but the remaining 2% appears to be sufficient to maintain and spread the populations of *Ulex.* The seed fly, *Euaresta aequalis,* was introduced to Australia to control *Xanthium strumatium* but according to Holloway (1964) almost total seed destruction will be required to control the weed. Both of these insect introductions were made after the plants had multiplied to the status of weeds but were still expanding their populations.

Post-dispersal predation

After seed has left the parent plant it is available as food for a wide variety of animals including small mammals and birds. The most detailed studies have been made of the predation of forest tree seeds because it is often suspected that the regeneration of forests may be hindered by seed predation. One such study was made (Gashwiler, 1967, 1970) to determine the fate of seeds arriving on the soil surface in a clear-cut area of a forest in western Oregon. Observations were concentrated on 2 years with comparatively heavy seed falls (the fate of seeds in years of light seed fall may of course be different). Three tree species contributed to the seed fall which was estimated by placing seed traps in the clear-cut area. Over the 2 years studied the seed fall averaged 315 seeds/m² for *Pseudotsuga douglasii,* 313 seeds/m² *Tsuga heterophylla* and 545 seeds/m² for *Thuja plicata.* Seed fall continued over many months so that three phases, seed fall, the period the seed spent on the soil surface, and the onset of germination, all overlapped. In Table 15/I two values for seedfall are given, one before and one after the germination period. Exclosures were erected in the forest clearing with a mesh size large enough to allow seeds to enter but sufficiently small to keep out all birds and mammals. Three types of exclosure were set up: (i) an open exclosure supported 10 cm above the ground so that there was easy access by predators but any physical changes in the environment due to the presence of an exclosure might be imitated; (ii) a closed exclosure, excluding predators; and (iii) open-closed, i.e. (i) until the start of the germination period and (ii) after the start of germination. In addition to determining the fate

Table 15/I

The fate of seeds falling in an Oregon clear-cut.
(From Gashwiler 1967)

SEEDS	Pseudotsuga douglasii	Tsuga heterophylla	Thuja plicata
Entering seed traps prior to germination (a)	307	302	520
Entering seed traps to end of germination (b)	315	313	545
Open ground samples (c)	97	181	471
ESTIMATED SEED LOSSES (%)			
$\dfrac{(b-c)}{(b)}$ x 100	69*	42	13
SEEDLINGS			
Closed exclosures	234	146	339
Open quadrats	35	57	82
Open/closed exclosures	40	99	337
Open exclosures	37	97	359

* 62% estimated due to birds and small mammals, 7% due to other causes.

of seeds in these exclosures, measures of the seed population remaining in unenclosed areas near the traps were made to estimate the number of seeds remaining after exposure to all kinds of mortality through the period from seedfall to germination (Table 15/I).

31% of the seeds of *Pseudotsuga douglasii* survived all causes of mortality in the open and judging from the different seed numbers in the open and closed exclosures some 62% of the seeds were eaten by predators and 7% destroyed by other causes. In the closed exclosures 57% of the seeds survived and the losses due to "other causes" were now up to 25%. This is an interesting example of compensating factors: if one cause of mortality is removed, another operates with greater severity.

In the open exclosures, 12% of the original seed population produced seedlings, which, had all survived, would have given an establishment of 3.7 trees per m². The causes of mortality are summarized in Table 15/II. Apparently much of the seed lost by *Pseudotsuga douglasii* and *Tsuga heterophylla* was taken by small mammals and birds. A study of the seed remains found near the trap confirm this view. *Thuja plicata* clearly did not suffer from predation, and showed relatively high seed and seedling

Table 15/II

The causes of mortality of filled seeds (a) from the start of seed fall to the start of germination and (b) during germination. (From Gashwiler, 1967)

	From seed fall to the start of germination			During germination			Combined total		
	Birds and Mammals	Other factors	Total	Birds and Mammals	Other factors	Total	Birds and Mammals	Other factors	Total
P. douglasii	62	7	69	1	18	19	63	25	88
Ts. heterophylla	15	25	40	1	28	29	16	53	69
Th. plicata	0	9	9	0	26	26	0	35	35

survival. However, considerable numbers of seeds escaped even in the most predated species.

Which seeds survive on the ground to germinate or to be incorporated into the seed bank depends in part on micro-events, such as whether a seed is quickly covered with a fallen leaf or falls into a crevice where it is hidden from searching predators. The seeds of *Yucca brevifolia* are eagerly sought by rodents as soon as they are shed and the rare survivors which germinate do so quickly and usually from protected cracks and crevices (Went, 1957).

The wild turkey (*Meleagris meleagris*) travels at about 2 miles per hour on a fairly regular daily circuit collecting seeds, particularly acorns (*Quercus* spp.) en route, eating about half a pound a day and digesting these thoroughly. The impact of this on the seed rain is not only to reduce the seed density on the soil surface but to reduce it in a non-random fashion: seeds that fall outside the normal feeding route of the turkey escape predation (Schorger, 1966). In the determination of plant population size the nature of such feeding activities becomes very important. There is evidence that some bird species (perhaps many) leave a particular food source when its density falls below a critical level at which the work done in further searching is not compensated by the additional food found (e.g. Murton *et al.*, 1966; Klopfer, 1962). If this is a general phenomenon it implies that seed predators do not exhaust the stock of seed but only reduce it to a density that may be quite adequate for the continued propagation of the plant species.

Many small mammals and birds collect and store seeds in caches and eat from the cache during periods of general food shortage. This habit poses a number of problems in the population biology of plants. Some seed is left *in situ*, some seed is eaten before caching, some is eaten from the cache and some may be left in the caches uneaten and able to germinate. An important question is whether such activities, taken all together, increase or decrease the chance that some seeds will form plants. In some cases there is little reason to suppose that seed collection and caching bring any benefits to the plant. The acorn woodpecker (*Melanerpes formicovorus*) stores acorns individually in neatly bored holes in pine or oak trees, and as many as 50 000 acorns may be placed in one tree (Bent, 1939). This is an active seed dispersal into a place wholly unsuitable for germination and represents a cost to the seed parents with no apparent benefit. Sometimes quite large seed deposits may be made in nests: the seeds of *Thelycrania alba* were found in abundance in the nests of redstarts (*Phoenicurus phoenicurus*). The seed deposits may eventually fall to the ground (Buxton, 1950) but it is very doubtful if this habit of the bird is in any way beneficial to the reproduction of *Thelycrania*.

If the germination and establishment of seeds from caches is relevant to the perpetuation of the plant, the position of the caches is clearly important. The behaviour of some seed-caching birds is difficult to interpret as part of a subtle coadapted process between predator and prey: for example during 3 months of the autumn, the Swedish nutcracker (*Nucifraga caryocatactes*) carries nuts from hazel thickets (*Corylus avellana*) and buries them in localized territories within spruce forests (*Picea abies*), often travelling as much as 6 km to do this (Swanberg, 1951). It is hard to see an adaptive advantage to hazel resulting from seed placement in spruce — this reverses any normal successional trend!

The seed caching habits of the jay (*Garrulus glandarius*) may be much more significant. The jay carries acorns (*Quercus* spp.) up to a distance of 1 km for burial. A population of 30—40 jays may carry away and bury 200 000 acorns in a season (Chettleburgh, 1952). The seeds are buried in groups of 2—5(+) and are rediscovered by trial and error in the autumn and winter. From April to August a search image develops among the birds based on the appearance of leaves (cotyledons in the case of *Fagus*) and the cotyledons are then searched for and eaten. The plants may survive removal of their cotyledons at this stage. A number of seedlings commonly survive and the activities of the jay clearly contribute to the dispersal of the oak (Turček, 1966). The seeds of *Quercus robur* rapidly lose viability on the soil surface and the act of burial by the birds may be vital for the regeneration of this species in some habitats (Watt, 1919).

Many small mammals collect, bury and then recover seeds, particularly the large seeds of many forest trees. Seed of *Pinus strobus* was isotopically labelled by soaking in scandium[46] chloride, and a mixture of about 1560 seeds (40 labelled, 1520 unlabelled) was exposed on the forest floor in special feeders. The seed was collected and buried mainly by white-footed mice (*Peromyscus leucopus*) and redbacked voles (*Clethrionomys gapperi*). These caches were detected by means of a gamma scintillation probe, and were examined daily (Abbott and Quink, 1970). The caches contained an average of 24 seeds each, buried 5 to 10 cm from the surface between the H and A1 soil horizons (Fig. 15/3); 35 caches were located, and of these 10 had been completely destroyed and eaten

Fig. 15/3. Seed cache of *Pinus strobus* collected by *Peromyscus*. (Photograph by A. L. Quink)

before the end of autumn. Half of the remaining caches were visited by the small mammals during the winter. Nine more caches were destroyed in May, leaving only 14 containing some whole seeds, and of these only six developed seedlings (49 seedlings altogether). Mice uprooted and killed the seedlings in three of the remaing caches and at the end of the first growing season only one of the caches bore any germinating seedlings! In a subsequent year, 92 caches were located, and only three of these produced surviving seedlings. The rodent activity had involved moving seed from its original position (distribution), depositing it in local hoards (concentration), at a distance from the source (dispersal). Subsequently, the population of seeds and seedlings was reduced by the recovery activities of the rodents (thinning). Pirating by species other than those that made the stores also occurred. One effect of this sequence of events was that a *few* tree seeds were placed in environments particularly suitable for germination: this was achieved at the cost of a massive seed loss, which might be envisaged as paying for the act of dispersal. Certainly at first sight the cost seems outrageously high, until one remembers that only one successful seedling on average is expected to establish from a whole lifetime of seed production by the parent. It may be that it is the act of seed caching that determines which seed among millions is the one that produces the descendant tree.

Seeds that have been cached, if they are not eaten, tend to germinate in clumps (see Fig. 15/4). Tree seedlings thin during the course of development but shrubs may form from a group of genets and what appears to be

Fig. 15/4. Seedlings of *Pinus strobus* germinating from a seed cache. (Photograph by A. L. Quink)

a single establishment may be a multiple colony. This situation is believed to occur in *Purshia tridentata*. About 50% of the shrubs of this species could be attributed to seed caching by small mammals in a study made by West (1968) in Central Oregon. About 15% of the establishments of *Pinus ponderosa* was also thought to be from seed caches.

Seeds are not the only materials that may be cached by mammals. Seeds, bulbs, tubers and corms may all be found in caches. The bulbs of *Oxalis cernua* are cached by the mole rat (*Spalax ehrenbergi*). The rat makes two types of cache, one associated with the nesting chamber and further stores as reserves. The female crops the young shoots as they appear from bulbs in the nesting chamber but bulbs in the reserve stores may remain uneaten and the plants establish readily and spread vigorously from these caches (see Fig. 15/5) (Galil, 1967). Some regenerating organs of plants are cached but are unable to establish from the caches. Galil notes *Oxalis cernua* as a species that derives some advantage from the caching of its bulbs but there is a long list of species that are unable to establish from the caches of the mole rat unless the stores are very shallow (e.g. *Cyperus rotundus*).

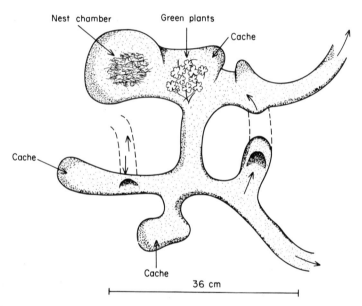

Fig. 15/5. The arrangement of tunnels and larders of the mole rat (*Spalax ehrenbergi*) which collects and stores bulbs of *Oxalis cernua* as well as a variety of seeds tubers and corms. (From Galil, 1967)

Escape from predation in time

Prey may escape from a predator by dispersal in space. A different aspect of predator escape involves disperal in time. An unreliable supply of a food can be a defence against predation just as crop rotation is a means of controlling host-specific pests and pathogens in agriculture. During the years when a specific crop is missing from a rotation the pest either moves elsewhere or dies of starvation. Rotational control of pests and diseases is effective if the organism concerned lacks a long facultative resting period; it is ineffective for the control of animals like eelworms with cysts that may remain dormant for many years and break dormancy in response to the presence of the host. Parasitic flowering plants like *Orobanche* and *Striga* require exceptionally long periods of crop rotation for effective control because the dormant seeds remain in the soil for decades, but germinate under the influence of root exudates from the right host plants when these are sown again (Sunderland, 1960).

Even relatively slight variation in seed output from year to year may be important for a plant if its predator is dependent on a particular species for its supply of food. This is the case particularly with seed-caching animals and birds. A successful species is one whose ancestors collected *more* than enough to tide them over the *worst* winters in their evolutionary history. An animal species of which the members collected only enough for the average winter would have starved to death in about half the winters — it would be extinct. On average, therefore, there will be seed left over after the winter and the survival of the plant depends on the difference between what the animal collects and what it eats. Thus the seed-caching habit does not require any specialized coadaptation, and the seed cached will on average exceed predation. The greater the unpredictability of the seed crop, the greater will be the average excess cached and left to germinate.

Most tree species, particularly in temperate regions, have "mast" years in which seed production is very high indeed, and such a year is followed by a group of years in which the tree behaves as if exhausted by this effort and produces relatively few seeds. The size of a population of predators is unlikely to be able to adjust quickly to such changes in the winter food supply. In "mast" years a particularly high proportion of the seed rain escapes predation, although it may be cached.

Unpredictability of seed production can be seen as a defence against predation (Murphy, 1968). The phenomenon of mast years is particularly strongly developed in woody perennials and is best known as a feature of many temperate trees, whereby years in which the seed crop is very large are interspersed more or less regularly with groups of years in which seed production fails or is very low (Salisbury, 1942; Smith, 1968; Sharp, 1958; Sharp and Sprague, 1967). Mast years are usually synchronized between individuals of the same species in a forest and consequently seed crops of glut proportions are interspersed with groups of famine years. The predators of seeds in temperate forests are mainly birds and small mammals with life cycles far too long for their populations to adjust to changing levels of food supply. Consequently much of the seed produced in mast years escapes predation and it may be only in these years that a plant leaves descendants. An unpredictably hard year for the predators, such as the spread of a disease among them, can have the same effect as a glut season for food; more seed is left for regeneration (Barrett, 1931; Parnell, 1966) and even-aged batches of tree seedlings may develop as a result.

The evolution of "mast year" phenomena is not easy to understand. It seems at first sight to require the operation of group selection: the advantage of escaping predation by synchronous mast years brings an advantage to the group and not, in its early stages, to the individual. Probably the mast year phenomenon has a major cause in the physiological consequences of occasional good seasons for growth. Seed number is determined before the season in which the seeds will be ripened. A heavy seed crop may be initiated in a good season and place a heavy drain on plant resources during the course of seed filling in a less good season (see Chapter 21). The setback to growth is reflected in a low seed output during subsequent years of recovery. In this way the environment effects a coarse timing of mast years. Non-conformist trees in the population (those that do not fit the crude rhythm of the majority), will be excessively penalized by predation: e.g. *Pinus ponderosa* trees that produce seed crops during non-mast years are excessively predated (Smith, 1970). The predator then acts as a fine-tuning device selecting for trees that conform to the environmentally determined response of the majority.

The significance of seasonality in the predation of seeds and fruits

The concept of "predator escape in time" has been developed particularly by Janzen (1970a, 1974) to explain the relations between seed predators and their prey in the tropics. Biologists unfamiliar with the tropics emphasize their climatic stability, while tropical biologists place a great emphasis on the seasonal rhythm of biological events. Seasonality of biological activity is very conspicuous even in humid equatorial environments. A study was made by McClure (1966) and followed up in more detail by Medway (1972) of the flowering, fruiting, and leaf formation of

trees and vines at a site near Selangor, Malaysia (3° 21' N). Climatic
records were available from Kuala Lumpur (3° 08' N) where day length
varies by only *ca* 20 min through the year and the annual mean monthly
temperature range is no more than 1.0–1.5°C (cf. a diurnal range of
7–11°C). There is a slight seasonality in rainfall but this is very variable
both between seasons and years.

Observations of plant phenology were made from a platform constructed 43 m above the ground at the point of branching of an emergent
dipterocarp. McClure recorded animal feeding activities in the canopy
and Medway's study overlapped with and added much more botanical
detail to McClure's records. The data have been arranged in a diagrammatic form in Fig. 15/6 to show the periodicities of feeding by the main
predators on 11 species of tree and liane. There is an extraordinary pattern of food-web complexity. The fruiting periods of most of the species
are strictly seasonal with the exception of *Ficus sumatrana* which fruited
about three times a year but with an apparently endogenous rhythm,
not synchronized with the calendar cycle. 11 of the 13 predators used
this fig as part of their diet, and no other fruit was as widely used. The
seasonal feeding habits of the predators involved an annual cycle following
the seasons of fruit production of specific tree species. For most of the
predators, food appears to have been available and eaten from the canopy
throughout the year. In some cases there were gaps in the annual cycle
of food for a predator which was probably feeding on a different species
out of sight of the observed canopy. A mosaic of predators moved through
a mosaic of fruit producers in a seasonal rhythm. None of the predators
was a specialist (*Arachnothera* appears from the diagram to feed only on
Aeschyanthus but as this was available for only a fraction of the year
Arachnothera was presumably feeding on other species in other parts of
the forest when *Aeschyanthus* was not in fruit). Two of the *Ficus* spp.
and *Santiria laevigata* were eaten by a considerable variety of animals
but there were also some clear specializations – differences in choice
between the predators which are polyphagous but appear to have different orders of preference.

The list of predators in McClure's study includes only birds and the
larger mammals and does not include fruit eaten on the forest floor. The
full food web must be much more complex. The study reveals, better
perhaps than any other set of published data, the role of seasonality in
herbivore–plant relationships yet this is an almost season-free climate.

Lord Medway's study of the phenology of the forest extended the

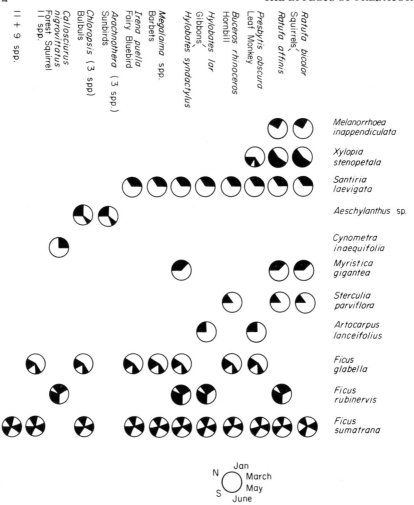

Fig. 15/6. Animals feeding on the fruit of trees at various times of the year at Selangor, Malaysia. Diagram drawn from data of McClure (1966)

number of trees under observation from McClure's platform to include 61 canopy trees representing 45 species. 10 of the species flowered in a seasonal and recurrent rhythm, and in the remainder the rhythm was still essentially seasonal but individuals failed to flower or to set fruit in some years or groups of years. Over the period 1963–69 only 10 of the species flowered every year and only six (including two figs) fruited every year. Yet in every year at least 20 species flowered and 12 produced fruit. It cannot be assumed that all the individuals of the same species in the forest were fruiting or failing to fruit at the same time: this information is not available and would be needed before detailed arguments can be developed about the survivorship of specialist predators. The community as a whole showed a more regular cycle of flowering and fruiting than the individual species within it.

The production of new leaves followed seasonal rhythms and these differed between species. Three species produced new leaves in an almost continuous cycle of leaf renewal, 21 species had an annual cycle of leaf renewal often with two periods of activity and 5 species had irregular twice-a-year cycles of leaf production. Four species were deciduous annually and four were deciduous at longer or less regular intervals. The phenologies of the various trees are completely wrong for any theory of escape from predators in time. Instead the flowering and fruiting patterns assure a very reliable diet for species of animal that can move between trees and treat the canopy as a fine-grained food resource. The system offers a remarkably steady and reliable source of food, and there is little opportunity for predators to be satiated at times and experience acute starvation at others. In many respects this appears to contradict Janzen's models of seed predation in tropical forests. The critical difference is that the predators in this forest are taking not seeds but *fruits*. The predators are eating fleshy fruits but dispersing the seed. This is not a form of predation that harms the plant and reduces its reproduction, it is a predation that is encouraged, paid for in the provision of fleshy, attractive, usually sugary fruits and rewarded for the act of seed dispersal. The consequences in natural selection of predators that eat fruit and disperse the seed are diametrically opposite to the consequences of predation on seeds. To maintain a population of animals as dispersal agents it is essential that food supply be available for the predators throughout the year, particularly if they are specialists on fruit. Moreover, each species of fruit-producing tree is a potential competitor for dispersal agents. It is not surprising if the seasonal rhythms of fruit production become, at the same time, constrained within seasons so that interspecific competition for dispersal agents is minimized, but also spread out over the year so that fruit is continuously available to supply the diet of the predators.

An extreme case of the non-synchrony of fruiting time occurs in the 18 species of the genus *Miconia* found in the Ariba valley, Trinidad, which between them completely fill the year with fruiting seasons (Fig. 15/7) (Snow, 1964). A new species of *Miconia* establishing and being

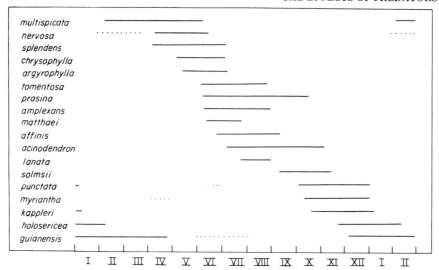

Fig. 15/7. The fruiting seasons of 18 species of *Miconia* in the Arima valley, Trinidad. Dots indicate occasional out-of-season fruiting (From Snow, 1964)

dispersed in this area by fruit-eating birds would need to find a gap in the season when fruit-eating birds were not food-saturated, or to bear fruits more attractive as food than those of existing species.

When predators take and destroy seeds the selective pressures are likely to favour synchronous ripening and individuals (and in consequence species populations) will be at an advantage if their seed is produced during glut periods. Thus one might expect that where *seed* is at risk from predators, whole groups of species would tend to have synchronous seed ripening and the season of seed ripening would tend to be too short. The wind-dispersed seeds of Central American tree species fit this ideal very closely (Janzen, 1967).

Seeds and fruits are very different food resources for an animal. Edible fruits are usually watery, fleshy and sugary; seeds are usually a concentrated supply of proteins and carbohydrates or oils. A diet of fruit requires a large intake to satisfy dietary demands and is apparently inadequate for the rearing of nestlings — very few birds feed their young with fruit. In the tropics, birds may maintain their own activities on fruits leaving time in which to search for insects for their nestlings (Morton, 1973). There are certainly examples in which energy-rich food is packed into pericarp tissue (the olive and the avocado) but these are very much the exceptions and in these cases the energetic costs of attracting the disperser must be very high. (It may be that energy is a superabundant resource for some plants and that we should be careful before making the assumption that the economy of a plant is governed by the efficiency of its energetics.)

"Predator escape" in space

For a small, rapidly multiplying pest or pathogen, a plant is a coarse grain of a highly localized, concentrated resource within the environment. When the grains themselves, e.g. the single species dominants of temperate forest, are close together the opportunities for a pest or pathogen to multiply are enormously increased. Search distances are small in such conditions and population explosions (epidemics) can spread readily. The more isolated from each other are the grains or islands of food resource, the slower will be the spread of the pest; isolation (and dispersal) bring protection. The situation is very similar to the chances of spread of a population between a group of islands. The more isolated the islands the slower is the spread of an organism between them and the lower is the overall density. The closer packed the islands into archipelagos, the more rapid the spread, the higher the overall density (see the discussion of the colonization of islands and archipelagos in MacArthur and Wilson, 1967).

The idea of "predator escape by isolation or dispersal" has been developed mainly in the work of Janzen (e.g. Janzen, 1970). He relates some of the characteristic features of seed production and dispersal to the search behaviour of predators — particularly the insect predators of seeds. Beetles of the Bruchidae oviposit in and the developing larvae destroy the seeds of various tree species of the Leguminosae in the tropics. Individual trees are usually both the source of seed and the centres of spread and search by ovipositing beetles. Most seeds tend to fall close to the parent plant (see Chapter 2) and the number of seeds falling per unit area (I) declines rapidly with distance from the parent tree. At the same time the probability that a seed will be missed by the searches of the species-specific predator (P) increases. The product of the I and P curves yields a population recruitment curve (PRS) with a peak at the distance from the parent where a new adult is most likely to appear (Fig. 15/8).

Two contrasted patterns of search behaviour can be envisaged which have very different consequences on the shape of the Population Recruit-

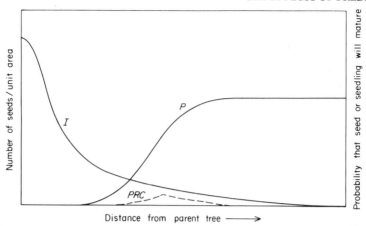

Fig. 15/8. A model showing the probability of maturation of a seed or seedling at a point as a
function of (1) seed-crop size, (2) type of dispersal agents, (3) distance from parent
tree, and (4) the activity of seed and seedling predators. With increasing distance
from the parent, the number of seeds per unit area (I) declines rapidly, but the probability (P) that a dispersed seed or seedling will be missed by the host-specific seed
and seedling predators, before maturing, increases. The product of the I and P curves
yields a population recruitment curve (PRC) with a peak at the distance from the
parent where a new adult is most likely to appear; the area under this curve represents the likelihood that the adult will reproduce at all, when summed over all seed
crops in the life of the adult tree. In most habitats, P will never approach 1, due to
nonspecific predation and competition by other plants independent of distance from
the parent. The curves in this and Fig. 15/9a, b are not precise quantifications of empirical observations or theoretical considerations, but are intended to illustrate general
relationships only. (From Janzen, 1970)

ment Curve. If the activity of the predator is determined by the distance
from the parent tree, from which they emerge and range, the highest
density of survivors is scarcely affected by the size of the seed crop, or
by the amount of predispersal predation (see Fig. 15/1). This situation
is illustrated in Figs 15/9a and 15/10. Janzen calls this distance-responsive predation. In contrast predators may act in a quite different way,
ranging almost independently of the distance from the parent tree but
cropping the dispersed seeds until they are more than some defined distance apart. The size of the seed crop (or the amount of pre-dispersal
predation) then becomes very important (Fig. 15/9b). This he calls
density-responsive* predation.

* The terms density-dependent, density-responsive, distance-responsive, are sometimes confusing.
The fixed positions of plants makes it easier to think about distance, meetings and neighbours
than it is with animals. Neither plants nor animals react to the density of their populations but
to the effects of the number and proximity of neighbours. The term "density" is an abstraction
removed from the level of cause and effect.

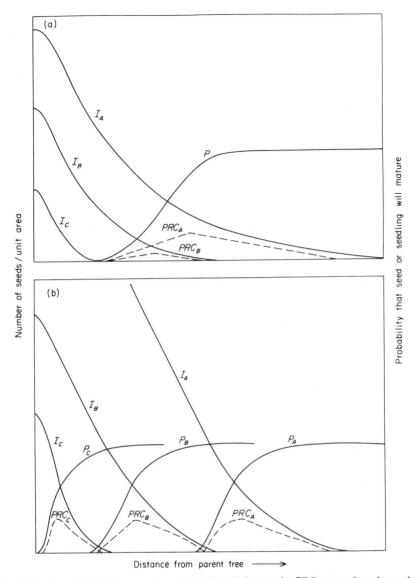

Fig. 15/9a. The effects of increased predispersal predation on the *PRC* curve when the predators are distance-responsive. The seed crop of I_B is about one-half of that of I_A, and the seed crop of I_C is about one-ninth that of I_A. This figure should be contrasted with Fig. 15/9b, where a reduction in seed crop affects the *PRC* curve quite differently when the predators act in a density-responsive manner. (From Janzen, 1970)

Fig. 15/9b. The effects of increased predispersal seed predation (progression from I_A to I_C) on the *PRC* curve when the predators are density-responsive. The PRC_C curve is slightly less peaked than would be the case if the density-responsive predators were identical for all three I curves; it is assumed that some density-responsive predators would be completely absent for the small I_C curve because there are not enough seeds or seedlings to attract them in the first place. (From Janzen, 1970)

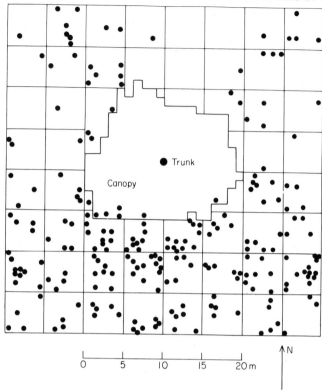

Fig. 15/10. The distribution of young oak trees at Silwood Park, Berkshire, England in an area
surrounding an old oak tree. An example of distance-responsive establishment.
(From Mellanby, 1968)

   Janzen interprets some of the remarkable diversity of tropical forests
in the light of these models. If a variety of species are able to tolerate a
particular environment but individuals of the same species are unable to
establish close to a parent or to each other, a high level of floristic diversity can easily be maintained and no single species can achieve local
dominance. Examples of this effect of predators are mainly anecdotal
but very numerous e.g.

> In second growth vegetation, . . . *Leucaena glauca*, a mimosaceous shrub . . . loses
> upward of 90% of its seed crop to bruchid beetle predation in central America,
> and occurs there as a scattered adult with rare seedlings. In Puerto Rico, no
> bruchids attack this species, virtually every seed is viable, the adults are surrounded
> by dense stands of seedlings and intermediate-aged juveniles, and are extremely
> common.
>
> (Janzen, 1970)

For example, the crown on a mature woody vine, *Dioclea megacarpa*, in lowland deciduous forests in Costa Rica harbors a large population of apparently host-specific erebine noctuid larvae that feed on shoot tips. They harvest as much as 50% of the new branch ends. There is a steady but slow rain of these caterpillars on the forest floor; most return to the crown or wander off to pupate. However, they will feed on any intact shoot tip of a seedling of *D. megacarpa* they encounter. The young plant has sufficient reserves to produce about three main axes; further decapitation kills the seedling (the previous decapitations slowed its development, which probably would also be fatal over a longer period). For this reason, there is no survival of seedlings directly under the parent. But seedlings more than about 5 m from the edge of the parent's crown show only slight damage from these caterpillars.

(Janzen, 1971)

... in evergreen primary forest on the Osa Peninsula in southwestern Costa Rica, the large and winged seeds of *Huberodendron allenii* (Bombacaceae) are heavily preyed upon by numerous rodents on the forest floor. Any seed placed near the base of the parent, sterile or fertile, is invariably eaten within two nights. Seeds placed more than 50 m from adult *H. alleni* are found much more slowly, some lasting at least 7 days.

(Janzen, 1970)

Numerous forest-floor mammals may subsist almost entirely on fallen seeds and fruits at certain times (e.g. peccaries, agoutis, pacas, coatis, deer, rats, etc.) and they tend to concentrate on the fruit under a particular tree (e.g. Kaufman 1962). That they will wander off when full, but return later, creates nearly the same effect as though they were obligatorily host-specific to that tree species.

(Janzen, 1970)

Janzen uses the phrase "escape from predation", and indeed, the seed that is widely dispersed from other members of its own sort gains essentially the same freedom from its specific pests and pathogens as the invader of an entirely new territory — an island or a continent. There is a period of unpredated life before the recluse is ultimately discovered (see discussion in Gillett, 1962 and Bullock 1967). Janzen (1968) described host plants as islands in evolutionary and contemporary time and many of the aspects of island biology are mimicked in the colonization of plants by pests particularly fungal and bacterial pathogens and those insects that multiply on a host plant, e.g. aphids.

The hypothesis of predator escape in space has been put to experimental test by Wilson and Janzen (1972). They placed nuts of the palm *Scheelea* at various distances from the parent tree. This did not affect the chance that a nut would be discovered by a bruchid and for this example the hypothesis is rejected. However there was a clear effect of seed density: the larger the pile of freshly cleaned nuts that was exposed, the more likely was it to be discovered by bruchids.

Selection by and evolutionary consequences of the predation of fruits and seeds

Much of the mortality of most organisms occurs in the juvenile stages but it does not follow that it is an important determinant of the size of plant populations. If, during the post-juvenile phases, a population of plants experiences density stress that thins the population to some low level, mortality that occurs before this stage may be irrelevant in the control of ultimate population size (Fig. 15/2). At the same time, the early mortality may have differentiated between genotypes so that what is left after juvenile mortality is a selected part of the original seed population. In this way a predator taking seeds may be a vital selective force although irrelevant in determining the size of populations.

Modifications among fruits that increase their attraction to dispersal agents are likely to be favoured if a dispersed seed is more likely to leave descendants than one that falls close to the parent. The reds and orange colours of fruits lie within the visual range of colour sensitivity of birds and, like the sweetness of pulp, develop only when the seed is ripe and ready for dispersal. In contrast any modification that makes a seed unattractive to a predator will be of selective value if it increases the chance that the seed will produce a mature plant compared with one that lacks the modification.

There is a little evidence of selection by predators between the seeds of members of the same species. The bullfinch (*Pyrrhula pyrrhula*) takes large quantities of seed from ash trees (*Fraxinus excelsior*) in Britain, stripping some trees of all their seeds while leaving others untouched (Newton, 1967). The explanation of such discrimination is apparently not known and it may or may not have a genetic cause. Most of the hints about the evolutionary consequences of seed predation come from the less satisfactory comparisons between species.

Species of three *Papaver* spp. found in Britain, *P. rhoeas, P. dubium* and *P. lecoqii,* shed their seeds from the capsule as soon as they are ripe. The seeds are too small to be worthwhile prey for most predators but a capsule full of seeds is a valuable article of diet. These three species of poppy have large capsule pores and the seeds escape easily. A ripe capsule is scarcely ever found full of seeds and birds rarely predate the capsules of these species. In contrast, *Papaver hybridum* and *P. argemone* which live in the same weedy habitats as *P. rhoeas,* retain seeds within the capsules, sometimes throughout the winter. The capsule pores of these species are small relative to the rather larger seeds and the seeds escape slowly over a period of time. The capsules of these species bear spines and are again seldom eaten by birds. However, forms with spine-less capsules occasionally occur and these are eaten by birds. The oil seed poppy, *Papaver somniferum,* has smooth capsules and in the wild forms the seeds escape quickly as in *P. rhoeas.* The cultivated form has closed capsule pores and the seed is retained; the capsules are a very popular food for birds. The poppies illustrate three aspects of protection of the seed crop against predators: (i) the seed is very small and scarcely worth collecting after dispersal; (ii) a capsule full of seeds is attractive to pre-dators and the more quickly the seed is shed the safer it is; (iii) if the seed is retained in the capsule, long spines and hairs are a defence against pre-dation (Harper, 1966).

It would be surprising if, during the course of evolution, the predation of seeds had not had selective influences on both plant and predator. It is not easy to recognize coadaptation with any certainty. The separation of proximate (ecological) and ultimate (evolutionary) effects is beset with philosophical difficulties. Coadaptation is most clearly suggested when different species and their prey can be compared and a group of multiple correlations can be discovered. The relationship between squirrels and pines may include elements derived from a long period of co-evolution (Smith, 1970). East of the Cascades in British Columbia fire has been a major environmental hazard and *Pinus contorta* among other species has evolved the habit of serotiny, i.e. the cones remain attached to the tree for many years after ripening and are sealed with resin so that the seeds are not released. After fire the resin seal is broken and seed is dispersed and germinates in a flush of seedling establishment. The serotinous habit means that a seed source is available on the tree for predators all through the year. The squirrel is a major predator of the seeds of *P. contorta* although it prefers the seed of *Pseudotsuga douglasii. P. douglasii* has very unpredictable seed crops with mast years and periods of crop failure and the squirrels then depend on *P. contorta.* In the Eastern Cascade region the squirrel population is rather stable because of the reliable fall-back food supply in the serotinous cones of *P. contorta.* In contrast, west of the Cascades, *P. contorta* is less common, is not serotinous and has mast years and crop failures. The squirrel population correspondingly fluctuates very greatly according to food supply. The serotinous cones east of the Cascades are hard with strongly developed apophyses and carry fewer seeds per cone than the soft cones of *P. contorta* west of the Cascades. The squirrels east of the Cascades have a stronger jaw masculature and different skull configuration from those to the west. Smith interpreted this series of differences as a series of coadapted responses of pine and squirrel. The evolutionary sequence is seen as fire $\longrightarrow$ serotiny $\longrightarrow$ all year round food supply $\longrightarrow$ high and constant squirrel predation $\longrightarrow$ evolution of hard cones in *P. contorta* + few seeds per cone in *P. douglasii* (increasing the work to be done per squirrel per seed extracted) $\longrightarrow$ changed musculature of the squirrel. Smith pointed out that squirrels have probably interacted with conifers for a long period of evolutionary time. The modern genera of pines appear in the late Palaeocene–Eocene Epochs of the Cenozoic coinciding with the appearance of rodents and other placental herbivores.

The coevolution of predators and their prey and the variety of evolved strategies has scarcely been studied in cases where the prey are plants and the predators are animals. The sophistications of animal–animal and of plant–pathogen co-evolution have been deeply explored. Botanists seem frightened by a search for adaptive interpretations of plant behaviour, perhaps as an over-reaction to "... the brothels and gin palaces of unbridled hypothecation" (Large, 1940) of the late nineteenth and early twentieth century. One of the reactions to the acceptance of Darwin's theories was a readiness to jump at facile adaptive interpretations. This is no reason why serious scientific enquiry should be abandoned; merely a reason for framing hypotheses in a form that can be tested.

16

Pathogens

In the order of subjects chosen for this book pathogens appear as an appendix to a rather generous allocation of space to predators. The order could just as well have been reversed. The essential principles involved are very similar. A plant represents a complex of food niches that have been exploited with extraordinary parallelism by pathogens (mainly fungi) and pests (mainly insects) — see Fig. 16/1. It is an open question whether, in the population biology of plants, pathogens are more or less important than predators. There are some differences however between pathogens and predators which need defining and illustrating.

There is nothing really comparable to search and choice in the discovery of a plant by a pathogen. A fungus may, in the course of its growth, meet a new host and in this way an infection may spread through a population. For example the rhizomorphs of *Armillaria mellea* may grow from food bases and establish a new infection across a gap between hosts, although the distances involved are apparently usually small (Garrett, 1956, 1960). Most of the spread of mycelial pathogens that is effective in causing disease seems to depend more on the growth of the host's root system than on independent exploration by the pathogen. In general the spread of pathogens is passive: the movement of spores, like seeds, is determined by the agent of transport — wind, water or animals.

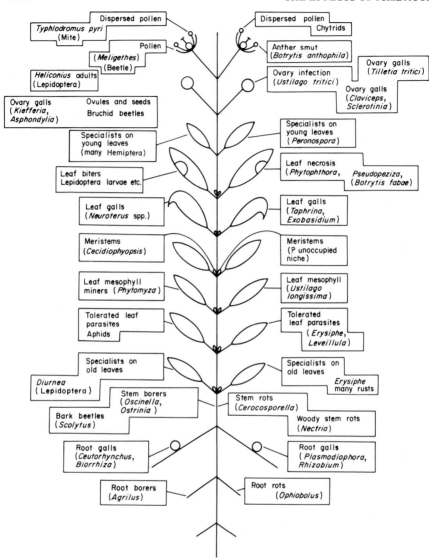

Fig. 16/1. The higher plant as a series of specialized niches occupied by insect predators and
        fungal parasites. Comparable, though rather less diverse, diagrams could be drawn
        for nematodes and pathogenic bacteria. Some of the subtlety in niche occupancy
        is omitted for clarity — for example *Neuroterus* spp., producing leaf galls on oak,
        are believed to specialize at a subspecific level with respect to tip, middle or base of
        the leaf as well as to age of leaf (Askew, 1962; Hough, 1953). Note also that some
        species may alternate between two highly specific niches on the same plant, e.g.
        Biorrhiza pallida forms root galls and meristem galls on the oak at different seasons.

Disease escape in space

The shape of the curve that relates the number of infections to the distance from an infected source varies between pathogens and reflects the way in which epidemics develop in a population of plants. If the curve of log infections against log distance from the source is steep — with a slope of > 2(2 = inverse square law) a pathogen is likely to spread as a wave through a population of plants or as an advancing horizon. This will be the case particularly for pathogens that spread through the soil where the medium is resistant to dispersal. If the curve of log infections against distance is shallow (< 2) the spread of the pathogen is likely to appear as local spots of infection within the population, not as a horizon. The slope of the curve is likely to be shallow when the disease is spread by some widely ranging animal. For example, the transport of *Endothia parasitica* by woodpeckers is believed to have played an important part in the early stages of spread of chestnut blight in the forests of North America. This interpretation of the relationship between dispersal pattern and infection is based on the ideas of van der Plank (e.g. 1960) (see Fig. 16/2) which may well prove a stimulating addition to ecological theory.

It is to be expected, on the basis of van der Plank's theories, that the escape of a host through space (in the sense used by Janzen: see Chapter 15) may be effective for diseases with a steep curve of infection against distance, particularly soil-borne diseases, but rather ineffective against diseases with a shallow curve. The dispersal range of wind-dispersed seeds may often be greater than that of soil-inhabiting pathogens and the wide interspersion of host trees of different species in a tropical rain forest will tend to protect them (and may partly result) from the activities of specialist soil-borne pathogens and others with a short dispersal range. To be effective in spreading within a diverse community, in which the hosts are separated and interspersed, a pathogen would need a shallow dispersal curve, obtained for example by specific attachment to a discriminating insect or higher animal vector. Even where there is a highly developed relationship between a pathogen and an insect the infection/distance curve will be shallow only if the insect itself has a wide range of search for its host plants. Dutch elm disease (*Ceratostomella ulmi*) is carried by bark beetles (mainly *Scolytus*) which have a very limited search range; consequently this disease has a steep gradient of infection/distance and infected trees are rarely found more than a few hundred yards from a

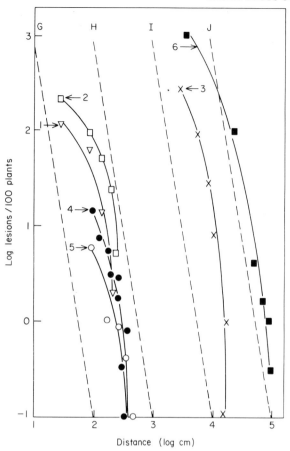

Fig. 16/2. The amount of potato blight (*Phytophthora infestans*), curves 1–5, and of *Pereno-spora destructor,* curve 6, at varying distaces from the inoculum. Curves 1, 2, 4 and 5 are based on a transformed value of the number of lesions. The broken lines G, H, I and J are inverse fourth power lines shown for comparison with the observed results. (From van der Plank, 1960)

source of infection (Zentmyer *et al.*, 1944; van der Plank, 1960) (see Fig. 16/3).

The fact that a disease spreads slowly in a plant population from an advancing front does not mean that it will not ultimately reach epidemic proportions. van der Plank has emphasized how wrong is the view that an epidemic requires "an aggressive parasite that multiplies fast, spreads far and swiftly, and is not particularly selective in its requirements". Swollen shoot disease of cacao is caused by a virus that is spread by flightless mealy-bugs. The main spread is between trees that make physical contact with each other. The virus is endemic to indigenous West African trees in the families Sterculiaceae and Bombacaceae. Nevertheless this disease spread over a period from 1922 to reach epidemic proportions in the early 1930s, killing cacao over great areas and destroying many plantations.

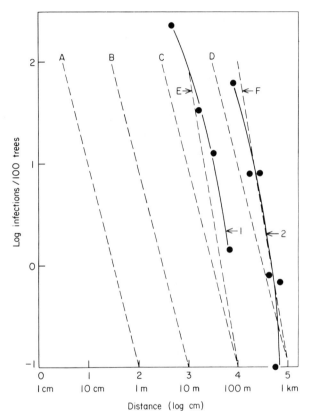

Fig. 16/3. The amount of Dutch elm disease (*Ceratostomella ulmi*) at varying distances from the source of inoculum shown as transformed numbers of infections. The broken lines A, B, C and D are inverse square lines and E and F are inverse cube lines. (From van der Plank, 1960)

The spread of epidemics of swollen shoot disease of cacao (as with many other diseases of forest trees) is a function of the spatial relationships between individuals. A pure stand of host plants provides the perfect conditions for disaster and it is easy to show that, for a pathogen that spreads from an infection front, the speed of spread is a function of the density of host plants. There is a clear relationship between the density of pine seedlings and the incidence of damping-off disease caused by *Pythium* (Gibson, 1956). The activities of *Pythium* spp. in seedling populations are very convenient models for studying epidemiology — "true epidemics in microcosm" (Garrett, 1970). The spread of infection by *Pythium irregulare* in a population of cress (*Lepidium sativum*) was followed by placing infected seedlings at the edge of populations and measuring the progression of the disease (Burdon and Chilvers, 1975). The higher the density of plants the faster was the movement of the infection front; the infection rate by *Pythium* was linearly related to the mean interplant distance (Fig. 16/4).

Diseases can on occasions reach epidemic proportions even when the host plants are widely scattered (*Puccinia malvacearum* had spread throughout Europe within 5 years of its introduction although it attacks only wild and garden hosts (Gäumann, 1946)) but it is a general rule that the proximity of susceptible hosts brings with it greater risk of epidemics. van der Plank (1960) generalizes this observation in relation to birth rates and death rates of an infection. He distinguishes pathogens with a high "birth rate" (the rate of appearance of new lesions or new systemically infected hosts) from those with a low "death rate" that occupy a host for a long period, taking a long time to kill or weaken it without killing. The proximity of plants increases the "birth rate" without much affecting the "death rate". There is therefore for each pathogen some critical density (under a stated set of environmental conditions) when the relationship between birth and death rates is crucial and the population becomes vulnerable to epidemics.

The idea of birth rates and death rates of a disease is illustrated by three diseases of trees. The rosette virus of peach spreads slowly and kills the host rapidly. The pathogen has a very low birth rate and a high death rate — the disease is locally suicidal and eradicates itself. In this case the epidemic crowding point is usually imaginary (though there have been occasional cases of whole orchards being affected). The swollen shoot virus of cacao has a slightly higher birth rate than peach rosette virus and it kills the host more slowly. It has a low epidemic point and it was only after cacao cultivation had developed intensively that epidemic spread occurred. The third example, Tristeza virus of sour oranges, has a high birth rate because the vector is abundant and efficient and a rather lower death rate — the host trees go into a decline that lasts for several years. In countries where the vector is present the virus spreads

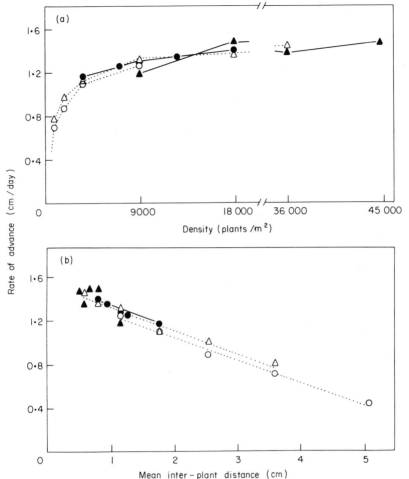

Fig. 16/4. The rate of advance of *Pythium irregulare* into populations of cress (*Lepidium sativum*) from infected seedlings placed at the edge of the population of plants, expressed in relation to (a) the density of plants and (b) mean distance between plants. (From Burdon and Chilvers, 1975)

quickly on all susceptible species of citrus. The disease has a low epidemic point, epidemics being strongly favoured by crowding. Van der Plank argues that in cultivated crops (and by analogy in plants in the wild) the bringing of populations nearer together increases the chance of epidemics of systemic virus diseases faster than local lesion diseases.

Continuous populations of a single species occur in nature in northern temperate forests that have been selectively felled; these are classic sites

for epidemics both of fungal and of insect pests. Plantation cropping
has also demonstrated the hazards of monoculture. Coffee rust (*Hemeleia
vastatrix*) has been absent until recently from the Western Hemisphere
and plantation cropping has been feasible in a way impossible in the East.
Rubber (*Hevea brasiliensis*) can be grown as a monoculture in south-east
Asia where South American Leaf Blight is absent, but suffers badly
from this disease in areas of Latin America where the fungus is endemic.
It is a well known hazard in the life of the experimental ecologist, that
plant species which appear to be free from disease in nature develop remarkable collections of epidemic diseases as soon as they are grown in
pure stands in experimental gardens!

The host-specific pathogen (like the monophagous herbivore) may be
an important agent in the diversification of vegetation. The presence of
a host-specific disease on a parent plant will usually increase the chance
that others in the immediate neighbourhood will be damaged by the
same disease. The plant that disperses its seeds widely increases the chance
that some of the progeny will escape from the hazard. If colonizers are
at risk near a parent of the same species the chance of other species establishing is correspondingly increased and an interspersion of individuals
of different species is encouraged.

It may be that many natural populations of plants in which groups of
species are intermingled, owe their diversity to disease (and pest attack).
Disease will develop in many species when neighbours become too close;
diversity in plant communities may therefore reflect the failure of pure
stands. It has been suggested that fungal parasites may contribute to
maintain the rather puzzling mixtures of species of *Eucalyptus* in Australia. Pure stands are rare and characteristic assemblages of species are the
rule (Pryor, 1959). Fungal parasites cause an average of 20% "effective
leaf loss" in *Eucalyptus* forests and host-specific pathogens appear to
penalize a population that becomes dominated by a single species (Burdon and Chilvers, 1974). It would be of the greatest interest to observe
the consequences of deliberately creating local pure stands of native
species in a variety of plant communities to determine what hazards
they then experience.

Interspersion of resistant and susceptible plants reduces r, the intrinsic
rate of natural increase of a disease. In the "native" potatoes of the
mountains of Basutoland* "the chief vector of the potato viruses, *Myzus
persicae,* is abundant, and its winter host, the peach, almost ubiquitous.

*now Lesotho

The varieties have no great resistance to virus disease, and succumb quickly when they are planted in garden plots. But when they grow — as they ordinarily do — as odd plants in the corn fields or in the grass near roadsides, there is little evidence of virus disease, and the varieties have persisted for more than a century in good health" (van der Plank, 1960). Plant pathologists have recognized that the spread of disease can sometimes be controlled by interspersion of different species within a pure stand — for example narrow rows of barley around seed beds of cauliflower reduced infection by mosaic virus from 3.0 to 0.6%. (Broadbent, 1957).

With a few exceptions, plants free themselves from systemic diseases when they pass through the phase of seed production. Passage through the sexual cycle cleans the cytoplasm of viruses and cytoplasmic variants (Mather, 1973). The new colonist arising from seed starts afresh, free from many of the systemic diseases that accumulate in older populations.

Disease escape in time

Most pathogenic fungi (and some bacterial pathogens and nematodes) have dormant resting stages: these may remain dormant for periods of time quite different from the diapause or winter dormancy of insects and small mammals. The long agricultural rotations that are required for the elimination, or even the partial control, of some fungal diseases witness the ability of many fungal pathogens to survive periods in which the host is absent. The ability to persist in the absence of the host plant characterizes soil-borne pathogens, which have rather poor dispersal in space but are excellently dispersed in time. In contrast foliar pathogens tend to have a relatively shorter life as dormant spores: they are well dispersed in space and poorly in time.

It is clear that if epidemics occur in a plant population there will be some selection for forms that complete their life cycles before the main force of the disease strikes. This must always be a means of escape in time for some members of a population of hosts. Precocious development may also mean that conditions for the growth of the pathogen are poorer — but the effect of such selection would usually be to force the host also into a less ideal phenology. So long as all individuals in a plant population grow synchronously, the opportunities for the development of epidemics are maximized and escape by precocity only works for the few that develop in advance of the many.

Coevolution of plants and pathogens

The idea that there is continuous genetic feedback between populations
of plant hosts and their pathogens is long established in plant pathology,
in contrast to the situation in ecology where the idea of evolution as a
continually operating force in ecological interactions seems to be accepted unwillingly. The rapid changes in the genotypic composition of
cereals and their rust races, and of potatoes and the races of *Phytophthora infestans* illustrate the speed and complexity of natural selection
where resistance, tolerance and susceptibility to disease are involved.

Epidemics are most common when either the pathogen or the host
have been introduced into a new area and have not had the opportunity
to coevolve. Epidemic diseases (like epidemic outbreaks of pest populations) are not seen in natural vegetation — except after some major disturbance. " . . . there does not seem to be a single record of natural infection (let alone an epidemic) by a harmful virus of any plant species that
had achieved ecological dominance and is growing in its natural environment . . . there have been epidemics of plant virus diseases only because
we have taken into cultivation a lot of annuals and other plants with no
record of ecological dominance in nature" (van der Plank, 1960). A
continuous display of genetic variation is of course necessary for a coevolutionary response to disease and it is not surprising that epidemics
are most common in self-pollinated plants and those that are propagated
vegetatively " . . . no disease control measures yet put into practise equal
in efficiency the natural ability of a cross-pollinated crop to protect itself through variation" (van der Plank, 1960).

It is not necessary at this point to document the evidence that coevolution occurs between plants and their pathogens — the very high
degree of specificity of hosts and parasites is sufficient evidence. This
specificity can only have arisen from an evolutionary history of coadaptation. The result of coevolution is (as with pests) the existence of
highly host-specific specialist pathogens (the rusts, smuts, mildews etc.)
doing relatively little harm to their obligate hosts; widespread but in
nature usually held well below epidemic level; not lethal but held in check
both by natural host resistance and by the spatial isolation of hosts within the natural vegetation. In contrast there is a group of generalist pathogens that have escaped the rut of specialization, "Jack-of-all-trades"
organisms with wide host ranges, often lethal to young plants, often
capable of living on dead organic matter and at their most damaging when

the hosts are unhealthy through some other cause. These are the pathogens that kill young seedlings, that afflict plants that are unadapted to waterlogging, that convert sub-lethal damage done by other causes into lethal damage. These are the organisms that may ultimately be the causes of the death of trees weakened by the activities of defoliators, of the overgrazed grass, the trampled clover leaf, or the plant that is subsisting at minimal net assimilation rate in the shade of others. The fungi that come into this broad category of relative generalists include species of *Pythium, Rhizoctonia, Nectria* and many *Fusarium* spp.

Invasions by pathogens into new areas

Diseases have on rare occasions made spectacular entry into natural vegetation. The classic case is the epidemic development of *Endothia parasitica* in the forests of the northern United States. The disease is believed to have come in the first place from China where the pathogen is relatively harmless against the native species of *Castanea* (presumably the host and pathogen had coevolved). *Castanea dentata,* which had been the dominant tree in large areas of North American forest, was almost completely eliminated by the disease. This is probably the largest single change in any natural plant population that has ever been recorded by man. The disease has not brought the host to extinction but *Castanea dentata* is now found as isolated young or suckering plants and is no longer a dominant. It is conceivable that in the long term *Endothia* and *Castanea* may coadapt in North America and reach a *modus vivendi* more in favour of *Castanea* (and of the pathogen, which is now rare!) but this process must be very slow because the disease has largely prevented the reproduction of *Castanea* by seed and so slowed down its potential rate of evolution.

The presence of a disease may thwart attempts to introduce a higher plant to a new area and it seems likely that natural invasions by angiosperms may sometimes have failed because of the pre-existence in the area of pathogens that had not coadapted. Introductions of *Pinus strobus* to Europe have failed because of their extreme susceptibility to *Cronartium ribicola.* Plants of *Pinus strobus* carried to the United States from Europe bore the pathogen and it caused epidemics in North American forests. Spread in America was made the more effective and rapid by the presence already in the forests of nine wild species of *Ribes,* the alternate host of the fungus (Gram, 1960).

Deliberate introductions of fungi have rarely been made but in one instance a fungus has been introduced for the biological control of a weed. Skeleton weed, *Chondrilla juncea,* is a common component of the weed flora of Mediterranean regions; it entered the flora of Australia and has become a dangerous weed of cereal crops. The biology of *Chondrilla* was studied in the Mediterranean region where the densities of the plant are strongly influenced by cultivation practices (Wapshere *et al.,* 1974; Hasan and Wapshere, 1973; Wapshere, 1970, 1971). Of various organisms attacking *Chondrilla*, the rust *Puccinia chondrillina* destroys seedlings and rosettes, heavily damages flowering shoots and reduces the seed output. Two powdery mildews (*Leveillula taurica* and what is probably a form of *Erysiphe cichoracearum*) and a gall mite may also cause some damage. The rust is the most important of the species that attack *Chondrilla* in the Mediterranean area, despite the fact that it in turn is attacked by a polyphagous parasite of rusts *Darluca filum*. Table 16/Ia shows the mortality in populations of *Chondrilla juncea* in a site in southern France and one in Spain where both rusted and non-rusted plants were present: the risk of death of infected plants was 2.0–4.6 times greater than of uninfected plants. *C. juncea* is a perennial, regenerating from underground parts and Table 16/Ib shows the effect of rust infection on the regeneration potential. In areas studied in the Mediterranean region, populations of *C. juncea* can become very high (100–300 plants/m^2) under conditions of cultivation that inhibit the development of *Puccinia chondrillina*. If cultivation stops, the rust immediately attacks most of the plants of *Chondrilla* and kills many. The populations of *Chondrilla* then decline over a few years to densities of 1–10 per m^2, and stabilize at this level. The proportion of rusted plants also declines, which suggests that the populations now consist mainly of rust resistant plants. Though some individuals continue to be heavily attacked, the process of selection of the host population by the pathogen is a continuing process. The similarities with the effects of an insect attack are striking: the host population settles down to a new low density at which damage is rarely seen and the agent that is the true determinant of the plant's population density ceases to be an obvious element in its biology. This example illustrates, as well as any, the way in which the population dynamics of a plant may be regulated by a host-specific predator or pathogen in which the host and pathogen play an interminable game of co-adaptive evolution. The introduction of *P. chondrillina* to south-eastern Australia has apparently been a success in biological control and densities of the weed

Table 16/Ia

Mortalities of *Chondrilla juncea* plants rusted and non-rusted with
Puccinia chondrillina at two natural stands between November
1968 and June 1969

	Les Campeaux S. France	Playa de Pals Costa Brava, Spain
Rusted plants		
No. infected during period	36	9
No. disappearing during period	23	5
Mortality (%)	63.8	55.5
Non-rusted plants		
No. remaining uninfected during period	93	33
No. disappearing during period	28	4
Mortality (%)	30.1	12.1
Mortality differential: infected/uninfected	2.1	4.6

Table 16/Ib

Regeneration of rosettes from *Chondrilla juncea* plants rusted and non-rusted
with *Puccinia chondrillina* at end of season (October, 1969) at Playa de Pals,
Costa Brava, Spain

	Rusted plants	Non-rusted plants
No. of flower shoots	15	22
No. plants with rosettes at base	2	19
Regeneration (%)	13.3	86.4

Hasan and Wapshere, 1973.

have been reduced in cereal crops from 200—400 plants per m^2 to densities of *ca* $10/m^2$ (A. J. Wapshere, pers. comm.).

The nature of host damage by pathogens

In general, those pathogens that have been trapped in the coevolutionary rut of host specialization cause relatively little damage to the host. A reliably high intrinsic rate of natural increase of a pathogen is best obtained from a host that continues to live and support the pathogen's growth: the diseases that cause epidemics are nearly all non-lethal specialists. In some cases host plants suffer little from a pathogen because it develops mainly on the older modules of the plant, tillers that have completed flowering, and mature leaves (e.g. *Erysiphe* spp. and many rusts). Like precocious development of the host, delayed development of the pathogen may reduce the severity of a disease and represent one pathway towards a co-evolved mutual tolerance. Specialized parasites tend to have their greatest development late in the life of the host and *Puccinia chondrillina* is probably rather exceptional in causing some early death. Infected plants tend, however, like grazed or insect damaged plants, to be weakened in competition with healthy individuals and competition exaggerates any differentials within a population that are caused by disease or any other form of damage.

There is one intriguing group of exceptions to the rule that a plant attacked by a pathogen is weakened. Some systemic pathogens prevent the host from forming seeds, e.g. *Epichloe typhina* sterilizes its grass hosts by preventing the emergence of an inflorescence. Such eunuch plants have high vegetative persistence in pastures and apparently greater vegetative vigour. It may be that some of the resources diverted from flowering are available for the more rapid production of tillers (Bradshaw, 1959). Such infected plants have lost all ability to form and disperse progeny and been converted by the pathogen into permanent perennials. Some smut diseases of grasses have the same effects (e.g. *Ustilago hypodytes* on *Zerna erecta*); the anther-sterilizing pathogen of *Trifolium pratense* (*Botrytis anthophila*) also seems to confer greater expectation of life upon infected plants.

Pathogens have rarely been studied as part of an ecological system in nature, and their relevance to population biology and population genetics is mainly unexplored territory. This chapter has quoted at length from the work of van der Plank which seems acutely relevant to the development of any general theory of population biology. The theory of epidemics is essentially the same as that for the growth of any biological population; it is concerned with births, deaths, emigrations and immigrations — with the ecological theatre and the evolutionary play. The stage is ripe for bringing the ideas and techniques of plant pathology from its traditional area of concern with crops and man-managed forests to the exploration of natural vegetation.

17

The Role of Predation in Vegetation

It is relatively easy to produce a list of anecdotes that describe the effects of predation or pathogens on plants. It is much more difficult to answer the question how important predators are in determining the number of plants and the character of natural vegetation. Hairston *et al.* (1960) maintained that herbivores are mainly controlled in numbers by the activity of their predators and parasites rather than by a limited food supply. Clear cases of the rape of vegetation by animals occur, but in relatively specialized circumstances: plants in natural environments are seldom seen to suffer from animal damage. "The earth is always green" — plant food usually appears to be superabundant and it can easily be argued that when this is the case animals are not very important to the vegetation. " . . . we contend that heavily grazed plants are rare in natural systems" . . . " . . . Floras as wholes are resource limited. In particular, this applies to dominant component species" (Slobodkin *et al.,* 1967). This view was strongly challenged by Ehrlich and Birch (1967) who pointed to the profound effects of introduced animals on vegetation, the effectiveness of some insects in the biological control of weeds and the disasters to vegetation caused by introduced pests and diseases. There has been vigorous controversy which depended like many other arguments in science on the opponents taking extreme positions in a continuum of truth. Slobodkin *et al.* emphasized the balance of nature in natural habitats, while Ehrlich and Birch emphasized the catastrophes and disasters in disturbed communities. The controversy was in some

497

ways a continuation of an older argument between Nicholson (1933, 1958: Nicholson and Bailey, 1935) and Andrewartha and Birch (1954; Andrewartha, 1957) about whether the numbers of animals are regulated by competitive interactions or determined by the rates of recovery after disasters.

What we observe in natural systems are complexes of partially interdependent animals and plants which are the products of past ecological interactions and evolutionary change. The components of natural vegetation are the result of long-term evolution of plant species and their predators to which new additions have occasionally been recruited by dispersal into an area. In such a system we will not see the failures, because they will be extinct and we will not have seen how they were extinguished. We will see the proof of the successes, it is their descendants that now populate the area. We will tend to see systems that are stable, if only because, by definition, the unstable systems are more likely to have disappeared. If epidemics have occurred we will see descendants of the successful survivors, probably with changed genotypes, the prey now controlled in their density by some relatively stable interaction with the predator or pathogen that once may have caused an epidemic. We will see species with a genetic composition that reflects past forces of predation and parasitism: mutually evolved species relationships, highly specialized and often monophagous links in food chains. The reason why we do not see the animal as a vital part in plant activity is that most of the exciting phases of invasions and the vigorous interactions that occur in the early stages of a predator–prey interaction have been played out. Long-evolved undisturbed natural communities (a rare phenomenon) are the stalemates in the existential game that is the struggle for existence (Slobodkin, 1968).

Janzen's view of the nature of predation in a tropical rain forest is well in line with this interpretation (Chapter 15). A very diverse system of intermingled dominant tree species suffers little from predation in the adult stage. Disease spreads rapidly through pure stands — epidemics of insects are devastating when the host plants are close together and within easy search range of each other. When such pests and diseases have taken their toll the residual populations of each species of plant, its predators and pathogens, are likely to be at low density, and the interactions will rarely be obvious. Neither species in a predator–prey system is likely to be of the same genetic composition as it was before the interaction. Genetic modifications that result from the convulsions of early

epidemic encounters tend to be in the direction of (i) greater host resistance, (ii) greater parasite and pathogen specificity, (iii) interactions less
damaging to either component. This is the classic route, from facultative
and lethal to obligate and tolerable relationships, between co-evolving
predators and prey, between parasites and their hosts.

If a complex system has had a long time in which to settle down, the
events within it are unlikely to be dramatic. We are unlikely to see in
action every day the more violent repercussions of events that occurred
in the early stages of the system. However, two natural perturbations
may disrupt the equilibrium: there may be genetic change within one
species, or a new species or genotype (predator or prey) may immigrate,
emigrate or die. Any of these events may trigger repercussions along a
food chain as the longer-established populations are selected by the new
changed presence amongst them. Such events will, almost by definition,
be extremely rare in "natural vegetation" since most of the common
possible mutations, recombinations and immigrations will already have
occurred. A long-evolved system is unlikely to be seriously upset by the
normal recurrent hazards such as fire, hurricane or season of extreme
cold. Past experiences are likely to have left their mark within the genotype and each subsequent experience is therefore likely to produce less
reaction.

It is perhaps a valid generalization that if in natural vegetation a predator or pathogen is important in the population dynamics of its prey,
it will rarely be seen to be so. At the stage at which we are likely to observe the system, the feedbacks within it will be operating at a fine
scale of adjustment which is rarely obvious. It is only after a disturbance,
a perturbation of the "natural vegetation", that the force of the feedback processes becomes vigorous enough to see.

The interpretation given above does not deny in any way that most
vegetation is probably resource limited, though its character may be
determined by subtle co-adapted interactions in which predators and
parasites play a major part. Undisturbed, stable "natural" vegetation is
probably the worst possible place in which to look for the processes by
which its character is maintained. It is there that they are least likely to
be seen. *The nature of a balancing system is best explored by disturbing
the balance.*

Perturbed systems

The activities of man as a colonizer and exploiter have provided massive perturbations in natural systems. He has introduced a new order of magnitude into distances of dispersal both of plants and (usually following later) of animals. The most striking effects of immigration have occurred where the distances involved were greatest and the chances of natural dispersal had been least. The flora of New Zealand is now composed of *ca* 35% introduced species, all of which arrived since European invasions (Good, 1974). The floras of Australia and of North America have received large injections of flora from climatically analogous regions in the Mediterranean. The islands of the Pacific have recruited largely from the continent of South America. The massive shifts in vegetation consequent on these immigrations were already apparent to Darwin: "Cases could be given of introduced plants which have become common throughout whole islands in a period of less than ten years. Several of the plants now most numerous over the wide plains of La Plata covering square leagues of surface almost to the exclusion of all other plants have been introduced from Europe" (Darwin, 1859). This is the type of perturbation that gives the opportunity to see vegetation in flux and to discover the ways in which new equilibria are established. This is ideally the chance to determine the role of animals in the structure of vegetation.

Unfortunately, wherever man has introduced new species into existing communities, whether by accident or design, he has also changed the communities in other ways. It is doubtful whether there is any indisputable example of a plant introduced by man that has successfully invaded a natural community, not because it cannot happen, but because the habitat has also been changed just before or at the same time as the introduction. In Australia, New Zealand, and California where floras have been particularly enriched, man has brought the plough and grazing animals. The newly established rangelands and arable lands have been the "home from home" for most of the invaders. European man has usually brought new species with him on his inter-continental migrations but has also hunted native animal populations and distorted existing balances between predators and prey. The escape of introduced animals like the rabbit, brought by the Normans to Britain, and by early settlers to Australia, has had profound influences on the nature of existing vegetation. Too many changes happened too quickly and it is tragic that the earliest phases of colonization have rarely been studied.

Biological control of plants by animals

The most completely documented studies of the role of animals in the control of plant populations come from the literature of biological control. Animals have been introduced deliberately to control weeds. This biological control of plants has usually involved the choice and release of a host-specific pest or pathogen into an area where an introduced plant has become a nuisance. The procedure for biological control has then involved: (i) search for a natural enemy of the weed in its country of origin; (ii) cultivation of the enemy in bulk and in quarantine to free it of any parasites or diseases that might reduce its activity in the country of introduction; (iii) extensive testing to ensure that it will not feed on useful plants and that it is very unlikely to adapt to these. Legislative restrictions have been very tight because of the risk that a supposedly specific predator might turn to attacking an economic crop. It is both the strength and weakness of a biological control agent that it multiplies — pesticides do not.

The three most spectacular examples of biological control of weeds by introduced animals involve *Opuntia* spp., *Senecio jacobaea* and *Hypericum perforatum*. There are many others and detailed accounts have been given by DeBach (1964) and Huffaker (1971). Most accounts of biological control are more concerned with the behaviour of the control agent than with the plant that is controlled and, unfortunately, changes in the plant populations that result from these introductions have rarely been studied in depth.

Opuntia

Many species of *Opuntia* have become noxious weeds. The genus is naturally restricted to the New World and has either escaped from gardens or spread from areas where it had been deliberately planted for hedging. An introduction to Australia of *Opuntia vulgaris* was made in 1788 but it did not become a serious pest. The main problem with *Opuntia* in Australia was started by the introduction of *Opuntia stricta* from Texas or Florida in 1839. This species spread rapidly to occupy 4×10^6 ha in 1900 and 24×10^6 ha by 1920. It probably reached its peak rate of spread in 1925 at which time it was colonizing *ca* 400 000 ha a year. In 1925 half of the total infested area bore stands of *Opuntia* so dense as to be practically impenetrable by man and larger animals. Photographs of infestations (Dodd, 1940) show populations so dense that it is difficult to believe that they are not resources limited. Various insect pests were introduced including *Dactylopius indicus* as early as 1903.

This species failed to establish but was successfully reintroduced in 1913 and gave a degree of control of *Opuntia vulgaris*. In 1925, 2750 eggs of the moth *Cactoblastis cactorum* were imported from Argentina and the larvae were reared on *Opuntia stricta*. The spread of the insect was extremely rapid and between 1927 and 1929 three billion eggs were available from collections in the field. By 1930 huge areas of *Opuntia* had been killed: ". . . in this process of food supply depletion there was a time when millions upon millions of starving larvae were in search of food, and it is most interesting that this already proved specific insect, under great stress for survival, was not observed attacking any plants other than *Opuntia*." (Holloway, 1964). The moth population fell drastically after the decline of *Opuntia* but *Cactoblastis* was able to increase sufficiently rapidly to prevent a major resurgence of the weed. *Opuntia stricta* has now declined to a sparse population of occasional isolated plants and a few larger patches. At present there are no obvious ups and downs in the populations of either predator or host. Neither is common and the plant density (of *Opuntia*) is well below any resource limits (A. J. Wapshere, pers. comm.). The areas occupied previously by *Opuntia* are now mainly in agriculture use. It appears that the density of *Opuntia stricta* is now controlled at a level determined by the search range of the moth.

Other examples of the attempted biological control of *Opuntia* species (Holloway, 1964) have seldom been as spectacular as in Australia. In some areas the cochineal insect (a scale, *Dactylopius tomentosus*) has reduced the densities of *Opuntia* spp. In South Africa the cochineal insect was hindered in an otherwise very successful control programme because it itself was attacked by an indigenous coccinellid *Exochomus flavipes* and the imported *Cryptolaemus montrouzieri*. However, a large proportion of the *Opuntia megacantha* populations was destroyed. In this case mechanical control methods were used to supplement the attack of *Dactylopius*. This case is interesting because the presence of a third trophic level (a predator on the herbivore) reduced the influence of the herbivore on the plant. It seems likely that the density of *Opunita megacantha* may be regulated at three different levels: (i) a resource-limited population size in the absence of *Dactylopius*; (ii) a herbivore-limited population size determined by the search range of *Dactylopius* in the absence of its predators; and (iii) a herbivore-limited population size determined by *Dactylopius* in the presence of its predators.

In Hawaii the control of *Opuntia megacantha* was attempted with both *Cactoblastis cactorum* and *Dactylopius tomentosus*. The cactus has been effectively controlled but *Cactoblastis* has become the dominant control at upper and *Dactylopius* at lower elevations.

Hypericum perforatum

This plant, which is relatively unimportant to agriculture in Europe, has become widely distributed in other temperate parts of the world. The seeds are very small and are readily dispersed in hair, hide and clothing. In Australia *Hypericum* spread rapidly after an introduction in about 1880. By 1916 it occupied an estimated 74 500 ha in the state of Victoria. Three species of chrysomelid beetle (*Chrysolina* spp.) were introduced and of these *C. hyperici* became established and spread slowly. Later *Chrysolina quadrigemina* and *Agrilus hyperici* were introduced and later still a gall midge *Zeuxidiplosis giardi*. The plant has been controlled in localized areas but *Chrysolina* prefers unshaded areas for oviposition and much of the infestation is in shade. The gall midge may be more useful in shady areas.

On the west coast of the United States (California, Oregon and Washington) *Hypericum perforatum* spread very rapidly in rangelands. In California it was first reported in 1900 and by 1944 occupied more than 800 000 ha. The plant is a serious weed not only because it is a vigorous aggressor towards useful grasses (in places *Hypericum* formed almost pure stands) but also because it photosensitizes the skin of white-skinned animals and acts as an irritant in the mouth. In 1944, three species of beetle that had already shown promise in Australia were introduced to California: *Chrysolina hyperici*, *C. quadrigemina* and *Agrilus hyperici*. The species all passed stringent feeding tests to ensure that they were unlikely to feed on agricultural crop plants and the two species of *Chrysolina* were readily adjusted to synchronize their life cycles to the northern hemisphere. Difficulties with *Agrilus* delayed its release until 1950. The gall midge *Zeuxidiplosis giardi* was introduced in 1950. The behaviour of the four introduced insects was very different and between them they provide perhaps the most exciting experiment in the whole of the science of plant—animal relationships.

After the release of *C. hyperici* in the spring of 1945 and of *C. quadrige-mina* in the following spring, both become established. An original colony of 5000 *Chrysolina quadrigemina* multiplied so fast that in 1950 more than 3 000 000 adults were collected for redistribution. Both *Chrysolina* spp. feed on the vegetative rosettes of *Hypericum* during the autumn, winter and spring. The timing of the insect's life cycle is crucial. Both *Chrysolina* species aestivate (as does the host) and the aestivation is broken by autumn rains. *C. quadrigemina* requires less rainfall to break aestivation and starts to lay eggs earlier in the autumn. It is therefore more critically damaging to *Hypericum* than *C. hyperici* which does not begin to lay eggs until December by which time rosettes are already well developed. Feeding by *Chrysolina quadrigemina* prevents *Hypericum* from flowering and setting seed but *Hypericum* regenerates readily from rhizome buds. Three years of repeated attack, plus competition from neighbouring plants, is usually sufficient to kill *Hypericum*. One of the effects of repeated shoot damage is that the root system is seriously reduced (see Chapter 12) and the plants are less able to tolerate hot dry summers. The plant is much more tolerant of defoliation in areas where irrigation is practised.

"Within a decade after release of the insects, Klamath-weed was reduced from an extremely important pest of range lands to a road-side weed, with a resultant improvement in land-carrying capacity and land values" (Holloway, 1964).

Agrilus is a root borer and the insects proved easy to establish — but " . . . they were overwhelmed by *Chrysolina quadrigemina* which ultimately moved into the colonisation area". *Zeuxidiplosis* — the gall midge — requires succulent foliage during the summer months and this severely limited its activities except along irrigation ditches. The differences in biology of the four organisms that have been established in California determine that *Chrysolina quadrigemina* has had by far the largest impact. It is not possible to determine what each might have accomplished on its own. The timing of the life cycle of *C. hyperici* does not mean that it would have failed ultimately to reduce the density of *Hypericum* — only that it might have taken much longer. It does not follow that the insect that spread most rapidly and was most damaging to high density populations of *Hypericum* is the same species that will ultimately affect it most at low densities. Much will depend on the relative search ranges of the species.

Chrysolina beetles are reluctant to oviposit in the shade, with the result that *Hypericum* has been drastically reduced in open habitats but not in shaded areas. This same characteristic has reduced the value of the species for *Hypericum* control in Australia where quite large areas of infestation were in woodland. The result is that *Hypericum* is now seen in the open range lands of California only as occasional plants at a density determined by the search range of the insect; it is much more abundant in the shade. The beetle is rarely seen, nor is its damage.

> It is believed that in the absence of previous knowledge of this programme, and unless he made specific studies, an entomologist or ecologist viewing the current picture would conclude that what we know to be the key insect species, *Chrysolina quadrigemina*, is not a significant influent of the stand of vegetation and that the few plants of Klamath weed seen here and there are not primarily limited by this insect. He might also erroneously conclude that this plant is a shade-loving species, since the beetle checks it much less effectively under shade, hence more survive there.   (Huffaker, 1964)

It is indeed quite likely that the ecologist would seek to explain the "shade-loving" character of *Hypericum* by growing it in growth chambers at a range of light intensities under various climatic regimes and sooner or later would discover something that the plant did better at low light intensities. He can then be imagined shouting "eureka" as he rushes to publish his false causal explanation."

Senecio jacobaea — ragwort, tansy ragwort

Senecio jacobaea is another European plant that has spread under the influence of man. It has established as a serious weed in rangelands in Australia, New Zealand, South Africa, South America, north-western U.S.A. and Canada. The plant is perennial, though commonly monocarpic, dying after flowering. In the first year or years it forms a rosette and this ultimately flowers. Flowering may under ideal conditions occur in the second year but it can persist vegetatively for a number of years before flowering. The plant is capable of regenerating new shoots from root buds, mainly from fine rootlets distant from the parent plant. It is also capable of regenerating from the crown after damage, particularly if the damage occurs before seed set. It is as if the plant uses up some limiting resource in the act of seed production and dies, but if seed set is prevented the plant maintains a perennial habit and multiplies clonally (Harper and Wood, 1957).

In Europe, *Senecio jacobaea* is the host plant of a large number of insects and fungi. Cameron (1935) surveyed the natural enemies and out of at least 60 species recorded from *Senecio* in Britain, selected a moth *Tyria jacobaeae* L. and a seed fly *Pegohylemyia seneciella* as possible

agents for biological control of the weed in New Zealand. The caterpillars of *Tyria* defoliate the plants. Introductions of *Tyria* to New
Zealand were not as effective as hoped for, since the species was heavily
predated and parasitized. Close relatives of *Tyria* were already present
in New Zealand and parasites of these attacked *Tyria*. The fly was
more successful and in 1954 was reported as destroying 98% of the
flowers. The fly lays eggs in the capitulum and the larvae destroy part
of the seed crop. The faeces of the larvae encourage fungal growth in the
capitulum and destroy the remainder of the seeds (Cameron, 1935;
Poole and Cairns, 1940).

   Tyria has behaved in different ways in different places after its
deliberate introduction for biological control of *Senecio jacobaea*. Establishment was not successful in Australia. It was introduced into the
United States in 1959 (Frick and Holloway, 1964; Hawkes, 1968) and
populations of *Senecio* were dramatically reduced in a dense stand of
ragwort on a coastal flat at Fort Bragg, California. By 1965 an area of
about 5 ha was affected and the highest density of larvae occurred
on an advancing front of invasion with lower numbers ahead and
behind. Larval attack produced complete defoliation and the number
of flowering stems declined to less than 20% of the original density. The
spread of the moth and its damage was remarkably slow, about 450 m
between 1963 and 1967, perhaps because the plant density was so high
that females did not search far afield (see account in Andres and Goeden,
1971).

   Polyphagous animals are obviously dangerous introductions: even
though they may prefer a weed as diet and are perhaps more likely to
drive it to extinction than a monophagous specialist, they may subsequently turn to eating something valued by man. However, if a polyphagous herbivore is strictly confined by its ecological tolerance to a
specific habitat, introduction may be safe. Thus aquatic herbivores,
although generalists, have been introduced to control water weed problems. Manatees, carp and the snail *Marisa cornuarieties* have been introduced for aquatic weed control, apparently with success, but the results
have not so far been written up in detail (Huffaker, 1975).

   From the existing cases of effective biological control by animals
some generalizations emerge, though the present speed of exploration
of control methods may break these generalities.

   (1) Introduced animals have profoundly altered the densities of plant
species but there has not been extinction. Rather, the pest and its host

have settled down to densities that are low; the activities of the predators are rarely seen. In many cases it appears that the density of both components in the interaction is determined ultimately by the search range of the animal and the dispersal range of the plant.

(2) Most plants that have been successfully controlled are perennials, either herbaceous or woody. It is clear that a perennial habit, conferring some ability to regenerate after damage, is one guard against the predator going to extinction. Annuals may be capable of shedding a specialist predator which eats growing plants and dies in famine after which the annual regenerates from buried viable seed.

There seem to be few highly specific monophages of annuals in nature. The reason may be the same as in (2), that predator escape through time is too easy.

(3) In most cases of successful biological control by an introduced animal, the attacked plant has suffered from competition from other plants in the area. The reduction in numbers of the host plant may be attributed in part to the direct effects of the pest, but in part to its reduced aggressiveness to other vegetation after it has been attacked.

(4) Examples of biological control of weeds show highly specific monophagous herbivores in action. Polyphagous species that *preferred* the noxious plant might be even more effective in regulating its population, and perhaps lead to its extinction.

(5) All cases of explosion of plant population followed by successful biological control have involved dispersal across trans-oceanic barriers.

(6) Cases of successful control have involved divorcing the predator from its own dependent food chain of hyper-predators and -parasites. This divorce is normal for any species invading a new territory. If the food chain is composed of monophagous feeders, each trophic level must arrive in turn: plant, then specialized herbivore, then its specialized predator and parasites. A weed invasion followed by a biological control agent models what may be a common course of events in a process of colonization. In some cases of biological control we can see the later stages of a model, the evolution locally or late invasion by the agent that can control the herbivore (e.g. the adaptation of parasites of *Tyria* relatives in New Zealand to introduced *Tyria jacobaeae*).

The experiments of biological control are observed crises in the relationship between pairs of species. The opportunity for detailed follow-up of the behaviour of populations ought not to be missed. Wherever biological control of a plant has been successful, a specialized series of selective forces has operated on both plant and predator. We may expect (on analogy with e.g. myxoma virus) that new genetic combinations in plant and predator will continually appear and that what we saw originally as a game played between two species will, sooner or later, refine itself into a more subtle game played between selected races of the two opponents. Seeds of all weeds that are the object of biological control should be put into genetic storage, so that at later stages we have original genotypes against which to judge the evolutionary changes that have occurred.

Epidemics of pests and diseases are the uncontrolled equivalent of biological control. Usually the outbreak of a pest epidemic can be ascribed to an introduction (e.g. the pathogen *Endothia parasitica* Ch. 16) or to a change in the habitat. Pest epidemics can often be ascribed to some activity of man that has brought a floristically diverse ecosystem closer to a monoculture, e.g. the selective removal of some species from a forest leaving a floristically poorer community. This is thought to explain epidemics of several forest pests, such as the spruce budworm *Choristoneura fumiferans* in north-eastern American forests. The favoured host of *Choristoneura* is balsam fir (*Abies balsamea*). To judge from the present status of this fir in unlogged forests, it was apparently a minor and occasional component of the forests of the region. Cleancut logging practices and forest fires have favoured the regeneration of balsam fir and vast areas of nearly pure stands have become established in which the budworm has devastating effects. *Choristoneura* is present in the natural mixed forests but does not produce epidemics. This is either because, under such circumstances, its natural enemies control it or because the interspersion of other trees hinders the spread of the insect among its favoured host (Graham, 1929; Tothill, 1958, van den Bosch and Telford, 1964). Much as the development of an epidemic of a pathogen may be slowed down by interspersing its favoured host with a different species it has been suggested that the appropriate ecological control of *Choristoneura* might be the planting of hardwood barriers 30 miles or more wide (Heimburger, 1945).

All vegetation consists of islands in space and time. Sometimes the islands are small and discrete, like the trees in a mixed tropical forest, sometimes extensive and continuous (almost continents) like the single species stands of northern coniferous forests. The islands are subject to gain and loss of species and each unit of vegetation represents some stage in a process of adjustment to the recruits and losses. The animals in such islands of vegetation are part of the adjusting complex, damaging some elements in the population of plants and contributing to the adjustment. Impacts are likely to be commonest in the early stages of entry. Prey species will usually enter before their specific predators and in the initial stages of an invasion a new plant entrant may fail in the face of attack by generalist predators that already exist in the area. If it succeeds there may be profound repercussions on the existing vegetation and the new entrant may perhaps extinguish some original occupant. The spectacular entries of new weedy species demonstrate this clearly. At some later stage a specialized predator (or pathogen) of the introduced plant may invade, evolve locally or (in biological control) be introduced deliberately. The stage is then set for a new confrontation as the densities of both prey plant and specialized predator adjust to new densities. The numbers of both prey and the predator are probably likely to be determined primarily by the search range of the predator, and genetic feedback between prey and predator may continually modify the relative abundance of each. The predator will have had to survive the generalist hyper-predators or parasites of the island, but sooner or later its own specialist enemies may arrive, evolve or be introduced (as in the biological control of an insect by an insect) and a new wave of repercussions stretches down the food chain and affects the competitive relationships and the *status quo* in the plant population. As time passes there may be a succession of new predators or parasites arriving as the food web becomes increasingly complex.

It may be that during the phases of colonization by a predator, the plants that dominate the area are continuously limited by resources of light, water and nutrients; the arrival, disappearance or fluctuations in one plant species being continually absorbed by the expansion of the territory of another. Although it is convenient to describe this process as one affecting the presence and abundance of species the process must be essentially similar for gene complexes and the immigration to an "island" of alien genotype of an already present species (e.g. as pollen) may affect colonization or cause extinctions within the population of genes.

The argument between Hairston *et al.* (1960) and Ehrlich and Birch (1967) about the role of herbivores in the regulation of vegetation seems to become irrelevant in the context of island theory. The important matter is how frequently do invasions occur, and how often do we see the recurrent disturbances? If we are preoccupied with long-established vegetation largely free from risk of new colonizations we shall see an environment that appears to be stable, in which predators seldom do much damage (although they may indeed be the true regulators) and in which resources are clearly limiting. Only occasionally, if we are fortunate in the point of time when we observe, may we see a new species (or gene) naturally dispersed into the area, starting a chain of events that ricochets through the ecosystem. If we live in New Zealand or Australia or are primarily concerned with disturbed habitats, our impression of nature is entirely different. The speed of new introductions has become temporarily very great owing to man's activities as a dispersal agent and most of the events that we witness are part of a process of change. Predators are seen in action before they are limited by the shortage of prey and before they have co-adapted with the prey.

The famous argument about whether animals are determined by their natural rate of increase or by shortage of resources boils down to a question of when, and how often, are numbers determined by one factor and when and how often by the other. Similarly the argument about whether plants are regulated by herbivores boils down to a question of how often. The answer to both questions lies in island theory, for it is the degree of isolation in space and time that determines the opportunities for new invasions.

It is amusing to reflect that it has been the destructive effects of man and his role as a dispersal agent and creator of new environments (new "islands") that has speeded up the process of evolution of communities to a rate at which the processes involved in the undisturbed systems can begin to be understood. In an insect—plant relationship we must remember that what we see in the "undisturbed" situation may be the result of up to 200 million years of coevolution.

It is important to recognize that the prime cause of death in a population need not be a regulator; it need not be density-dependent. Similarly, the regulator in a population need not be the most obvious cause of death. It seems very probable that, proximally or ultimately, animals may frequently be the regulators of plant populations without its being at all obvious that this is the case. The problem becomes a technical one — how do we determine if animals are significant regulators in the life of a given plant population. Probably there is no simple answer but two perturbation approaches are as likely as any to deliver hints.

(1) *Deliberate reduction of a predator population* may reveal that it had

effects within the plant community that were not apparent before. The exclosure of potential predators is one technique that has already been widely used. It is not easy to exclude the smallest mammals and invertebrates by a physical barrier. Chemical exclusion may sometimes serve the same end.

Cantlon (1969) applied insecticides repeatedly to the ground in a Michigan woodland. The effect was surprising — a rapid increase in the population density of *Melampyrum*. He was able to trace the cause to

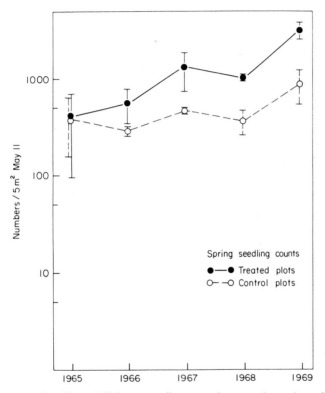

Fig. 17/1. Numbers of seedlings of *Melampyrum lineare* on the approximate date of maximum population density. The treated plots received an insecticidal treatment of *ca* 1 kg aldrin per ha each spring. In the first 2 years the treated plots also received a light foliar spray every week form mid-May to mid-August with a 1:1 mixture of malathion and DDT. (From Cantlon, 1969)

the exclusion of a beetle which is apparently partial to the seedlings of *Melampyrum*. It is very doubtful whether the role of the beetle could have been exposed by any classically oriented study of the autoecology of *Melampyrum*. Foster (1964) made a somewhat similar experiment in which he applied insecticides, fungicides and molluscicides to experimental quadrats in a permanent grassland in which *Bellis perennis* was a conspicuous species. The effect of the combination of pesticides (but not each acting alone) released *Bellis* from some form of population control and its numbers increased markedly. The cause was not identified.

(2) *Deliberate increase of the prey population.* A predator may control the numbers of its prey down to a level at which they are limited by the predator's search range. The search distance may be great or interspersion of prey amongst other plants may hinder search. The predator and its damage will then be rarely seen, unless the density of the prey is deliberately increased. Then there may be a chance to see the regulator in action as it returns the prey to the "safe" density. Both agricultural practice and forestry do this: a species is planted in stands more extensive and more concentrated than are found in nature and pest and disease outbreaks are the consequence. Deliberate ecological experimentation might do the same but this type of study has apparently never been attempted. For the autoecologist it is common experience that experimental pure stands often fall foul of pests and disease but this has been regarded as a nuisance, hindering the study. The liability of experimental pure stands to epidemics may be the critical clue to the real regulating factors that are responsible for maintaining the densities of plant populations in nature.

The Natural Dynamics of
Plant Populations

18

Introduction: Annuals and Biennials

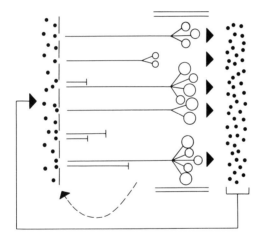

Plants stand still and wait to be counted. It is odd therefore that the great part of our knowledge of population dynamics has come from studies of animals where numbers can usually only be estimated and this requires sophisticated marking and trapping techniques. There are two main reasons why botanists have been scared of counts: plants vary enormously in size even within even-aged populations (see Chapter 6) and, because many species spread clonally, the genetic unit and the physiological unit may be different. For most animals the genetic

unit of the individual corresponds to the physiological unit of function:
a count of the number of rabbits in a colony measures the number of
genotypes represented within it, and also gives an estimate of the biomass
of rabbit present, the capacity of the colony to produce young and the
demands that it will make on environmental resources. Conversely, given
the biomass of a rabbit population it is possible to make a rough estimate (within a factor of say 10) of the number of individuals present
within it. In a plant population the nature of the individual is much
less sharply defined — the physiological unit of function may be the
whole or part of a genet and the parts may have lost all physical connection. A genet of e.g. *Trifolium repens* may be a short stolen bearing
two or three leaves, the whole not more than 2—3 cm long, or it may
be a fragmented population of parts extending over distances of *ca*
20 m (Harberd, 1963) with an immense biomass. If the plant is small
this may be because it is very young but growing fast, or very old,
most of the genet having died. Individual genets of the fern *Pteridium
aquilinum* have been estimated to extend over distances of 500 m
(Oinonen, 1967b) and may weigh hundreds of tons. A count of the
number of genets of a species therefore gives minimal information
about biomass, potential reproduction or the demands made on the
environment. Similarly a measure of biomass gives negligible information
about the number of genotypes. The number of individuals in a plant
population can mean two different things. The number of fronds of
bracken gives some measure of the activity of the species in an area, of
its demands on resources and of its status in a hierarchy of other species,
but it gives no information that could be useful to a geneticist or evolutionist who is interested in the populations as a source of variation. On
the other hand the number of genets in the population might be counted
(techniques of electrophoretic study of isoenzymes begin to make this
possible) but the size of the population determined in this way, though
relevant to the geneticist and evolutionist, is virtually irrelevant to the
production ecologist, the forester or the agronomist.

It was pointed out in Chapter 1 that, although it is often convenient
to describe an animal population in terms of its numbers (N) and to discuss population change as the change from N_t to N_{t+1} a proper description
of a plant population needs to be expressed as $N\eta$ where N is the number
of genets present and η is the number of structural modules that compose a genet. The production ecologist may be interested only in the
product, $N\eta$, and be unconcerned with the number of genotypes repre-

sented in a population. The geneticist may be interested only in N or in the variations in η because they roughly parallel the individuals' reproductive capacity.

In practice it is easy to count genets and to determine N in populations of annuals and in trees that do not spread clonally. The real problem arises with clonally spreading perennials where the clones intermingle and fragment. It is then impossible to determine N without recourse to genetic markers, electrophoretic typing or long term mapping.

At first sight, the fact that plants are not mobile suggests that counts (of $N\eta$ if not of N) should reveal the essentials of the population dynamics. In fact this is not the case because populations may gain and lose individuals at the same time. Figure 18/1 shows the population of plants of *Ranunculus acris* in 1 m² of a permanent grassland in which each individual seedling was mapped as it appeared and its fate followed, at approximately 2 week intervals over 2 years. The population present was very constant — but the apparent stability was the result of a balance between births and deaths and disguised the fact that the population received and lost within 2 years more than 5 times as many new genets as were present at any given time.

Between two censuses of a population new recruits may appear and individuals may die. For this reason, some form of marking of individuals is needed for a proper census. The most convenient way of doing this is to map the population with sufficient precision so that two neighbouring seedlings, one of which appears and quickly dies and is replaced by another close by, will be recorded as two plants and not as one. Even when a population is mapped at intervals a seedling may appear one day and be gone tomorrow (e.g. eaten by a snail) leaving no trace, so that maps made every few weeks may still underestimate the true flux within a population.

The population of plants present on an area of land is represented by those that are in an active state and those that are dormant. An investigator gains a biased sample if he counts only the plants that are growing and neglects the population of buried viable seeds and vegetative buds. The buried seed population may represent an accumulation of seeds from many previous generations (see Chapter 4) and the flux of losses and gains from the seed bank can be estimated only if cohorts of new additions can be distinguished from the bulk accumulated from the past. Mark and recapture methods may have to be used, thus making a census as difficult as with a highly mobile animal! It is quite realistic

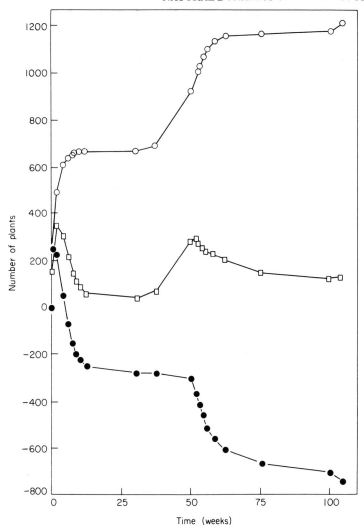

Fig. 18/1. The population flux of *Ranunculus acris* within a plot of 1 m² in permanent grassland. ○ = cumulative gains; ● = cumulative losses; □ = actual population present. (From Sarukhán and Harper, 1973)

to envisage a population of an annual weed (e.g. *Chenopodium album*) in a disturbed site, developing from a mixture of seeds that span 400 years in the times at which the zygotes were formed (Ødum, 1965). There is nothing remotely comparable to this "evolutionary memory" in animal populations except in such simple animals as nematodes, protozoa and rotifers and the degree of its development is one of the important differences between the demographies of different plant populations.

The dynamics of the populations of annual plants

An annual is a plant that completes its life cycle and dies within 12 months, though the life span may overlap two calendar years (Fig. 18/2).

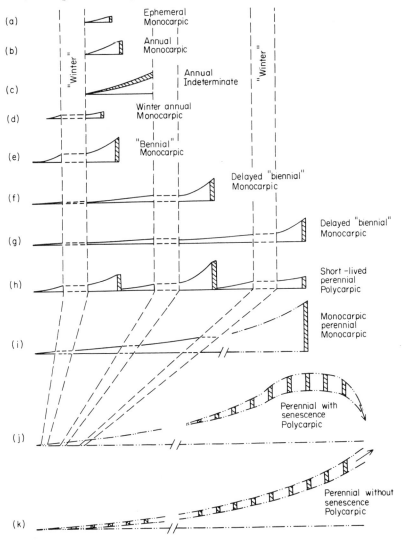

Fig. 18/2. Diagrammatic comparison of the fecundity schedules found among flowering plants. The period of seed production is shown shaded.

There are two categories of annual life cycle, that which is ended more or less abruptly with flowering and seed set and that which has a potentially indefinite length of life that is usually ended within a year by a climatic event. It is characteristic of plants in the first category that a phase of growth is followed by an episode of flowering and that the main meristems are used in the formation of the inflorescence — effectively putting an end to further potential for vegetative growth. In this category are many of the annual grasses of Mediterranean climates and of arable land. Most of these annuals have a precisely triggered transition from the vegetative to reproductive phase depending on photoperiod. They are species with life cycles that are synchronized with recurrent seasonal events. They are exemplified by the weeds of arable land: life for an annual in a crop is relatively safe until harvest and the safe period is used in a continuous cycle of growth; reproduction is the event that ends the growth period.

In contrast to the determinate annuals, some species start to flower when small and continue to grow, flower and set seed until they die from some extrinsic cause, e.g. frost or drought. This is an opportunist life: if they die young they will have left some progeny but they may sometimes be able to exploit a very long growing season (the first killing frost may be late) and they are able to exploit this by producing more and more seeds as the season passes. Such plants die with flower buds unopened. Their main meristems are not used up in flowering — they tend to produce their first flowers when very young and flowering and seed production continue simultaneously. This category of annual plants is illustrated by many species of *Veronica*, by *Poa annua* and *Senecio vulgaris*; these species are common as weeds in horticulture where the crops in the rotation have widely different growing seasons and the weeds may be killed by cultivation in almost any season of the year. Some of these indeterminate annuals root at the nodes and form clones within their single season of growth (e.g. *Veronica persica*). The details of population dynamics have been studied only for annuals in the first — determinate — category of growth forms.

The population dynamics of two annual grass weeds of cereal crops

Alopecurus myosuroides is a weed of cereal crops in Europe, particularly on heavy soils. It develops to high population densities in agricultural systems in which wheat is grown year after year. Some seeds germinate in autumn and a second crop of seedlings usually appears in spring, so that the mature population is often recruited from two differently aged cohorts. The plants from autumn-germinating seedlings flower in May when about 30 weeks old and those from spring germinating seedlings in June when about 10 weeks old. The autumn seedlings usually make some growth and start to branch before winter. The basal branches, "tillers", are the convenient modules for describing the growth of grasses as a population of parts. Plants enter the winter with 2–5 tillers and the greater the number of tillers the greater the chance that the genet will survive the winter (Barallis, 1968). Further tillers are produced in spring and the mature genets may be composed of up to 80 tillers. The spring plants tiller much less. Each tiller will usually form its own root system and either produces a single inflorescence or dies.

The population of seed of *A. myosuroides* accumulates in the soil and the flux in this buried seed bank is an important part of the population dynamics of the species. In order to study this flux, seed was collected from plants at three times during the season of seed production and marked by applying differently coloured fluorescent paints in an aerosol to the dispersal units. Marked seed was added to field plots in which natural seed production was occurring and from the ratios of marked to unmarked seeds it was possible to calculate the size of the total seed bank (Naylor, 1972a). The seed population in the soil estimated by the mark and recapture method was 23×10^3 seeds per m^2 which agrees closely with the value of 26×10^3 seeds per m^2 obtained by very tedious direct counts of the buried seed population. After germination the seedlings still carry marked glumes and seedlings derived from the experimentally added seeds can be recognized. Naylor used this technique to calculate the fraction of plants of *A. myosuroides* that was derived from seeds less than 1 year old. Seed from the previous year's crop represented only 35% of the total buried seed population but contributed about 66% of the next year's seedlings. Clearly the seedling recruits were not a representative sample of the whole buried seed population. If the buried seed bank is relevant as a genetic memory it is a biased memory.

Less than 50% of the spikelets of *A. myosuroides* contain a fully developed grain (caryopsis). This may be because this annual rather surprisingly (see Chapter 24) is an obligate outbreeder (Naylor, 1972b); failures in seed set may be the result of selfing but, whatever the cause, this magnitude of reproductive failure is an interesting feature of the population dynamics of this species.

The population dynamics of *A. myosuroides* in one year of study in a few carefully chosen "representative" wheat fields in eastern England is shown in Fig. 18/3. The seed production by the plants in year t (6.5 × 10^3 per m^2) corresponds closely with the seed production of the plants

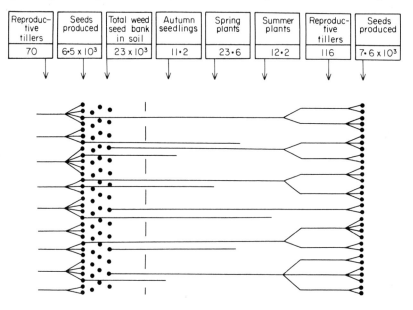

Reproduc-tive tillers	Seeds produced	Total weed seed bank in soil	Autumn seedlings	Spring plants	Summer plants	Reproduc-tive tillers	Seeds produced
70	6·5 × 10³	23 × 10³	11·2	23·6	12·2	116	7·6 × 10³

Fig. 18/3. The population dynamics of *Alopecurus myosuroides* in representative wheat fields in Eastern England in the season 1968–1969. (From Naylor, 1972a)

in year $t + 1$ (7.6 × 10^3 per m^2). The populations did not increase significantly (× 1.17) compared to the potential multiplication of × 620, the average number of seeds produced per plant. The number of individuals of *Alopecurus myosuroides* present as viable dormant seeds per m^2 was about 500 times the number present as mature plants — not an unusual situation in a weedy species.

Avena fatua is another annual grass weed of cereal crops with (in northern Europe) two peak periods of germination, a major peak in late March or April and a lesser peak in the autumn. Populations are maintained or increase on land that is cultivated in autumn or early spring, but a late spring cultivation destroys most of the seedlings and if regularly repeated can keep the weed under control. One of the very few studies of population changes in an annual plant over a sequence

of years was made on *Avena fatua* in plots in which barley was grown
year after year (Selman, 1970). Figure 18/4 shows the rate of increase
of a population in crops that were cultivated early. The number of
plants present increased *ca* 2.7-fold per annum, and, although there
was some year-to-year variation in the rate of increase, there was no
sign that the populations were reaching a saturation density. The same

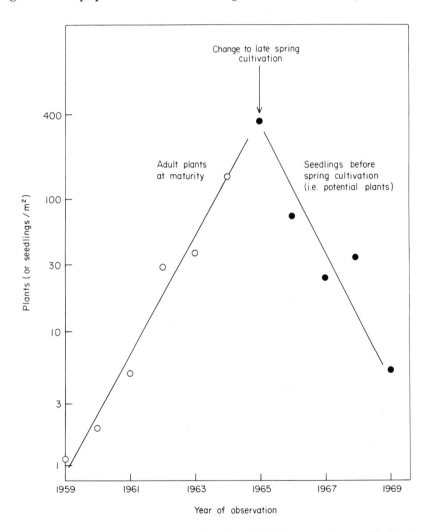

Fig. 18/4. Changes in populations of *Avena fatua* in a field with repeated crops of barley showing the effect of a change in the time of cultivation of the soil. (From data in Selman,
1970)

populations were followed for 5 further years in which barley was again sown repeatedly but the cultivation and sowing were deliberately delayed; the populations declined over 5 years to about a hundredth of the starting density. (During the phase of increase (1959—64) the population was measured as the number of mature plants present at harvest and in the decrease phase as the number of seedlings present

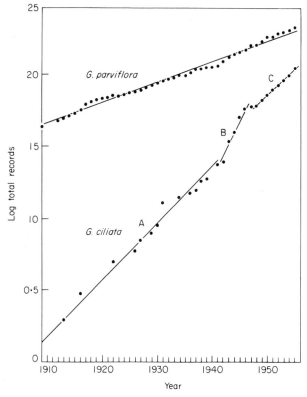

Fig. 18/5. Logarithm of cumulative total number of records of *Galinsoga parviflora* and *G. ciliata* in the British Isles from 1909 to 1955. Fitted regression lines are shown. The lines A, B and C for *G. ciliata* are for the three periods 1909—41, 1942—47 and 1948—55 respectively. (From Greig-Smith, 1957, after Lacey, 1957)

before the late spring cultivation, i.e. the potential population.) There can be few better documented records of changes in population growth associated with a single controlled alteration in the environment. The time required for a population to double under the favourable conditions of the first 6 years was 0.6 years and the time for the population to halve under the unfavourable conditions was also *ca* 0.6 years, but perhaps this was no more than coincidence.

Only rarely have plant populations been studied during invasions into new territory when the growth rate of the population can be measured in the absence of density stress from its own abundance. Lacey has recorded the spread of two related annuals, within the vice-counties of Britain. *Galinsoga ciliata* was introduced in 1909 and *G. parviflora* in 1860 (Fig. 18/5). The number of vice-county records is a very indirect measure for detecting population growth and gives no information about actual numbers of plants. However, both species increased their number of vice-counties exponentially from 1910. The rate of increase of *G. ciliata* changed twice between 1910 and 1956 but remained exponential.

Minuartia uniflora and *Sedum smallii* — two annuals of natural granite outcrops in the south-eastern United States.

A comparison of the biology of two species usually yields much more meaningful information than the sum of two independent studies. Sharitz and McCormick (1972) studied the population dynamics of *Minuartia uniflora* and *Sedum smallii* in "island" communities on granite outcrops in North Carolina and Georgia, U.S.A. These two annuals together with *Viguiera porteri* dominate sharply defined depressions that are filled with a soil derived from the weathered granite and organic debris. The islands vary from less than 1 ha to more than 100 ha. Both the species are winter germinating annuals and occur in dense populations. *Sedum smallii* is a primary invader of the shallow soils of the depressions and *Minuartia uniflora* is a secondary invader of deeper soils. The species tend to occur in concentric zones.

Five island communities were studied and two permanently marked transects were placed in each at right-angles to each other. The population density of the species was recorded at 2 month intervals in quadrats along the transect, the times of recording coinciding with the phases of seedling establishment, rosette formation, flowering and seed production. The soil depth in each quadrat, and the dry weight of plants at the rosette and flowering stages were also measured. The reproductive potential of each species was determined from 250 mature plants in each of the Georgia populations. The data were subjected to Life Table analysis — a procedure widely used for animal populations but which has scarcely been used for plants (see Harper, 1967). The results of the analysis are shown in Table 18/1.

Table 18/I

Life Tables of populations of two annuals, *Sedum smallii* and *Minuartia uniflora* (in parentheses) on granitic outcrops in Georgia, U.S.A. (from Sharitz and McCormick, 1972).

Life cycle stage		Seed produced	Available	Germinated	Established	Rosettes	Mature plants
Length of the "Life cycle stage" in months	D_x	4	1	1	2	2	2
Age of the population at this stage	A_x	0–4	4–5	5–6	6–8	8–10	10–12
Percentage age (i.e. age at beginning of stage as % of mean length of life of population)	A'_x	−100 (−100)	−10 (+52)	+13 (+90)	+35 (+128)	+81 (+204)	+126 (+280)
Survivorship (i.e. number of individuals surviving to the start of next life stage)	l_x	1000 (1000)	840 (210)	210 (64)	33 (11)	24 (8)	14 (5)
Senescence (i.e. number of individuals dying during the life cycle stage)	$d_x = l_x - l_{x+1}$	160 (790)	630 (146)	177 (53)	9 (3)	10 (3)	14 (5)

Mortality rate (1000 × q_x fraction of individuals alive at start but dead at end of the life stage) $= q_x = d_x/l_x$	160 (790)	750 (695)	843 (828)	273 (273)	417 (375)	1000 (1000)
Stationary population $L_x = \dfrac{(l_x + l_{x+1})}{2}$	920 (605)	525 (137)	122 (38)	28 (10)	19 (6)	7 (2)
Residual population life-span (plant-months remaining to individuals alive at the start of the stage) $T_x = D_x L_x + T_{x+1}$	4436 (2632)	756 (212)	230 (74)	109 (37)	52 (18)	14 (5)
Life expectancy (plant-months remaining to the population per plant alive at this stage) $e_x = T_x/l_x$	4.4 (2.6)	0.9 (1.0)	1.1 (1.2)	3.3 (3.4)	2.2 (2.2)	1.0 (1.0)

There is a worrying arbitrariness about the time from which survivor-ship is to be measured; a decision that all population biologists have to face. Strictly, the life of a genet begins with the formation of the zygote and the loss of individuals from a potential population starts with the abortion of embryos: this is an extremely difficult stage at which to make counts. The position taken by Sharitz and McCormick as the start of independent life is the ripe seed. This is the point at which the seed of *Minuartia* is released from the parent and hazarded to the en-vironment but in *Sedum* the seed is retained, protected, in the capsule for 4–5 months. Others (e.g. Sarukhán and Harper, 1973) have used the time of germination as the start of life for calculating a survivor-ship curve on analogy with mammals in which birth is the time at which the developed embryo escapes from the investing maternal tissues. When survivorship curves are compared it is vital that they start at the same developmental stages. For the geneticist and evolutionist it is the survivorship of zygotes that matters and this point must be emphasized if only to encourage more studies of this difficult stage. In *Alopecurus myosuroides* as much as 50% of genet mortality occurs before the seed ripens and these embryonic deaths would not usually figure in survivor-ship curves.

The survivorship curves for *Minuartia uniflora* and *Sedum smallii* are shown in Fig. 18/6 and the number of seeds produced by the two species is shown in Table 18/II.

Table 18/II

The components of reproductive activity of *Sedum smallii*
and *Minuartia uniflora*

	Seeds per flower	Flowers per plant	Seeds per plant	Plants per dm^2	Seeds per dm^2
Sedum smallii	12.0	9.5	114.1	41.2	4698
Minuartia uniflora	12.5	24.4	305.2	9.6	2946

The seed output is very high for plants that are so small and the seed populations suffer very heavy mortality. The mortality rate ($1000\ q_x$ in Table 18/I) is very different for the two species. *Minuartia* disperses its seeds as soon as they are ripe and they are immediately at risk. One major hazard is the washing of seeds out of the islands by swirling water after rains. The seeds of *Sedum smallii* float and would also be lost in

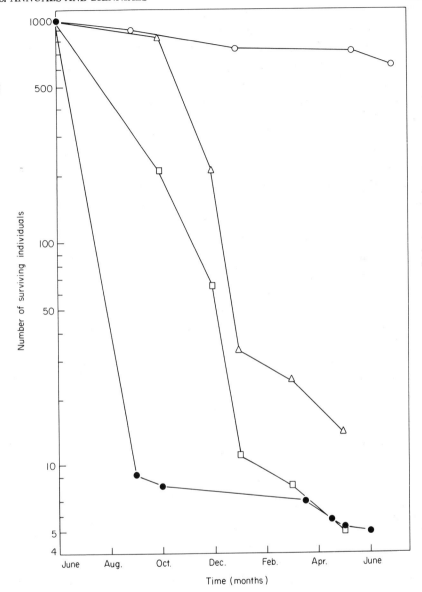

Fig. 18/6. Survivorship curves for natural populations of four species of annual plants.
○ *Vulpia membranacea* (from Watkinson, 1975) on fixed dunes in N. Wales;
● *Spergula vernalis* (from Symonides, 1974) on fixed dunes in the Torun Basin,
Poland; □ *Minuartia uniflora* and △ *Sedum smallii* (from Sharitz and McCormick,
1973) on granitic outcrops, Georgia, U.S.A.

this way if they were not retained within the capsules. The difference in q_x during the period of seed production (160 compared with 790) reflects the difference between retaining or dispersing seeds (N.B. in some species there may be a greater risk from retention than dispersal: see Chapter 15). In subsequent life-cycle stages the mortality risk of the two species is comparable — the germination phase is the most dangerous and the establishment phase the safest in the life cycle. After seedling establishment the risk again increases as the rosettes grow (and presumably interfere with each other); the remaining plants die after the seed is ripened.

The population dynamics of two annuals growing on sand dunes

A number of winter-germinating annuals are found in stable coastal dunes. In Britain *Cerastium atrovirens* is a member of this group, together with *Vulpia membranacea* (= *V. fasciculata* (Forskal) Samp.), *Mibora minima*, *Phleum arenarium* and *Aira praecox*. All of these species germinate in the autumn, over-winter as small plants and flower and set seed in spring before the habitat becomes too dry and is dominated by the perennial inhabitants *Festuca rubra*, *Galium verum*, *Ononis repens* and *Thymus drucei* (Pemadasa and Lovell, 1974).

For the study of *Cerastium atrovirens* 12 permanent quadrats were marked in stable dunes on the North Wales coast by driving metal pins deep into the sand to serve as permanent markers for the corners of quadrats. A removable counting frame could be fixed accurately on the pins. The plants of *Cerastium atrovirens* are very small and population dynamics can be studied at an extremely fine scale. The number of individual plants was counted every 2 weeks within 5 x 5 cm squares in the quadrats; the populations in some of the 5 x 5 cm squares were mapped to obtain co-ordinates of every plant position and these were checked every 2 weeks (Mack, 1976). The first seedlings appeared in September and new individuals continued to appear until the end of October; there was negligible mortality during this phase. The plants overwinter as tiny rosettes and some death occurred in this period, mainly because the root systems of some of the seedlings dried out after they were exposed by movement of the sand. Some seedlings were buried by sand but these usually re-emerged.

The plants reach anthesis in late March to mid-April, seed matures 3 weeks later and the plants are dead by mid-May. About 40% of the

plants present at the start of the winter died before flowering. The survivorship curve for these populations is shown in Fig. 18/7. During the
winter months the rosettes scarcely changed in size and presumably
made little demand on the nutrient resources of the habitat. When a

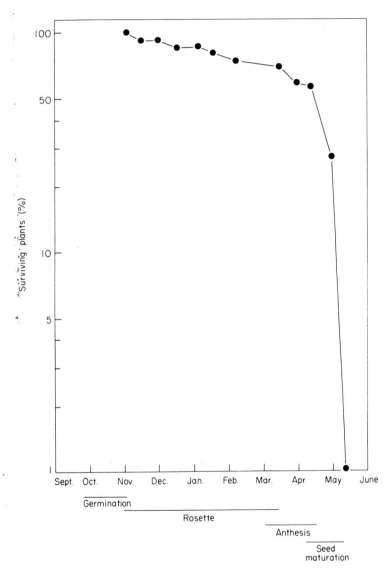

Fig. 18/7. Survivorship curve for *Cerastium atrovirens* on fixed dunes at Aberffraw, North
Wales. (From Mack, 1976)

new phase of growth started in spring the individuals were probably forced into active competition, not only with each other but also with the perennials, e.g. *Festuca rubra*. The major death risk occurred during this phase, when survivors were growing rapidly and preparing for flowering.

A number of seedlings were transplanted in January into dune sand; they were spaced 5 cm apart and grown to maturity in an unheated glasshouse for comparison with plants in the crowded conditions of the field (Table 18/III).

Table 18/III

The survival and reproduction of natural crowded plants and transplanted spaced plants of *Cerastium atrovirens*

	Plants in natural crowded population	Transplants under spaced conditions
Probability of a seedling surviving to maturity	0.64	0.98
Seeds produced per plant	44	388
Potential multiplication of the population in 1 year (R_0)	28.16	380.24

These values suggest that the field population might multiply 28-fold in a single season though in fact the populations are known to be very stable from year to year (Pemadasa, 1973) which points to a considerable loss from the populations at the seed and early germination stages. The transplants had a higher expectation of surviving to maturity and a much greater seed output per plant, suggesting that interference between plants in the dense field populations was a powerful regulating force.

In very marked contrast with *Alopecurus myosuroides,* populations of *C. atrovirens* have no reserve of dormant seeds in the soil. Soil cores taken in January produced only a few seedlings and none emerged from soil samples in March. Apparently the populations are wholly dependent on the seed output in year t for the population in year $t + 1$.

Vulpia membranacea is a small annual grass in the same community of the fixed dunes as *Cerastium atrovirens*. Like *Cerastium* it lacks an effective seed bank. Batches of seed were collected in the field in July

and sown in the same month within circular plots (10 cm diameter)
within the natural habitat of the parents. All potential seed parents
in and around the plots had been removed before they flowered so that
there was no natural input of seeds to the area. On seven dates from
August to the following May the soil was removed from the plots (from
an area 15 cm in diameter so as to include seeds that might have moved
from the original sown area). The seeds were grouped into classes:
(i) those that had germinated, (ii) those that were in enforced
dormancy (i.e. germinated readily in laboratory tests), (iii) those that
were in a state of induced or innate dormancy (i.e. were alive but did
not germinate in standard laboratory tests) and (iv) seeds that were
dead (i.e. reacted negatively to the tetrazolium chloride test (Iseley,
1952)). The changing pattern of distribution of the seeds between the
classes is shown in Fig. 18/8. The innate dormancy of the seeds had
been completely lost by late summer. About 5% of the seeds died during
the late summer and early autumn. All the remainder germinated between mid-August and late December with the exception of 0.5% that
remained viable and dormant in the soil. Soil samples taken in the field
after the flush of germination was completed, confirmed the almost
total absence of a reserve of buried viable seeds in the soil. This contrast with the weeds of arable land is very remarkable and is presumably associated with the reliability of the fixed dune habitat. Not all
annuals are fugitive species of unreliable habitats and the fixed dune
provides, over periods of many years, an environment that is highly
predictable compared to that of the weedy annuals.

The dune annuals are badly dispersed not only in time but also in
space. Dispersal in space was measured by lightly spraying infructescences
with aerosol paint and searching for the marked seeds after natural dispersal. Of all the grains ripened 98% were shed and only 2.7% of the
total seed output was unrecovered at the end of the experiment. The
release of seed from the parents continued over 3 months but 94% were
dispersed in July. The distance of dispersal depended on the height of
the infructescence: 73% of the seed from the small plants fell within
5 cm of the parent. Only 12% of the seed from tall inflorescences fell
within 5 cm, but overall, 79% of the dispersal units (1—4 caryopses are
dispersed together as a spikelet unit) fell within 10 cm of the parent
plant and the maximum distance of dispersal was 36.0 cm.

The fate of dispersed seed was measured by labelling them with the
radioactive tracer scandium-46 applied in nail varnish to the proximal

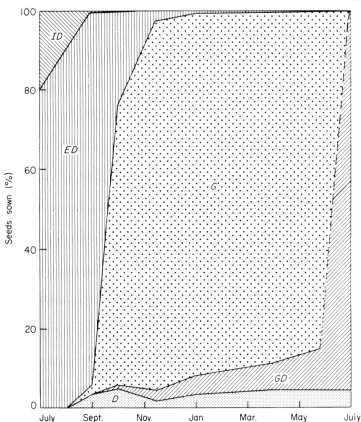

Fig. 18/8. The fates of seeds of *Vulpia membranacea* sown in the natural sand dune habitat of
the species in July.
ID = innate dormancy, *ED* = enforced dormancy, *G* = germinated, *D* = died as seed,
GD = died after germination. (From Watkinson, 1975)

end of the spikelet. Seed was sown in 10 cm diameter plots at five sites
on the fixed dunes. A portable contamination monitor was used to find
each seed in the field and this was sufficiently sensitive to detect a marked
seed through 30 cm of sand and 100 cm of air. Where there was significant vegetation cover (fixed dunes and slack) the seeds did not move
far after sowing: 57—80% remained within the sown area and the greatest distance moved by any seed was 37.7 cm. On the yellow dune,
where there was very little vegetation, the seeds moved in all directions
and only 10% remained within the sown area after 9 days. Most of these
seeds moved 60—70 cm down the open slope and became lodged and

buried by sand near clumps of *Ammophila arenaria*. The death of marked seeds occurred mostly in the seedling stage (mainly wind drag of seedlings with badly anchored root systems) and on one site predators left the shredded remains of the marked spikelets (see Table 18/IV).

Table 18/IV

The fates of spikelets of *Vulpia membranacea* labelled with scandium-46 and sown in the natural dune habitat of the species. (From Watkinson, 1975)

| | Site | | | |
	1	2	3	4
Dispersal units recovered (%)	99.0	98.0	98.0	94.0
Seedling mortality (%)	18.8	11.7	23.8	27.0
Predation (%)	0.0	0.0	5.0	0.0
Maximum distance moved by a dispersal unit outside sown area (cm)	21.2	7.7	37.7	7.2

The dynamics of natural populations of *Vulpia* were followed by repeated detailed mapping of naturally established individuals within fixed quadrats. From these censuses survivorship curves could be drawn (Fig. 18/6). On average each plant of *Vulpia* produced 1.7 mature seeds of which 90% germinated, and 69% of the seedlings survived to flowering. This is in contrast with species like *Cerastium atrovirens* and other annuals that have a high seed output and a very high death risk. *Vulpia* is somewhat like the small desert annuals that have a very low seed output but a high probability of survival (Went, 1973). The annual grasses *Avena fatua* and *A. barbata,* which are common in the Californian rangelands, averaged 6—36 seeds per plant (Marshall and Jain, 1967) and on some of the sites studied the overall mortality between seed and adult stages was again as low as 41%.

The dynamics of natural populations are unlikely to be understood simply by observing and recording — a population needs to be perturbed if the factors that determine its characteristics are to be exposed. The natural densities of *Vulpia membranacea* were altered on 37 permanent plots on fixed dunes at Aberffraw, Anglesey. Low densities of 100, 200 and 400 plants per 0.25 m² were obtained by thinning natural populations in November and January. Seed was added to other plots to give densities of 800, 1600, 3200, 4800, 6400 and 8000 seedlings per plot. Mortality was no greater on the denser plots and virtually all the response of populations to density was a plastic growth response by individuals. The relationship between the number of seeds per plant and density is shown in Fig. 18/9. Below a density of 100 plants per 0.25 m² the number of spikelets per plant was apparently independent of density but, above this level, seed output per plant was dependent of their distance apart and spatial arrangement (see Chapter 8). The generalized regression for the relationship between density and seed output at Aberffraw was

$$\frac{N_s}{N} = -1.406 \log N + 6.012$$ where N_s is the number of seeds and N the

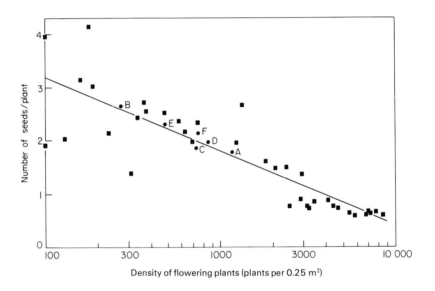

Fig. 18/9. The relationship between the density of flowering plants of *Vulpia membranacea* and the number of seeds produced per plant in the natural habitat. ● A—F = natural densities. ■ = thinned or supplemented populations. (From Watkinson, 1975)

number of plants per 0.25 m². If there were no mortality and seed production was regulated according to this equation, the equilibrium population density would be 3572 plants per m². This prediction is 1.33-fold higher than the highest natural density observed in the field and 3 times the highest density observed in the permanent plots. Clearly, density-dependent seed production cannot account wholly for the range of plant densities found in the field. However, density-independent mortality also occurs and Williamson (1972) has shown that an interaction between density-dependent and density-independent factors can lead to equilibrium population densities.

The equilibrium density of 3572 plants per 0.25 m² with no mortality is halved if the probability of plant survival is 0.7 and quartered when the probability is 0.54. The probability of survival to maturity that would be required (given density-dependent seed production) to equilibrate populations at the densities found in the field ranges from 0.39 to 0.59 for the populations at Aberffraw. The real values of survivorship measured in the field range from 0.50 to 0.67 (Table 18/V). It is peculiarly satisfying to discover that the population density of a species in nature can be so wholly accounted for by measured variables and indeed the variations in population density of *Vulpia* between plots at Aberffraw and in a nearby series of dunes at Newborough can also be largely explained by differences in the levels of the various types of mortality (Table 18/V).

Two causes of mortality are particularly interesting — some seed is taken by invertebrate predators (up to 15% in one plot) and both developing and flowering inflorescences are taken by rabbits: the losses to the seed crop may well vary from year to year. There is some evidence that rabbits may have a large effect: the rabbit population on some sand dunes in Devon and Somerset was drastically reduced in and after 1954 by myxoma virus and the populations of *Vulpia membranacea* subsequently exploded from a few individuals to millions in the 1960s (A. J. Willis, pers. comm.).

Watkinson's study of *Vulpia membranacea* (Watkinson, 1975) exposed some delightful subtleties in the interactions occurring within a population. The later a seed germinated the greater, in general, was the chance that the plant would survive to maturity (Table 18/VI a). However, the later the germination the fewer spikelets were produced per plant (Table 18/VI b). Rabbits were selective in their grazing and the greater the number of spikelets per plant the greater was the chance that the plant would be eaten (Table 18/VI c). The number of grains ripened per spikelet was higher when plants were growing near a rabbit warren! Thus (i) late germination favoured survivorship, (ii) early germination favoured reproductive output, (iii) high reproductive output

Table 18/V

The probability of a seed of *Vulpia membranacea* surviving various phases of development. (From Watkinson, 1975)

Period of mortality	Cause of mortality	Plots											
		Aberffraw						Newborough					
		A	B	C	D	E	F	1	2	3	4	5	6
(a) Seed dissemination – germination	Loss of viability	0.95	0.95	0.95	0.95	0.95	0.95	0.95	0.95	0.95	0.95	0.95	0.95
(b) Seed dissemination – germination	Seed predation	1.00	0.94	1.00	0.97	1.00	1.00	0.85	0.87	0.98	0.87	0.95	0.92
(c) Post-emergence –pre-establishment	Wind drag Desiccation	0.88	0.72	0.80	0.80	0.73	0.80	0.76	0.81	0.81	0.76	0.72	0.76
(d) Post-establishment – flowering	Various (see text)	0.83	0.78	0.87	0.90	0.88	0.91	0.95	0.83	0.88	0.76	0.61	0.91
(e) Inflorescence development	Rabbit grazing	0.96	1.00	0.90	1.00	1.00	0.95	1.00	1.00	1.00	0.98	1.00	0.99
(f) Flowering	Rabbit grazing	1.00	1.00	0.99	1.00	1.00	1.00	0.95	0.92	0.69	0.72	0.79	0.78
Complete life cycle		0.67	0.50	0.59	0.66	0.61	0.66	0.55	0.51	0.46	0.34	0.31	0.47

(a) The risk of death of seedlings germinating at different times. (Data from two sites in North Wales, from Watkinson, 1975)

	Germination period							
	22 Aug. – 31 Aug.	1 Sept. – 6 Sept.	7 Sept. – 25 Sept.	26 Sept. – 11 Oct.	12 Oct. – 5 Nov.	6 Nov. – 8 Dec.	9 Dec. – 24 Jan.	25 Jan. – 27 Feb.
Aberffraw	0.14	0.33	0.09	0.11	0.30	0.36	0.69	0.80
Newborough	0.09	0.29	0.12	0.10	0.23	0.35	0.45	–

(b) The number of spikelets produced by plants that germinated at different times (data corrected for the effect of rabbit grazing). (From Watkinson, 1975)

	Germination period							
	22 Aug. – 31 Aug.	1 Sept. – 6 Sept.	7 Sept. – 25 Sept.	26 Sept. – 11 Oct.	12 Oct. – 5 Nov.	6 Nov. – 8 Dec.	9 Dec. – 24 Jan.	25 Jan. – 27 Feb.
Aberffraw	4.09	3.09	3.34	2.44	1.93	1.74	1.17	2.00
Newborough	3.47	2.51	3.14	2.64	1.88	1.50	1.48	–

(c) The probability of a plant being eaten by a rabbit at flowering time in relation to the number of spikelets per plant. (Data from Newborough, North Wales, from Watkinson, 1975)

	Number of spikelets per plant						
Plot	1	2	3	4	5	6	7
3	0.01	0.07	0.36	0.63	0.56	0.67	0.67
4	0.03	0.18	0.33	0.17	0.45	0.67	0.50
5	0	0.05	0.17	0.17	0.43	1.00	1.00
6	0.02	0.14	0.13	0.15	0.35	0.40	1.00

encouraged predation but (iv) the presence of the predator encouraged reproductive output!

Desert annuals

In desert communities annuals may be an important element in the flora — rain storms may permit germination and a limited growth period in which very small plants can complete their development cycle. In extreme desert conditions, e.g. the centre of Death Valley, California, with an average yearly rainfall of *ca* 40 mm, the number of annuals appearing after rains is often very small and in many years the number of ripe seeds produced is barely sufficient to maintain a population (see Table 18/VII). The equilibrium population is obviously finely balanced and the excess seed produced in the exceptional good season is added to the bank of seeds in the soil and compensates for years in which the seed production is low or absent. In such an environment an annual with no buried viable seed population would surely become extinct.

Table 18/VII

Germination of annual plants and seed production in a 1-m^2 plot in the centre of Death Valley, California (from Went, 1973).

Year	Number of seeds germinated	Number of new seeds produced
1966	17	11
1967	30	31
1968	37	9
1969	32	274
1970	0	0
1971	19	35
1972	1	0

In less dry parts of Death Valley, Went (1973) surveyed 12 plots, each of about 0.2 m^2, in the periods January to June 1968 and 1969. Of 2893 seedlings 41% survived, flowered and set seed and 33 951 mature seeds were formed. The breakdown of the population is given by habitats and not by species (Table 18/VIII). No survivorship curves were determined but "Some seedlings disappeared because they were eaten by rodents or insect larvae (*Oenothera clavaeformis* by *Altica*

Table 18/VIII

Number of seedlings germinated after early winter rains in 0.1- to 1-m² plots in Death Valley, the number that survived to the flowering and fruiting stage, and the number of ripe seeds produced per plot

	Altitude in m	1968			1969		
		Germinating	Surviving	Seed	Germinating	Surviving	Seed
Valley Bottom	−80	37	5	9	32	19	274
Road to Beatty	−40	303	42	270	63	50	897
Bennett's Camp	−80	148	69	3093	—	28	1086
Bennett's Well	−80	33	11	210			
Jubilee Pass	100	149	74	140	—	48	2778
Titus Canyon	200	35	12	372	63	33	1553
Weir	600	186	51	375	80	72	150
Grapevine SW	800	222	47	612	170	62	2830
Grapevine NW	800	145	13	82	124	61	736
Grapevine SE	800	102	16	686	61	45	500
Grapevine NE	800	144	45	1005	104	33	600
Entrance	1000	194	168	6000	151	83	6258
Scotties	1100	272	80	3335	75	11	100
All plots	—	1970	663	16 189	923+	545	17 762

from Went (1973)

torquata), but most died in the early stages of germination when their roots did not penetrate properly into the soil. But, once established, the seedlings survived for practically 100% to flowering". Although there is negligible mortality after seedling establishment natural selection will of course act within such a population through the differential fecundity of individuals.

At first sight it might appear that annual plants are the ideal material for analysing population growth because the generations do not overlap. In fact the situation is not so simple because where there is a buried seed population there is a hidden overlap of generations. The dune annuals are exceptions in lacking a persistent seed population and are probably the simplest of all plant populations for a model study. At the same time, they are best regarded as idealized exceptions to the general rule that annual plants have depended on seed longevity to buffer their population against environmental unpredictibilities.

The dynamics of the populations of biennial plants

The category of biennial plants is well known to the horticulturist. They are species which must be reared as vegetative plants in one calendar year for flowering in the following year, since they die after flowering. It is, however, very doubtful whether this category of behaviour is clearly defined in nature: the biennial habit is only an extension of the phenology of the winter-germinating annual which also makes some growth in the late autumn of one calendar year and resumes and completes its growth cycle in the spring of the second year (Fig. 18/2). Most biennials in cultivation are germinated in year 1 and grown on for transplanting at the end of the year or in the following spring. They are planted as spaced plants and under these ideal conditions flower in year 2. Such biennials require a seasonal stimulus (vernalization and/or photoperiodic induction) before they will flower but there may also be a critical size that must be achieved before flower initiation will occur. In nature the growth rate is usually slower and may be much slower than in cultivation in gardens; inter- or intraspecific density stress may slow the growth rate still further and delay flowering into year 3, 4 or later. It is a serious error to regard the biennial habit as anything but an ideal in horticultural practice. The critical feature of the so-called biennial habit is that the plants usually die after seed set. However, death does not necessarily follow flowering and plants of e.g. *Digitalis purpurea* regenerate from rosette buds if the inflorescence is damaged before the seed is ripened. It is often possible to extend the perenniality of a biennial by continually preventing seed formation, and natural enemies of the plant in the field may sometimes extend life in this way. Seed production rather than flowering is the lethal event. Biennials of this sort are best described as perennial monocarpic species with the *potential* for completing the life cycle in two growth seasons. A group of so-called biennials in cultivation are really perennials, but under conditions of cultivation reach their peak reproductive activity in year 2 and decline in vigour subsequently. They are short-lived perennials, not monocarpic, and include such garden plants as *Cheiranthus cheiri, Antirrhinum major* and *Dianthus barbatus*.

Three perennial monocarpic species with a potential for completing their life cycle in 2 years have been studied in some detail as field populations: *Digitalis purpurea* (the foxglove), *Dipsacus fullonum* (the teasel) and *Daucus carota* (the wild carrot). These are all species of disturbed habitats but annual disturbance is lethal. Normally, they are completely excluded from agricultural and horticultural systems by the frequency of cultivation — they are normally excluded from late phases of successions because they possess small seeds and seedlings and require full light for their early establishment as rosettes. Most of the species have poor dispersal of seed in space (e.g. *Verbascum*, see Fig. 2/2f) but good dispersal in time, i.e. the seeds have great longevity in the soil. They are slow to reach new habitable sites but lie in wait for favourable conditions to return. The species produce very large numbers of seeds in a "big bang" of reproduction before death. These are species which may appear in very dense populations after a local disturbance, e.g. *Digitalis* after clear felling of woodland (Oxley, 1977), and it is a curious feature of the dynamics of these populations that when they are conspicuous enough to be chosen for study they are usually already in a condition of decline! These are opportunist species which exploit rare occasions that are suitable for seedling establishment and capitalize on these establishments with a more or less extended growth cycle ending in a single episode of reproduction (see Chapter 7). The species may then become rare and disappear in the locality until a disaster affects the vegetation, when populations appear suddenly and dramatically as if from nowhere. The population dynamics of *Digitalis purpurea* has been studied in artificial communities and this is described in Chapter 7).

Daucus carota is often an invader of early phases in old field succession, i.e. the vegetational sequence on land that is allowed to recolonize naturally after a period of cropping. It is a monocarpic species that may complete its life cycle in 2 years (sometimes in 1 year). Holt (1972) studied a section of a former cornfield that had been farmed for about 100 years. A sequence of stages in succession was obtained by ploughing and then leaving plots to colonize naturally. The plots quickly became dominated by *Agropyron repens, Poa compressa, P. pratensis, Melilotus alba* and *Rhus typhina.* Seed of *Daucus* was collected from near the study area and sown at a range of densities in areas that had last been ploughed in the early summer of 1964 or 1966. One set of quadrats was denuded of vegetation before the seed was sown and was kept clipped during the first growing season. Delaying the input of *Daucus*, relative to the succession, reduced the number of seedlings that established, delayed the onset of their reproduction and reduced seed yields. Some seeds were washed out of the plots during storms — about 10% was lost in this way from the younger stage and none from the older stage in the succession. The densities of seedlings that became established were proportional to the number of seeds sown but in the unclipped quadrats were positively correlated with the percentage of bare ground ($P < 0.01$). The number of seedlings emerging in the second year was negligible although 42—80% of the originally sown seeds was unaccounted for by emigration and germination. The mortality of seedlings was much greater in herbage than on bare ground (14% and 3% respectively) but the survivorship curves subsequently ran closely parallel (Fig. 18/10). The plots were completely defoliated in the summer of 1967 (probably by grasshoppers). In the first growing season 3.5% of the plants flowered (behaving as annuals) but most of the inflorescences were destroyed by the predators. The precocious flowering occurred only on the young fallow, not in the 2 year successional stage. In 1968 (year 2) 37% of the plants flowered in the young fallow and 7% in the older phases producing *ca* 21 000 seeds per m² in the young successions and 1540 in the older phase (Fig. 18/11). This experiment illustrates many of the general characteristics of so-called biennials. The environment was suitable after ploughing for the establishment of the carrot but, as succession proceeded, establishment became more difficult. Within the succession the presence of associated vegetation delayed the onset of flowering, perhaps by as much as 4—5 years. Populations of *Daucus carota* in such a successional community represent a cohort of plants

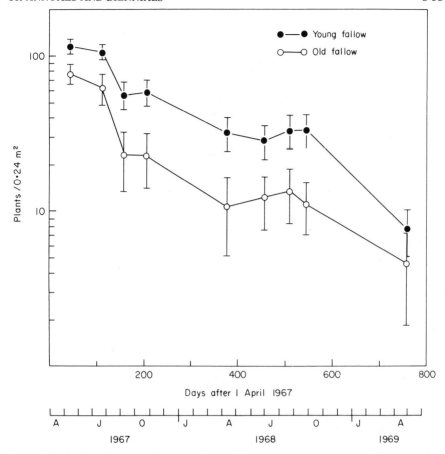

Fig. 18/10. Survivorship curves for populations of *Daucus carota* sown into old field successions
in the first season after cultivation (young fallow) and two years after the last cultivation (old fallow). (From Holt, 1972)

recruited when the community was young and of which the numbers
progressively decline as individuals flower and die. The denser the
community, the longer the life of the population of *Daucus* and the later
and more extended its period of seed production.

Studies of *Dipsacus fullonum* in old field successions were made by
Werner (1975a). Seeds were sown into 2- and 3-year-old successions and
the germination of these was spread out over at least 2 years. No rosettes produced flowers in the first year, a few did so in the second year
and the majority in the third. In the fourth year a few plants flowered

but a few very large 4-year-old rosettes were still present in the population and these showed no sign of flowering. Werner could find no relationship between rosette size and liability to flower except that rosettes of less than 15 cm diameter did not flower (see Chapter 7).

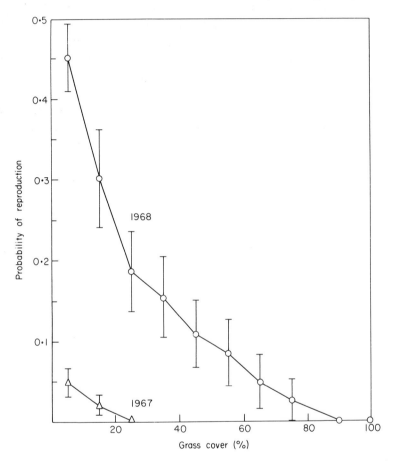

Fig. 18/11. The probability of flowering in populations of *Daucus carota* sown in December 1966 in old field successions, Michigan. (From Holt, 1972)

   The population dynamics of so few annual/biennial plants have been studied that it is dangerous at this stage to attempt too many generalizations. The annual growth habit in particular includes a wide variety of seasonal growth strategies that grade imperceptibly into the category traditionally considered biennial so that it is unrealistic to think of it as a natural biological grouping. The annual habit is a continuum of behaviour patterns that occur within species as well as between them. An annual cornfield weed such as *Papaver dubium* contains genetic variants which determine autumn or spring germination so that the single species and a single population includes winter annuals and spring annuals. Somatic polymorphism and perhaps also genetic polymorphism in *Alopecurus myosuroides* and *Avena fatua* determine that these species also have autumn- and spring-germinating forms. Biennials such as *Daucus carota* include genetic variants that confer an annual habit and the biennial habit itself is extended phenotypically into perennial (though usually still monocarpic) life cycles. The genetic variation for an annual habit occurs amongst some biennials and has been exploited by horticulturists. Annual forms of *Digitalis purpurea* and of *Althea rosea* are now in cultivation. The study of the population biology of annual and biennial species immediately reveals that the traditional classification of the group is unsatisfactory and a more subtle method for describing the continuum of life cycle strategies that are represented amongst short-lived plants is clearly needed. We need a methodology for the quantitative description and comparison of what has previously been part of anecdotal natural history.

19

Herbaceous Perennials

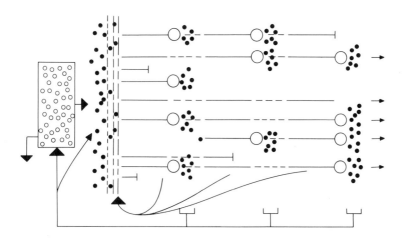

There are no natural breaks in the continuum of plant life cycles —
separation into the categories of annual, biennial and perennial is rather
arbitrary. The category of herbaceous perennials includes plants with a
very short expectation of life after first flowering (e.g. *Rumex crispus*,
in which Canadian ecotypes may flower in the same year as they germinated, whereas other ecotypes may flower only in the second or later
years) and species that include life cycle variants such as *Trifolium
pratensis* which in subalpine meadows may live for 20 years — beginning

to flower at 5–10 years old and continuing to flower for several years, and forms present on flood plain meadows that usually flower only once and die within 2 years of germination (Rabotnov, 1960).

The category of herbaceous perennials includes forms which, like a tree, accumulate ageing or dead soma and an increasing body of dependent tissue that may be a respiratory burden (Kershaw, 1964) lowering the net assimilation rate and producing effective senescence. The category also includes forms that renew the vegetative body every year and slough off the older tissues (e.g. many bulbous species, some forms with corms — e.g. *Ranunculus bulbosus* — and most stoloniferous species). These plants spend their perennial life in a state of perpetual somatic youth. There is probably no sense in which such plants can be said to senesce except that they may accumulate viruses and somatic mutants that may ultimately reduce their vigour. There is again a continuum of forms that extends from those plants that retain no tissues more than a year old to species like *Iris, Zingiber,* herbaceous bamboos, *Polygonatum* in which the rhizome persists for many years although eventually decaying. Perennial plants vary in the amount of their past that they carry with them.

A few populations of herbaceous perennials have been studied to discover the essential features of their population dynamics but the species chosen are not a representative sample of the variety of perennial life forms. Most of the species studied are inhabitants of temperate grasslands and woodlands.

A critical difference between the population biology of perennials and annuals is that generations may overlap, although in some circumstances, for example in the early recovery stages of vegetation after a disaster such as fire, even-aged populations may become established. It is a matter of much interest to determine whether natural populations of plants derive from single massive acts of even-aged recruitment or a continual process of death and replacement. If overlap of generations occurs it has consequences in natural selection because individuals of different ages will usually represent the survivors from slightly different selective processes.

In populations of higher animals the age structure is critical because the activities of an individual change in an age-determined pattern. Most animals pass through phases of juvenility to reproductive maturity and then to senescence and death. The age structure of the population therefore determines its dynamics: a population dominated by old animals has a very different future from one that is mainly young. For this reason zoologists put much effort into determining age-specific fecundity schedules for a species. Taken together, three measures of an animal population are needed for an appreciation of its population dynamics: these are (i) the number of individuals present, (ii) the age distribution of the population and (iii) the fecundity schedule of the species (see e.g. Birch, 1948). It is most irritating that such a relatively simple approach can rarely be adopted for plants because fecundity is only loosely related to age. Even with animals the concept of a fecundity schedule that is species-specific breaks down if it is applied too rigidly to organisms like fish which are relatively plastic in their developmental phases.

The problem of defining fecundity schedules for perennial plants is that the plasticity of growth is so enormous. A seedling in the ideal conditions of an uncrowded flower bed may grow fast and some perennials may reach flowering size in their first year but the same species growing in dense stands or mixed with aggressive neighbours of another species may grow slowly and spend many years reaching a reproductive condition. A clone of *Trifolium repens,* for example, as it invades a new habitat, may pass through a phase of rapid growth and vigorous flowering, decline in size as the habitat deteriorates in the presence of other species and then make a new vigorous growth after the habitat is disturbed. Indeed, parts of the same clone may be growing vigorously, other parts passing through a phase of intense flower production while other parts are dying under some vigorous neighbour. For this reason Rabotnov and his colleagues of the Russian school of population biologists have taken the view that the real age of a plant is not a particularly useful concept. They base much of the study of plant demography on a classification of "life states" (e.g. see Uranov and Smirnova, 1969). They describe a plant population as a spectrum of 10 ontogenetic states (Uranov et al., 1970), viable seeds in the soil, seedlings, several intermediate stages and senile plants. These states differ in their response to environmental factors, their influence on the microclimate of each other and in their reproductive activity. Classification of individuals into these age states makes it possible to classify populations themselves, for example (i) invasive populations that do not have all the states yet present, (ii) normal type populations with all stages present and (iii) regressive populations that have lost the ability to reproduce by seed. Quite unlike a normal age distribution, such a succession of life states is reversible, in that senile perennial plants may revert to an active

reproductive state if environmental conditions become favourable
(Rabotnov, 1969).

A population analysis by the Russian method is an extremely sensitive means of detecting environmental change and has been applied to
a number of perennial herbs (e.g. *Galeobdolon luteum*: Smirnova
and Toropova, 1972; *Ranunculus acris* and *R. auricomus*: Rabotnov
and Saurina, 1971; Saurina, 1972) and many others (see Harper and
White, 1974, for a fuller bibliography). The method has not yet
been used outside Russia although a rather similar approach was taken
by Summerfield (1972) who described a population of the bog
asphodel *Narthecium ossifragum* which was no longer recruiting new
seedlings nor producing seed but persisted in its habitat in a state that
he called "biological inertia". In the Russian terminology this might be
described as a population of regressive type with a spectrum of life states
including only senile plants. In his study of the population biology of
Ranunculus spp. Sarukhán described quadrats within permanent grassland in which continuous recruitment of new individuals of *Ranunculus
acris* occurred. Each flush of new seedlings died before any individuals
reproduced but the flushes overlapped so that the species was always
present, but only as juvenile plants (Sarukhán and Harper, 1973). Again
this is a population that is better described by the Russian concept of
life states than by a formal age structure.

In general, age is a poor predictor of size or reproductive activity
among plants. The Russian procedure accepts this fact and uses life
states as an alternative to fecundity schedules. Although age is a poor
predictor, it is not useless: among a group of plants of different sizes
it is virtually certain that the largest will be old. The problem arises with
the small plants, which may be young plants or very old but suppressed
individuals. The Russian procedure is no use if the knowledge of a population's biology is to be used in the study of selection, genetic change or
microevolution; for these ends it is important to know the real ages of
the individual plants that are present. Is the genetic composition of a
population changing fast or are a few old genets waxing and waning in
their position in the community? Life states may be the most useful
predictors of the future development of a changing population but are
of little use in tracing its past. It may be that the real strength of the life
state concept will be its use in combination with more formal analyses of
real age.

It is sometimes possible to determine the age of a herbaceous perennial by examining its morphology. If old rhizomes or stems decay slowly there may be sufficient record of annual growth increments to allow individuals to be aged. In a very few cases herbaceous perennials develop annual growth rings (e.g. the corms of *Cyclamen* and *Liatris*) and the plants can then be aged as accurately as temperate trees. One such attempt at ageing the plants in a population was made for dicotyledonous herbs in a Finnish meadow (Linkola, 1935). 1-year-old plants were the commonest plants in each species that was studied (with the exception of *Trollius europaeus* in one of the 2 years in which counts were made). 1-year-old plants accounted for 27% of the populations of *Geum rivale*, 37% of *Ranunculus acris*, 67–75% of *Ranunculus auricomus* and 67–81% of *Polygonum viviparum*. Linkola made his estimates in 1931 and 1932 in different plots. In 1931, 1-year-old plants formed 48% of the population of *Trollius europaeus* but in 1932 there were very few new recruits, only 4% of the total population. There had been nearly total failure of this population to recruit new seedlings into the population in 1932 and 2-year-olds predominated.

Figure 19/1 shows the age distributions for some of the populations in Linkola's study. By the side of each age structure diagram is plotted the logarithm of population density in each age group. If one assumes that recruitment is constant from year to year (which is clearly untrue in the case of *Trollius*) these curves can be interpreted as survivorship curves. The graphs are in most cases nearly linear and if they were true survivorship curves could be interpreted as an exponential decline of numbers with age — a death risk that is independent of the age of plants. Following this interpretation, half-lives (the time taken for a population to fall by 50%) can be calculated for the species. *Ranunculus auricomus* and *R. acris* appear to have a higher death risk in the first year and then to settle down to a steady death-rate. Such an interpretation depends entirely on the assumption of a constant rate of recruitment; a bad or a particularly good year for seedling establishment would exaggerate the representation of one age group in the population and completely alter the estimate of death rates.

The danger in using the age structure of a population to deduce its past history is illustrated in Fig. 19/2 in which the same theoretical population is plotted to support quite different theses: (i) the age structure may represent a population that recruits a constant number of new individuals each year and the death risk is independent of age — the figure is then a survivorship curve; (ii) the age structure may represent

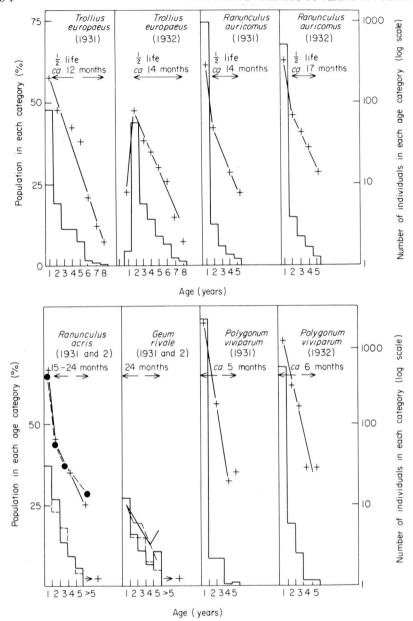

Fig. 19/1a, b. The age structure of some perennial herbs in a meadow in Finland — from data
of Linkola (1935) obtained by determining the age of plants from their morphology. The data are plotted as % of plants in each age category (left-hand scale) and
as the logarithm of the numbers present (right-hand scale). Approximate values are
shown for half-lives.

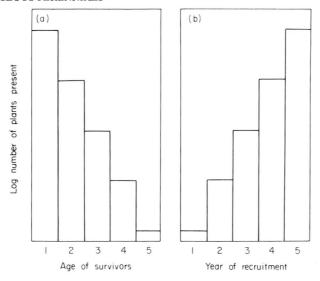

Fig. 19/2. The composition of a population of plants of different ages plotted (a) to support the hypothesis of constant death risk (exponential decay) *or* (b) of constant rate of growth (exponential increase).

a population that recruits new individuals in exponentially increasing numbers in succeeding years and in which there is no death. This would occur if a seed parent was growing nearby and increasing its seed output exponentially or the species is a perennial that seeds in its first year. The age structure then described the curve of growth of the population over time. Clearly such a population age structure could also be obtained by some mixture of (i) and (ii) or if the chance of seedling establishment is increased by the presence of older plants. An age structure must be supported by other real evidence of the population's history before it is used to discover past history. In the meadows studied by Linkola the past history of management had been constant and it seems reasonable to suggest that most of the species were suffering constant death risks.

Another community of plants that was studied in the same way is a perennial vegetation of the arctic fjaeldmark of Greenland (Wager, 1938). Tissue decay is slow in this environment and Wager could use growth and decay zones on the perennating organs to determine age. His records have been plotted in Fig. 19/3 as if they represented survivorship curves. The communities are long established and had not been disturbed by man: if we assume constant annual recruitment, the half-lives of the four species are very similar and of the order of 1 year. Wager noticed

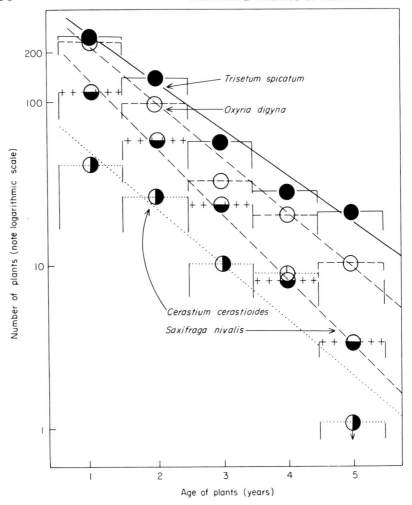

Fig. 19/3. The age structure of four perennial herbs in an alpine "fjaeldmark" community. The data of Wager (1938) was obtained by morphological examination of the plants and is shown here plotted on a logarithmic scale.

that there was "a tendency for young plants to be grouped together, but no tendency to get adults in that position". This suggests that the mortality was concentrated among the more closely packed individuals.

   Climax prairie vegetation consists largely of herbaceous perennials and in an area in Lake County, Illinois an area of undisturbed prairie contained high densities (48–108 plants per m²) of *Liatris aspera*. This species produces annual growth rings on its corms and most of the

plants can be aged very accurately. Kerster (1968) determined the age
structure of a number of populations (Fig. 19/4a, b). The younger categories of plants were very much under-represented in most of the
quadrats (e.g. Fig. 19/4a), and this suggests that the recruitment of
seedlings was inadequate to maintain the population, though, as individuals may live at least 34 years, the decline of the species would be
slow. One quadrat showed evidence of a resurgence in the population
(Fig. 19/4b) and this was on a soil profile that had been buried by
wind-blown sand. Evidence from elsewhere suggested that the establishment of *Liatris* was favoured by this condition. It is of course impossible
to be certain that the small numbers of plants in the 7–11-year-old
category was due to poor establishment or to a high mortality risk. The
youngest flowering plant was 2 years old but flowering was rare in plants
less than 9 years old. Kerster suggests that *Liatris* is a long-lived colonizing
species, eventually doomed to local extinction but with colonies migrating within the general prairie structure.

Three species of *Liatris* (*L. aspera*, *L. cylindracea* and *L. spicata*) were
present at Zion, Illinois together with their interspecific hybrids. The
mean age of the species populations was greater than that of the hybrids (significantly so in four out of the nine possible combinations)
but the hybrids appeared to flower precociously (Levin, 1973). In these
populations there were a few plants of 40 years old. In gardens *Liatris
aspera* reaches flowering condition in 2 years and then has corms larger
than many of the older plants in the field. Clearly in their natural environment the plants are very much delayed in their reproduction and
growth compared with what is possible in unstressed conditions.

It is a snare and delusion to believe that age structures offer an easy
short cut to understanding population dynamics. Levin concludes from
his study of *Liatris* that a full life table analysis would be needed properly to understand what was happening in the populations. An alternative approach is to record the demographic details of marked populations over a period of time — this is extremely laborious but is the only
really satisfactory solution. In the case of perennial plants with a long
expectation of life, a census may need to be continued for many years
before the real trends emerge. The most significant long-term study of
any plant population was made by Carl Olov Tamm in Swedish meadows
and forests. He started observations of marked quadrats (1 m² or
1/4 m²) in 1943 and continued to record annually; the most recently
reported results include the years up to 1971. The quadrats were per-

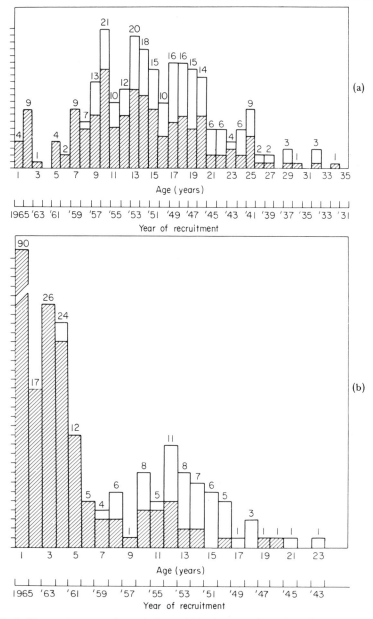

Fig. 19/4a, b. The age structure of populations of *Liatris aspera* determined from growth rings
   on the corms. Fig. 19/4a is characteristic of a number of the prairie communities
   studied at Zion, Illinois. The relatively low numbers of young plants suggest that
   the populations are not recruiting sufficient new members to maintain the population. Fig. 19/4b shows a population that has recently rejuvenated. Shaded
   columns and the numbers on top are vegetative plants. Unshaded columns are
   flowering plants. (From Kerster, 1968)

manently marked and the plants present were counted within the sections of a fine wire quadrat. There are problems that arise in this kind of census, for example "doubts can sometimes arise, as to whether an individual is identical with one observed earlier or has grown up since the last revision on the site of a dead one observed earlier . . . seedlings may well be overlooked in one or more years." Where there was room for doubt, Tamm compared records from year to year with extreme care

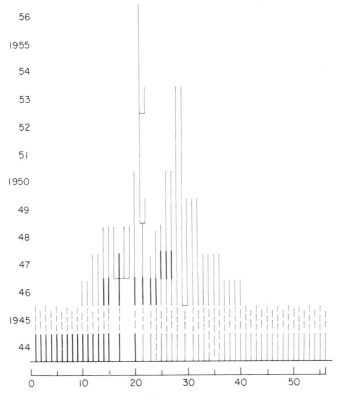

Fig. 19/5a. The behaviour of *Centaurea jacea* in a declining population.

Fig. 19/5. Each vertical line represents one individual, straight for unramified ones, and branched where the plant has ramified. The heavier lines are for years when the specimens flowered, and broken lines indicate that the plant was not seen that year. Group A includes specimens of large or intermediate size at the first inspections, group B small or rather small ones, and group C those appearing in 1944 or later, presumably from seedlings. When first observed they were as a rule smaller than those of group B. On this plot the observations of the flowering are not entirely reliable for some of the years (1946 and 1948 in particular), when the inspections were made in unsuitable seasons (too early or too late). (From Tamm, 1956)

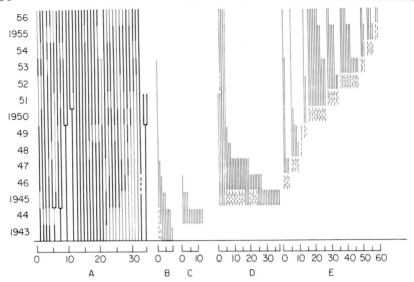

Fig. 19/5b. The behaviour of *Anemone hepatica* in a forest. Groups C, D and E include different crops of seedlings.

and it is unlikely that there is significant error in the study. Tamm has presented this work in a series of important papers (Tamm, 1948; 1965a, b, 1972a, b). Examples of the kind of record obtained are shown in Fig. 19/5a—c.

The age structure of the populations present at the beginning of the study was quite unknown. The plant units that were counted were the above-ground shoots, which may have arisen as the branches of clonal growth or as seedlings. Only when new seedlings arrived during the period of study was it certain that a new genet had been added to the population but from such seedlings it was sometimes possible to determine the length of life, the survivorship curve and the life expectancy of genets. A number of plants were observed to form clonal branches but this was not common. Seedling populations were often quite dense but survivorship was low. The general picture that emerges is of rather long-lived plants and when Tamm made his first report after 13 years' study, the populations were largely composed of the same plants that had been present at the beginning. Flowering in all the species was a very unpredictable event and there is very little sense that can be read into the pattern of flowering. There is no suggestion of anything like the mast year phenomenon of trees. Flowering was generally confined

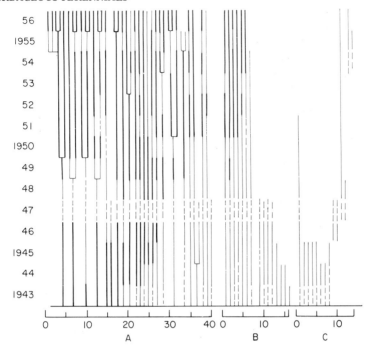

Fig. 19/5c. The behaviour of *Anemone hepatica* in a meadow.

to long-established individuals.

The recruitment of seedlings into the population was not at all regular. *Anemone hepatica* was recruited as new seedlings in large numbers in 1945 but 1946—49 were poor years and this same irregularity was true for most of the species. Moreover the size of the seedling crop did not determine the number of plants recruited into the population: some years were suitable for the appearance of many seedlings but survivorship was poor — in other years successful long-lived plants survived from very poor seedling crops.

Some changes in management occurred during Tamm's study of the meadows. They had been used regularly for hay-making and then grazed until 1956. After that time hay-making was stopped and the herbage became increasingly dense. One species, *Centaurea jacea*, declined steadily from a population of 56 plants in a quadrat in 1944 to one survivor in 1956 (Fig. 19/5a). During the last phases the plants continued to produce ramets but with a short life. The one plant that was present in 1944 and persisted to 1956 had flowered in 1946, produced a ramet in 1949 (which died after a year) and a further short-lived ramet in 1953. The depletion rate of the population can be examined by plotting the number of survivors as logarithms against time (Fig. 19/6). The depletion curve is remarkably linear, implying that the death risk within the population was constant with time. The population behaved in the same way as a radioactive isotope and the original population declined with a half-life of 1.9 years. No new seedlings were produced and the populations moved relentlessly towards extinction.

The term "depletion curve" has been used to describe the rate of loss of individuals from a whole population of unknown age structure. This is not a true "survivorship curve" and this latter term has been reserved for the rate of loss of plants from a particular age group — a particular cohort of plants recruited within a defined period.

Other species present in Tamm's plots also show similar linear depletion curves over the 13-year period (see account in Harper, 1967). For example *Sanicula europaea* had a depletion rate with half-life of about 50 years and *Filipendula vulgaris* a half-life of 18.4 years. These populations were continually recruiting new members as seedlings and the size of the whole populations did not change very much, as recruitment approximately compensated for mortality.

The implications of such linear depletion curves are important. The impression given by the data of Fig. 19/6 is of extraordinary constancy in the death risk within populations that had experienced the full range of seasonal and annual fluctuations in climate for 13 years. There is no sign of the good years and the bad — depletion rates differed between species but not appreciably between years.

Depletion rates may vary for the same species in different habitats. Tamm (1972b) followed the fate of plants of *Primula veris* in permanent quadrats in three sites. Site 1 was a grove of *Fraxinus excelsior* with some hazel (*Corylus avellana*). The population of primroses remained very constant from 1943 to 1957, and although it was continually losing individuals the number of rosettes present was kept up by the formation of new ramets. This population did not recruit new seedlings and the genetic diversity of this "stable" population was declining all the time. There was a large break in the observations between 1958 and 1962 and after this period the population of rosettes began to decline rapidly, though with a linear depletion curve giving a half-life of 2.9 years. The last lingering members of the population still flowered. (Note that the depletion curve is based on the number of unit rosettes originally pre-

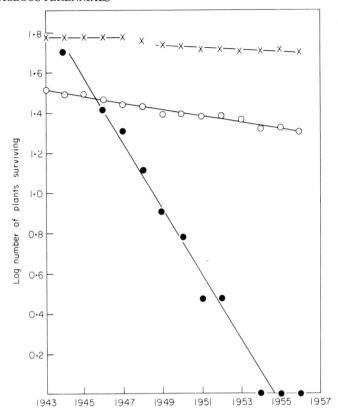

Fig. 19/6. The depletion curve for populations of *Sanicula europaea* x (with half life >50 years) *Filipendula vulgaris* ○ (half-life *ca* 18.4 years) and *Centaurea jacea* ● (half-life *ca* 1.9 years). (From Harper, 1967, calculated from data of Tamm, 1956)

sent in 1943; if a rosette produced a ramet after that time it was still counted as one individual for the preparation of the depletion curve.)

Site II was a meadow close to trees of *Quercus robur* and *Betula verrucosa*. Site III was a dry meadow and until the late 1930s both had been lightly grazed. From 1943 to 1956 the areas were regularly mown and after that time the mowing was stopped over the area as a whole, though the plots themselves were usually mown in summer. During the period of the observations the nearby trees had grown and increased their influence on the vegetation. Site II was not mown during the period 1943–47 although the surrounding vegetation was cut. Voles found the tall vegetation of the plots attractive and did a

great deal of damage. Tamm ascribes the sudden fall in the numbers of
plants in 1946—47 almost entirely to the voles. The populations on plots
28 and 29 in this site subsequently declined still further (one to extinction, one to a single survivor) but in another plot (30) all the original
population had disappeared but new seedlings had been recruited and
two of these flowered in 1968—70. In Fig. 19/8 a linear depletion curve
has been set to these data but it is obviously unsatisfactory, partly because of the gap in the records between 1957 and 1967, partly because
the initial fall in numbers occurs as a clear step in the curve and has
a well defined special cause (the voles).

In the dry meadow (Site III) the populations of rosettes declined
slowly and regularly (Fig. 19/7). The initial population of about 40
rosettes had fallen to 32 in 28 years. The population present at the
beginning had a depletion curve with a half-life of about 50 years (assuming again that ramets produced during the period are not offset against
original plants that were lost). At the end of the period of 28 years,
the population of *Primula veris* was composed of 33 rosettes:
two were new genets, rosettes formed from established seedlings, while
six rosettes were ramets of other plants. If we assume that the 33 plants
originally present were all different genets (the outside estimate) the
population had lost 18% of its original diversity of genets and gained
6%. From a genetical point of view this is a progressively impoverished
population despite its very long half-life. One original plant was represented by three ramets in the final population, another was represented
by four ramets in 1957 but by 1971 the line had become extinct. This
suggests that underlying a very constant depletion rate in the population
of rosettes there was a considerable shuffling in the representation of
the original genets.

The population dynamics of some orchids

A number of orchids were present in Tamm's permanent plots including
Dactylorchis incarnata (in a wet meadow by a small lake), *Dactylorchis
sambucina* (in a meadow on shallow soil near a pine tree which increased
its shading of the quadrat during the period), *Orchis mascula* (in a mesic
wooded meadow) and *Listera ovata* (in a meadow shaded by *Fraxinus
excelsior* and *Alnus glutinosa* which increased their shading during the
period of observation). The vegetation was changing slowly in the
habitats because of the alterations in mowing and grazing practice

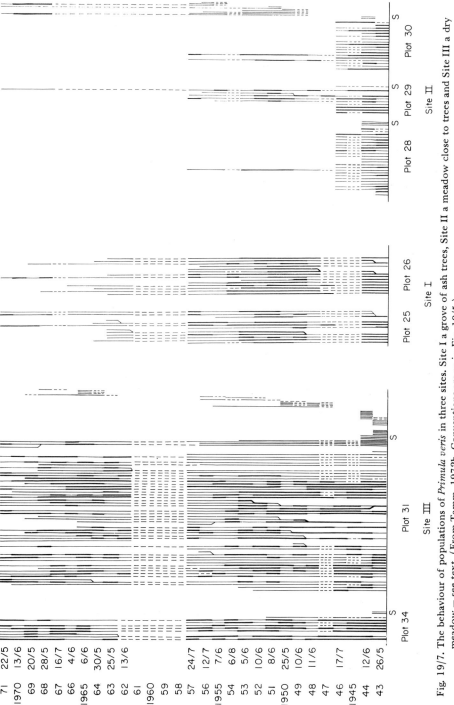

Fig. 19/7. The behaviour of populations of *Primula veris* in three sites. Site I a grove of ash trees, Site II a meadow close to trees and Site III a dry meadow — see text. (From Tamm, 1972b. Conventions are as in Fig. 19/5.)

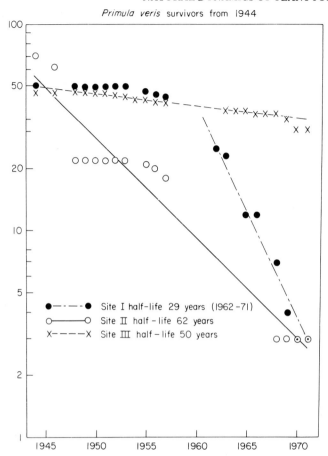

Fig. 19/8. Depletion curves for the three *Primula* populations of Fig. 19/7. Note logarithmic scale. The "individuals" in the diagram are the *Primula* rosettes observed in 1944, without regard to possible subterranean connections. All rosettes formed later by branching from one 1944 individual are considered as belonging to the same individual. (From Tamm, 1972b)

described earlier. The linear depletion curves observed for other species are not found with the orchids over the whole observed period, though episodes of exponential decline can be recognized (Tamm, 1972a).

The populations of *Dactylorchis incarnata* remained stable from 1943 to 1952 and then declined rapidly. A new population appeared in 1967 and declined rapidly with a nearly constant half-life of *ca* 2 years like that of the earlier population (Fig. 19/9a). Populations of *Dactylorchis sambucina* remained stable for about the same period and

then started to decline in a very irregular manner and the same behaviour can be seen in *Orchis mascula* (cf. Figs 19/9b, c) though the decline
started earlier. *Listera ovata* was apparently favoured by the increasing
shade or some correlated change in the quadrats. The original population continued to recruit new members throughout the 27 years of

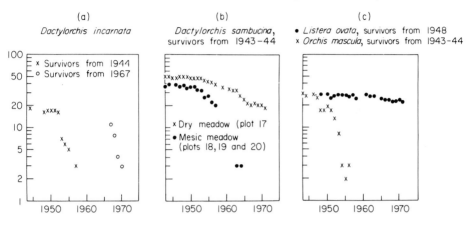

Fig. 19/9. Population depletion curves for various orchids in permanent quadrats. (From Tamm, 1972a)

study (Fig. 19/10). Tamm believes that most if not all of the new recruits were ramets but it was impossible to be sure without excavating
the plots and destroying the continuity of records.

An odd feature of the depletion curves for the orchids is that the
number of survivors appears to go up as well as down! Clearly the number of survivors can never increase. The explanation is that the orchids
appear to be capable of disappearing from the above-ground population
for a year, or perhaps two. Tamm believes that the mycorrhizal habit
of the orchids may explain the ability of these plants to opt out of the
above-ground struggle for existence for a period. The phenomenon has
been noticed in other orchids (e.g. *Spiranthes spiralis*: Wells, 1967). It
may be that this habit is more common than we know. *Delphinium*
species have also been reported to be driven into a dormant state by
unfavourable conditions at almost any stage of the life cycle and to
remain dormant for several years before making new aerial growth
(Epling and Lewis, 1952). It is very unlikely that this phenomenon will
be detected unless populations are repeatedly mapped in detail.

The flowering habits of the orchids are very erratic and this is a well
known feature of these plants (Curtis and Greene, 1953; Wells, 1967)

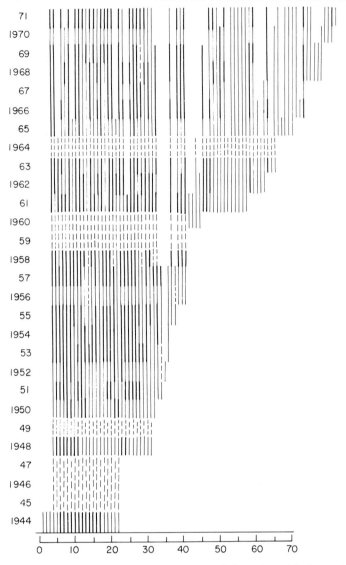

Fig. 19/10. The behaviour of *Listera ovata* in a mesic wooded meadow with almost closed canopy. (From Tamm, 1972a.) Conventions are as in Fig. 19/5.

but on closer analysis it may not be more marked than in many other perennial plants. The whole area of study of the control of flowering of perennials in nature is ripe for deeper study.

The extraordinary studies of Tamm make it possible to give accurate minimal ages to genets in the populations. Both *Primula veris* and *Listera ovata* are certainly capable of living for 28 years and *Dactylorchis sambucina* for 30 years. The age structure of a population of *Listera ovata* in 1971 is shown in Fig. 19/11, calculated from Tamm's data. 30% of the plants were more than 27 years old and could not be aged. The remainder had been recruited at odd periods throughout the years of study, becoming more frequent in the last 10 year period. There is no suggestion that recruitment was a steady process. It is unfortunate in some ways that the number of new recruits that appeared in Tamm's populations was too small for proper survivorship curves to be calculated for any species.

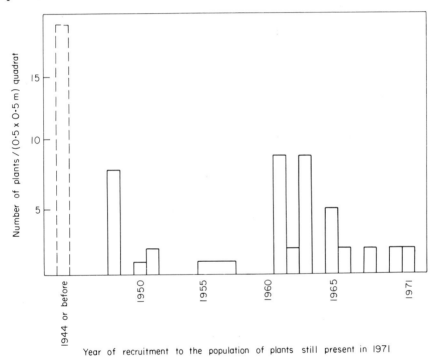

Fig. 19/11. The age structure of a population of *Listera ovata* in a mesic wooded meadow. (Calculated from data of Tamm, 1972a)

Ranunculus species and other grassland herbs

Three very different forms of perennial habit are found in species of *Ranunculus* which are commonly found growing together in grazed meadows in Europe. *Ranunculus repens, R. acris* and *R. bulbosus* differ slightly in their habitats within such pastures and when the three are present together *R. bulbosus* tends to be most abundant on the well drained sites, *R. acris* on intermediate sites and *R. repens* on poorly drained positions (see Fig. 5/10a). However, even where there are sharp differences in drainage the three species tend to overlap.

Ranunculus repens forms spreading clones. Stolons develop from buds in the axils of the older rosette leaves and ramets are formed which may in turn produce daughters. In open ground, a single rosette may develop up to 200 ramets of first, second and subsequent orders in the course of a single season and spread over an area of 2 m diameter. In a grassy sward the number of ramets is much smaller and often only a single ramet is produced, close to the parent rosette which then dies. The life of a ramet is less than a year and the plants accumulate no ageing tissue.

In contrast, *R. acris* has an oblique rhizome which may carry 2 or 3 years of older tissue (? + viable buds). The above-ground parts again form a rosette and occasionally an axillary bud may develop a new rosette close to the parent. The connection may ultimately be lost but the ramets remain close to the parent rosette, forming a tight clump. In this species the clones are small and compact whereas in *R. repens* they are potentially very large and diffuse.

Ranunculus bulbosus bears a corm and the plants pass the mid-summer in a dormant state with no above-ground parts (the other species normally bear leaves all the year round). In autumn a rosette of leaves is initiated from the corm and the base of the new shoot begins to swell to form the new corm which grows to mature size in the spring. After flowering and seed set the plant re-enters the dormant phase. Only in very exceptional circumstances does the parent corm produce more than one daughter. The old corm rots as the daughter develops and the plant accumulates no ageing tissue — the new corm is a replacement, not a means of clonal growth. The multiplication of rosettes of this species is entirely by the addition of new seedling recruits. The rosettes of the three species are not dissimilar and those of *R. repens* and *R. bulbosus* are easily confused. However, the population of rosettes in a pasture will be quite differently composed — all the rosettes of *R. bulbosus* will be genets, most but not all of *R. acris* will be genets but a large fraction of the rosettes of *R. repens* will be clonal modules of successful genets.

The three species differ in their palatability to livestock. *R. bulbosus* contains up to 1% by weight of ranunculin — a glycoside which releases the toxic vesicant protoanemonin when the tissues are damaged. *R. acris* contains the glycoside in lower concentration and *R. repens* contains scarcely any of the compound. *R. repens* is readily eaten whereas the other species are avoided (Harper, 1957).

The contrasts in biology of these three closely related and cohabiting species makes them ideal for a comparative study of their population dynamics. A study was made by Sarukhán in a field of permanent grassland in North Wales. The field had been grazed by sheep, cattle and occasional horses for at least 50 years and the only other management had been the regular cutting of thistles (*Cirsium arvense*) and the application of farmyard manure. The community of the sward contained 50 species of flowering plants with *Lolium perenne* as the most common grass together with *Agrostis stolonifera*, *A. tenuis*, *Cynosurus cristatus*, *Festuca rubra* and *Holcus lanatus*. In addition to the buttercups other dicotyledonous species were present including *Achillea millefolium*, *Bellis perennis* and *Leontodon autumnalis*.

The population dynamics of the buttercups was studied in 1 m² quadrats; 21 quadrats were arranged in triangular groups of three, each of which could be accurately identified within a large metal frame, which was designed so that it could be attached to metal rods which had been fixed into concrete cylinders in the ground. This procedure made sure that the fixed points were well away from the observed quadrats and that nothing protruded above ground between observation periods that might encourage beasts to visit the sites out of curiosity, and at the same time allowed extremely accurate positioning of the triangular frame and the quadrats within it. The plants were mapped by means of a pantograph (see Fig. 19/12) and each plant and seedling was assigned a number and a punched card. Consecutive maps could be superimposed to identify new recruits and individuals that died between observations. Maps were usually made at approximately 15-day intervals but were less frequent during periods when changes were small. The sites of the triplets of plots were chosen deliberately to include areas in which at least one of the species was common and areas containing mixed populations. From the accumulated data it was possible to calculate depletion curves from the pre-existing population, survivorship curves for new recruits and some elements of life tables (Sarukhán and Harper, 1973).

Fig. 19/12a. Metre square plots delimited by a removable frame for demographic study of pasture species. The study plots are fixed in position within the triangle which is itself temporarily attached to permanent pins in the ground.

Fig. 19/12b. A pantograph in position for mapping of plants within a sward.

The depletion curves for populations of the three species are shown in Fig. 19/13. It is important to remember that these are records of the survivors from all the plants present at one point in time, that the populations were composed of mixed ages and that the units counted were rosettes and so included ramets. The useful module of growth in *Ranunculus* spp. is the rosette and the counts are of $N\eta$ (see Chapter 1) where N is the number of genets and η is the number of ramets per genet. η is 1.0 in *R. bulbosus,* slightly more than 1.0 in *R. acris* and much more than 1.0 in *R. repens*. The curves are remarkably linear over the period of study and only *R. acris* deviated noticeably from linearity in the first few weeks which coincided with the recent arrival of a large number of seedling recruits. The half-lives of the populations were 23–25 weeks for *Ranunculus repens*, 87, 54, 77 and 19 weeks for *R. bulbosus* and

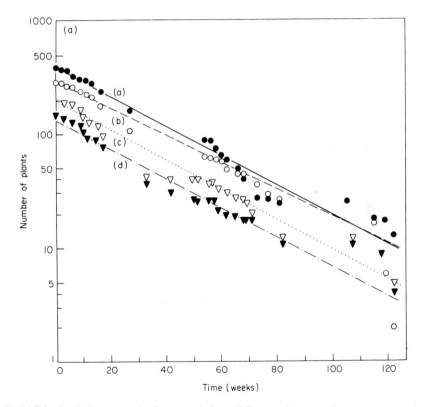

Fig. 19/13a. Depletion curves for four populations of *Ranunculus repens* in permanent grass-land sites of 1 m²; time in weeks after first observation (April 1969). (From Sarukhán and Harper, 1973)

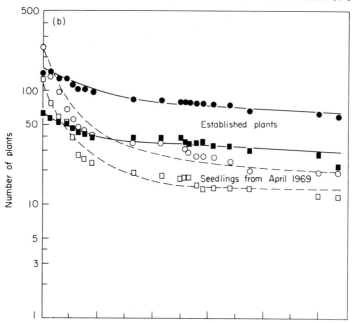

Fig. 19/13b. Depletion curves for two populations of *Ranunculus acris* in 1 m² permanent grassland sites; time in weeks after the first observation (April 1969). (From Sarukhán and Harper, 1973)

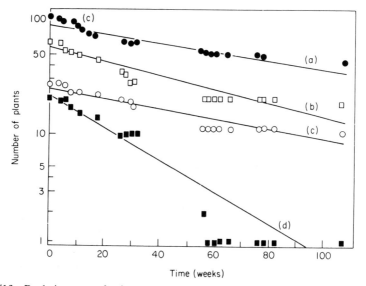

Fig. 19/13c. Depletion curves for four populations of *Ranunculus bulbosus* in 1 m² permanent grassland sites; time in weeks after first observation (April 1969). (From Sarukhán and Harper, 1973)

102 and 196 weeks for *R. acris*. This latter half-life can be compared with
estimates made on populations of *R. acris* from data of Rabotnov
(1958) who recorded, but did not analyse, survivorship of this species
in a meadow in the Oka River basin near Moscow. The half-life in
Rabotnov's population was about 99 weeks (Sarukhán and Harper,
1973). The populations of the same species studied by Linkola in a
Finnish meadow (Fig. 19/17) give values (based on the age structure)
of 65–100 weeks. The half-lives are remarkably similar and suggest
that big differences in the climatic regimes between North Wales,
continental Russia and Finland are not very important in the population dynamics of this species.

The depletion curves for Sarukhán's populations contain a clear
seasonal rhythm. This is shown for *R. repens* in Fig. 19/14 as the rate of
death per week. The plants emerge from winter with a low death risk
which accelerates sharply with the start of the spring flush of growth.
The risk of death then declines to a very low value during the flowering
period when the growth rate is low, increases again after flowering and
falls again to a very low death risk during the autumn and winter. *The
greatest risk of death is when the survivors are growing most rapidly.*

The seasonal cycles of mortality are shown in Fig. 19/15 for the
three species of *Ranunculus*. The cycle for each species is distinctive
and in each species the mortality period for adult plants coincides with
the periods of growth. The recruitment of seedlings also has a seasonal
rhythm, the seeds of *R. bulbosus* germinating in the autumn and the others
in the spring, first *R. acris* and then *R. repens*.

Age-specific death rates can be calculated when large populations of
seedlings appear nearly synchronously. The age specific death rate is
$$\frac{\text{number of plants dying in age class } x}{\text{number that attain age class } x}$$ and Fig. 19/16 shows the values
for one population each of *Ranunculus acris* and *R. repens*. Seasonal
rhythms are obvious, though not so clear as in population death rates
because the latter include the larger flowering plants whereas the cohorts of new recruits are in the pre-reproductive phase.

It is commonly asserted that juvenile phases are risky and that most
deaths occur when plants are young; these statements are not the same.
If a population is exposed to a constant risk of death, more young
plants will die than old plants because there are more young plants
to experience the risk. If the mortality risk itself decreases with age
an even greater number of plants will die young. The critical question

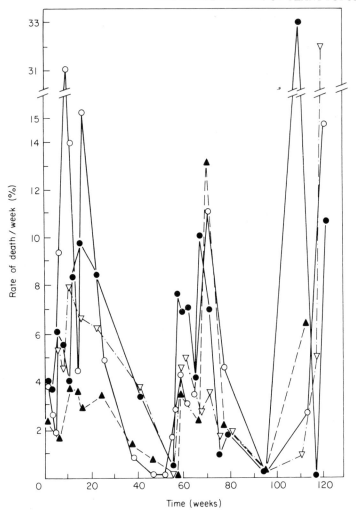

Fig. 19/14. Rate of death per week of mature plants of *Ranunculus repens*. Site A1, high density, intense grazing (●————●; site A3, low density, intense grazing (▽·————·▽);
site C3, low density light grazing (▲ — — —▲); site C2, high density, light
grazing (○————○). (From Sarukhán and Harper, 1973)

is whether a plant has a greater chance of dying during a week in its
youth than at other stages in life. Figure 19/17 shows the survivorship
of cohorts of even-aged plants of *R. repens* arising from seedlings and
from ramets. A seedling is counted from the time that its cotyledons
appear above ground and a ramet from the moment that an emerging
stolon is visible in a leaf axil. The two curves are very different — the

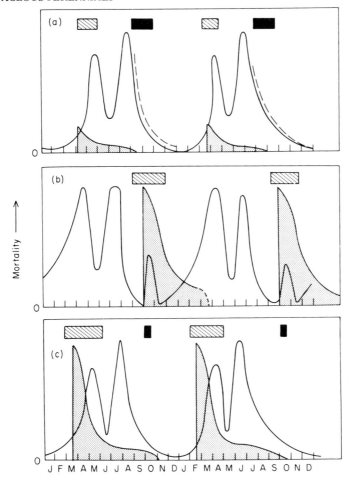

Fig. 19/15. Diagrammatic representation of the rates of mortality in populations of *Ranunculus repens* (a), *R. bulbosus* (b) and *R. acris* (c) at different times through the year. The continuous line represents mortality of mature plants; the shaded areas seedling mortality and the broken line the mortality of newly born vegetative propagules. Blocks show time of seedling recruitment (filled) and vegetative recruitment (cross hatched). (From Sarukhán and Harper, 1973)

death risk for a seedling declines continuously from birth whereas that of a ramet is constant. The difference between the survivorship curves of ramets and seedlings may have two origins. It may be that the small size of a seedling and the risks involved in changing from heterotrophic nutrition based on seed reserves to autotrophic life make this stage

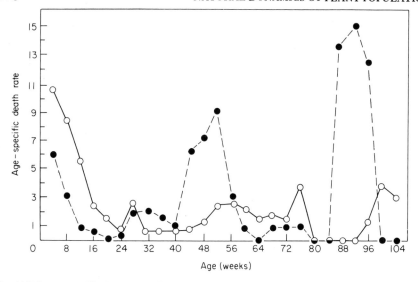

Fig. 19/16. Age-specific death rates for one population each of *Ranunculus repens* and *R. acris* observed in two different 1 m² permanent plots during 2 years. (From Sarukhán and Harper, 1973.) • = *R. repens* ○ = *R. acris*

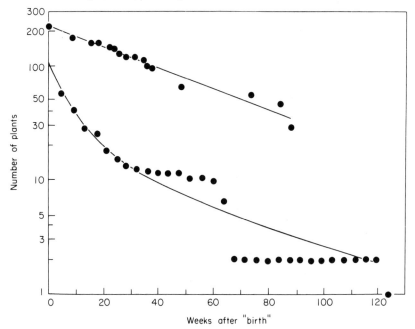

Fig. 19/17. Survivorship curves of populations of *Ranunculus repens*.

Lower curve, plants originated from seed at different dates during a 3 year period of study

Upper curve, plants originated from stolons from the time of first appearance as emerging axillary buds. (From Sarukhán and Harper, 1973)

especially hazardous. The ramet, by contrast, develops with a continued supply of resources from the parent rosette during establishment (Ginzo and Lovell, 1973a, b). A second origin of the high mortality risk of seedlings is that this may be the period when the genetic load of the population is discarded. *Ranunculus repens* is an outbreeding species and seedlings represent new genetic recombinants — evolutionary experiments. Some of the ill-adapted genotypes will presumably have died in the post-zygotic and pregermination phase but many of the real tests of an effective genotype must arise first when the seedling starts independent life. Ramets are of course only somatic branches of genets that have already proved themselves.

The juvenile phase is only one of the high-risk states in the life cycle — the other is the seasonal hazard when the survivors are growing fast. The period of rapid growth is the time at which individuals make their heaviest demands on environmental resources and are most likely to interfere with their neighbour's activities. This would explain the high mortality risk at this time and also the fact that the risk is species-specific. Each plant is most likely to suffer from the presence of a neighbour of its own species because their phenologies will be synchronous. The idea that death is brought about by the vigorous growth of survivors is of course the conclusion from experiments with controlled populations of single species where survivorship and growth are clearly correlated (see Chapter 6). If the growth rate of neighbours is a main cause of death we would expect to find a relationship between density and the risk of death. Such an effect is shown for the life expectancy of ramets of *Ranunculus repens* (Fig. 19/18).

Sarukhán noticed that quadrats that contained *R. acris* but not the other species of *Ranunculus* tended to increase in population density but where *R. acris* was present in mixed populations its numbers tended to decline. This is the only evidence for any interaction between the species but each may play a part in the control of others. Where other species of *Ranunculus* were present the turnover rate of populations of *R. acris* was high and in one plot there were four complete turnovers of the population within 2 years.

During the period when the buttercup populations were being censused above ground, an attempt was made to follow the dynamics of the seed reserve in the soil (Sarukhán, 1974). Mark-recapture methods proved difficult because the seedlings throw off their pericarps when they germinate, so that paint marking is useless for identifying batches

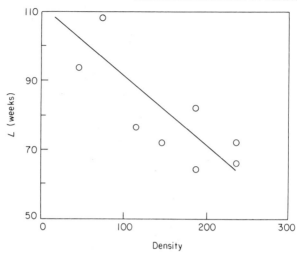

Fig. 19/18. The relationship between the life expectancy (*L*) of vegetative propagules of
    Ranunculus repens and the density of this species. Densities are averages of the
    number of plants/m² observed at each site in April 1969, 1970 and 1971. (From
    Sarukhán and Harper, 1973)

of seeds. Instead seeds were added in groups of 100 to marked areas
3 cm in diameter close by the censused plots. Natural seeding into the
area was prevented by removing flower heads within a 2 m wide strip.
At intervals seed was sampled by taking soil cores to include the central
sown zone and extracting the seed by washing, sifting and sorting by
hand. The seeds were separated into categories: (i) germinated; (ii) in a
state of enforced dormancy — these were seeds that germinated readily
when sown in soil in the greenhouse; (iii) in a state of induced dormancy —
these did not germinate in subsequent tests but were viable judged by
their reaction to the tetrazolium test for viability; (iv) dead seeds —
empty, rotten or negative in the tetrazolium test; (v) predated — seed
that had been eaten by voles and could be counted easily because the
animals neatly slice open the achenes and leave the half pericarps. The
added seed population was censused in this way seven times during a
period of 14 months and the results are shown in Fig. 19/19. Observations made subsequently by Ian Soane in the same field show that the
element of predation is very variable and depends on the placement
of the seeds, whether grouped or scattered. The proportion of seeds
lost to predators may much exceed the values shown in the Fig. 19/19.
The decay rates of unpredated seed appear to be roughly exponential

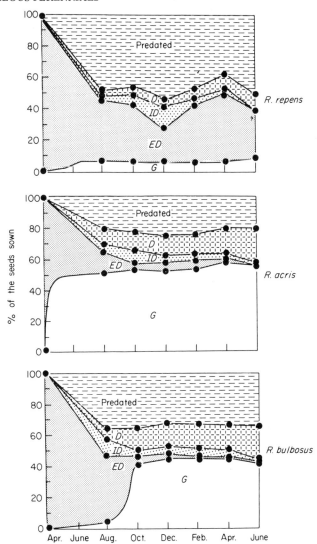

Fig. 19/19. Variation in the size of the different fractions of the seed population of *Ranunculus*
spp. with time. *G*, seeds observed to germinate *in situ*; *ED*, seeds in enforced dormancy; *ID*, seeds in induced dormancy; *D*, fraction of the population that decays.
(From Sarukhán, 1974)

like horticultural weeds (see Fig. 4/11). The half-lives of the seed are
very different for the three species of *Ranunculus*: *ca* 16—20 months
for *R. repens*, 8 months for *R. bulbosus* and *ca* 5 months for *R. acris*.
   The population dynamics of the three species of buttercup in the

field in North Wales can be summarized in diagrammatic life cycles
(Fig. 19/20a, b, c). These diagrams include estimates of seed production by the three species and a measure of the extent to which the
rosette populations were maintained by recruiting seedlings or
ramets.

Plants of *Ranunculus bulbosus* produced about 15 seeds per plant,
adding 565 new seeds to a seed bank in the soil that already contained
ca 300 seeds/m². The seed bank declined steadily until autumn when
it was rapidly depleted by germination. The population of 43 plants
present in the spring had fallen to 32 plants per m² and was suddenly
boosted to 127 plants by this flush of seedlings. Most of the mortality
occurred among these new recruits and the established plants had a
high expectation of life. There were no ramets, but a large number of
seeds were produced.

The life cycle of *R. acris* contrasts in several ways. About 10% of the
rosettes produced ramets, repeating the representation of proven genotypes among the rosette population. The number of plants was high
(*ca* 100) but the average seed number per plant was 10 adding 1000
seeds per m² to a seed bank of only 100. The seed bank decayed
rapidly and a large fraction was lost by germination in the spring. The
expectation of life of established plants was lower than for *R. bulbosus*.

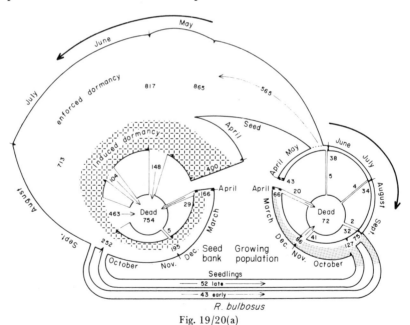

R. bulbosus

Fig. 19/20(a)

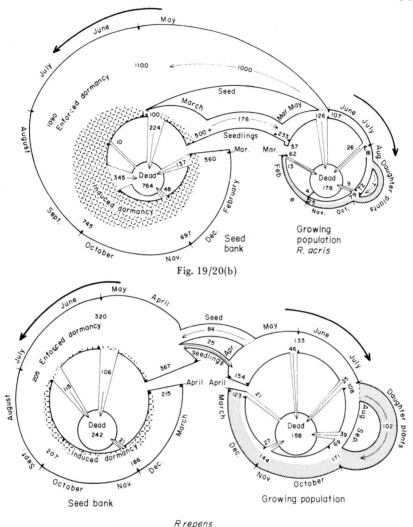

Fig. 19/20(b)

R. repens

Fig. 19/20. Diagrammatic life cycles of (a) *Ranunculus bulbosus*, (b) *R. acris* and (c) *R. repens*. The diagrams represent population fluxes within average quadrats of 1 m² of permanent grassland in a site in North Wales (see text). (From Sarukhán, 1971)

The most extreme contrast is with *R. repens*. A population of 108 rosettes per m² produced 102 ramets and the expectation of life of rosettes was short. Most of the turnover in the population was through the birth and death of ramets. The plants produced on average less than 1 seed per plant adding 84 seeds to a seed bank that already contained about 280 seeds per m^2. The seed bank decayed very slowly and a very small proportion (7%) germinated.

The greatest contrasts are (i) between the species that specialize in clonal growth and those that concentrate on seed production and (ii) between those with a long ecologic and genetic "memory" in the form of buried viable seeds in the soil, and those with a short half-life as seeds.

A diagrammatic and scaled representation of a life cycle is one step towards the quantification of life-cycle phenomena. A further step was attempted (Sarukhán and Gadgil, 1974) by fitting quantitative values from the field census data to flow diagrams. The year was divided into five periods in which major events occurred, such as seed germination, flowering, the production of ramets. Figure 19/21 shows the flow diagrams for the three species of *Ranunculus,* and the census data can be fitted to such a flow chart to describe the number of plants involved at each step. This procedure recognizes that the changes that occur in the population are not simply a function of the number of plants present but also of the structure of the population. Thus a seed, a seedling and a mature plant have different roles in determining the next stage of growth or decline of the population. Clearly a population of seedlings has an immediate future quite different from one composed entirely of mature plants (see Lotka, 1925 and Birch, 1948, for discussions of this problem in animals). The flow diagram recognizes the existence of different sub-sets within the population and defines their relationships formally.

The flow diagram forms the basis for the construction of a population matrix (for a general treatment of matrix applications to population biology see Leslie, 1945, and Usher, 1973). An imaginary population might have the following categories of members: 0 = seed, 1 = ramet, 2 = non-flowering adult, 3 = flowering adult and 4 = flowering and ramet-producing adult. Various transitions between the categories may take place in a time step: (i) a seed may remain a seed ($a_{00} = P_0$); (ii) a seed may germinate to become a non-flowering adult ($a_{20} = G$); (iii) a non-flowering adult may develop to a flowering adult ($a_{32} = P_2$); (iv) a ramet may become a flowering adult ($a_{31} = P_1$); (v) a flowering adult may become a flowering and ramet-producing adult ($a_{43} = P_3$); (vi) a flowering adult may produce seeds ($a_{03} = F_3$); (vii) a flowering and ramet-producing adult may remain in that category ($a_{44} = P_4$) or (vii) it may produce seeds ($a_{04} = P_4$) or (viii) it may produce ramets ($a_{14} =$

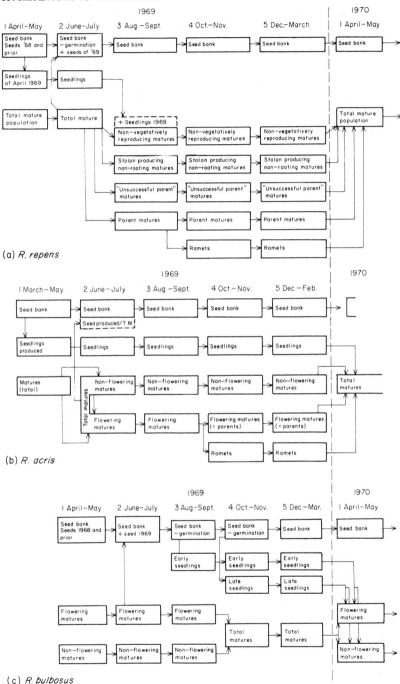

Fig. 19/21. Flow diagrams describing the life cycles of *Ranunculus repens, R. acris* and *R. bulbosus.* (From Sarukhán and Gadgil, 1974)

V). The projection matrix for such a system may be represented in this way:

$$
\begin{matrix}
P_0 & 0 & 0 & F_3 & F_4 \\
0 & 0 & 0 & 0 & V \\
G & 0 & 0 & 0 & 0 \\
0 & P_1 & P_2 & 0 & 0 \\
0 & 0 & 0 & P_3 & P_4
\end{matrix}
$$

The transition matrix defines the transition from one time step to the next and if the year is divided into seasons, the projection matrix can be specified for each season of the year. The matrices may then be used to determine the rate of growth or decline of the populations, given the values found in the field censuses. It is known that, for a fixed schedule of mortality and fertility rates, a population generally achieves a stable age distribution irrespective of its original state. The assumption is made that the populations have a constant growth rate

$$N_{(t+1)} = \lambda N_{(t)} = N_{(t)}e^{m}.$$

λ is used as an estimator of population growth rate and when $\lambda = 1$ or $m = 0$ the population is constant. The computed values of λ are shown in Table 19/I. It must be remembered that N in these calculations is the number of rosettes and in the terminology of this book is really $N\eta$.

Table 19/I

Population growth rates for three species of *Ranunculus* over two years of observation on 15 x 1 m^2 quadrats (from Sarukhán and Gadgil, 1974)

	Ranunculus repens	R. acris	R. bulbosus
Mean	1.04	1.167	1.603
Variance	0.089	0.264	3.583
Coefficient of variation	0.287	0.455	1.181

The species with strongly developed clonal growth (*R. repens*) appeared to be the more stable in population growth rate and *R. bulbosus*, which produces most seed per plant and is poorly buffered by the seed bank had the highest coefficient of variation in population growth rate.

An alternative way of looking at the life-cycle strategies of the species, still using the matrix formulation, is to compute what would happen if the rates of clonal growth and reproduction by seed were modified. Figure 19/22 shows the predicted rates of population increase if either

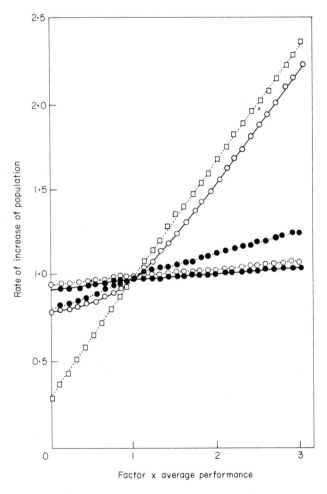

Fig. 19/22. Theoretical rates of increase attained by populations of *Ranunculus repens* (○), *R. bulbosus* (□) and *R. acris* (●) when their population growth by seed (······) and ramets (———) is modified. Rates of increase have been normalized by dividing the resulting rates at each value of the factor by the rate of increase of the "normal" population. (From Sarukhán and Gadgil, 1974)

of the two mechanisms of population growth were increased or decreased by constant factors. *R. repens* is intensely sensitive to variations in ramet production and very stable in the face of altered rates of reproduction by seed. One can argue that, if seed production by this species was inhibited entirely, the population of rosettes would be maintained for a long period. *Ranunculus bulbosus* is extremely sensitive to changes in its reproduction by seed and would apparently drift fast towards extinction if seed production were inhibited. *R. acris* is more sensitive to alterations in reproduction by seed than in ramet production and is as insensitive to alterations in its ramet production as *R. repens* is insensitive to alterations in reproduction by seed.

This comparison of three closely related species was made in one clearly defined habitat. It shows that three sharply different life-cycle strategies can be successful side by side. The differences may indeed reflect just those divergences between the species that permit them to cohabit without a relentless struggle for existence in which one proves superior and ousts the other from the habitat.

I know of no other studies that have followed the population dynamics of both the active and dormant fractions of a plant population at the same time. There are however several similar detailed censuses of the active populations of perennial plants. Sagar (1959) followed a population of *Plantago lanceolata* over 2 years in a heavily grazed grassland near Oxford, England. The data allow a depletion curve to be calculated (Fig. 19/23) for the populations that were present in the spring of 1957. The half-life was about 8 months and just as with *Ranunculus* spp. there was a clear seasonal rhythm in the depletion curve. The greatest death risk occurred in May–June, the period of most active growth. Sagar calculated many of the components of a life table for this plant (see Harper, 1967). A very similar study of two related species of *Plantago*, *P. major* and *P. rugelii* was made in Ontario, Canada (Hawthorn, 1973) and again there was a very similar story of a generally exponential depletion curve for the established plants. In this environment there is a long winter with snow cover but the populations generally emerged, when the snow melted, at virtually the same densities as at the start of the winter. The mortality risk developed when active growth started and also at the end of autumn: mortality was generally negligible during the midsummer periods of drought and the winter period of intense cold. Deliberately sown seeds were used to obtain cohorts of seedlings of even age. These suffered heavier mortality risk

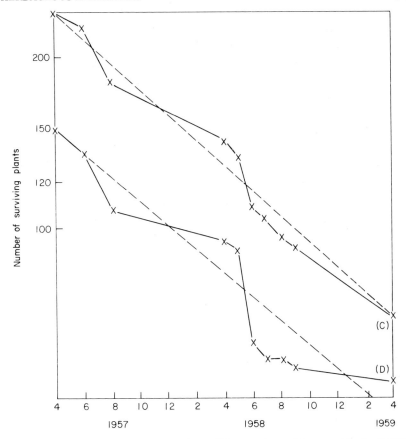

Fig. 19/23. Depletion curves of two populations of *Plantago lanceolata* present in April 1957
in replicate quadrats 1 m² in a field of permanent grassland at Oxford, England. (From
data of Sagar, 1959)

than the established plants but the risk of death occurred at much
the same time as for the adults.

The half-life of established plants of *P. major* was about 16.2 weeks
in a newly sown pasture and 42 weeks in a ploughed but unseeded
area. In contrast, *P. rugelii* had a half-life of 4.1 years in an old pasture
and about 1.1 years in a recently disturbed pasture. In the pasture
P. major behaved virtually as a monocarpic species, flowering once and
dying, but *P. rugelii* was interoparous (polycarpic) and could persist in
a vegetative state for several years after flowering.

The population dynamics of some grasses

A long series of observations has been made of grasses on rangeland in Arizona (Canfield, 1957). Over a period of 17 years the position of grass clumps was mapped using a pantograph in areas subject to year-long grazing and in exclosures. Each clump represented an original seedling and the graphs shown in Fig. 19/24 of surviving plants against time are

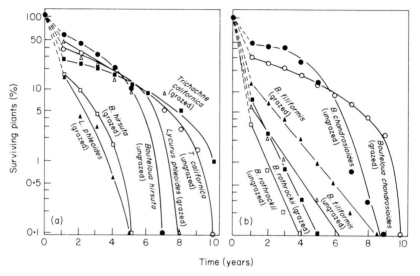

Fig. 19/24. Survivorship of (a) primary and (b) secondary range grasses in south-eastern Arizona. (From Sarukhán and Harper, 1973, drawn from data of Canfield, 1957)

true survivorship curves. The shape of the curves in the first year is not known. Two species of *Bouteloua* (*B. filiformis* and *B. rothrockii*) had a constant annual death risk with half-lives of 9—10 months and 4.5—6 months respectively. The remaining species followed remarkable patterns of survivorship with a low risk of death in middle age followed by a high death risk as they grew older. This is a survivorship curve of the type described by Deevey as Type I and characteristic of laboratory populations of *Drosophila, Hydra, Microtus,* wild populations of Dall mountain sheep, a rotifer (*Floscularia*) and man (Deevey, 1947; Macfadyen, 1957); they differ from any other populations described for plants except the dune annuals such as *Cerastium atrovirens* and *Vulpia membranacea* (Figs 18/6 and 18/7) in which the shape of the curve is formed within one season of growth, whereas the grasses studied by Canfield are tussock-forming perennials with a theoretical potential for indefinite life. It may be that the tussock habit brings an increasing death risk as the plants grow larger. As a clump becomes dense and the tussock accumulates undecayed remains, there is increasing difficulty in the rooting of tillers. A similar high risk of death in old age occurs in *Corynephros canescens,* a perennial of sand dunes in northern Europe and in this case the tillers on old tussocks often fail to produce any roots.

Another group of perennial grasses was studied in a semi-arid environment in Australia. Williams (1970) followed the survivorship of cohorts of seedlings that established in different years (Fig. 19/25): these sur-

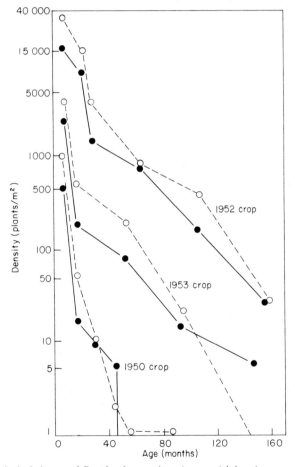

Fig. 19/25. Survival of clumps of *Danthonia caespitosa* in grazed (o) and protected (•) disclimax grassland for the 1950, 1952 and 1953 crops in semi-arid Australia. (From Sarukhán and Harper, 1973, redrawn from Williams, 1970)

vivorship curves appear to be negatively skewed (Deevey Type III). The most intriguing aspect of this population dynamics is that the survivorship curves are clearly different for different cohorts. It is as though the death risk to a population is determined at the time it is formed rather than by the characteristics of the years in which its members die.

A long-term study of the survivorship of perennial grasses was made at Deniliquin in the middle of the geographical range of *Danthonia caespitosa*. Life tables were prepared for three species, *Stipa variabilis, Enteropogon acicularis* and *Danthonia caespitosa*, in plots grazed by sheep. The survivorship curves of different cohorts of seedling recruits are shown in Fig. 19/26a, b, c, together with the depletion curve of the

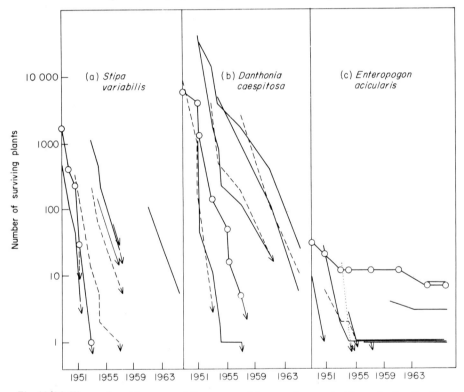

Fig. 19/26. The survivorship curves of three perennial grasses in a semi-arid grassland at Denili-quin, south-western New South Wales, Australia. The areas studied were grazed by sheep. The curves marked o————o are depletion curves of the population present at the start of the study — the age distribution of these populations is unknown. The remaining curves are survivorship curves for cohorts of seedlings establishing at different times. The broken and continuous lines are used to avoid confusion between curves — they have no other significance. (Drawn from data of Williams and Roe, 1975)

population present at the start of observations. The three species illustrate quite markedly different age structures. The populations of *Enteropogon* were small but the individuals were often very long lived. The
population present at the beginning of the experiment was depleted
with a half-life of 4—5 years. Some large individuals outlived other
plants in their cohorts by 10.8—17.5 years. This is a species that is
favoured by exclosure but nevertheless had a very high expectation of
life in the grazed plots. *Stipa variabilis,* in very marked contrast, is a
short-lived perennial with a half-life measured in months. There were
frequent recruitments of quite substantial populations in which each cohort moved rapidly and inexorably to extinction. There were no individuals that had long survivorship as was the case with *Enteropogon.*
Danthonia was intermediate in behaviour with a half-life of a little less
than a year but the recruitment of seedlings was often on a massive
scale. Williams (1970) and Williams and Roe (1975) suggest that death
is due to the severity of the climate though when the data are plotted
on appropriate logarithmic scales there is remarkably little difference
in the mortality risk between years. Certainly the differences between
species are much greater than the differences between cohorts over
years or between years over cohorts.

It is an all but impossible task (perhaps completely impossible) to
follow the dynamics of a grass population in a continuous sward
unless the dynamics of tillers are studied instead of genets. Individuals
intermingle, connections between tillers are rapidly lost and the sward
has to suffer so much manipulation from the observer that it rapidly
ceases to be an undisturbed population. Only where the vegetation is
very open, as in semi-arid grasslands, is there a reasonable chance of
tracing the fate of genets. *Anthoxanthum odoratum* is a common member of closed swards but also colonizes some open habitats. There is a
great deal of ecotypic differentiation within the species and the life
cycles are not immune from the trend to local adaptation. In Britain
it is described as a perennial (Clapham, Tutin and Warburg, 1952) but
in the south-eastern United States as "perennial or winter annual".
Böcher (1961) grew plants from 27 European populations and one of
these behaved as a biennial, seven populations lived for three years and
the remainder lived longer. Antonovics (1972) mapped populations of
Anthoxanthum odoratum on a disused mine in North Wales where the
soil had a high concentration of zinc and was also contaminated by
lead and copper. The area had probably been exposed for colonization

between 1873 and 1898. Plant density is low in such sites and restricted to species and ecotypes tolerant of the contaminating metals. The
mapping method involved using linear transects: a plant was recorded
as present if its tillers came directly below a tape measure stretched between two fixed points. The fates of cohorts of plants recruited in different years was followed (see Fig. 19/27a, b) and two striking features

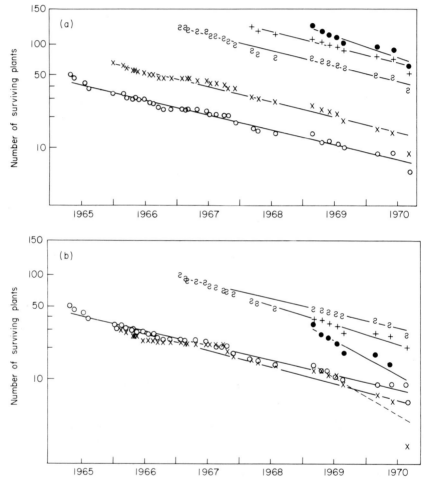

Fig. 19/27. The depletion and survivorship curves of populations of *Anthoxanthum odoratum*
on a colonized old mine working in North Wales. (From Antonovics, 1972)
(a) Depletion curves for the population present in the spring of each year from
1965 to 1970.
(b) The survivorship curves of annual cohorts of plants of *Anthoxanthum odoratum*
recruited to the populations in the years 1965–1970.

emerge — both the population depletion curves and the survivorship curves of specific cohorts are clearly linear but the recruitment of new individuals to the population was most erratic from year to year. There were good and bad years for seedling establishment but the depletion curve remained linear. The depletion curves give a half-life for the populations of 2.05 years, i.e. 71.3% of the population present in one year survives through to the next. The survivorship curves for each cohort of recruited seedlings also have constant decay rates but differ between cohorts (Fig. 19/27 b), thus the population recruited in 1966 had a half-life of 2.17 years but that recruited in 1968 had a half-life of only 0.95 years. Just as with the populations of *Danthonia* (Fig. 19/25) the survivorship of a population seems to be determined at the outset rather than by the conditions at the time of death. There is also a seasonal trend in the death rates of *Anthoxanthum*, in that the death risk is low in winter and high in summer.

Only 2.9% of the plants lived for more than 5 years and a large number of the short-lived plants never flowered. Every plant that lived more than 3 years flowered at some time during its life. The data provide an intriguing footnote to the comments made earlier about the relation between age and size in plants (Chapter 18). Among the long-lived plants of *Anthoxanthum* the six that persisted throughout the study period showed a slight (non-significant) decrease in diameter and towards the end of their life span individuals had often become smaller! Any foolish observer who imagined that the size distribution in such a population represented an age distribution would be greviously in error. This assumption is often made quite happily for forest trees, where it is just as fallacious (see Chapter 20).

The use of regression techniques for analysing the dynamics of plant populations

Regression techniques have been much used by entomologists to determine which factors are most important in controlling the numbers of an insect from generation to generation. For example Morris (1959, 1963a, b) related N_{t+1} as a dependent variable to N_t, the independent variable. There would appear to be some potential in this method applied to plant populations and an initial attempt was made by Foster (1964) in a study of the daisy (*Bellis perennis*) in grasslands. Putwain *et al.* (1968) extended Foster's methods to determine what contributions

were made to populations of *Rumex acetosella* by ramets, and seedlings
from the buried viable seed in the soil. This plant is dioecious and
spreads clonally from root buds. It is a common species in heathland
and acid grasslands in Europe. In a pasture in North Wales, Putwain
marked out large adjacent blocks for three different treatments: (i) control, in which natural seed dispersal was allowed to occur, (ii) all female
inflorescences were removed, and (iii) the female inflorescences were all
bagged so that normal seed ripening might occur but without seed dispersal. In addition he deliberately introduced seed which was sown at
densities varying from 50 to 5000 seeds per 30 x 30 cm plot. Natural
seed production was estimated on the control plots and the number of
seedlings and mature plants (including ramets) was counted at intervals
over a period of 5 months. He then fitted the multiple regression equation (see Table 19/II which also shows the results for control plots).

The analysis showed (i) that the contribution of seeds to maintaining
the population was quite negligible — most of the plants were recruited
as ramets and (ii) that the denser the population at the start the slower
was its rate of increase. The experiment was short yet revealed these
two features which otherwise could only have been demonstrated by
much more cumbersome procedure. There is probably much potential
in this method in which deliberate perturbation of the population is
coupled with regression analysis.

The series of examples described in this chapter reveal some unexpected generalities in plant behaviour. Perhaps most important are (i)
the remarkable smoothness of survivorship and depletion curves (which
suggest that climates and their variations are less important than might
have been imagined among the mortality risks; (ii) the occurrence of
density-dependent regulation within populations and the coincidence of
high death risks with periods of high growth rates among survivors (which
suggest that the most serious hazards in the life of a plant may often
be too many neighbours, especially of the same species); (iii) the very
long life of a few individuals within a continuing flux of deaths and recruitments; (iv) the demonstration in two quite different habitats that
two quite different grass species have survivorship curves that are determined at the time of recruitment rather than at the time when death
comes — this last point is very puzzling; (v) the surprisingly long periods
spent by perennial plants in a vegetative stage compared to the early
flowering of the same species in culture, which suggests that most plants
in the communities studied were living under considerable stress; (vi)

Table 19/II

The relationship between the population size of *Rumex acetosella* and (i) the number of mature plants present on 24 March 1965 (x_m) (ii) the number of seeds sown (x_s) and (iii) the number of inflorescences produced (x_t). (From Putwain et al., 1968)

The table shows the regression coefficients in the equation

$$Y_t = a + b_1 x_m + b_2 x_m^2 + b_3 x_s + b_4 x_s^2 + b_5 x_f + b_6 x_f^2 + b_7 x_m x_s + b_8 x_m x_f + b_9 x_s x_f$$

Regression equation — The independent variables and their respective regression coefficients which were selected out and which made a significant contribution to the size of the total populations

Date of recording	Probability of regression equation	Constant a	x_m	x_m^2	x_t	x_f^2	$x_m x_s$	$x_m x_f$
15 April	<0.001	+0.24 × 10^1	+0.10 × 10^1		+0.61 × 1	−0.12 × 10^{-1}		
	<0.001	+0.38 × 10^1	+0.99 × 1					
12 May	<0.001	+0.57 × 10^1	+0.11 × 10^1					
3 June	<0.001	+0.78 × 10^1	+0.28 × 10^1	−0.29 × 10^{-1}				
29 June	<0.001	+0.19 × 10^2	+0.15 × 10^1					
	<0.001	+0.16 × 10^2	+0.15 × 10^1					
20 August	<0.001	−0.47 × 10^1	+0.83 × 1			−0.41 × 10^{-1}		
	<0.001	−0.21 × 10^1	+0.87 × 1			−0.32 × 10^{-1}	+0.15 × 10^{-3}	
	<0.001	−0.26 × 10^1	+0.99 × 1					+0.66 × 10

the large number of plants that may persist for several years and then die without ever producing seed — evolutionarily sterile members of the population, consumers of resources that stay in the struggle for existence for a long time and have no measurable fitness at all; and (vii) the generally small contribution to the maintenance of populations that is made by seeds (*Ranunculus acris* and *R. bulbosus* are exceptional in this respect). Presumably we would need to study dynamics of populations on a larger scale if we were to detect the critical role of the seed within what are always dynamic communities with local invasions and extinctions.

20

Woody Plants

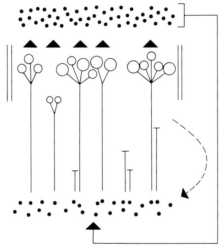

"A tree is, properly speaking, a family or swarm of buds, each bud being an individual plant." Studhalter

The great advantage of a woody habit of growth is that it can give perenniality to height. The advantages of height are clear — leaves may overtop and shade neighbours, flowers may be more obviously displayed to roving pollinators or to the wind and seeds may be shed over greater distances (see Chapter 2). The woody plant is not dependent on the growth made in a single season to gain these benefits because the meristems themselves can be held high on the perennial woody stem. Offsetting the advantages of height are a number of very serious disadvantages. By its nature the woody perennial accumulates dead structure and dependent living tissues. The great lengths of cambium, phloem and wood parenchyma are a respiratory burden, one of the costs of being tall. Accumulated dead tissue may serve as a food base for fungal colonization and attack by pathogens. Height increases the risk from some hazards such as fire, wind, lightning and hurricane damage from which the herb is protected. The early periods of life of the woody perennial are usually devoted to gaining height and reproduction is delayed; a long juvenile period is a characteristic feature of trees (see Chapter 22). Hence the rate of colonization by trees tends to be slow and so also is the rate at which populations recover from disasters. The tree has to pass through growth stages appropriate to the ground layer of vegetation, the field layer and the shrub layer before it emerges as a proper tree in a canopy. In a sense the tree has to master all trades — to be successful in a variety of life stages and to meet the hazards of each layer of the vegetation that it penetrates.

The study of trees is a study of short cuts; the long life and large size of trees makes many of the conventional methods of plant biology impossible or unrealistic (see Chapter 7). This is certainly true of population dynamics and there is no reported study of a forest population that can be compared with the detailed censuses over time that have been made for perennial and annual herbs. There are three short cuts to discovering the population dynamics of a forest or woodland:

(i) *A detailed census procedure can be applied to a strictly defined age-class in a population.* In practice this usually means concentrating on the behaviour of seedlings and saplings which can be mapped and often aged by counting apical bud scars: the information from such an age-class can sometimes be projected to account for some of the behaviour of older trees in the community.

(ii) *The assumption can be made that size reflects age.* It is easy to obtain a measure of the size of trees in a population, e.g. the diameter at breast height. If the assumption is true, it becomes easy to determine the age structure of forests. In fact the assumption is false and dangerously so. Within an even-aged forest, marked differences in the size of individuals develop quickly, as in populations of other sorts of plants (see Chapter 6). The frequency distribution of tree sizes in such even-aged stands frequently becomes exponential. Even-aged pure stands of *Abies sacchalinensis, Betula platyphylla, B. ermanii, B. mazimowiczii* and *Cryptomeria japonica* were shown to have a weight distribution that conformed very closely to the exponential form (Hozumi and Shinozaki, 1970; Hozumi *et al.,* 1968). It is very easy to misinterpret such a distribution of sizes as a distribution of ages. The use of size as a measure of age is only justified if the relationship has been formally proved by the use of some independent accurate measure of age. When such a correlation is looked for in multiple-aged stands it usually turns out to be very weak (see e.g. discussion in Chew and Chew, 1965; Fig 20/18). Spaced forest plantings of even age sometimes develop differences in size that are multimodal: this is, for example, a well known feature of *Pseudotsuga douglasii* where it may reflect genetic differences within a stand. The temptation to use size as an index of age is very strong in the tropics where annual rings are not formed and no other method of ageing a stand seems available. There is no reason to suppose that age is more closely related to size to the tropics than it is in temperate regions. No confidence can be placed in the results of using this method.

There is, however, an important way in which the size distribution of trees in a forest can play a part in population biology. Just as it is possible to describe the condition of a population of herbs in terms of life states (Chapter 19) this is also possible for a forest. The size structure of a population of trees is some measure of its future and is useful to a forester who needs to make predictions about yield or to plan thinning operations – the real age of the trees may then not be very important (see e.g. Buell, 1945; Leak, 1964; McGee and Della-Bianca, 1967; Meyer and Stevenson, 1949; and Meyer, 1952). A number of measures of tree size can be used to obtain a size distribution and the easiest short cut is to measure diameter at breast height. The French forester de Liocourt suggested that the quotients between the numbers of trees in successive diameter classes remained essentially constant over the range of diameters found in the forest. Just as the idea of a balanced age distribution can be applied to an animal population so a balanced size distribution can be envisaged as a comparable condition in a forest. Meyer (1952) suggested that the forests over any large area will approach a balanced condition in which the quotient (q) between the numbers of trees in successive diameter classes approaches a constant value. In practice most forests do not have such a balanced size distribution though they may always be converging towards it. In a study of a 1050 ha tract of hardwood forest in New Hampshire, U.S.A., the diameter distributions were analysed on 22 compartments of the forest, each of about 16 ha (Leak, 1964). The quotients of the number of trees in successive diameter classes are shown in Fig. 20/1. The relationships are effectively linear (provided that large classes are taken, cf. Fig. 20/1a and b) and the slope describes the departure of the size distribution from the "ideal" balanced condition. It may be that this procedure could be applied usefully to a number of other plant forms besides trees and, like Rabotnov's life classes for herbs (Chapter 19) may offer an alternative to the more classical age-related methods for describing population structure. The main argument for measuring age distributions in population biology is that reproductive behaviour is age specific. In fact much of the reproductive behaviour of perennial plants appears to be more specific to size than to age, and it may be that predictions about the future of a population are best obtained by studying the size rather

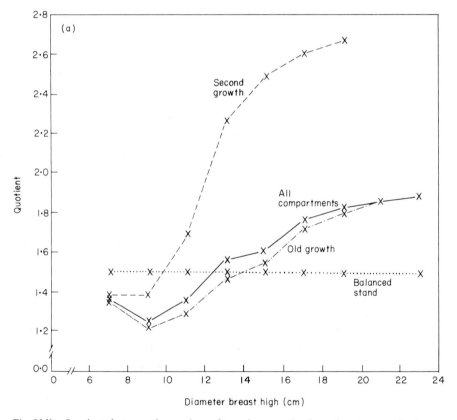

Fig. 20/1a. Quotients between the numbers of trees in successive 5 cm classes over midpoints between DBH classes for second-growth, old-growth, all compartments (Bartlett Forest), and a hypothetical balanced stand ($q = 1.5$).

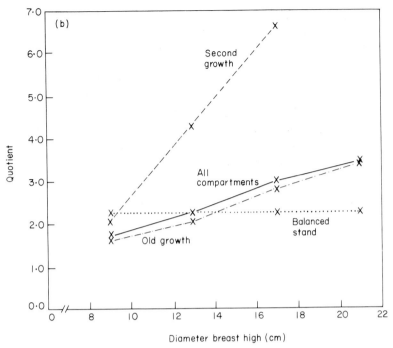

Fig. 20/1b. Quotients between the numbers of trees in successive 10 cm classes over midpoints between DBH classes for second growth, old growth, all compartments (Bartlett Forest) and a hypothetical balanced stand (average quotient for 10 cm classes equals 2.25). (From Leak, 1964)

than the age distribution. Unfortunately, this approach is completely unhelpful in tracing past history (it can project forwards but not back) and it is of no use to the student of selection or of evolution.

(iii) *The past population dynamics of a forest may be inferred from the age structure of the present population.* The formation of annual growth rings in the wood of temperate trees means that an individual can be aged and the characteristics of the growth rings can sometimes reveal more details about the past history of the individual such as the times when it was suppressed by or released from the influence of a neighbour (Fig. 20/12), attacked by a pest or disease, suffered from drought or some other hazard. There are some problems in the method. A ring-count reveals the age of the tree at the level at which the trunk is cut or a core is taken, not the age from germination. A ring-count made at ground level can give an accurate age but is technically difficult.

In the forests of the North American Appalachian Mountains the three "climax" species, *Abies balsamea, A. fraseri* and *Picea rubens* spend some 40 years reaching sapling or transgressive size (*ca* 7 ft) and much of this age would be missed in a core sample taken at breast height (Oosting and Billings, 1951). An accurate estimate of age can be made by felling a tree but non-destructive methods involve taking cores and it is not easy to ensure that a core passes through the centre of the old wood. In some environments, e.g. tundra, false growth rings may develop and confuse the dating (Barrow *et al.*, 1968).

Woody species that form clones introduce a special problem into demography, in that the age of the tree may not be the age of the genet and this makes it very difficult to derive a real age structure for stands of species like *Populus tremuloides.* A similar problem arises with trees that regenerate from stump sprouts like *Sequoia sempervirens,* the coastal redwood.

The biggest problem in interpreting age distributions made at one point in time is the impossibility of determining how much mortality has occurred in the past history of the stand. Occasionally relics of past forests can give a clue about catastrophes that have occurred but in general this is not possible. The age structure at a point in time is the age structure of survivors — it is past births minus past deaths — and a number of assumptions have to be made about past recruitment and mortality before an age structure can be interpreted.

The population structure of invading pines and juniper

A community of pinyon juniper

Woodland dominated by pinyon pine (*Pinus monophylla*) and juniper (*Juniperus osteosperma*) in U.S.A. is intimately associated with *Artemisia* sage-bush communities. The woodland usually lies at altitudes of 975—2560 m with an annual precipitation of 25—63 cm, above the zone of desert scrub and below sub-montane scrub or forests of *Pinus ponderosa.* Pinyon-juniper woodland occupies about 40×10^6 ha in U.S.A. In a region of east-central Nevada both species have invaded communities of black salt bush and analysis of the age-structure of the populations makes it possible to reconstruct the population dynamics of the invasions (Blackburn and Tueller, 1970). The ages of survivors were determined in five types of woodland: (a) *"Open"*, with few seed-lings and saplings in a well developed sage-bush population; (b) *"Dispersed"*, with abundant seedlings, young saplings and maturing trees but no old trees; (c) *"Scattered'*, with abundant juniper and pine seed-lings, saplings plus a few old trees, and a well developed understorey of *Artemisia*; (d) *"Dense"*, a population with a full range of age classes present and still with an understorey of *Artemisia*; and (e) *"Closed"*, communities with no understorey. The age-structures of these five communities are shown in Fig. 20/2a and b.

If we ignore any mortality that has occurred in the history of the populations these graphs can be interpreted as extremely elegant curves of population growth, representing five stages in a cycle of invasion. Juniper was the earlier colonizer and in the community that was described as "closed" in the 1950s there were more than 100 trees that had established before 1725. The dense and scattered communities had already been colonized by juniper at that time, though there were fewer survivors from that age still present in the 1950s. No plants of such age were present in the communities described as "dispersed" and "open". Colonization by pinyon pine was later and followed very similar patterns of invasion and population growth, as judged by the individuals that survived. Again ignoring mortality, we can compare the rates of growth of the populations at different periods. A convenient way to do this is to compare the time taken for a population to double. The data have been plotted on a logarithmic scale to emphasize growth rates (Blackburn and Tueller used linear scales which made interpretation more difficult); the rates of population growth may then be read as the tangents to the growth curves. The values obtained are very crudely calculated from eye-fitted curves based on the figures quoted by the authors rather than on their graphs (see Table 20/I). The history of colonization by juniper in the "closed" communities involved a long period of very slow population growth that accelerated with the passage of time, particularly in the period 1850—70. The rate then fell. Essentially the same picture emerges from the "dense" population which reached the same population density as the closed community in about 1882. The scattered community reached the same density but not until 1956. The dispersed and open communities were still increasing rapidly, though the rate of growth of the populations was falling steadily. Table 20/I shows clearly that in every period examined the fastest population growth was in the least dense communities. The

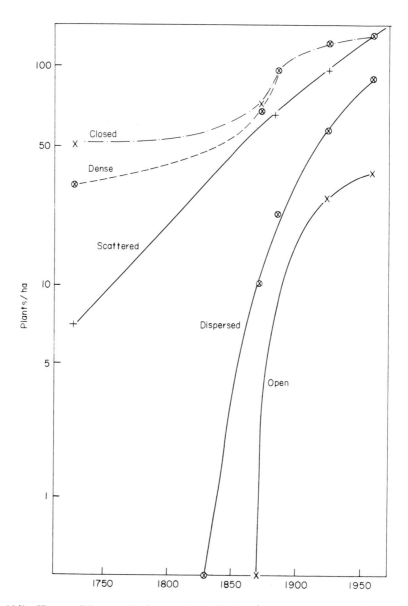

Fig. 20/2a. History of the growth of populations of juniper (*Juniperus osteosperma*) as indicated by the ages of survivors in 1956. (Plotted from data of Blackburn and Tueller, 1970)

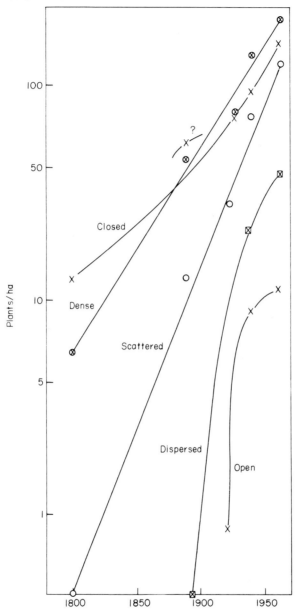

Fig. 20/2b. History of the growth of populations of pinyon pine (*Pinus monophylla*) as indicated by the ages of survivors in 1958 (see text). (Plotted from data of Blackburn and Tueller, 1970)

same was true of pinyon pine, though there were curious changes in
population growth rate at the end of the period when the densest
population appeared to be accelerating its growth and the "open"
community slowing down.

Table 20/I

The time taken by populations of pinyon pine and juniper
to double. Note the populations are those of survivors
present in 1956–58 (Values in years).

	1700	1750	1800	1850	1900
Juniper					
Closed	>1000	250	92	108	
Dense	245	188	66	90	
Scattered	45	45	45	56	
Dispersed	—	—	12	26	
Open	—	—	—	24	
Pine					
Closed	—	60	54	41	
Dense	—	31	31	31	
Scattered	—	20	20	20	
Dispersed	—	—	—	8	

There is no question that in these cases we are seeing population
growth mirrored in the age-structure (cf. Fig. 19/22). The question
remains how far we are justified in neglecting mortality and variations
in the rate of recruitment. The data cannot answer the question whether
the decline in the population growth role of juniper with time and with
density was due to differential mortality or differential birth rates. If
the differences were in the death rates the true curves of population
growth might have been differently shaped. It is extremely interesting
that the *apparent growth rate* of juniper populations in the scattered
communities remained effectively constant for 150 years!

Despite the problems of interpretation that are posed by this sort of
study, there is probably no more beautiful set of population growth
curves in the whole of plant biology. The treatment adopted here depends on the use of logarithmic transformation of the data, which is
almost always appropriate for the analysis of biological populations

(Williamson, 1972) because, amongst other reasons, it displays large-scale trends and reduces to their proper scale the minor fluctuations that can appear so important in a linear analysis.

Communities of *Pinus taeda*

The loblolly pine *Pinus taeda* is an early colonizer of old field systems on the Piedmont, North and South Carolina and Georgia, U.S.A. The dynamics of population growth are strongly affected by the position of seed parents. When colonization is from an established stand, for example when an old field regenerates at the edge of a forest, stands of graded density become established. Areas close to a seed source may become fully stocked in one year so that the population grows up as an even-aged stand. Further away it will take longer for the area to become fully stocked and the resulting population will then contain an age spectrum representing several years' recruitment into the area. Still further away occasional seeds may have landed and a very sparse population develops over time (McQuilkin, 1940). Free-grown individuals of *Pinus taeda* are precocious reproducers and can start bearing cones when 7 years old. These trees then serve as secondary foci for "infection" of the land with *Pinus taeda*. (The term infection is used deliberately to draw attention to the great similarities between the phases of colonization of land by such a plant and the development of epidemics. The pathologist's concepts of infection fronts and horizons, pockets of infection and local foci of spread (see e.g. van der Planck, 1960) have exact equivalents in the population biology of trees!). When a population develops from scattered pioneers, followed by secondary colonization from their descendants, the age structure will show a few large old trees and a secondary peak of numbers about 8–10 years younger. Patterns of seedling establishment are illustrated in Fig. 20/3 and an idealized diagram of the age-structures of populations near to and far from a seed source is given in Fig. 20/4.

One old field population of *Pinus taeda* has been studied in detail (Spring *et al.*, 1974): the field had last been cultivated in about 1951. 268 trees including seedlings were accurately aged and the age distribution of the population is shown in Fig. 20/5. The few trees of 15–21 years old were advance colonists and the age group 1–11 years represent the progeny of these colonists.

The fate of young plants growing in a fully stocked population is

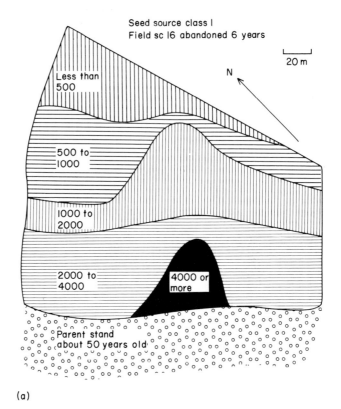

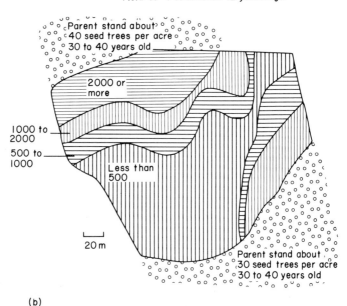

Fig. 20/3. Maps of seedling density in old fields near seed sources of *Pinus taeda*. (From McQuilkin, 1940)

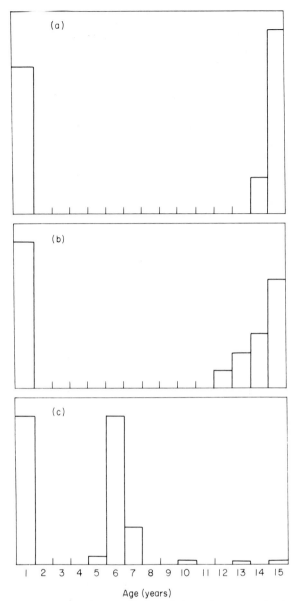

Age (years)

Fig. 20/4. Idealized age structure diagrams for populations of a tree species that requires at least 7 years pre-reproductive growth.

(a) Population close to a dense source of seed. Population fully stocked in the first year of recruitment — becomes an even aged population except for doomed one-year-olds.

(b) Population further from a source of seed. Several years are needed before the population of seedlings is fully stocked. The age structure is nearly even aged except for doomed one-year-olds.

(c) Population even further from a seed source. A few recruits appear early but dispersal into the area is insufficient to stock it fully before the first colonists start to set seed. The trees of age 5, 6 and 7 represent progeny of the first colonisers and bring the community to a fully stocked state. Again there is a population of doomed one-year-olds.

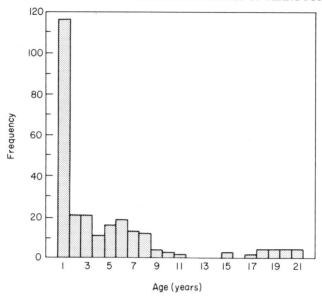

Fig. 20/5. The age distribution of loblolly pines (*Pinus taeda*) in an old field succession. (From Spring *et al.*, 1974)

usually to die young. *Pinus taeda* is extremely sensitive to shading whether from its own species or other trees and shrubs. Every seedling of *Pinus taeda* was marked within a population regenerating after a pure stand had been destroyed by fire in 1925 (Chapman, 1945). The first census was made at the time of marking in 1933 and the age of each seedling was determined. The population of seedlings was followed through to 1944. The seedling establishment had not been very dense after the fire and only 50 of the seedlings that were present in 1933 had been established in 1928–29, yet of the 55 trees present in 1944, 36 were survivors from this first wave of immigrants. Of the 303 seedlings recorded as appearing in later years only 19 survived to 1944. A very large crop of seedlings appeared in 1929 but by 1944 every one of them was dead, and neither were there any survivors from 1930 or 1931. Each of the seedlings of *Pinus taeda* was scored according to its experience of shade cast by other colonizing trees and shrubs. In many cases a seedling of *P. taeda* established in full light but was subsequently shaded by the growth of neighbours. Such shaded pines lingered for a few years and then died. Death occurred, on average, 5.2 years after a seedling was shaded. In the early years of the study it was mainly the smallest seed-

lings that died but, as the stronger trees grew and extended the range
and height of their shade, larger and larger saplings were killed. It is impossible ever to gain a proper picture of the dynamics of such a tree
population without marking and recording or mapping every individual
repeatedly. Seedlings and saplings that died left no trace after 2 years
and only a repeated census could reveal that they had ever been present.

Pinus taeda is a pioneer colonizing tree and, left without disturbance,
gives way to hardwood stands. One very old stand of the pine contained
mainly individuals 100–175 years old (Stalter, 1971). In a series of
populations of different ages *P. taeda* can be seen gradually losing dominance, though a relatively few 150-year-old trees that remained in the forest
continued to contribute disproportionately to relative dominance in
the canopy (Table 20/II).

Table 20/II

The relative importance of *Pinus taeda* to hardwoods in old forest
successions on lowland soils (from Stalter, 1971)

Average age of *Pinus taeda* (years)	15	18	34	45	90	150
Relative density	92	89	30	22	15	13
Relative dominance	99	99	88	64	76	48

A somewhat similar study was made of the dynamics of young populations of holly (*Ilex aquifolium*) in the New Forest (Hampshire,
England). The forest had no continuous history of regeneration but there
were two periods when active establishment, particularly of beech
(*Fagus sylvatica*), occurred: 1650–1760 and from 1850 to the present
(Peterken and Tubbs, 1965). Regeneration was active when browsing
pressures were low. The fate of marked holly seedlings was followed in
a number of quadrats placed in different parts of the forest mosaic. The
seedlings could not be aged accurately without destroying them, and
they were therefore classified into growth stages: C (with cotyledons)
present), S (small seedlings without cotyledons) and Y (young plants).
Ages were determined by ring counts for a number of plants off the
quadrats. Class C included plants of ages 1–2, class S 2–4, and class Y
4–13 years (Peterken, 1966).

The annual turnover among the holly seedlings was calculated as a
percentage —

$$\frac{100(\text{Gains}_{t_1 - t_2} + \text{Losses}_{t_1 - t_2})}{\text{Total}_{t_1} + \text{Total}_{t_2}}$$

and an expression of the age structure was given by

$$\frac{\Sigma S + \Sigma Y}{\Sigma C + \Sigma S + \Sigma Y}.$$

The measures of turnover and age structure are given in Table 20/III for the various quadrats grouped according to the state of regeneration.

Table 20/III

The age-structure and annual turnover of holly seedlings in 20 x 1 m^2 quadrats in a variety of woods in the New Forest, Hampshire, England. (From Peterken, 1966)

Quadrats	Regeneration	Y plants per m^2	Age-structure 1962	1963	% annual turnover C	S	Y
2–6	Active	30.6	0.88	0.87	66	23	14
14–18	Active	9.4	0.88	0.90	60	39	19
7–13	In check	6.4	0.68	0.68	72	38	33
23–26	In check	4.8	0.61	0.79	73	48	53
27–30	Absent	0	0.10	0.00	78	–	–
19–22	Absent	0	0.37	0.00	94	91	–

Table 20/IV

The percentage survivorship of seedling holly to the fifth year. (From Peterken 1966)

Community type	% survival
(a) Open pine woodland	7.26
(b) Clearing in beech woodland	3.23
(c) Open beech woodland with scattered holly	0.57
(d) Thin oak woodland with scattered holly	0.053
(e) Deep shade below holly	0.008

The density of seedlings was greatest under holly trees, but regeneration was not commoner under female than under male trees. Much of the seed is taken by birds and presumably the density of the seed rain is determined by the positions that birds choose to defecate.

The annual turnover was fastest where regeneration was failing and lowest where there was active regeneration. The age structure was higher in the regenerating sites. Table 20/IV shows the percentage survival of seedlings to their fifth year in different parts of the mosaic of habitats.

The best chance of survival was in the clearings and open sites — not under holly, although (or perhaps because) seedling density was highest under holly. Survival was clearly lowest in the shaded plots, although seedlings of holly have a very low compensation point of photosynthesis. The causes of death were very hard to identify except in special cases. A fire killed all the seedlings in one quadrat, a mole uprooted the seedlings in another and in a third a deposit of horse dung killed about 20 seedlings. These deaths are not obviously related in any way to light intensity which was the environmental feature that correlated (negatively) with mortality risk. A large number of seedlings were grazed and it seemed that although the *potential* regeneration of holly was determined by the frequency of patches of high light intensity the real regeneration was controlled by grazing animals.

There are a few studies based on forest inventory that show the dynamics of successional change in terms of the numbers of trees present instead of the more qualitative description that dominates most ecological treatment of the subject. 27 years of the history of a middle-aged mixed hardwood forest have been described for a forest on Sugar Island, St Mary's River, Michigan (Spurr, 1964). The data are shown graphically in Fig. 20/6. The aspens, birches and red maple declined steadily but whereas the surviving maples had increased greatly in size and increased the basal area that they occupied from 6.8 to 7.8 m² per ha, the birches and aspen had lost basal area during the same period. The populations of sugar maple and red oak lost a few individuals during the period but the majority of these trees had continued to grow vigorously.

In many analyses of forest structure the younger stages and the smallest plants are neglected. A few studies have concentrated on this fraction of the total population. An undisturbed pine forest was chosen for one such study (Hett and Loucks, 1968). The forest, which was in Ontario, Canada, had no history of cutting and appeared to have originated by colonization after a fire about 200 years before. The forest was predominantly pine and the major seedlings present were *Abies balsamea*, *Pinus strobus* and *Acer rubrum*. A restricted population (those individuals less than 137 cm tall) was analysed and these "seedlings" were aged by counting terminal bud scars and annual growth rings. Two-year age classes were easily recognized up to age 8 but the remainder included individuals with a maximum age of 25 years and an approximate median age of 17. It was assumed that there was a constant annual input of seed-

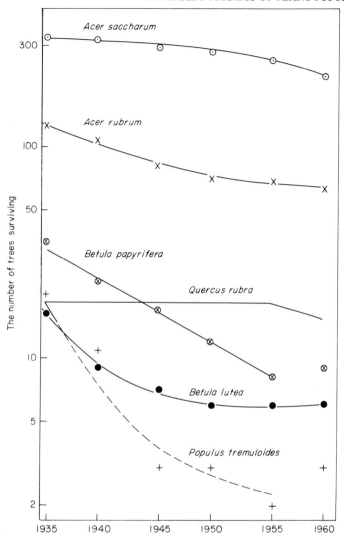

Fig. 20/6. Population changes in a mixed hardwood forest over a 27 year period. (From data of Spurr, 1964)

lings into the population; the age distribution was then treated as a survivorship curve (Fig. 20/7) and the percentage survivorship was calculated for each age class (Fig. 20/8). Both pines and red maple populations declined exponentially over the first 8 years and the survivorship percentage, judged by the rather dubious 17-year class, then fell — *Acer rubrum*

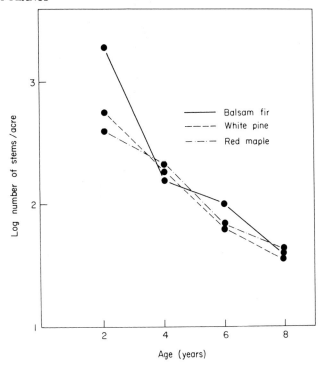

Fig. 20/7. The age distribution (logarithmic scale) of seedlings and saplings in a pine forest in Ontario, Canada. (From Hett and Loucks, 1968)

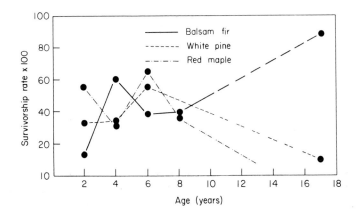

Fig. 20/8. The survivorship percentage by age class of seedlings and saplings in a pine forest in Ontario, Canada. (From Hett and Loucks, 1968)

had disappeared entirely from this class. Both *Acer rubrum* and *Pinus strobus* have a reputation for shade intolerance. In contrast *Abies balsamea*, after an initial period of low survivorship, gradually increased its potential for survival until at the median age of the oldest class (17 years) the probability of living to the next age class had approached 0.99.

A rather similar analysis of the dynamics of the juvenile phase in a forest was made on *Acer saccharum*, one of the major species in the sugar maple/basswood forests of southern Wisconsin (Hett, 1971). This species has very variable seed production from year to year. The size distribution of individuals in the forest has a clear negative exponential distribution (Knuchel, 1953). Curtis had started a study in three stands and recorded a crop of seedlings in 1953 that was followed until 1956. Hett made a further analysis in 1965. The initial losses from the population gave a good fit to a negative exponential decline, i.e. a risk of mortality that was independent of age (Fig. 20/9). The shape of the curve changes when the 1965 data are included and is then best fitted

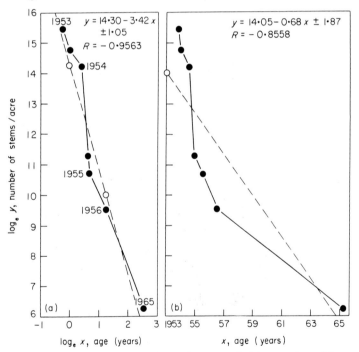

Fig. 20/9. The age class distribution of seedlings of sugar maple (*Acer saccharum*) in sugar maple/basswood forest, Wisconsin. (From Hett, 1971)

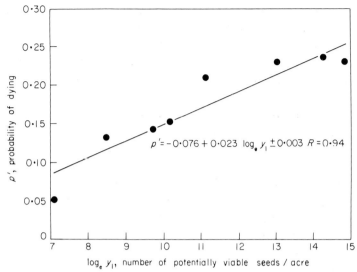

Fig. 20/10. The relationship between the density of potentially viable seeds of sugar maple and the probability that a seedling will die during its first year. (From Hett, 1971)

by a power function (a linear relationship between log numbers and log age). It is unfortunate that there is a 9 year gap in the records; the evidence for a declining death rate rests on the observations made in 1965 after this gap in the records and after a change of observer. The declining death rate of *Abies balsamea* in the Ontario forest (Fig. 20/8) likewise rested on the ages of one very anomalous age group. Taken together these studies of young stages in forests suggest that the dominants have a declining death rate and it would be interesting to see this generalization widely tested.

Records were made, in the same Wisconsin forest, of the size of the seed crops in different years and Fig. 20/10 shows the relationship between the number of seeds shed and the probability of death. There is a striking density-dependent effect that helps explain why, although seed crops were so variable, the number of plants recruited into each age class was remarkably constant from year to year.

The demography of the cherry (*Prunus serotina*) in oak forest

Oak forests are relatively recent developments in the vegetational history of southern Wisconsin. Few of these forests (which are dominated by *Quercus macrocarpa* and *Q. alba*) are more than 125 years old, though they contain open-grown individuals, 300 years old, which remain from before European settlement when the vegetation was maintained as prairie and oak savannah by fire. The black cherry, *Prunus serotina* accounts for from half to all the plants in the sapling and small tree strata in these forests. The age structure of a population of cherry was determined by ring counts (see Fig. 20/11). There were four distinct

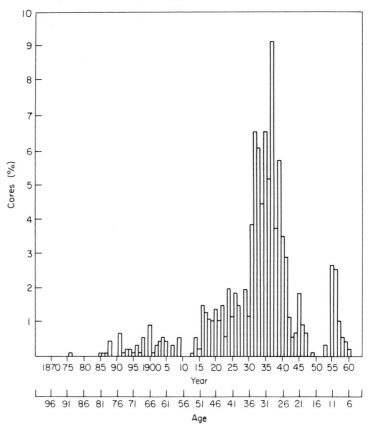

Fig. 20/11. The age class distribution of *Prunus serotina* in a Wisconsin oak wood — based on 854 live stems. (From Auclair and Cottam, 1971)

periods of establishment. (i) The oldest stem was from 1876, and there followed a period of slow recruitment. (ii) After 1916 a new, higher, though rather constant rate of recruitment took place. (iii) 58% of the total population present in 1961 came from a sudden period of rapid recruitment in 1931—41. (iv) A further apparent flush of recruitment in 1955—60 is thought by the authors (Auclair and Cottam, 1971) to be a sampling artefact.

The periodicities in cherry recruitment can be tied quite closely to changes in land management in the area. In particular the increases in cherry populations after 1915 coincide with a period of vigorous farming activity in the area when wood lots were used for grazing. During periods of drought the farmers are thought to have cut tree foliage as browse for their livestock, thus opening up the canopy for regeneration. Other factors than the activities of man in the region also had repercussions on the behaviour of the cherry. Oak wilt disease was spreading in the area and the annual growth increments made by some cherry trees show a great increase in width when an overshadowing oak tree was attacked by wilt and died (Fig. 20/12).

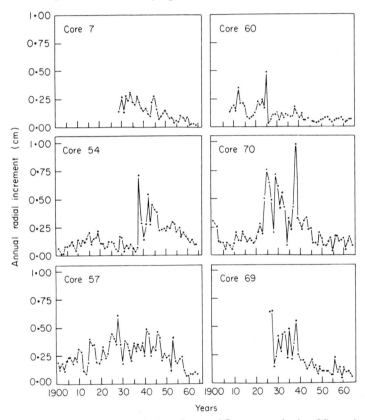

Fig. 20/12. Annual radial increments of selected trees of *Prunus serotina* in a Wisconsin oak-wood. Core 7 and 60 are from understorey trees. Cores 54 and 70 are from trees that show a sudden burst of rapid growth after the death of the overstorey oaks from wilt disease. Cores 57 and 69 are codominant trees of seed and sprout origin. (From Auclair and Cottam, 1971)

The demography of old spruce—fir forests

In the eastern, non-montane regions of North America, forests dominated by spruce and fir reach their southern limits in Maine and Nova
Scotia. These forests have been regarded as a physiographic climax on
the shallow acidic soils and their age structure, like other putative climax
or virgin forests, is of great interest. A series of 14 stands was selected
in which at least 50% of the trees in the upper canopy were red spruce
(*Picea rubens*) and/or white spruce (*Picea glauca*) and/or balsam fir
(*Abies balsamea*). Seventy sampling units of 10 x 10 m quadrats were
distributed within the 14 stands. Increment borings were taken from
individuals in each 10 cm diameter class and age structure diagrams
were constructed. Among the various forests three major types of age
structure could be recognized, and the history of the forests could be
partly reconstructed (Fig. 20/13a, b, c).

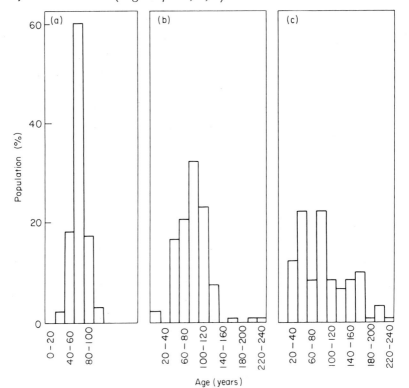

Fig. 20/13. Age structure diagrams for populations of *Picea rubens* in spruce—fir forests of the
Maine coast (U.S.A.) drawn from data of Davis (1966), see text. Data include only
trees of DBH > 2.5 cm.

Shade-intolerant hardwoods usually dominate the early development of forest in this area (e.g. after a large fire) after a herb—shrub stage. *Betula papyrifera* and *Populus tremuloides* are the usual hardwood dominants. The three conifers enter in proportions that are a function of the distance from seed parents and either replace the hardwood species or establish themselves directly on open sites. When the conifers are 20—45 years old and 20—35 ft tall they form a dense thicket of their dead lower branches. At about 50 years old they are 30—45 ft tall and 10—20 cm DBH. Seedlings, especially of *Picea rubens* appear but die when very small (2.5—10 cm). The stands are effectively even-aged, dominated by a single age class (Fig. 20/13a). Sapling-sized trees in the population are nearly as old as the largest upper-storey trees. Such a forest is commercially mature and would be harvested for pulp at this stage.

Later in the development of the forest, trees may be lost from the upper canopy by wind damage or other causes and red spruce and balsam fir start to colonize the openings (Fig. 20/13b). Such secondary age classes probably always begin to enter the forest before it is 100 years old (Davis, 1966). Only one of this series of forests was classified as "virgin". The number of annual rings on the trees of *Picea rubens* that formed part of the upper storey was 163 ± 7 (at breast height). This forest was multi-aged (Fig. 20/13c) though the distribution of ages was not at all uniform. There were no young recruits of less than 20 years old (but trees of d.b.h. < 2.5 cm were excluded from the samples) and it looks as if two big periods of recruitment (or the establishment of plants with a low mortality risk!) occurred in the 40—60 and 80—100 year classes.

It is seldom possible to age the trees in a forest so accurately (or in sufficient numbers) that discrete year classes can be separated and, in most such data, groups of years are taken together to make an age class. This hides annual variations in mortality and recruitment; the disparity in Fig. 20/13c between age classes 40—60 and 60—80 presumably hides even bigger disparities between individual years. It is usually necessary to take age classes with a much larger group interval if a forest is very long lived and the sample size is small.

A population of coastal redwoods (*Sequoia sempervirens*)

The coastal redwood is long lived; the age-structure of a "virgin" redwood population of 12 ha (Fritz, 1929; Roy, 1966) included one individual

1380 years old. The age-distribution of this forest is shown in Fig. 20/14 but the picture is biased against the younger age classes since the sample did not include trees less than 30 cm in diameter (there were apparently as many trees less than 30 cm in diameter as there were larger individuals).

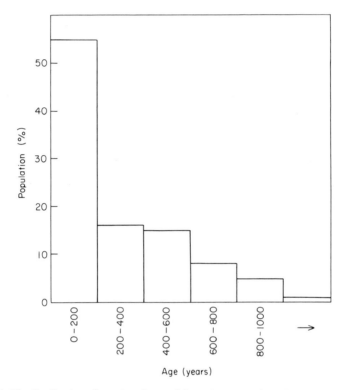

Fig. 20/14. The distribution of ages in a forest of *Sequoia sempervirens*. Data did not include trees of < 30 cm. d.b.h. although about half the population was in this category. (From Fritz, 1929)

An attempt was made to model the population dynamics of the redwood using a matrix derived from this age-structure (Bosch, 1971). Unfortunately there were serious errors in the use of this procedure (Brussard *et al.*, 1971; Halbach, 1971, Diem and McGregor, 1971) which was applied to predict the effects of commercial felling on the future life of the forest. The data were misleadingly used to suggest a policy of felling juveniles and matures while leaving old trees. It was unfortunate that one of the first serious attempts to use age-structures in a modelling process was faulty, both biologically and mathematically. Ecology, if it is to gain maturity as a science, must become predictive and the use of matrix methods gives one clear way in which predictive ecology can develop (see e.g. Usher, 1966).

The coastal redwood is one of the many tree species that can regenerate from stump sprouts. In one locality stumps produced an average of 72 sprouts from the bank of buds normally dormant at the base of the trunk. The age of a redwood trunk may therefore not be the age of the genet, which may be very much older. Bosch (1971) says that "typically 8 per cent of new trees arise from stump sprouts" though it is not easy to see how such a figure can be arrived at or how much variation there is between different types of redwood stand.

Reconstruction of the history of a "virgin forest"

The dynamics of a forest can sometimes be reconstructed for longer periods than the life of the surviving trees. Wood fragments can be used to estimate the diameters and ages of trees long dead. When a living tree is growing on the remains of a dead one it is sometimes possible to determine when the dead tree fell. A rare example of an old growth forest stand, undisturbed by man, was studied in south-western New Hampshire, U.S.A. (Henry and Swan, 1974). One plot of 0.04 ha was studied in great detail to include the place and time of germination and the diameter growth of every tree more than 15 cm tall. The ageing was done by taking cores at 23 cm above ground level and applying a correction factor so as to obtain the number of annual rings at ground level. (The correction factor for a 23 cm difference in height was 6 years in the case of *Tsuga canadensis*.) The width of the annual growth increments was used to discover when each tree was suppressed or released. Fallen trees were mapped to record the direction of fall: if trees are felled by a hurricane groups fall at the same time and in the same direction making it possible to correlate time of fall, direction of fall and the time of known hurricanes. Each stem was treated as a special case and all the available evidence was pieced together. For example, if the date of fall of one tree is known, all down stems beneath it must have fallen before and all stems above must have fallen later.

The reconstructed history of this forest fragment shows that it was destroyed by two catastrophes, a fire in about 1665 and a hurricane in 1938. The fire in 1665 may have developed after an earlier catastrophe

as two of the stems of trees present in 1665 had wide growth rings near
their centres and narrow rings further out; this suggests that they made
their early growth in released conditions and were suppressed later.

Five old conifers blown down in 1938 had charcoal beneath their
place of germination and had germinated in 1665, 1673, 1677, 1678
and 1687. All the old growth of *Pinus strobus* in the area had germinated between 1665 and 1687. The fire had initiated a completely new
forest. A hurricane had left a storm track in the region in 1635 and
much of the forest may have fallen before the fire. A series of later
hurricanes left their mark in the remains of fallen trees (each hurricane
causing a fall in a characteristic direction) in 1898, 1909 and 1921. The
final destruction of the forest was by a hurricane in 1938 when 55% of
the old growth canopy was of trees that had germinated between 1665
and 1702.

The composition of the forest plot in 1967 is shown in Fig. 20/15 together with the times of the most recent hurricanes. The conclusion from

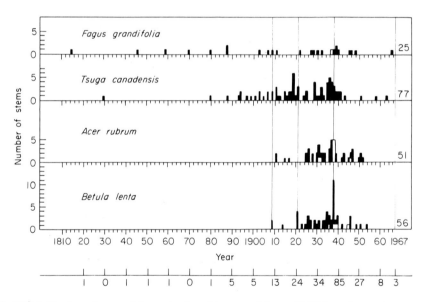

Fig. 20/15. Number of successful germinations (those surviving to 1967) in each year for the
four common tree species on a one-tenth-acre (0.04-ha) forest plot, Harvard Pisgah
Tract, SW New Hampshire. The numbers beneath the years are the number of stems
established in the preceding decade (e.g., 34 stems were established 1920–1929,
85 stems 1930–1939). To the right of each histogram is the total number of stems/
species. White areas in histogram are sprouts, black areas seedlings. (From Henry and
Swan, 1974)

this study is that, far from representing a climax or near-climax community, this old forest was always recovering from the last calamity.

The virgin forest

The concept of a virgin forest varies between authors but is usually taken to be one undisturbed by major disasters* (felling, fire, hurricane, disease etc.) and in which the unrelenting processes of natural succession have produced some sort of asymptotic condition. The ecological ideal of a climax community stimulated a search for vegetation types that might, more or less nearly, be regarded as virgin. In practice, as in the forest of New England described above, recurrent disasters are common in some regions and the so-called "virgin" stands are in fact recovery stages from the most recent disasters. Nevertheless, forests have been studied in which the idea of a virgin stand seems to be vindicated. In an analysis of the forests of the northern and southern Appalachian mountain systems of North America multi-aged populations are found. These forests are dominated by *Picea rubens* and *Abies balsamea* (in the north) and *P. rubens* with *Abies fraseri* (in the south). In general the largest trees in these forests are the oldest, though not always. Individuals of shade-tolerant species that had been overtopped lived 50 years or more (cf. *Pinus taeda,* an intolerant species) and when "released" by the death of an old tree, grew very rapidly. Thus trees of the same size were often of widely differing ages and, even in the overstorey, the age of individual trees differed by as much as 200 years (Oosting and Billings, 1951). Unfortunately, the existence of a mixed age-structure does not mean that a forest is virgin; it may have passed through phases in which some individuals have survived a disaster that killed others; then some old trees will persist in a regenerating population. It is indeed rather difficult to know by what characteristics a virgin stand might be recognized. Spurr (1964) took the attitude of an extreme sceptic:

> . . . we conceive of the virgin forest as being simply an unharmed old-growth

*I distinguish (as the ends of a continuum) disasters and catastrophes. A disaster recurs frequently enough for there to be reasonable expectation of occurrence within the life cycles of successive generations — e.g. hurricanes at *ca* 70 year intervals are "disasters" in the life of a forest tree and the selective consequence may be expected to leave relevant genetic and evolutionary memories in succeeding generations. A "catastrophe" occurs sufficiently rarely that few of its selective consequences are relevant to the fitness of succeeding generations. The selective consequence of disasters is therefore likely to be to increase short-term fitness and the consequence of catastrophes is to decrease it.

forest. Such stands simply do not exist. . . . Interferences with normal growth and development are common, and it is a meaningless semanticism to try to distinguish between "natural" disturbances and "artificial" disturbances caused by man. To the tree, it makes little difference if a fire is set by lightning or by a human incendiary, if the soil is upturned by a plough or by a tree uprooted in a storm, if a leaf is eaten by an insect who came by wind or one that came by airplane. . . . Disturbance to tree development and growth are normal, instability of the forest is inevitable, and the changeless virgin forest is a myth.*

There is little in the study of the population biology of trees that supports the idea of stable asymptotic systems. Rather it would seem that disasters are the main determinants of community structure, or if not disasters, then bad patches in a sea of favourable conditions, good patches in a sea of bad. One of the consequences of the development of the theory of vegetational climax has been to guide the observer's mind forwards. Vegetation is interpreted as a stage on the way to something. It might be more healthy and scientifically more sound to look more often backwards and search for the explanation of the present in the past, to explain systems in relation to their history rather than their "goal".

There is an intriguing parallel between the history of population biology in the hands of zoologists and botanists. A great argument raged amongst zoologists as to whether the numbers of animals were determined by the stress of limited resources (Nicholson, 1933; Andrewartha and Birch, 1954) or by the time lapsed since the last disaster. The parallel argument about plant succession is that communities represent the working out of the effects of past disasters *or* stages in approach to a climax state. The disaster theory is put strongly in the writings of Raup (e.g. 1957) and Olson (1958) and the climax theory by Clements (1916), Oosting (1956) and Daubenmire (1952). For both arguments (in both kingdoms) the stated positions are extremes. The important question is, do we account most completely for the characteristics of a population by a knowledge of its history or of its destiny? In forest systems it seems that we are unlikely to find communities so old and in such stable environments that the effects of the last disaster have been obliterated. It may be that, when more is known of the population dynamics, and particularly the age-structures, of trees, it will become possible to define the state of a stand much as one can define the state of an ideal population of laboratory organisms, where the transition from r- to K-dominated processes can be expressed formally in mathematical terms (see Chapter

*Stephen H. Spurr — "Forest Ecology". Copyright © 1964 The Ronald Press Company, New York.

1). A comparable measure for the development of a community might be:

$$\frac{dC}{dt} = s \left(\frac{C' - C}{C'} \right)$$

| The rate of change of community C | = | the early rate of unimpeded succession | (a function of (the climax state (function of the climax state | — | a function of) community C)) |

Such a concept demands a measure of rates of change in communities and this would need to take account both of changes in species and of their age-structure. The function needed must be quantitative and is unlikely to be anything as simple as productivity or biomass.*

The dynamics of some disaster-prone populations of trees

Many of the hazards that affect the life of trees do not really count as disasters. A plague of seed-eating mammals, birds or insects may prevent regeneration for a year or two, but be relatively unimportant in the life of a tree that may live for 200—1000 years. The most serious hazards are those that kill a tree which may have taken 1—200 years to reach maximal reproductive activity. Fire, hurricane and devastating plagues of defoliating pests and diseases are probably the major disasters of forest systems.

The Canadian Forest Service can apparently find no forests that are older than 150 years; fire is the catastrophe that ends the life of each forest. Fires may vary in frequency in different areas and 80—90% of the "virgin" forest of Minnesota can apparently be traced to a post-fire origin, the oldest stand dating back to 1595 (Heinselman, 1971). Where fire has been a repeated part of the hazards of life there have been evolutionary consequences. The evolution of serotiny in pines ensures that seed is stored in cones on the tree and only released to germinate after a fire. A hazard has become a requirement. (This is another evolutionary parallel to the evolution of sulphonamide-requiring strains of bacteria among populations of pathogens that have repeatedly suffered disasters from sulphonamide application!) Many species of *Eucalyptus* appear to need fire to stimulate regeneration. *Eucalyptus* spp. are generally tolerant of fire but they also produce abundant fuel as fallen leaves and bark that burns readily; large areas of forest are burnt regu-

*Horn (1976) and others have considered models of forest successions as Markov chains — predictable from present probabilities of replacement and independent of history. The models prove very effective predictors in many cases, thus removing some of the mystique surrounding the Clementsian concept of succession.

larly without harm to the trees. The regeneration of *E. regnans*
is apparently dependent on fire (Gilbert, 1959) and the regeneration of
eucalypts with wet sclerophyll understoreys apparently also needs fire
(Ashton, 1956; Cunningham, 1960). Forests of these types are invariably even-aged to step-aged and each age group probably represents a
historical episode of fire (Mount, 1964).

Eucalypts, like many other tree species of hazard-ridden environments, regenerate easily from basal buds. The lignotuber, a buried mass
of stem tissue, is capable of putting out repeated crops of shoots after
damage and in the low shrubby eucalypts known as "mallee" this form
of regrowth may be more significant than seedlings as a means of recovery after fire. The regenerative capacity of eucalypts is extraordinary:
1—1.5 year old seedlings of some species bear lignotubers that can survive 26 successive removals of all leaves before they die (Chattaway,
1958). The ages of the shoot systems can be determined in a mallee
community though it would be very hard to age lignotubers and obtain
the ages of genets. Two sites were selected in a mallee vegetation that
had not been grazed commercially for 25 years (Holland, 1969), one
site containing one species, *E. incrassata* and the other containing two,
E. dumosa and *E. oleosa*. The study was designed to measure productivity in the communities but, as part of the programme, ages were
determined and age class distributions were fitted statistically (Fig.
20/16). The site containing *E. dumosa* and *E. oleosa* was believed to
have been burned about 60 years before the study was made. The age-
class distribution of the shoots was uneven and the calculated mean
age of the shoots of *E. dumosa* was about 34 years and of *E. oleosa*
36 years. Occasional shoots had evidently escaped the fire and one shoot
of *E. oleosa* was about 105 years old. In this population the wave of ages
obviously does not represent a sudden regrowth after the fire but a
build-up of a population of shoots over a number of years after the fire:
then the formation of new shoots was either inhibited or they died
when very young in the established dense stand. The site that contained
only *E. incrassata* had been burned during the summer of 1945—46 and
a limited number of ring counts was made. Of the 15 stems counted 11
had between 18 and 21 growth rings, suggesting very rapid re-establishment.

A disastrous defoliation can have repercussions right through the
age structure of a forest. In spruce—fir forests, after severe defoliation
by the spruce budworm (*Choristoneura fumerana*), seedlings of *Abies*

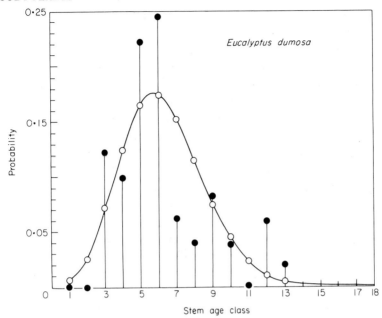

Fig. 20/16. The age distribution (●) and the calculated probabilities (○) for the age classes of *Eucalyptus dumosa*. The area was burned approximately 60 years before the measures were taken. The values are for stems — most or all of which represent regrowth from lignotubers. (From Holland, 1969)

balsamea establish in great numbers (Ghent, 1958), outnumbering the spruces, *Picea mariana* and *P. glauca,* by from 10 to 200 times. A population of regenerating plants was aged by making ring counts at ground level and the age frequency in the different height classes is shown in Fig. 20/17. The seedlings present in the forest after the budworm attack were not new colonists but an accumulation of seedlings that had been building up for 20–30 years! When a canopy was damaged by the budworm the effect was to release this large population of suppressed seedlings and the regeneration depended on this reserve population that was lying in wait. Not only is fresh seed unimportant in regeneration; seed production was in fact drastically reduced by the budworm attack and female cones were not produced.

Before the attack by budworm the balsam seedlings had been growing extremely slowly, about 2.5 cm per year. The seedlings continued to be suppressed during the epidemic because the larvae of *Choristoneura* damaged the leading shoots. As soon as the epidemic was over and the

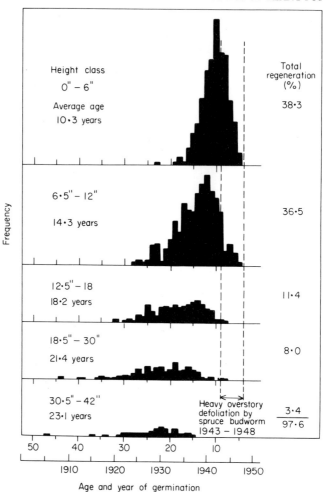

Fig. 20/17. The distribution of ages within different height classes of balsam fir seedlings (*Abies balsamea*) regenerating after a severe defoliation of the overstorey by spruce budworm (*Choristoneura fumerana*). (From Ghent, 1958)

overstorey had started to die, the seedlings grew rapidly.

The strategies of forest regeneration after disasters are quite clearly different — depending both on the nature of the disaster and the species.

(i) *Regeneration from locally produced seed* — There are few forest trees that have significant seed longevity — as a rule tree seeds do not accumulate in a seed bank in the soil (see Chapter 4). There are a few specialized exceptions. The black cherry (*Prunus serotina*) appears to

be dependent on a buried seed bank for its persistence in forest and
the serotinous seeds of pines accumulate over many years, though on
the parent tree, not in the soil. In general, regeneration from locally
produced seed seems to depend on the disaster following fairly closely
on an act of seed dispersal so that fresh seed is available for immediate
establishment of a seedling crop (as in the case of Eucalypts).

(ii) *Regeneration from seed dispersed from elsewhere*. This is the type
of regeneration described for *Pinus taeda* and produces age distributions
that are a function of the proximity of seed parents (Figs 20/3 and 4).
The effectiveness of this type of colonization depends, of course, on the
dispersal powers of the species and after a catastrophe the speed of re-
colonization may be limited simply by the absence of dispersed seed.
Mora, a tropical forest tree with cannon-ball-like seeds, may still be
expanding its range from the time of its evolution (!) and clearly is
poorly fitted to recolonize after a catastrophe.

(iii) *Regeneration from dormant buds*. Tree species vary greatly in
the extent to which they accumulate a reserve of dormant buds near
the base of the trunk. The red maple (*Acer rubrum*) develops stump
sprouts from both buried and external buds within 6 weeks after the
main stem is cut. In this case the buds have accumulated over many
years and have grown with the growth of the stem and bark, remaining
protected but with continued vascular connection to the main stem.
In the Harvard Forest (Massachusetts, U.S.A.) the great majority of red
maple trees have originated from stump sprouts (Wilson, 1968). In
Eucalyptus spp. the lignotuber is a specialized bank of reserve buds and
in clonally growing species such as *Populus tremuloides* the ability to
form shoot buds on the root system serves the same end, providing
means for the regeneration of a genet as well as extending its range. In
these cases we do not know true age-structures of genets.

(iv) *Regeneration from a reserve of established seedlings*. Seedlings
may establish and remain suppressed (e.g. balsam fir) until release after
a disaster. This may be highly effective after a hurricane or disease
has damaged the canopy but is obviously ineffective as a means of regeneration after a fire. The reserve population of seedlings that wait,
suppressed, for release may derive from a phase of seedling establishment long past. In a 70-year-old population of *Populus tremuloides*
with balsam fir as sub-dominant, the fir had been producing flowers
for 37 years (including some non-flowering periods and a serious attack
by budworm). The estimated seed production during this period was

19 500 000 seeds per ha (55 seeds per m² of forest floor per year). This
seed fall had produced about 12 000 seedlings per ha, most of them less
than 120 cm high (i.e. 1 seedling per 1500 seeds). 67% of these seedlings
had established during the first 19 years of the flowering history of the
firs (when only 28% of the seed production had occurred). During the
early period of reproduction, one seedling must have become established
for every 600 seeds produced and the peak of seedling establishment
had passed by the time the forest was 50—55 years old. The regenerating
potential of this forest depended on an ageing residue of tiny old
"seedlings" (Ghent, 1958).

Darwin noticed that a population of very tiny pine "seedlings" could
remain alive for years though repeatedly grazed in a pasture — but serv-
ing as a source for very rapid regeneration as soon as the grazing animal
was removed (Darwin, 1859).

The age structure and dynamics of forest systems — a resumé

The examples of forest age-structures that have been studied are not
representative of the forests of the world. We know next to nothing
about the dynamics of systems in the tropics where annual rings are
not produced. For the forest systems that we do know something
about, some generalizations (mainly negative) begin to emerge.

(i) It is wholly unrealistic and very dangerous to assume any relation
between the size of trees and their age, other than the vague principle
that the largest trees in a canopy are likely to be old. However, it cannot
be argued conversely that small trees are likely to be young: they may
be as old as the main occupants of the canopy. If a tree is very young
it is likely to be small, but if it is small it may be any age.

Even-aged populations of trees develop skewed distributions of size
with the passage of time and it is easy to imagine (quite wrongly) that
such a size distribution represents ages.

(ii) The age-structure of a population of trees may reflect the distance
of the population from a seed source.

(iii) Once an even-aged population of trees has become established
it is difficult for new age classes to enter the canopy until the colonists
die from accident, disease or old age. The new age classes may accumulate as an ageing population of very small plants in the case of so-called
"tolerant" species, or may be continually recruited and die young in
the case of intolerant species. It may be very difficult to determine true

ages for trees that have spent a long period in the suppressed stage because the height at which it is convenient to estimate ring number will
usually miss all the early rings.

(iv) After a wave of colonization, a forest may be dominated by a
restricted age class. New entrants to the canopy depend on the death
of this class and the next wave of recruits is likely to be less clear. Over
generations, the after-effects of the first wave may become blurred but
presumably a tree species that spends 250 years in the canopy after its
first invasion will need at least four generations before the secondary
waves have disappeared entirely from its age structure.

(v) It is an open question what sort of age structure we might expect
in an "equilibrium" forest. Some evidence suggests that death risks become constant with age and that the numbers in successive age groups
decline exponentially. No mature forest has yet been studied that reveals a situation that can be interpreted as equilibrial; even old "virgin"
forests reveal phases of recruitment or mortality that suggest past
disasters.

(vi) The life of a generation of trees is long compared to the evolutionary time scale of its major enemies. It is extremely doubtful (even if
climate remains unchanged) that the biotic components — the pests and
predators — remain genetically invariant. The life of most forests appears
also to be short compared to the frequency of major disasters such as
fire and hurricanes and even to the time scale of climatic shift.

(vii) The grouping of age classes hides year-to-year variations in mortality and recruitment. For plants that live long it is essential to have
sample sizes that adequately cover the structure of the population. A
dominant tree is a very large patch in an environment and it controls
the fate of potentially dense populations of seedlings beneath it. The
age structure of trees in a forest is the sum of the age structures in the
various mosaic patches of the forest and the mosaic must be adequately
sampled. Descriptions of the mosaic may indeed be a better clue to the
dynamics of the system than an overall average. The idea of a forest as
a mosaic of regeneration cycles (Watt, 1947) may give a clue to more
efficient methods of determining dynamics. In the past the emphasis
has lain on what happens in an area of land — a convenient but perhaps
misleading abstraction for the population biologist. A "plant's eye view"
of a forest might best be represented as a spectrum of the experiences
of individuals: what life is like in the various phases of the mosaic. Of
all the studies of the population biology of trees reported in this chapter,

perhaps the most completely revealing are the "plant's eye views" of *Pinus taeda* and of holly. Both involved marking and following the fate of individuals. Any population biology that has hopes of explaining and predicting events must record the life of individuals: the statistics come later.

Shrubs

Shrubs are, for the most part, as difficult subjects for population biology as the most difficult trees. Those that have clonal growth present the worst problems because like "mallee" eucalypts, though usually on a smaller scale, they may have two quite different age structures — that of the above-ground shoots and that of the underground system. There may indeed be three different age-structures, for the age of the below-ground parts may not represent the whole life of the genet. Those shrubs that grow clonally pose problems to the demographer as acute as long-lived clonal herbs. The great majority of shrubs bear berries and the seed is dispersed in clumps by birds or small rodents. A shrub may grow up from a patch of seedlings and represent several genets. It is possible to measure the age of individuals only when the shrub has the single-stemmed habit of a miniature tree and unfortunately the smaller shrubs (e.g. *Thymus vulgaris*) do not form clear annual rings (it is annoying that many of the most interesting communities of dwarf shrubs in maritime and desert areas (e.g. *Salicornia* spp.) have the chenopodiaceous habit of anomalous cambial growth and cannot be aged).

Three detailed studies have been made of the population dynamics of shrubs in very contrasted communities: a study of *Larrea tridentata* the creosote bush in San Simon Valley, Arizona (U.S.A.) (Chew and Chew, 1965), of *Vaccinium myrtillus,* the bilberry, in Sweden (Flower-Ellis, 1971); and of *Symphoricarpos occidentalis* in Minnesota (Pelton, 1953).

The populations of *Larrea* existed in a mosaic of desert shrubs including *Flourensia cernua, Prosopis—Acacia* and *Yucca elata* with *Opuntia* spp. in an open cattle range. The study plots were protected from livestock in 1958. The age of a number of plants was determined by counting rings and the relationship between age and shoot volume was calculated (see Fig. 20/18). Subsequently, age structures were determined by measuring shoot volume and determining age from the correlation. Clearly

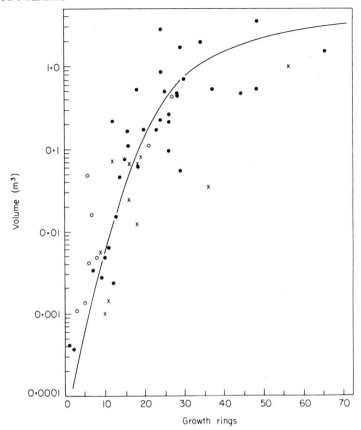

Fig. 20/18. The relationship between shoot volume and age in *Larrea tridentata* (creosote bush) in San Simon Valley, Arizona. (From Chew and Chew, 1965)

the relationship is untidy; for the original purposes of calculating the productivity of the community it was probably quite adequate but the variation is too great for accurate demography. Shrubs of the same volume may differ in age by 25–40 years, and shrubs of the same age may differ 60-fold in volume. Nevertheless even crude estimates of the age-structure are very welcome for a type of plant that has rarely been studied demographically. The age-structure calculated for *Larrea triden-tata* is shown in Fig. 20/19. The interpretation placed on this age-structure is that the population of *Larrea* rather slowly began to invade the area in 1893; the rate of population growth suddenly increased in the period 1940–50 mainly at the expense of *Flourensia* and then declined

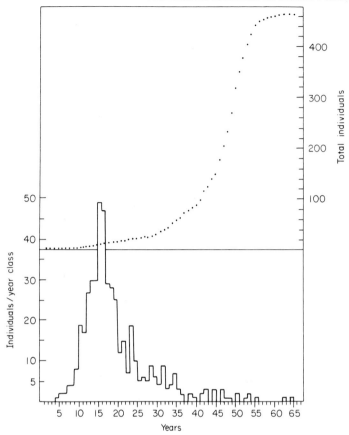

Fig. 20/19. The age frequency distribution of a population of *Larrea tridentata* in San Simon
    Valley, Arizona, based on measured shoot volumes and the correlation with age
    shown in Fig. 20/18. The calculated cumulative survivorship is superimposed. (From
    Chew and Chew, 1965)

equally fast. Classes of 1–4-year-olds were apparently absent. Observations of very old communities dominated by *Larrea* suggest that the
population might ultimately settle down to a "stable" density of *ca*
1000 plants per ha.

   Symphoricarpos occidentalis represents a very different picture of
structure and population dynamics. A spreading system of rhizomes
bears aerial shoots 75–110 cm height. The rhizomes live for up to 20
years and (like the shoots) can be aged quite accurately. The shoots in
a mature population have an average age of about 7 years. The rhizome
system remains undecayed through the life of the genet but the ramifi-
cations are so complex (see Fig. 20/20) that it would be a dreadful task

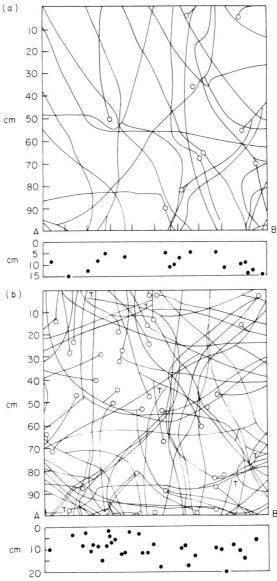

Fig. 20/20. The structure of the rhizome system of *Symphoricarpos occidentalis* within areas of 1 m², (a) a moderately open colony and (b) a very dense colony — both on loamy sand. The sections below each map show the depth of the rhizomes in the section A–B.

○ = live erect stem

● = dead erect stem

T = rhizome tip (From Pelton, 1953)

to disentangle it into components. Colonies develop an age-structure with the oldest rhizomes at the centre of a zone of spread (Fig. 20/21). Fire kills the shoots but new ones are regenerated from the rhizomes. Thus

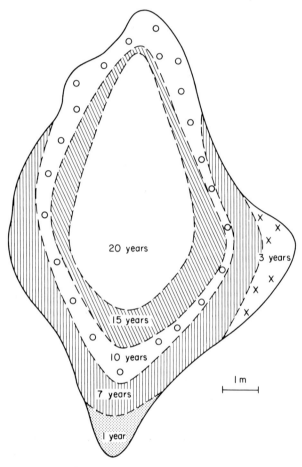

Fig. 20/21. The age structure of a spreading colony of *Symphoricarpos occidentalis*. The years refer to the oldest rhizomes found within a given zone. (From Pelton, 1953)

the age-structure of the shoot system reflects past disasters (Fig. 20/22). *Symphoricarpos* is an invader of open woodlands where it forms a shrub layer, of chapparal and of disturbed habitats. It is particularly successful as an invader of grassland and provides a shrubby environment in which pioneer trees such as *Populus tremuloides* can establish.

A woody shrub that grows clonally will be a complex of age-structures —

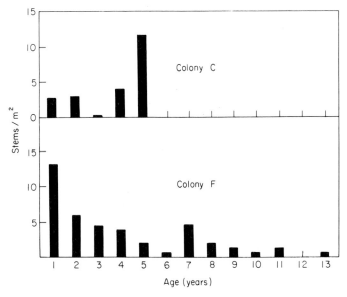

Fig. 20/22. The age distribution of stems of *Symphoricarpos occidentalis* in two nearby colo-
    nies one of which (C) had been burned five years before and the other (F) had
    escaped the fire. (From Pelton, 1953)

the distribution of ages found within the rhizome system may be quite
different from that found in the aerial shoots; the age-structure of the
viable buds may be different from that of rhizomes and shoots and the
leaf population will have yet a further different age structure. The most
complete analysis of such a system has been made for *Vaccinium myrtillus* (Flower-Ellis, 1971), in Swedish forests and in eastern Scotland. The
age of aerial shoots and of rhizomes could be determined by counting
annual rings, though at the end of its life a rhizome may persist for a
few years without laying down a ring. The structure of expanding clones
is very similar to that of bracken (*Pteridium aquilinum*) (see Chapter
19) with an advancing front leaving behind it a mosaic of growth phases.
This is in contrast to another low-growing ericaceous shrub, *Arctostaphy-
los uva-ursi,* in which growth is regularly centrifugal and in which decay
spreads regularly from the centre. The clones of *Vaccinium* expanded
radially at about 7 cm per annum, though some rhizomes grew much
faster. The rhizomes persisted longer than the aerial shoots so that at
one site the oldest rhizomes were 28 years and shoots were 10 years
old. Occasionally very old shoots were found — one of 34 years old
was still vigorous.

The older parts of the rhizomes of *Vaccinium* eventually die and decay and, although the age of living rhizomes can be measured, the length of life of the genet of which they are the parts can only be estimated. This can be done by assuming a constant rate of diameter increase by the clones. Individual clones were estimated in this way to be 40, 50 and 70 years old. Clones grow into each other's territory and intermingle but where a single clone can be studied in isolation there is a very clear relationship between productivity and the age of a rhizome unit (age is here taken to be the age of the rhizome, *not* the age of the genet) see Fig. 20/23. This curve is readily interpreted as an initially

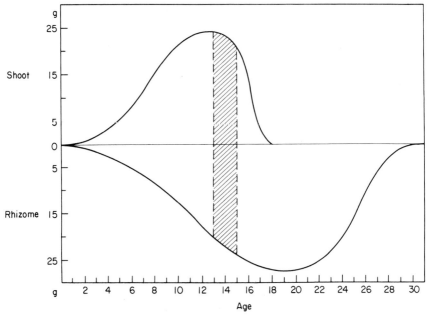

Fig. 20/23. Diagram of the changing productivity of a rhizome unit of *Vaccinium myrtillus* with age. The "rhizome unit" is not a genet but a length of rhizome without any remaining physical connection to a possible parent rhizome. (From Flower-Ellis, 1971)

vigorous assimilating system that accumulates an increasing respiratory burden of accumulating underground structures (cf. Figs 10/3 and 10/4). The aerial shoots achieve their maximal standing crop 8 or more years before the rhizome system.

The age structure was determined for several populations of aerial shoots (Fig. 20/24). There is apparently an increasing probability of death in the first 4 or 5 years and subsequently the death risk becomes

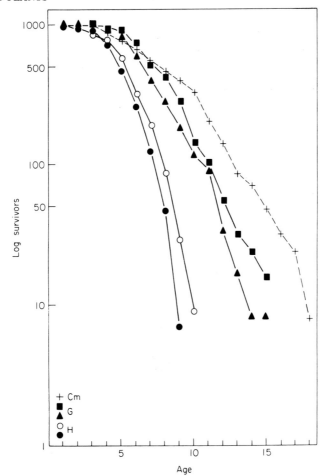

Fig. 20/24. The age structure of five populations of *Vaccinium myrtillus* based on the annual
ring counts of aerial shoots. Cm = Cammachamore, Scotland, G = Garpenberg,
Sweden, and H = Hamra, Sweden. (From Flower-Ellis, 1971)

almost independent of age. It is important to remember that this
conclusion infers the survivorship curve from the age distribution and
describes the parts of genets; it is not a survivorship curve of genetic
individuals.

Woody shrubs illustrate better than any other group of higher plants
the way in which the modular structure of plant form contributes to
determine the population dynamics, both within and between genets.
The interests of the population biologist might indeed catalyse a resurgence of interest in some parts of classical morphology.

Plants, Vegetation and Evolution

21

Reproduction and Growth

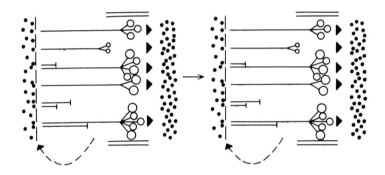

On average each plant during the course of its life leaves one descendant – one replacement. If this were not true the earth would soon be bare or covered with the progeny of a single plant or pair. Yet despite the general 1:1 ratio between parents and offspring we find a fantastic variation in the range of reproductive capacities of plants. In the course of its life a desert annual may produce 10^{0-1} seeds, a weedy annual grass (e.g. *Alopecurus myosuroides*) about 10^2 seeds, about the same number as the double coconut tree (*Lodoicea seychellarum*); and a coastal redwood tree will produce 10^{9-10} seeds, about the same number as a small herbaceous orchid! If the plant is only the means by which a seed produces more seeds the variations in efficiency of the process are extraordinary. The meaning of this vast variation in reproductive capacity is

647

a problem at the heart of any science of population biology and of the study of evolution. A great deal of analysis of the meaning of reproductive capacity has been done by ornithologists who have asked questions about the variation in clutch size in birds (e.g. Lack, 1954). Plants pose the problems even more sharply and are easier to study.

Three types of explanation are commonly given for variations in reproductive capacity, two of which are certainly wrong.

(a) *Differences in reproductive capacity reflect differences in the hazards of life.* In a crude sense this is true; if organisms with a large reproductive capacity, and those with small, both leave on average one descendant it follows that the chance of survival is greater among the progeny of the organism with the lower reproductive capacity. Where this interpretation is wrong is in its extension to the thesis that the reproductive capacity has evolved to match the extent of the hazards. Darwin (1859) was wrong when he wrote, "the real importance of a large number of eggs and seeds is to make up for much destruction at some early stage in life — and this period in the great majority of cases is an early one." The objections to this argument are, firstly, that all organisms that we know have had successful ancestors. Among the ancestors will be those that have experienced the worst calamities in the history of the species; their reproductive capacity proved sufficient for them to leave descendants. The objection can be phrased in a different way: it is impossible for an organism to evolve to a state in which it leaves enough offspring to replace itself from a condition in which it did not leave enough.

The second objection is that if the reproductive capacity of individuals has been adjusted by natural selection to balance the hardships of life we would expect ecotypes at the margins of a species' distribution to have a greater reproductive capacity than those at the more favourable centre. This is clearly not the case. Conversely, if the reproductive capacity at the margins of a distribution are adequate to replace the population, the reproduction at the centre of the distribution must be more than enough.

(b) *Differences in reproductive capacity are responsible for the differing abundances of organisms.* The objection to this view was put by Darwin (1859):

> The only difference between organisms which annually produce eggs or seeds by the thousand, and those that produce extremely few, is, that the slow breeders would require a few more years to people, under favourable conditions, a whole

district let it be ever so large. The condor lays a couple of eggs and the ostrich a score and yet in the same country the condor may be the more numerous of the two; the Fulmar petrel lays but one egg, yet it is believed to be the most numerous bird in the world.

It is easy to give comparable examples from the plant kingdom. The two rarest plants in the native British flora are *Cypripedium calceolus* and *Orchis militaris* yet they are in the class of those producing the most seeds. The abundance of a species has more to do with the abundance of habitable sites and the genotypic and phenotypic plasticity that permit a wide range of such sites to be occupied — it has nothing to do directly with the reproductive capacity of plants.

(c) *The third explanation of variations in reproductive capacity depends on Fisher's "Fundamental theorem of Natural Selection".* Natural selection operates on individuals in a population and selectively favours those individuals and those genotypes that contribute most descendants to subsequent generations. This process leads remorselessly to an increase in the population of those forms that contribute more descendants than their neighbours. The critical word in this argument is descendants — the contribution of parents to subsequent generations. *It does not follow that an organism that produces a large number of progeny will also leave a large number of descendants.* On this argument the crucial driving force in evolution is not a trend of "the species" towards some desirable end, it is not a trend to some optimal behaviour on the part of the species and its individuals — it is the never-ending game of leaving more descendants than others in the population. Variation in the number of progeny or seeds produced is just *one* of the ploys in this game. If a genetic change in one member of a population increases its number of progeny, leaving other attributes unchanged, then natural selection is bound to favour it over less fecund neighbours. But it is almost unimaginable that any genetic change that altered the number of progeny could occur while other features remained "equal". An increase in the number of progeny might be gained at the expense of their size — this need not increase the chance of leaving descendants and would often reduce it. An increase in the number of seeds left by a forest tree might be gained at the expense of a slower rate of growth in height, but again this would be unlikely to increase the number of descendants left by the tree because it would tend to lose in a struggle for existence with neighbours to reach the canopy.

* "The rate of increase in fitness of any organism at any time is equal to its genetic variance in fitness at that time."

It is important to remember that "fitness" is not an absolute measure — it describes the relative numbers of descendants left to future generations by one form compared with others. For this reason a plant might increase its fitness by reducing its number of progeny if a compensating increase in its vegetative vigour deprived its neighbours of the chance of leaving descendants.

The explanation of the differences in fecundity of organisms (on an interpretation based on considerations of fitness) is that the numbers of progeny produced by an organism is only one component of success in natural selection and that an emphasis on one component is likely to involve compromises in others. The variations in reproductive capacity that we see represent different adaptive compromises. Natural selection acts to *optimize* the form of the compromises in such a way as to *maximize* individual fitness.

If heritable properties that permit some individuals to leave more descendants than others are regularly favoured in evolution, it is not surprising that the problems of overpopulation continually recur in every living system. This is the reason why so much of this book has been devoted to the effects of density. It is doubtful if natural selection acting on individuals can ever do anything to redress overpopulation because even in a famine-stricken overpopulated environment the genotypes most frequently represented in the descendants will dominate the future character of the population. The conditions required for group selection to occur are extremely limited (see e.g. Maynard Smith, 1964; Williams, 1966; Lack, 1966) and there is rarely if ever any reason why what is good for the individual should also be good for the population, for the community or for the species (see Chapter 24). There is also no reason why what is 'good' for the population, the community or the species should necessarily be good for the individual. Yet to interpret evolutionary forces we have to look at the individual as it is he that transmits the genotype. It is an intriguing paradox that, on average, a plant or animal leaves but one descendant yet natural selection continually favours individuals that leave more.

Some of the compromises that are involved in the evolution of reproductive capacity arise because the seed is much more than the means of multiplication. A variety of other roles have become attached to the seed phase in the life cycle.

(i) The seed is usually the stage of the life cycle at which dispersal and the colonization of new areas occurs. It may not be a very effective means of dispersal in some species, e.g. the cannon ball-like seeds of some tropical leguminous trees (Richards, 1952). Other parts of the life cycle may be involved in dispersal such as the parts of fragmented genets which are dispersed as bulbils (e.g. *Allium*) or the rootable fragments of aquatic plants. In general, however, it is the seed that is relied on for dispersal.

(ii) The seed is often the organ that persists through recurrent unfavourable seasons (particularly in the case of annual plants) and through irregular disasters such as fire (for perennial plants). It does not always have a perennating role (the seeds of most tropical rain-forest trees have very short viability), but in most plants the seed is a perennating organ.

(iii) The seed usually contains a reserve of food which gives the embryo a temporary continuation of its maternal support both before dispersal and during the post-dispersal period while it establishes as an independent seedling. It may not be a food storage structure — the seeds of orchids and of parasitic flowering plants such as *Orobanche* are little more than a few dried cells, but the seed is usually a food storage structure and is therefore often a choice food object.

(iv) The seed is usually the stage at which new genetic recombinants are released. The sexual process precedes the formation of the seed and some recombinants are tested while on the placenta of the parent ovary — just as some unsatisfactory recombinants are eliminated from a population of mammalian embryos when they are still on the mother's placenta — but most of the selective elimination of new genotypes occurs after "birth" and the seed is the carrier of most of the genetic variation found among zygotes.

The four responsibilities, dispersal, perennation, food supply and the display of genetic variation, are not wholly compatible with each other or with the fifth role of multiplication. It is impossible to be light enough to be readily dispersed and yet carry a large food supply. It is difficult for a seed to carry a large food reserve and also to perennate because the risks of predation become so great. It is impossible for a seed accurately to multiply the parent's genotype and at the same time be a genetic variant.

The production of seed is seldom wholly compatible with the vigorous growth of the vegetative plant, yet fitness depends as much on the survival of the plant in the vegetative phase of growth as on its ability to produce seed after it has survived. It will certainly be a fitter strategy

to produce few seeds and to reach maturity than to have the potential
of producing a vast number of seeds but to fail in a struggle for existence
with more vigorous neighbours in an environment of limited resources.

Variations in fecundity as the result of a partitioning of resources

It has been suggested that the process of individual development represents a "strategic" allocation of energy or time to conflicting ends.
Cody (1966) developed this idea to explain variations in clutch size in
birds where the number of eggs laid appears to be greater in those species
or forms that live in relatively hazard-free environments. Cody envisaged
natural selection operating to move the genotypic composition of a
population on an adaptive surface defined by a series of variables that
are partly mutually exclusive. He envisaged the three most important
components of fitness for a bird as: (a) reproduction, (b) the struggle
for existence with competitors and (c) the avoidance of predators.
Each of these adaptations requires the expenditure of energy and
time which are in limited supply. The principle of strategic allocation is
that organisms under natural selection optimize the partitioning of the
limited time or energy available in a way that maximizes fitness. Thus
a population of birds that finds itself on an island free from predators
may be selected to spend relatively more time or energy on competition
and reproduction because there is less benefit from spending it on predator avoidance. It is easy to see how such a principle might operate in
plants. Individuals in open colonizing environments will face little interference from neighbours; the chance of leaving descendants will depend
much more on high fecundity. However in a crowded community, such
as a meadow or a forest, an environment of limited resources, the "fit"
individual is one that pre-empts to itself a disproportionate share of
the resources. Such an individual depends for its success on such attributes as an early vigorous start in seedling growth (large seed reserves)
a vigorous and early root system, foliage produced earlier and higher
than that of neighbours. Under such circumstances an individual would
be likely to be favoured if it sacrificed some fecundity and freed resources to the other more vital contributions to fitness. In the same way,
if predators are an important element in the life of a plant population,
fitness may be increased by spending limited resources on repellent substances (pyrethrins, alkaloids, glycosides, terpenes, phenolics) or prickles
and spines, even though this leaves less to be allocated to vegetative

vigour and fecundity.

The concept of allocation depends absolutely on the idea that different structures or activities are alternatives, that a gain in one as a result of selection must be offset by a loss in another. There is evidence for and against this view. Much of the evidence is in the form of demonstrations that the physiological stimulation of one structure or activity may depress another in an apparently compensating fashion. It may be that this form of evidence (based on changes in phenotypes — tactics) is inadmissible in an argument about evolutionary change (strategies). What is possible as phenotypic variation may be no real indicator of what is possible through altered genotypes.

Flowering as an alternative to vegetative growth

In an essay on the art of prolonging the life of plants *"Über die Kunst, das Leben der Pflanzen zu verlängern"*, Molisch (1922) showed that some characteristically annual plants could be maintained in active vegetative growth if the flowering primordia were continually removed. He managed to maintain *Reseda odorata* for several years as a vigorous perennial that grew to a small tree, 2 m high, but as soon as it was allowed to flower and produce seeds it died. Species of *Agave* in their native Mexico grow vegetatively for 8—10 years then flower once and die but in the Riviera or in glasshouses in the north the same species may develop vegetatively for 20, 50 or even 100 years before flowering and dying. These experiments show that the juvenile period of monocarpic plants can be delayed and the plants then grow bigger but this is not quite the same as showing that vegetative growth and flowering are competitive processes. More convincing is the frequent evidence from perennial plants with photoperiodic requirements for flowering, that the prevention of flowering permits more active vegetative growth. It is not so clear that the prevention of vegetative growth increases flowering. It is still a matter for argument among physiologists whether patterns of allocation between parts of a plant are determined by competition between the parts of a plant and how often they are imposed by hormonal control; probably both mechanisms operate and interact. Questions about the evolution of the allocation of resources may not be answerable in terms of the physiology of allocation. Evolutionary change may alter hormonal instructions in ways that are not available to a plant physiologist.

Perennial polycarpic (iteroparous) plants often show an inverse correlation between vegetative growth and the production of fruit and seed which suggests that fecundity and vegetative activity are not wholly compatible. The fruit grower encourages fruiting by using vegetatively weak or dwarfing rootstocks. He also knows that after a particularly good crop he may expect one or two years of poor crop and that by reducing the load of fruit in a bumper year he can increase the chance of a good crop in the following year. In forest trees, "mast" years of high seed production are followed by years with a poorer seed crop. The mast year is marked in the annual rings of growth by a narrow ring of summer wood: the annual ring increment may be only half of that in a normal year (Fig. 21/1). The production of a bumper crop of seeds

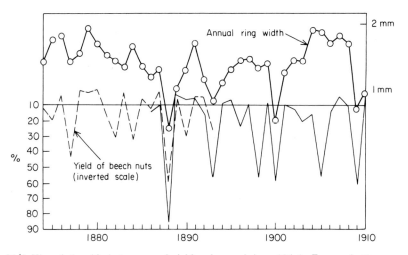

Fig. 21/1. The relationship between seed yield and annual ring width in *Fagus sylvatica* at Rohrbrunn. (Reproduced from Rohmeder, 1967)

leaves the same record in the annual increment as a heavy attack of defoliating caterpillars (see Fig. 12/4). The starch content of trees after a mast year is reduced to half or even a third of the values found after non-mast years (Hartig, 1889) and it is therefore not surprising that the year after a mast year may also show some reduction in the annual wood increment (Holmsgaard, 1955). In spruce (*Picea*), where seedless years are common, the annual growth increment in a seed year is on average 38% less than in a seedless year and in the second year after a mast year is still 20% less (Danilow, 1953).

A curious indirect piece of evidence showing the relationship between expenditure on reproduction and growth comes from dioecious plants

(Fig. 21/2). A grove of ash trees (*Fraxinus excelsior*) near Munich had
a complicated sex ratio, 40% having only male flowers, 4% only female
and the remainder bearing mixtures of bisexual, male and female
flowers (Rohmeder, 1967). The male trees exceeded the females in diameter growth by 14%, in height by 10% and in volume by 40%. One
interpretation of these data is that the female, burdened by the allocation
of resources to seeds, has less to spare than the male for vegetative
growth.

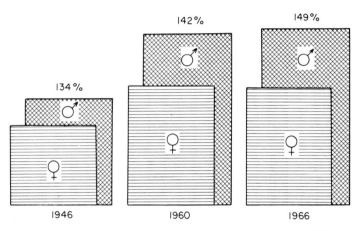

Fig. 21/2. The relative increments in stem volume of male and female trees of *Fraxinus
excelsior*. (From Rohmeder, 1967)

Cody thought that time and energy were the important resources to
be allocated in the strategy of a bird. Other resources may be even more
important and in a specific habitat one type of resource may be the
most crucially in short supply and its allocation between the activities
or parts of a plant will then be critical. Populations of moose (Botkin
et al., 1973) may be limited by the amount of sodium that they can
collect during a short season of aquatic grazing; the life cycle strategy
is likely in such a case to be influenced by the optimal allocation of
sodium between parents and offspring and between the various sodium-
demanding activities. Even if, as seems likely, a mast year is causal in
reducing the vegetative growth of trees it is not necessarily a shortage
of energy resources that provides the causal link. Other resources e.g.
nitrogen and potassium are depleted during a mast year. The flowers
and fruits of most forest trees are green and probably supply some (perhaps all?) of the carbohydrate demand created by seed production. The

production of seed is expensive in other respects than the energetic cost. The seeds of beech (*Fagus sylvatica*) contain 6 times as much mineral material per gram dry weight as does beech wood and a forest of beech uses 316 lb of nutrients per acre in a mast year, several times that used during the normal production of wood (Matthews, 1963). The green plant may indeed be a pathological overproducer of carbohydrates and the resource that needs critical allocation may often be something other than time or energy.

A further reason for believing that in plants the allocation of energy may not be the most critical issue is the now large body of evidence that plants normally function at a level of photosynthetic activity below that of which they are capable (see Chapter 15). The activity of the photosynthetic system seems to be determined in some way by the demands made by various other organs such as meristems and storage structures (Humphries, 1963; Sagar and Khan, 1969). The limited resources whose allocation matters are probably often minerals such as phosphorus, potassium, nitrogen and in some cases trace elements. It may be a useful exercise to compare the strategies of allocation of some limiting nutrients during the growth cycles of plants in their natural habitats. The "cost" of reproduction cannot really be measured without knowing what is the relevant currency. The currency may well differ between species and be a critical element in niche differentiation.

The phase of seed production in annual plants is often associated with the rapid death of leaves and the appearance of symptoms usually associated with mineral deficiency, as if the leaves have been "sucked dry" of some essential nutrient which has been translocated ("allocated") to the greater needs of the seeds. Where annuals grow in nutrient-deficient habitats, e.g. the annual grasses of sand dunes, it is common to find that each leaf withers and dies before the next is fully expanded and that when the inflorescence is formed the plant bears no remaining living leaves. It is as if a limited resource is shunted from place to place within the plant until ultimately it is allocated to the seed. One of the contrasts between a monocarpic and polycarpic habit of growth is that in the former a whole plant has to be constructed afresh for each seed crop, whereas in the polycarpic species it is possible for a maintenance nutrient budget to be continually recirculated from old to new tissues and (except for new growth) the mineral demand is only that for the seed crop. The mineral cycling system of the perennial *Eriophorum vaginatum* L. appears to be such a nearly closed cycle (Goodman and Perkins, 1959).

This may be one of the reasons why annual plants are ineffective in mineral-poor habitats.

The detailed patterns of allocation of resources are very difficult to follow in a plant; the technical problems are great because of the translocation of materials between different parts of the plant and it is difficult to decide which resource is the appropriate one to follow. Partly because it is easiest and partly following the example of Cody, the few comparisons of allocation have been made for energy or for dry weight. Even dry-weight determinations pose problems — whether to include dead parts and lost parts, and whether to make allowance for respiration. Crude estimates of the allocation of assimilates can be made by weighing the plant at maturity and determining the proportion of this "net assimilation" that is devoted to seed and ancillary reproductive structures (Harper and Ogden, 1970). Greater precision can be added by making calorific measurements and estimating energy budgets but the inaccuracies at other stages of such a study (e.g. the difficulty of obtaining an adequate sample of the roots) often makes such a refinement superfluous unless two very similar species are being compared.

Two patterns of allocation are shown in Fig. 21/3 for the annuals *Senecio vulgaris* and *Chrysanthemum segetum*. The patterns are remarkably similar. Allocation does not appear to be much affected by cultural conditions unless these are very extreme. *Senecio vulgaris* was grown as isolated plants in various sized pots of soil to give a very wide range of plant sizes (Fig. 21/4a). The allocation pattern was scarcely affected except in the most stressed conditions in which the plants failed to flower or managed to produce just one capitulum (Fig. 21/4b, c, d). Similar patterns of allocation can be extracted from a variety of sources in the agricultural and forestry literature and these have been assembled as a generalized diagram in Fig. 21/5. This is a convenient starting point in attempting to answer the series of questions put in a paper entitled "A Darwinian approach to ecology" (Harper, 1967):

1. Is the proportion of a plant's output that is devoted to reproduction higher in colonizing species than in those of mature habitats?

2. Is the proportion of a plant's output that is devoted to reproduction fixed or plastic? Is it changed by inter-or intraspecific density stress?

3. Does the proportion of a plant's output that is devoted to reproduction differ between plants of hazardous climatic conditions and those of more stable environments such as tropical rain forest?

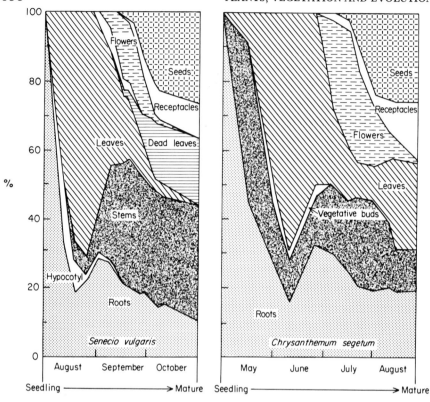

Fig. 21/3. Percentage allocation of dry weight to different structures throughout the life-cycles of *Senecio vulgaris* and *Chrysanthemum segetum*. (From Harper and Ogden, 1970 and Howarth and Williams, 1972)

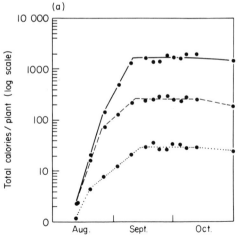

Fig. 21/4a. Growth curves for *Senecio vulgaris*. ————— "Low stress" treatment; — — — "medium stress" treatment; ·······, "high stress" treatment.

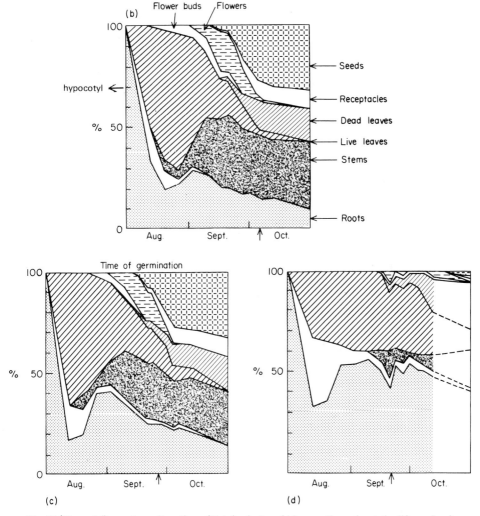

Fig. 21/4b, c, d. Percentage allocation of total calories of biomass throughout the life cycle of *Senecio vulgaris*. (b) "Low stress" treatment; (c) "medium stress" treatment; (d) "high stress" treatment. Vertical arrows mark dates at which maximum total calorific values were recorded. (From Harper and Ogden, 1970)

4. Do plants adapted to competitive environments devote a greater proportion of energy to non-photosynthetic organs (such as support organs)?

5. Do plants with clonal growth* sacrifice a proportion of the energy

*This is reworded from the original to eliminate a phrase that I now regret!

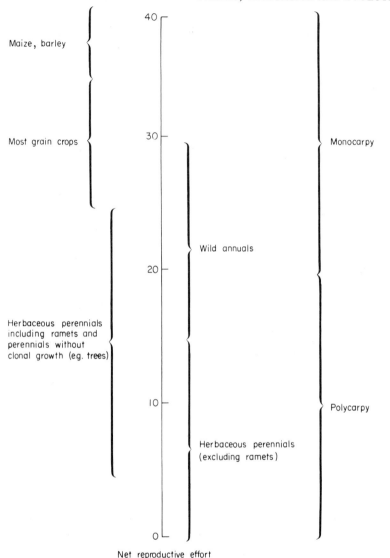

Fig. 21/5. The proportions of annual net assimilation involved in allocation to reproduction in different groups of flowering plants. (From Ogden, 1968)

which would otherwise be expended on seed? Are the processes of clonal growth competitive with those involved in producing seeds?

6. What is the relative energy expended in producing a ramet or tiller and that involved in producing a seed? Can this expenditure be related

to the relative risks of establishing a genet and a clonal module? Can this expenditure be related to the relative risks of establishment and the relative importance of local colonization as opposed to long-distance spread?

7. What is the expenditure on organs ancillary to the seed — the cost to the plant of attractive flowers, a pappus or the massive woody cone of some conifers? Is expenditure a measure of the selective advantages due to possessing these organs?

A further question should be added: what is the relevant resource to consider when comparing allocation patterns? Is it the same in different habitats and species?

There are some clear plasticities in reproductive allocation (Chapter 7): in polycarpic plants there may be whole years or groups of years in which vegetative growth occurs and no seed is produced (Fig. 22/8d, f). Variations in a strategy can be regarded as tactics; phenotypes are tactical solutions within a strategy that is set by a genotype. Where tactical changes in allocation occur it is possible that these may themselves be adaptive. Evolution may define both a strategy and the range of adaptive — tactical — responses to common conditions. In a study with golden-rod (*Solidago* species), the pattern of allocation of dry matter was followed in *Solidago nemoralis, S. speciosa, S. canadensis* and *S. rugosa* in a dry, heavily disturbed early successional site, a wet meadow site and a hardwood community (Abrahamson and Gadgil, 1973). Not all the combinations of species and sites could be studied but *S. speciosa* occurred in the dry and the woodland sites and *S. rugosa* in the wet and woodland site. The allocation of resources was expressed as the ratio of the dry weight of reproductive tissue to the total dry weight of the above ground tissue (see Fig. 21/6). The two comparisons between sites show a much greater proportion of biomass as seed produced by *S. rugosa* in the wet site than in the woodland site and for *S. speciosa* a greater reproductive allocation in the dry site than in the woodland site. In both cases the reproductive allocation was greater in the earlier successional community. It is not known whether the differences within the species were genetic or phenotypic.

The behaviour of the six combinations and sites is shown in Fig. 21/7a. Generally the allocation to reproduction decreased with increasing successional maturity of the habitat in the order *S. nemoralis* (dry) > *S. speciosa* (dry) > *S. canadensis* (wet) > *S. speciosa* (wood) > *S. rugosa* (wet) > *S. rugosa* (wood). A very similar result was obtained in a com-

parison of *Taraxacum* apomicts in disturbed and closed communities —
the forms in the open habitats put a greater proportion of dry matter
into seed (Gadgil and Solbrig, 1972). The partitioning of dry matter
between leaf, stem and the rest of the plant has also been studied in
Solidago (Fig. 21/7b, c). It is not easy to interpret the data for stems

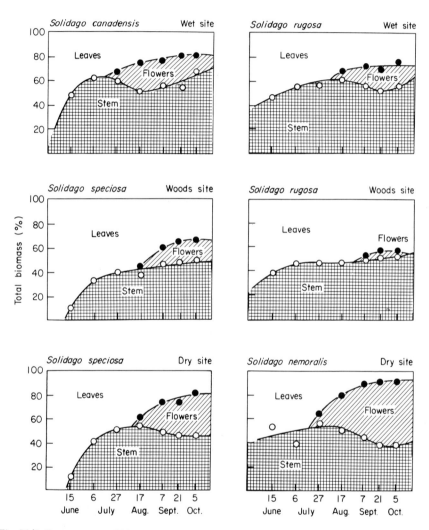

Fig. 21/6. Percentage total biomass of stem, leaves, and flowers (buds, flowers, pappi, seeds)
during the growing season for six populations of *Solidago* spp. Each point represents
the mean of the individuals included in a single population. (From Abrahamson and
Gadgil, 1973)

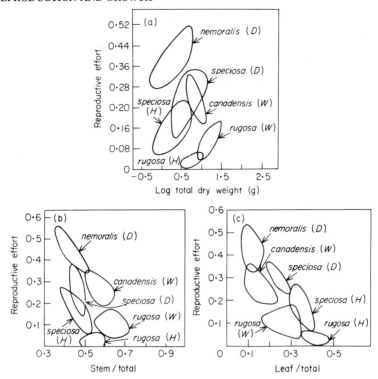

Fig. 21/7. Patterns of allocation in *Solidago* species, (a) proportional expenditure on reproduction in relation to log dry weight, (b) Proportional expenditure on reproduction in relation to the proportional expenditure on stem and (c) as (b) but for proportional expenditure on leaf. *D* = dry field site, *W* = wet meadow site, *H* = hardwood site. (From Abrahamson and Gadgil, 1973)

but there would appear to be a greater allocation to leaves in the more shaded habitats.

Another way of looking at the strategy of allocation in plants is to compare the number of seeds produced in closely related species where seed size is not a significant variable. The two grasses, *Agropyron repens* and *A. caninum* have very different patterns of growth. *A. repens* grows clonally by an extensive rhizome system whereas *A. caninum* has intravaginal tillers and forms a close tussock. The seeds of the two species are closely similar in weight and appearance but *A. caninum* has by far the greater seed production. In an experimental comparison of the two species grown as isolated plants, *A. caninum* produced an average of 258 seeds per plant and *A. repens* 30 in the first growing season (Tripathi and Harper, 1973). The clonal growth of *A. repens* involves the placing of buds along a very extensive rhizome system. This gives *A. repens* a potential for very rapid increase in the number of clonal modules compared with *A. caninum*. The number of seeds and rhizome buds produced per plant was:

	Q(genets)	q (rhizome buds)	Total $(Q + q)$
A. caninum	258	—	258
A. repens	30	215	245

These data strongly suggest that clonal growth and reproduction by seed are alternative processes. There is a great deal of ecotypic differentiation in *A. repens* and it would be interesting to know how resources are allocated between seeds and rhizome buds in ecotypes from the early and late colonizing phases of a succession.

The allocation of resources between seed numbers and seed size

The number of seeds produced by a plant (genet) is the product of three variables: the weight of the plant x the proportion allocated to seeds x the number of seeds per unit weight. Weight is a convenient measure but other components of weight may be more meaningful; thus the number of seeds per plant may be better described in some cases as the weight of nitrogen assimilated by the genet x the proportion allocated to seeds x the number of seeds per unit of nitrogen.

The weight of individual seeds varies over 10.5 orders of magnitude and examples illustrating this range are shown in Fig. 21/8. In some cases the "seeds" are really fruits and include the dry weight of investing tissues but on the scale of weights involved this difference is scarcely relevant — inclusion or exclusion of the investing structures will rarely alter the weight by as much as half an order of magnitude. The largest seed is that of the double coconut (*Lodoicea seychellarum*), Fig. 21/9, and the smallest seeds are those of the orchids and the parasitic flowering plants.

The seed is one of the least plastic organs on a plant (see Chapter 7); plants respond to stress phenotypically by varying almost every other component of yield before seed size is affected. Such homoeostasis suggests that seed size may be of much more crucial importance in evolution than seed number. Where variation in seed size does occur it is often an adaptive polymorphism (see Chapter 3). Many studies have been

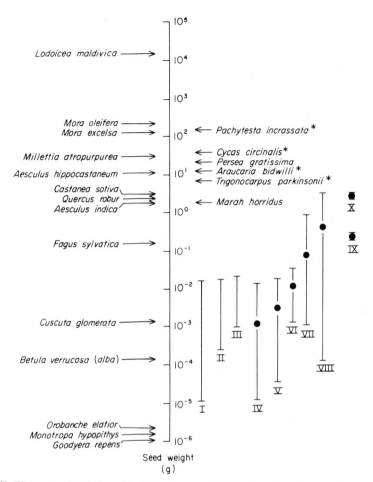

Fig. 21/8. The range of variation of individual seed weights in flowering plants and some gymnosperms.

I.–III. Seed weight ranges of N. American weeds. (Stevens, 1932, 1957)

I. *Rosaceae, Scrophulariaceae, Compositae, Ranunculaceae, Cruciferae, Gramineae, Cyperaceae, Chenopodiaceae, Caryophyllaceae.*

II. *Liliaceae, Polygonaceae.*

III. *Leguminosae, Umbelliferae.*

IV–VIII. Seed weight means and ranges of groups of species from various habitats in Britain (Salisbury, 1942)

IV. Open habitats, short grass and meadows.

V. Woodland margins.

VI. Woodland ground flora.

VII. Woodland shrubs.

VIII. Woodland trees.

IX and X. Seed weights (means and S.E.) of Central American woody legumes. (Janzen, 1969)

IX. Species subject to Bruchid attack.

X. Species not subject to Bruchid attack.

* Gymnosperm and fossil seeds — data from W. S. Lacey (pers. comm.).

(From Harper, Lovell and Moore, 1970)

Fig. 21/9. The double coconut (*Lodoicea seychellarum*) removed from the husk. Photograph
taken at Praslin, Seychelles. Copyright R. Frey-Hemmer, Zurich, Switzerland.

made of the effects of seed size on subsequent growth, particularly by
agronomists interested in obtaining maximal establishment from a sown
crop (see review: Harper *et al.*, 1970). *Trifolium subterraneum* produces an unusually wide range of seed sizes among the progeny of a
single plant and the effects of this variation were briefly discussed in
Chapter 6. Plants derived from small seeds were at no disadvantage provided that populations were at low density. When mixed populations
of large and small seeds were sown at high density the greater embryonic
capital gave the plants from large seeds an overwhelming advantage
in the developing struggle for existence. This experiment adds an
empirical dimension to the classic work of E. J. Salisbury, *The Reproductive Capacity of Plants* (Salisbury, 1942). He showed that species occu-
pying early successional habitats generally had small seeds: that larger
seeds were borne by plants in intermediate successional habitats and
that the species of woodland and forest tended to have the largest seeds.
Salisbury's seed size categories are shown in Fig. 21/8.

If a seed is small this implies that its embryo is small and that it carries few food reserves. The seedling from a small seed is dependent
from a very early phase in growth on its own independent assimilation.
In contrast the seedling from a large seed may have sufficient reserves

to continue growth for a much longer period or emerge as a more completely developed plantlet. The embryos of large-seeded legumes may contain already performed floral primordia and it is said that the seeds of some *Phaseolus* species can grow in the dark without nutrients or carbohydrate supply to a stage at which they not only flower but produce tiny viable seeds. Large embryos and large food reserves make it possible for a seedling to emerge from greater depth, survive for longer and grow to a more aggressive size in an environment that is starved of resources (see Fig. 21/10) (Grime and Jeffrey, 1965; Harper *et al.,* 1970; Harper and Obeid, 1967; Twamley, 1967).

The size of seeds is a very constant characteristic of a plant, of a variety and of a species yet there are important variations, e.g. the very exceptional case of *Trifolium subterraneum,* where a 17-fold variation may occur in the seeds from a single plant (Black, 1957, 1959). Whether the emphasis is placed on the constancy or the variability depends on the context. In a consideration of the reactions of a plant to stress it is remarkable how slight is the variation in seed size compared with the variation in other components of reproductive capacity. Thus in a density experiment with wheat the number of ears per plant varied 56-fold, the number of seeds per plant 833-fold, the number of grains per ear 1.43-fold but the mean weight per grain varied only 1.04-fold (Puckridge and Donald, 1967). Similar constancy in mean seed size has been observed in a variety of species (Harper, 1961). The comparative orders of magnitude of the variation in such an experiment emphasizes the constancy of seed weight. If experiments like those with *Trifolium subterraneum* and *Linum* are designed, the variations in seed weight can be clearly shown to have crucial adaptive importance. It is no contradiction to find that seed size is remarkably inflexible and at the same time to point to the very great significance of the flexibility (genotypic and phenotypic) that does exist.

One explanation for the constancy of seed size among plants under very differently stressed conditions is that most plants over-produce the number of seeds that they initiate and then, by abortion, reduce the number that develop. Presumably some internal regulation controls the number that develop and so controls their size. It is interesting that the pressure of intraspecific density stress rarely affects the size of seeds but changes are sometimes produced by interspecific interaction. In mixed populations of oats (*Avena sativa*) and barley (*Hordeum sativum*) the mean grain weight of the oat is increased, compared with pure oat

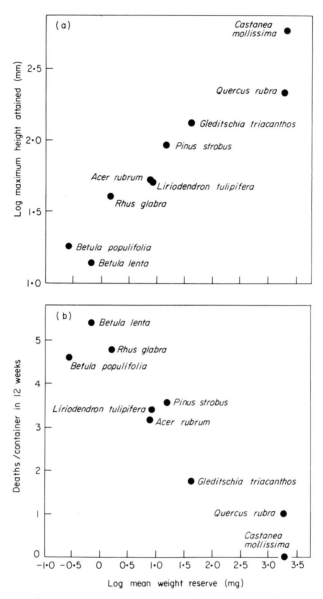

Fig. 21/10. The behaviour of seedlings of a variety of tree species grown in deep shade, (a) the maximal height attained in relation to mean seed weight. (b) the seedling mortality per container in relation to mean seed weight. (From Grime and Jeffrey, 1965)

populations (Fig. 8/15a) and in mixtures of *Linum usitatissimum* with *Camelina* the seed weight of *Linum* increases, compared with pure populations of *Linum* (Fig. 8/15b).

Seed size can be altered by surgery. If the flowers of *Vicia faba* are removed above the sixth node the mean seed weight in the pods that remain can be increased by 50% (Hodgson and Blackman, 1957a), and similar effects can be obtained by removing pods from *Phaseolus vulgaris* (Adams, 1967) and *Lupinus luteus* (van Steveninck, 1957) or by removing most of the grains from an ear of wheat (Bingham, 1967). These experiments all suggest that there is an integrated control of seed size, that seed number is sacrificed before seed size is reduced but that the integration can be less than perfect if the normal environment is distorted. The seed weight of *Vicia faba* can be increased by 7% if the population is thinned but only if this is done after the pods have formed and before they have filled (Hodgson and Blackman, 1957b).

The cultivated sunflower (*Helianthus annuus*) is often quoted as an example of the plasticity of seed size. It is exceptional among seed plants in responding to density stress by up to 6-fold variation in seed size (Clements *et al.*, 1929). The cultivars have been bred to bear a single apical capitulum and this leaves the plant with few sources of plastic variation. The wild sunflower, in contrast, branches freely and produces many capitula which develop in succession. A 156-fold variation in the density of wild sunflower produced only 1.25-fold change in mean seed weight (Khan, 1967; and see Fig. 7/5). Some variation in seed size occurs between different parts of a plant. In *Pisum sativum* proximal and distal positions within the pod tend to bear smaller seeds than those in the central region and seeds borne near the base of a capsule tend to be larger than those at the apex (e.g. *Spergularia* spp.: Salisbury, 1958; Sterk, 1969). Perhaps some of the emphasis on the constancy of seed size comes from the custom of quoting mean weight based on samples of 100 seeds. It is this value which tends to be very stable; perhaps there is greater variation between seeds in a sample than is commonly supposed. The weight of seed of *Rumex crispus* varies with position on the inflorescence: proximal seeds had a mean weight of 0.186 mg and distal seeds 0.161. The larger seeds were the last to be shed from the parent, germinated later than the small and produced larger cotyledons. Such variation may enable an individual to establish descendants in a wider range of conditions than one with invariant seeds (Cavers and Harper, 1966).

Variation in seed size is markedly bimodal in some species. Such polymorphism, whether genetic or somatic, can usually be taken to imply two distinct optima and the operation of disruptive selection. Seed polymorphism affecting size represents the allocation of resources into two distinct categories of low and high investment. In *Dimorphotheca* spp. large achenes are borne by the disc florets and small achenes by the rays of capitulum. In *Synedrella* the smaller achenes are borne by the disc florets (Salisbury, 1942). Seed polymorphism is common in the Compositae where the structure of the capitulum offers an opportunity for a division of labour between the florets; polymorphism is found in *Bidens, Calendula* and *Crepis.* An extreme form of seed polymorphism occurs in the desert annual *Gymnarrhena micrantha* which bears a very few (1—3) large achenes in axillary cleistogamous flowers below the soil surface. These large achenes lack a pappus but the aerial flowers produce large numbers of small achenes (0.37 mg) with a well developed pappus. The proportions of the two types of achene vary between seasons and in very dry years only cleistogamous flowers are produced (Koller and Roth, 1964). Seed polymorphism is associated with dispersal as well as with the allocation of resources (see Chapter 3).

Seed size is heritable: most of the evidence comes from crop plants. In the soybean (*Glycine soja*) recurrent cycles of mass selection for seed size in bulk populations produced a continuing response and the heritability of seed size was 0.93 (Fehr and Weber, 1968). A single cycle of selection in populations of *Bromus inermis* was sufficient to raise mean seed weight from 2.80 to 3.95 mg per grain (Christie and Kalton, 1960). Similarly cycles of simple recurrent selection for large seed in *Lotus corniculatus* gave increases of 20% and 6.25% per cycle in the cultivars 'Viking' and 'Empire' (Draper and Wilsie, 1965).

Not only can seed size be shown to be heritable but there is evidence that it is powerfully affected by natural selection. Twenty-five populations of the Lima bean (*Phaseolus lunatus*) were prepared by intercrossing varieties. The mean seed weight in these populations varied from more than 1000 mg to less than 250 mg. The populations were sown and managed under routine cultivations (in California) and no conscious selection was practised for seed size or any other character. After 6—8 generations the mean seed weight of the large-seeded populations had decreased to 600 mg and that of the small-seeded populations had increased to more than 350 mg. This experiment strongly suggests that there is an optimal size of seed for this species under this set of conditions and that seed size was subject to stabilizing selection (R. W. Allard, pers. comm.). There is evidence within single species of variation in seed size between populations. In *Arabidopsis thaliana* (a winter annual) differences between seed sizes within populations were very small but remarkable between-population differences occurred and were not obviously correlated with any environmental factor. In Britain populations were found with mean seed weight of 7.33 mg per 100 seeds ('Husbands Bosworth') and a population at Mickle Fell with a mean seed weight of 34.0 mg per 100 seeds. A population from Lodz (Poland) had a mean seed weight of 8.00, one from the Burren (Eire) 38.00 and one from Soria (Spain) 48.00 mg per 100 seeds (Evans, 1974). It would be extremely interesting to compare such populations side by side in a controlled experiment to discover whether there are other differences between these forms that compensate for such a great variation in individual seed weight.

Useful comparisons can be made between closely related species. A remarkable example is a study of six species of *Solidago* in an old field successional environment and in a prairie. Five of the species were to be found in both habitats, though it is not clear whether these represented differentiated ecotypes. There were large differences between the mean seed weight of species in the same habitat and within species between habitats (Table 21/I).

Table 21/I

Mean weight (μg) per achene of *Solidago* spp. in
"old field and prairie communities".
(From Werner and Platt, 1976)

	Old field	Prairie
S. nemoralis	26.7 ± 2.1	104.0 ± 8.3
S. missouriensis	17.6 ± 0.6	39.3 ± 3.2
S. speciosa	19.5 ± 1.6	146.3 ± 11.7
S. canadensis	27.3 ± 2.3	58.3 ± 11.1
S. gigantea	— —	50.8 ± 4.2
S. graminifolia	24.5 ± 2.7	10.6 ± 1.6

When the weight per seed of *Solidago* is plotted against the number of seeds per stem there is a clear negative relationship (Fig. 21/11). These data illustrate, better than any other, that seed number and seed size seem to be alternatives in the strategy of reproduction.

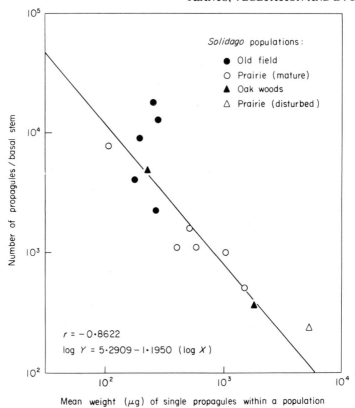

Fig. 21/11. The relationship between the number of achenes produced per shoot and the mean weight per achene for species of *Solidago* in a variety of habitats. (From Werner and Platt in press)

The significance of very small seeds

The plants that bear very small seeds are unusual in other respects. The group includes the orchids, saprophytes such as *Monotropa* and parasites such as *Orobanche*. All of these plants have anomalous nutrition. In the orchids a mycorrhizal association is developed at the earliest stages of germination; early growth appears to depend on this association and some species may spend several months or even years before they develop autotrophic tissues, some even remaining permanently saprophytic. The parasitic flowering plants do not all have dust seeds and the exceptions are interesting. The seeds of *Cuscuta*, *Bartsia*, *Rhinanthus* and *Melampyrum* are in the same categories of seed size as their non-parasitic neighbours in their habitats. The small-seeded parasites are all of species that require a germination stimulant from the host plant roots (Sunderland, 1960). These are species that only germinate when they are assured of an immediate food source. The larger-seeded parasites do not require a germination stimulant from the host and (even in the case of the chlorophyll-less *Cuscuta*) most grow in search of a host or as in *Viscum* and its allies (the mistletoes) must grow through host bark before becoming parasitic. The very small-seeded plants are apparently all forms in which *the germinating seedling is assured of an outside source of support from the moment of germination.* It is easy to see how natural selection might seize on such an advantage. The seed reserve is no longer very important, and the constraint on seed numbers that is imposed by the minimal necessary food stores to start independent life is therefore removed. Under such circumstances fitness can be increased by increasing the numbers of progeny and sacrificing seed size. This curious group of parasites and plants with a wholly or partially saprophytic stage have escaped from the constraint of a minimal effective seed size; seed size has decreased by an order of magnitude below that of the smallest "normal" plants and seed number has correspondingly become enormous. The common interpretation of the large seed output of these species has been that they "need" to produce many propagules to counter the peculiarly difficult task of finding a suitable host or mycorrhizal fungus. This argument is untenable (see pp. 648–649). Instead the group can be seen as opportunistic, having taken advantage of an alternative mode of embryo nutrition to reduce seed size to tiny dried bags of DNA and expand their reproductive capacity to a new limit.

The number of seeds as the product of successive allocation

During the lifetime of a plant, assimilates are allocated to a variety of ends, new assimilating organs, support structures and reproductive structures. What is allocated to reproduction may be divided between few large seeds or many small ones. The three variables, (a) the size of the plant, (b) the proportion devoted to seeds and (c) the mean seed weight, determine the total seed production by an individual. These variables can be expressed as the axes of a defined "adaptive space" (Fig. 21/12). There is insufficient information to define the position of many species within this adaptive space, not because of any shortage of information about seed size or difficulty in putting rough values

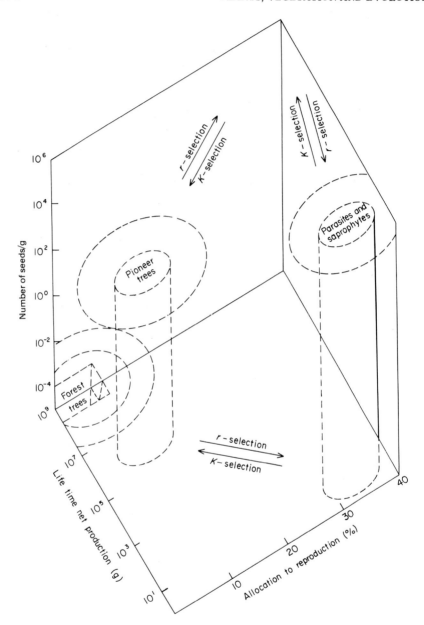

Fig. 21/12. The components of the reproductive capacity of plants shown as an adaptive space within which the forces of r- and K-selection are indicated and tentative positions for three major plant groups.

to the allocation of resources to reproduction. The problem arises in fitting values for the third axis — the net production in a plant's lifetime. There is no real meaning to this concept in the case of clonal species in which the lifetime may be quite indefinite and the net production, though it may be calculated for a year, cannot be realistically extended to the lifetime of a genet. The net productivity of a tree during its lifetime is more easily defined but most estimates of productivity made during the surge of effort during the International Biological Programme were specifically concerned with production per unit area of land, not per individual and so are evolutionarily largely irrelevant.

Gross evolutionary and ecological trends can be recognized in all three of the parameters that define the adaptive space that is the reproductive capacity of organisms. In an environment that is open (colonizable) the advantage will usually go to small-seeded species. They are generally more widely dispersed and the larger number of seeds means that they have a potential for rapid colonization; r-species — those that spend most of their time in acts of colonization and then give way to others in natural succession — will usually be forms with many small seeds. K-species — specialists in resource-limited environments where there is intense interference from neighbours — will tend to have larger seeds, devote a greater proportion of whatever resources are most limiting to ends that maximize individual survival rather than fecundity (perenniality, clonal growth) and devote resources to structures like stems, extensive roots or the ultimate in K-strategies, the trunk of the tree. These trends are shown as arrows in Fig. 21/12. However, this sort of allocation analysis treats all seeds produced by a plant as if they are equal in their contribution to fitness. In practice all seeds are not equal, even if they are identical in size, number and physiology, because progeny produced at different stages in the life of the parent have different values. For this reason the reproductive capacities of plants must also be compared as time schedules.

22

Reproduction—Life Cycles and Fertility Schedules

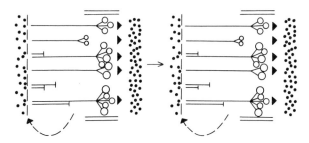

In order to obtain a distinct idea of the application of Natural Selection to all stages in the life-history of an organism, use may be made of the ideas developed in the actuarial study of human mortality. R. A. Fisher

Not all progeny produced by a parent have the same probability of contributing to descendants in subsequent generations. Fisher posed the question for humans:

We may ask, not only about the newly born, but about persons of any chosen age, what is the present value of their future offspring; . . . To what extent will persons of this age, on average, contribute to the ancestry of future generations? The question is of some interest, since the direct action of Natural Selection must be proportional to this contribution. (Fisher 1929)

Organisms of different ages have different "reproductive values", they make different contributions to the expectation of growth of the population as a whole, they contribute with different force to the genetic composition of future populations.

Two schedules define the behaviour of a population — the age structure (the life table) of the population and the fecundity schedule. The former describes how many organisms there are present in different age classes and the second describes the contribution made by individuals in the different age classes to future population growth. By combining the schedules, it is possible to calculate the present value of future offspring

$$v_x/v_0 = \frac{e^{mx}}{I_x} \int_x^\infty e^{-mt} I_t b_t \, dt.$$

I_x is the number of individuals living to age x
b_x is the rate of reproduction at age x
m is the Malthusian parameter of population increase
v_x is the reproductive value at age x.

The number m is implicit in any given system of rates of death and reproduction, and measures the relative rate of increase or decrease of a population when in the steady state that is appropriate to any such system. Fisher has an analogy which can be reworded in a botanical sense. If we regard the "birth" of a plant as the loaning to it of a life, and the birth of its "offspring" as a subsequent repayment of the debt, the method by which m is calculated shows that it is equivalent to answering the question — "At what rate of interest are the repayments the just equivalent of the loan?"

The value of an individual to the growth rate of a population and the value of that individual relative to its fellows within the population depends on the way in which the time course of the risk of death is related to the time course of producing offspring. It is to be expected that the reproductive schedule will be adjusted by natural selection in such ways as will maximize the number of descendants left. Life, and the risk of death, start when the zygote is formed. There is then a period of juvenile life (pre-reproductive) which itself can be sub-divided into the period of postzygotic development when the embryo is developing at maternal expense, a period of seed dormancy and a period of purely vegetative growth. A reproductive period follows which may occur as one lethal burst of activity or be spread as a repetitive series of episodes. A post-reproductive period may also be present in the life cycle. An idealized fecundity schedule is shown in Fig. 22/1a and a series of variations in the form of life cycles is illustrated in Fig. 22/1b—e.

The shape of a fecundity curve has important consequences in popu-

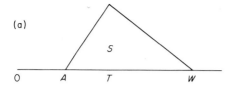

(a)

Generalized fecundity schedule: O = birth, A = age at the beginning of fecundity period, T = peak period of fecundity, W = age at end of period of fecundity, S = total number of offspring.

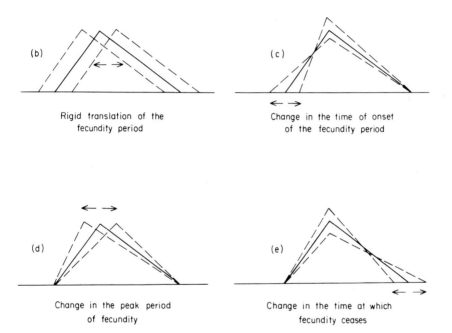

(b) Rigid translation of the fecundity period

(c) Change in the time of onset of the fecundity period

(d) Change in the peak period of fecundity

(e) Change in the time at which fecundity ceases

Fig. 22/1. Theoretical variations in a model fecundity schedule. (From Lewontin, 1965)

lation behaviour. In a famous essay, "The population consequences of life history phenomena", Lamont Cole (1954) showed that precocity of reproduction had a great influence on the potential growth rate of a population. Fisher had shown that the "reproductive value" of an individual was great during its early reproductive life; population growth is a compound interest phenomenon and the earlier an investment is made the greater is the interest accumulated. The point is made very simply by comparing two populations in one of which the organisms each produce two offspring at the end of the first year and then die: their descendants follow the same pattern. In the other population the organisms produce one offspring every year, the offspring do the same and the organisms live for ever. Both populations grow at the same rate — provided that there is no differential mortality. In both populations the numbers of individuals increase in the progression 2, 4, 8, 16 . . . even though in the first case the reproductive capacity of individuals is two but precocious and in the second is infinitely large but infinitely prolonged.

Cole showed for a variety of more realistic life cycle models that in general, for a growing population, the precocity of reproduction was a major determinant of the rate of population growth. Lewontin (1965) extended this argument for an idealized *Drosophila* life cycle (Fig. 22/1), which included a juvenile phase in which no eggs were laid (and includes the larval phase), a phase of increasing egg production to a peak rate and a phase of declining egg production as the fly ages. He used computer models to determine the rates of population growth when the life cycle was varied by (i) changing the length of the juvenile period, (ii) changing the time of peak egg-laying within the reproductive period and (iii) changing the number of eggs laid without changing the time schedule. The intrinsic rate of natural increase (r) rises from 0.501 to 0.565 if the total fecundity is doubled (from 5000 to 10 000 offspring), but an equal effect comes from reducing the juvenile period from 8.6 to 7.5 days. "Thus one day saved in development is worth a doubling of fecundity or 5000 eggs." When the fecundity is low the effects of precocity are even greater so that a doubling of fecundity from 150 to 300 eggs will increase r from 0.180 to 0.205, the equivalent of shortening the juvenile period by 3 days.

Lewontin summarized the findings from his model by considering the changes in life cycle that were necessary to increase r (the intrinsic rate of natural increase) from 0.300 to 0.330. This could be done by nearly doubling the number of eggs produced (780 to 1350), or by altering the shape of the fecundity curve without changing the number of eggs produced per fly. A rigid translation of the fecundity period of 1.55 days is needed to produce the same effect (Fig. 22/1b) *or* a decrease of 2.20 days in the time to sexual maturity (Fig. 22/1c) *or* a decrease in the turnover age of 5.55 days (Fig. 22/1d) *or* a decrease in the age at the last egg of 21.00 days (Fig. 22/1e).

The enormous effect of precocious reproduction in increasing the potential rate of population increase immediately poses two problems about the types of life cycles that are commonly found in nature. If precocity brings such large rewards why do so many plants have long juvenile periods (see Fig. 22/5) and, if the reproductive value of early progeny is so great, why do so many plants reproduce over long extended periods — why are all plants not monocarpic? Polycarpic (iteroparous) reproduction is a particular puzzle in view of Cole's demonstration that the effect of repeated reproduction is equivalent to adding just one extra to the "clutch" size of an individual that reproduces only once. The answers to these questions appear to lie in the way in which the risk of mortality varies through life (Gadgil and Bossert, 1970). The models of Cole include variations in the schedule of fecundity but not of mortality and Lewontin's models represent "the age-specific fecundity schedule adjusted for the probability of survival to age x". In general, if the risk of death increases with age there are even greater advantages to precocity than envisaged in Cole's model, but if the death risk declines with age, precocity is less rewarding. If the death risk remains constant throughout the life cycle Cole's argument remains unaffected.

If C is the probability that offspring will survive the first year and P is the adult survivorship, then "for an annual species the absolute gain in intrinsic population growth that can be achieved by changing to the perennial reproductive habit would be exactly equivalent to adding P/C individuals to the average litter size (Charnov and Schaffer, 1973).

The effects of different birth and death regimes can be illustrated by comparing the life cycles of an annual and a biennial plant. The genes are known that would permit an annual foxglove (*Digitalis purpurea*) to exist in nature but it is the less precocious biennial that occurs naturally. If we compare the population growth of an annual plant A that completes one generation in a 12 month period with the biennial B that requires 24 months to do the same — the number of seeds produced by the two forms would need to be in the ratio of $x_A^1 : x_B^2$. The populations of the two forms will grow at the same rate if the biennial produces the square of the seed output of the annual. A well developed plant of *Digitalis* produces about 300 000 seeds at the end of its second year. As an annual it would need to produce only 500—600 seeds to achieve the same population growth rate, provided there was no mortality or if the mortality was a constant, age-independent risk throughout life. This is not the case — the juvenile mortality risk is greater than that of adult plants (see Fig. 22/2). The foxglove is a fugitive species, invading temporarily open habitats such as woodland clearings and

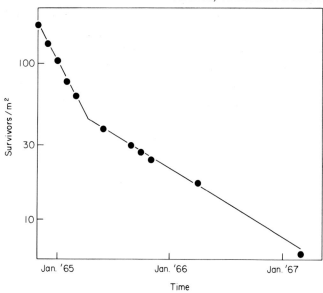

Fig. 22/2. The survivorship of populations of *Digitalis purpurea* over two years from germination in woodland sites in North Wales (data from E.R.B. Oxley).

then often failing to re-establish as other species invade and change the habitat. In such an environment it obviously pays to capitalize on a few successful establishments by a longer period of growth that culminates in a "big bang" of reproduction. The analogous life cycle in the animal kingdom is that of the salmon which has a hazardous juvenile life and the survivors capitalize on their escape from early death by growing to a similar big bang of reproduction (Gadgil and Bossert, 1970; Schaffer, 1974). The comparison between the annual and biennial strategies can be made by considering the power by which the seed output of the annual must be raised to give it the same intrinsic rate of natural increase if it became a biennial: this comparison is shown in Fig. 22/3 for different mortality risks that are confined to the first year of development. If there is no mortality the biennial would need the square of the seed output of the annual, but the greater the juvenile mortality the less seed is demanded from the biennial to match the population growth rate of the annual.

The physiological potential for very precocious reproduction certainly exists among plants. "The ideal speedy life cycle might involve a seed that germinates to expose a green flower that immediately proceeds to leave several seeds that germinate without delay. The flower would need to photosynthesize sufficiently to stock the new seeds with reserves and to support a root for the necessary mineral uptake" (Harper and White, 1974). This ideal is approached very closely by *Chenopodium rubrum* grown in 8 hr short days. The green flowers are produced immediately after expansion of the first pair of foliage leaves and are visible only 6 days after seed is moistened to start germination

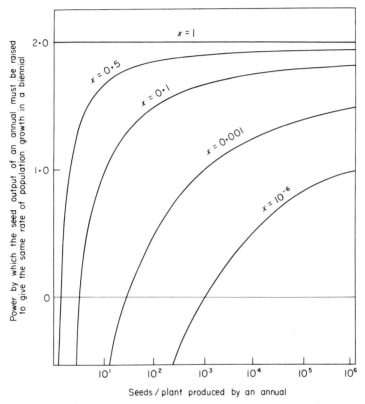

Fig. 22/3. The theoretical relationship between the seed output of an annual and a biennial for an equal intrinsic rate of natural increase at various values of density-independent mortality (*x*) where this risk is confined to the 1st year of development (data from E. R. B. Oxley). (From Harper, 1967)

(Cummings, 1959). The winter annual *Arabidopsis thaliana* can be persuaded to complete its life cycle from seed sown to first seed set in 6 weeks under cultural conditions and some weedy annuals such as *Poa annua*, *Senecio vulgaris* and small desert annuals may achieve similar speeds in nature.

Precocious reproduction brings no benefits to the rate of population increase if the seed, once it is formed, remains dormant. Precocity brings its advantage only when generation succeeds immediately on generation. It may be that some annual weeds such as *Senecio vulgaris* and *Poa annua* may repeat generations in this way though even these are suspected of having ecotypes that grow in different seasons so that the impression they give of rapidly repeating generations may be misleading. The precociously reproducing annual plants as a rule complete their life cycle in one short period of the year but spend the rest as dormant seeds. A potentially explosive rate of population growth is sacrificed in favour of a precisely adjusted seasonal rhythm of development.

The potential for precocious reproduction exists even in long-lived trees and can be demonstrated either by genetic selection or by physiological manipulation. The period between planting and first flowering in the oil palm has been reduced by plant breeders from 45 to about 30 months (Blaak, 1972) and three generations of selection for early flowering in *Betula verrucosa* reduced the juvenile period to 2 years (Rudolph, 1966). The juvenile period of *Sequoia* and *Sequoiadendron* can be reduced to 8–12 months by gibberellin treatment (Pharis and Morf, 1969). When physiologic and genetic potentials for precocity exist and precocious reproduction brings clear advantages in the potentially rapid growth of populations there must be powerful countervailing forces to precocity in nature. "The fact that the length of the juvenile period is heritable presumably implies that the length to which the species has become adjusted in nature represents an adaptive response" (Harper and White, 1974). Every delay in the process by which individuals leave descendants needs to be countered by some compensating contribution to fitness.

One interpretation of delayed reproduction is that some populations in nature do not necessarily experience the greater part of the forces of natural selection while they are in phases of population increase. The corollary of the selective advantage of precocity in expanding populations is a selective advantage of delay in declining populations. Most plants in nature spend their lives in successions in which they are doomed to local extinction and individuals that contribute progeny to new establishments elsewhere may well be those that have survived during periods of population decline. A long juvenile period may merely imply that in the course of the recent evolution of a species it has suffered more intense selection during its periods of population decline

than during its periods of expansion. Indeed, the almost obligate alternation between phases of population expansion and decline that occur
in most populations in nature is likely to mean that they experience
disruptive selection for the length of the juvenile period. "The fitnesses
of genotypes change as the result of changes in the overall rate of increase
of the population and in the complete absence of any genotype-specific
changes in environmental selective pressures" (see Giesel, 1972; Charlesworth and Giesel, 1972; also discussion in Hamilton, 1966).

The juvenile period is made up of three distinct phases (Fig. 22/4).
Time is spent after the formation of the zygote while the embryo develops, accumulates resources and grows at the expense of the mother
plant. This represents reproductive capacity sacrificed by one generation (which could presumably have produced a lot more tiny seeds instead of supporting the next generation as a parasite) to permit the precocious development of the next. The period of overlap of the two
generations is curiously variable, for it becomes entangled with another
curious delay in the reproductive process at the pre-zygotic stage, the
interval between pollination and the act of fertilization. This delay is
well known in gymnosperms, particularly pines, but there is much
variation amongst angiosperms (Table 22/I). In *Pinus sylvestris* the
whole process from pollination to production of ripe seed extends
over 2 years. In *Ulex gallii* seed is ripened in the year following flowering but in *U. europaeus* flowering and seed set occur in the same year.
The contrast within the oaks is very peculiar. *Quercus robur, Q. petraea*
and *Q. pubescens* complete the whole process of fertilization and
seed set in 4–6 months but the same process takes 12 months longer
in *Quercus coccifera*. The closely related *Q. ilex,* another evergreen,
requires 4–6 months and cohabits with *Q. coccifera* which makes the
difference even more remarkable.

The second part of the juvenile phase is spent by the plant as a dormant embryo, either awaiting a seasonal stimulus to germinate or in
enforced, more or less long-lived dormancy (see Chapters 3 and 4). The
precocity that is sacrificed at this stage of development has clear advantages in permitting accurate synchronization of the growth cycle with
seasonal events. If generations follow each other immediately, plants
must be adapted to live and grow in all seasons: precocity is sacrificed
to permit precise adaptation to a specific season of growth.

The third phase of the pre-reproductive period is the time spent in
vegetative growth. In the case of annuals this appears to be related

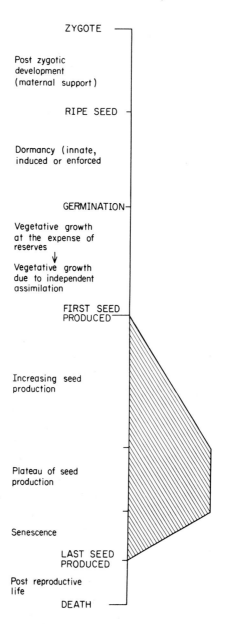

ZYGOTE

Post zygotic
development
(maternal support)

RIPE SEED

Dormancy (innate,
induced or enforced

GERMINATION

Vegetative growth
at the expense of
reserves

Vegetative growth
due to independent
assimilation

FIRST SEED
PRODUCED

Increasing seed
production

Plateau of seed
production

Senescence

LAST SEED
PRODUCED

Post reproductive
life

DEATH

Fig. 22/4. Idealized life cycle of a higher plant.

Table 22/I

The interval of time between pollination and
fertilization in various angiosperms. (From Coulter
and Chamberlain, 1909)

Lilium	65—72 hours
Fagus sylvatica	3 weeks
Arum	10 days— several months
Betula alba	1 month
Carpinus betulus	2 months
Alnus glutinosa	3 months
Corylus avellana	4 months
Quercus robur	4 months
Colchicum autumnale	More than 6 months
Quercus velutina	13 months

to the predictability or otherwise of the environment (see Chapter 18).
The two classes of annual habit are (i) those plants with an extremely
short vegetative period of growth followed by a long drawn out flowering and seed shedding period — a strategy appropriate for a growing
season of uncertain length and (ii) those plants with a growth period that
ends in a lethal burst of flowering and seed set — appropriate for a
reliable environment with a highly predictable growing season.

Among perennials the period of vegetative growth is immensely variable and often seems to be related to the time needed to achieve some
critical size necessary for reproduction. Even in semelparous (monocarpic) species the "big bang" of reproduction occurs when the plant
has reached a critical size (Chapter 18) though there may be other
factors that also operate in the odd cases of synchronous flowering
that occur in the semelparous bamboos. Figure 22/5 shows the relationship between the length of the juvenile period and the life span of a
number of herbs, shrubs and trees (data collected from a wide variety
of sources (Harper and White, 1974)). There is of course a perfect correlation between the two parameters in the case of semelparous species
(which have to conform!) but if they are excluded, a broad trend remains in which there is enormous variation. It is not very useful to put
clonal plants onto such a diagram because the concept of life span has
no very sensible meaning. The values given for trees are estimates of
the age commonly reached by individuals that reach a dominant position in their respective canopies.

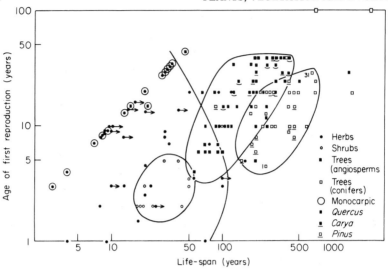

Fig. 22/5. The relationship between pre-reproductive life (or juvenile period) and total life span for perennial plants. The data give the known normal life spans and not exceptional examples of great longevity (due to White in Harper and White (1974) which includes identification and bibliography for each of the points on the diagram).

Some herbaceous perennials take an extremely long time to reach a flowering condition. *Listera ovata, Spiranthes spiralis* and *Cypripedium calceolus* require at least 13–16 years before flowering (Summerhayes, 1951) and *Orchis ustulata* does not even begin to form leaves until it is 10–15 years old. There are no annual orchids and there is an intriguing contrast between the length of the juvenile phase in these mycorrhizal plants in comparison with many parasites such as *Orobanche, Striga* which in many other respects (e.g. their small seeds) have similar reproductive strategies to the orchids. Sylvén (1906) measured the lengths of the juvenile period of many herbaceous perennials in culture and his records were supplemented by those of Linkola (1935) in Finland: *Trollius europaeus* will flower in its third year under garden conditions but takes much longer in the field: *Ranunculus auricomus* will often flower in its second year if conditions are good and in a garden *Ranunculus acris* sometimes flowers in its first year. *Potentilla erecta* and *Polygonum viviparum* can also flower in the first year. *Alchemilla vulgaris, Geum rivale* and *Chrysanthemum segetum* take 2 years before the first individuals flower in garden cultivation and *Cirsium palustre* at least 3 years. In all these cases, the recorded time to first flowering is that of the exceptionally advanced members of the sown populations, while the majority had even longer juvenile periods. In the field the juvenile period is usually much longer, though the occasional individual growing in a favoured spot may be precocious. There seems to be no reason why the juvenile period should not be infinitely long for clonal herbs lingering in a vegetative condition in a marginal environment. We might indeed expect genetic polymorphism in the length of the juvenile period.

The juvenile period in the life of geophytes has concerned horticulturists for whom it causes a serious delay to breeding programmes. Under cultural conditions designed to promote early flowering, the juvenile period of the tulip is 4—7 years and the daffodil 4—6 years. The shortest juvenile phases are in *Freesia, Brodiaea, Tritonia* among the corm-bearing species, and *Dahlia* and potato among the tuber-bearers, all of which may complete the juvenile phase in a year. The plants apparently have to attain a critical size for flowering and, not unexpectedly, the larger this critical size the longer is the juvenile period. There is an exception to the rule of a critical size for flowering in some species of *Allium*, e.g. *A. porrum* which flowers in the second year from seed even if it is a tiny plant grown under stressed conditions (J. Boscher, pers. comm.). There does not seem to be any obvious correlation between the length of the juvenile period and clonal growth before flowering. Many of the geophytes develop daughter bulbs, corms or rhizomes during the juvenile stage (*Tulipa, Iris, Crocus, Dahlia* and potato) whereas clonal multiplication occurs only after flowering in *Hyacinthus, Narcissus* (daffodil), *Gladiolus, Freesia, Tritonia* and *Brodiaea* (Fortanier, 1973). The very long juvenile phase in the tulip is explained (Fortanier, 1973) by the short season of plant growth; the period of assimilation is only 3 months: the growth rate of the bulb is slow and only large bulbs flower.

Among trees, the time taken from seed germination to seed production depends, at least as much as in herbs, on the conditions of growth. Isolated specimens in an arboretum will usually flower when much younger than the same species in a forest stand. The few trees in a forest that expand into the upper canopy flower much earlier than the suppressed individuals, which may never flower at all (Schmidling, 1969). Environmental conditions that encourage rapid vegetative growth tend to shorten the juvenile phase (Higazy, 1962; Visser, 1964, 1965, 1967, 1970).

The generalizations that emerge from Fig. 22/5 are that shrubs tend to have shorter juvenile periods than trees and that conifers tend to live rather longer but have the same order of length of juvenile period as hardwood trees. Most hardwood trees that live more than 200 years spend at least 20 years in the juvenile phase. A series of generalizations made by Molisch (1938) are generally supported by the data in Fig. 22/5: (a) plants with brief youth have short life, (b) plants with a long youthful period tend to have a long life and (c) a long youthful period is usually followed by a long and often very extended reproductive life.

Some trees can certainly be very precocious in reproduction (e.g. *Pinus taeda*, see Chapter 20). In the ideal conditions of an experimental garden in a research institute in California, 55 species, hybrids and varieties of *Pinus* produced ovulate cones at an average minimum age of 5.2 years and 39 produced staminate cones at average minimum age of 4.4 years. The juvenile period for female cones was on average longer than for male cones (Righter, 1939), but this is not always true even in *Pinus*: in a population of 1050 trees of *P. sylvestris* in Pennsylvania (U.S.A.) two carried ovulate cones at age 3 and 790 by age 8 whereas only 199 bore male cones at age 8 (Gerhold, 1966).

The conclusions from a wide literature review (Harper and White, 1974) were:

> In comparisons *between* species, vegetative vigour is associated with late flowering. *Within* the species the more vegetatively vigorous individuals flower first . . . shade intolerant, colonizing (*r*-type) species tend to have precocious reproduction associated with large seed numbers, small seeds and a high reproductive efficiency. In contrast, species that occupy later positions in forest succession and are shade-tolerant have fewer and larger seeds and a long juvenile period: all available assimilates in early life are apparently channelled into establishing canopy height and achieving competitive dominance.

These generalizations were made for trees, but they can be extended more widely to perennial plants as a class, except that in herbs the channelling of early resources is not into a tall trunk but may be to a vigorous root system or a perennating or storage organ underground. These are *among* the interpretations that can be offered as explaining (in terms of fitness) the various delays in the onset of reproduction in flowering plants and coniferous trees. To these may be added that it may be a fitter strategy to delay reproduction and remain dormant or vegetative if the chance of progeny finding a suitable site for establishment is increased in this way.

The reproductive period

Much effort has been put into understanding the significance of variation in life cycle strategies — mainly by modelling exercises. Most models assume that reproduction has a cost to the organism in terms of its future growth rate, expectation of life and chance of producing future progeny. Some of these costs have been illustrated in Chapter 21. The effort put into an act of reproduction can then be envisaged as a sacrifice of some future reproduction. The reproductive value of early pro-

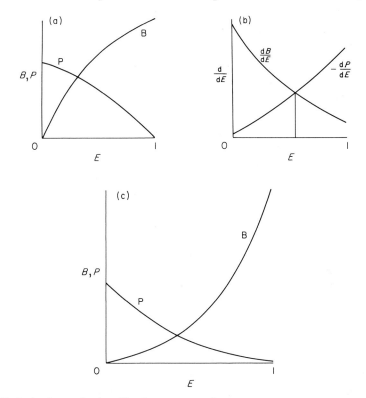

Fig. 22/6. Optimal reproductive effort in constant environments.

(a) Effective litter size B and post-breeding survival P plotted against reproductive effort E. The curves are concave and the value of effort E that maximizes $B+P$ (i.e. the yearly rate at which the population multiplies) intermediate between 0 and 100% selects for iteroparity (polycarpy).

(b) For concave B and P, E satisfies the relation $(dB/dE) = -(dP/dE)$.

(c) B and P plotted against reproductive effort but curves are convex. $E = 100\%$ — selects for semelparity (monocarpy).

(From Schaffer, 1974)

geny is greater than that of late in an expanding population and vice versa in a decreasing population, so it is not sufficient simply to set early progeny against late as equal contributors to the number of descendants — it is necessary to discount organisms produced at different stages in the life cycle according to their "value" in contributing descendants to subsequent generations. Thus ". . . optimal life history maximizes for each age class the expected fecundity at that age plus the sum of all future expected fecundities, each discounted by an appropriate power of e^{-m} where m is Fisher's Malthusian parameter" (Schaffer, 1974).

Schaffer (following Gadgil and Bossert, 1970; Schaffer, 1972) considers the relationship between reproductive effort and both fertility and post-breeding survival and growth. Figure 22/6 shows three idealized diagrams in which the cost of reproductive effort is shown in relation to fertility, post-breeding survival and growth. When both the functions are concave (Fig. 22/6a) a little reproductive effort brings a big fecundity reward at a low cost to post-breeding survival and growth. Under these circumstances, Schaffer shows that fitness is maximized by iteroparous (polycarpic) reproduction. In contrast, where both functions are convex (Fig. 22/6b), increasing reproductive effort brings more than proportionate rewards at decreasing cost. Under these circumstances, fitness is maximized by a semelparous (polycarpic) "single Herculean effort" (Schaffer, 1974), "a big bang" of reproduction (Gadgil and Bossert, 1970). Where there are concave—convex functions of greater complexity there can be alternative iteroparous or semelparous strategies, each of which represents a local maximum in fitness and in which the evolutionary choice taken by a population may be determined primarily by its genotypic composition at the start of the period of selection. For this reason, Schaffer feels that it is not surprising if iteroparous and semelparous strategies are found in closely related species.

Fecundity schedules are known for very few plants. Probably the normal course of events in polycarpic perennials is for the plant to increase its reproductive output over the years to a plateau and then decline. In *Citrus* species (*C. paradisi, C. reticulata* and *C. sinensis*) trees begin to bear fruit at about 3 years old, reaching a plateau at about 20 years and maintain their output (or slowly decrease it) over the next *ca* 30 years (Savage, 1966). The mango bears 15—20 fruits at age 5, 400—600 (in an "on" year) after age 10, about 1000 between ages 10 and 15, 2500 by the 20th year rising to about 5000 per year and beginning to decline after about 40 years (Singh, 1960). The fecundity schedule of several species of oak reaches peak reproductive activity at age 40—100 and they then start to die as portions of the crown decay (Goodrum *et al.*, 1971).

Although we have little information about the shape of the reproductive period over the lifetimes of genets there is a great deal of knowledge about variations within it. It is rare for the reproductive activity of a perennial to be a continuous process, even in the most constant of environments (see Chapter 19). Even the fig *Ficus sumatrana* which fruits repeatedly through the year in Malayan rain forest has cycles of on and off periods three times a year (Fig. 15/4). Usually the period of seed production is an episode within the annual cycle of growth and there is a lot of variation between years in the reproductive activity of an individual. In perennial herbs the years of flowering and seed production are often interspersed with periods of purely vegetative activity and there is usually no sign of synchronization between members of a population between years (although those that flower are usually highly synchronized *within* years). The records of flowering obtained by Tamm for herbs in woodland and meadow (see Figs 19/5—19/11) illustrate very clearly the sporadic flowering habits of most individuals. The underlying causes are barely understood; though the fact that most perennial garden herbs can be relied on to flower every year after the juvenile period, suggests that the ability to flower in nature is probably curtailed by density stresses from neighbours. Perhaps the great variation in reproductive activity of individual plants in nature could be explained by correlation with the growth and activity of the vegetation surrounding each individual.

The Russian population biologists of the school of Rabotnov have made the most detailed studies of the reproductive frequency of herbs in the field. They showed, for example, that individual plants of *Heracleum sibiricum* had a lifespan of 13—15 years and that in this period most individuals flowered and set seed three or four times, although a few plants died after the first or second flowering period (Rabotnov, 1956). Reproductive schedules have sometimes been determined for even-aged populations of plants: for example by agronomists concerned with the useful length of life of a grass sward grown for seed production. Such a schedule is illustrated in Fig. 22/7 for *Poa pratensis* (Evans and Canode, 1971). A peak reproductive activity in the third year is followed by a rapid decline, perhaps to a plateau. The fall in the rate of

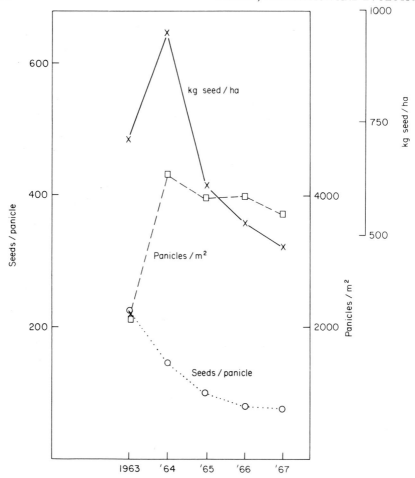

Fig. 22/7. The time course of seed production by a population of *Poa pratensis* sown in September, 1961. (From data in Evans and Canode, 1971)

seed production is accounted for in part by a reduction in the number of panicles per m² after year 4 but mainly by a reducing number of seeds borne per panicle. Essentially the same reproductive schedule is shown by *Bromus marginatus* (Stark *et al.*, 1949) where seed production per hectare declined from 1243 kg 2 years after sowing, to 467, 380 and 319 kg in succeeding years. The reproductive schedule of a population is a somewhat dangerous substitute for one made on individuals: the genets that contribute to the activity of the population as a whole may well have on and off years for flowering and during the years through which reproductive activity is followed fewer and fewer genets will be contributing to the seed produced by the whole population.

The most striking periodicities of flowering are found among bamboos. Species appear to have characteristic flowering cycles of 1, 3, 11, 15, 30, 48, 60 or even 120 years when all the plants from a given seedling (a whole genet) flower simultaneously (McClure, 1967). Most of the evidence for flowering periodicities in the bamboos comes from anecdotal information and "there is no positive basis for the generally held assumption that bamboos have in their character a flowering cycle of precise and invariable length" (McClure, 1967). It has recently been argued (Janzen, 1976) that the periodic flowering in the bamboos is a predator-escape strategy and Janzen has produced a very convincing body of evidence that major cycles of predator activity are generated by a bamboo flowering episode.

The mast year phenomenon is widely known in forest trees. Years of vigorous seed production are interspersed among years in which trees are sterile or nearly so. The advantages of irregular reproduction have been discussed in relation to predation in Chapters 13 and 15, and it was pointed out that the drain on resources of a period of intensive seed production may itself set back growth in that period and be reflected in reduced reproduction in one or more succeeding years. A tree may devote a large part of its resources to seed production or to vegetative growth but in many species the activities are not fully compatible. The mango illustrates this phenomenon very elegantly. Young trees of *Mangifera indica* produce a crop of fruit regularly every year but, after about 10 years of fruiting, reproductive activity becomes periodic (sometimes biennial). When a heavy crop of fruit is carried, the tree makes few new vegetative shoots in the same year. Inflorescences are borne on new shoots and so the tree has lost its potential for reproduction in the year following a large seed crop. Cultivars that fruit regularly every year produce about 30 fruits a year but the fully biennial trees bear 500–1000 fruits (Singh, 1960). It may be more than a coincidence that the biennial produces about the square of the number of fruits every other year that the annual bearer produces every year.

Periodicities in reproductive activity may or may not be synchronized between trees in a forest. A hazard that affects an individual in a stand may set back its vigour and throw it out of synchrony with the remainder. A hazard that affects a population as a whole — a disastrous

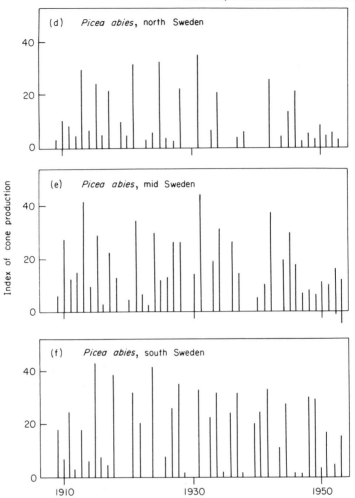

Fig. 22/8. Fluctuations in the cone crop of Scots pine (*Pinus silvestris*) and Norway spruce
(*Picea abies*) in north, central and southern Sweden. Based on cone counts made
by the Swedish National Forest Survey from 1954—62 and reports by rangers in
years 1909—61. (Drawn from data of Hagner, 1965)

epidemic defoliation or a failure of pollination or fertilization due to
a seasonal hazard — may synchronize all the members of a population
which then recover to a seed-bearing condition at about the same time.
The hazard may have a time lag of 2 years in species with a long period
of embryonic development (e.g. *Pinus sylvestris*), so that the same

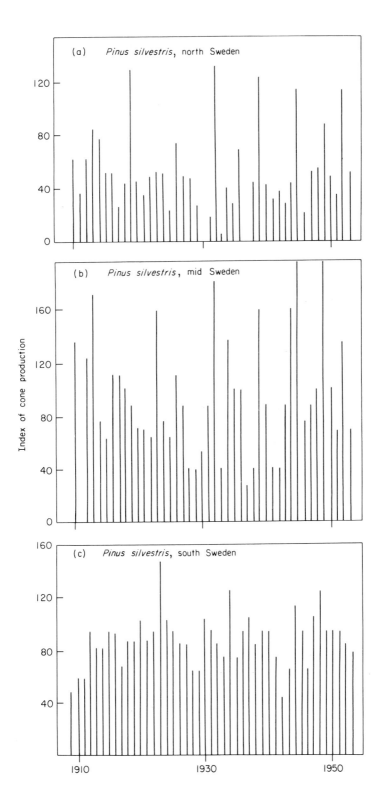

(a) *Pinus silvestris*, north Sweden

(b) *Pinus silvestris*, mid Sweden

(c) *Pinus silvestris*, south Sweden

Index of cone production

1910   1930   1950

environmental event may start differently timed periodicities in different species. Analysis of 45 years of flowering records for dipterocarps showed that the members of a population tended to flower gregariously every 2—5 years. Up to half the individuals flowered in any given year.

The mast year phenomenon is best known in *Quercus* and *Fagus* (e.g. Gysel, 1971) and in conifers such as *Pinus, Pseudotsuga, Abies* and *Picea*. The occurrence of mast years seems to be most strongly marked in species that produce large crops of large seeds and in which the reproductive organs are not provided with photosynthetic appendages. The phenomenon seems to be less marked in *Tilia, Carpinus, Ulmus,* in which there are conspicuous photosynthetic bracts associated with the flowers. An analysis of the periodicity in seed production was made on four trees of *Pinus monticola*, 24—33 years old, that had deliberately been freed from interference from neighbours so that their seed production was maximized. The trees were within 1 km of each other at elevations of about 1000 m in North Idaho (Rehfeldt *et al.*, 1971). A spectral analysis of these records showed major cycles of 4 years' duration and that the individual trees were in phase.

A great many attempts have been made to explain the mast years of tree species by correlation with climatic variables. The long process of fertilization and "gestation" in most pines means that a causal environmental factor may have operated 1, 2 or even 3 years before the consequences are seen in seed output. The following correlations have been claimed, among others:

(i) Mean temperature in early summer of $Year_{-3}$ negatively associated with cone development in *P. resinosa* (Lester, 1967).

(ii) Early summer irrigation and late summer drought in $Year_{-2}$ increased strobilus production in *P. taeda* (Dewers and Moehring, 1970).

(iii) Mean precipitation in early summer of $Year_{-2}$ was positively correlated with seed production in *Pinus pinea* and *P. taeda* (Pozzera, 1959).

(iv) Mean temperatures in $Year_{-2}$ were positively correlated with seed crops of *P. ponderosa* (Maguire, 1956).

(v) Mean temperatures in $Year_{-2}$ but for the midsummer period alone, were positively correlated with seed crops in *P. resinosa* (Lester, 1967) and *P. ponderosa* (Daubenmire, 1960).

(vi) Spring droughts in $Year_{-0}$ appeared to be associated with the abortion of young cones in *P. radiata* (Pawsey, 1960).

Among the various correlations that have been detected it appears that the climate in year$_{-1}$ is almost irrelevant.

It is always difficult to disentangle correlations involving meteorological variables and from the large amount of research on correlation betwen seed crops and the climate no single correlate emerges that adequately "explains" mast year phenomena in the genus *Pinus* — certainly there is no overwhelming correlate for trees in general. There seems to be wide agreement, however, that a large seed crop reflects a condition of vigour: the largest crops are borne by the largest trees, by the individuals that extend high into the canopy and have not recently suffered a major setback in growth (Larson and Schubert, 1970).

As the distribution of a species approaches the limits of its range, hazards to life are likely to be more frequent and plants are more often to be found recovering from the last disaster. This principle is well illustrated by the very extensive records of cone production made by Swedish foresters in pine and spruce forests in different parts of Sweden. Figure 22/8 shows the cone production of *Pinus silvestris* in northern, central and southern Sweden over a period of 60 years. The periodicity and the total variance of cone production were much greater in the northern forests. In the south the number of cones has higher predictability; there were no seedless years and the mast years were scarcely recognizable. Over much the same range (though not necessarily the same range of tolerance of the species) *Picea* showed very little difference in the predictability of the seed crop and, in comparison with the pine, was an erratic cone producer. Over the three regions for the 60 years there were two seedless years for the pines (both in the north). The spruce in contrast failed to produce seed in 22 of the site-years.

The spatially fixed character of plants places restrictions on the optimal reproductive strategies. Most seeds fall close to the parent plant — such seeds are unlikely to grow to maturity as long as the parent is alive and monopolizing the local resources. The chance that a plant will leave descendants is thus linked to the chance that the parent will die and this trend becomes stronger, the later the phase in a succession that is occupied by the species. The reproductively most valuable trees in an old forest may often be the ones that die in a last burst of reproduction because it is only then that their seed progeny have an opportunity to develop. A quite contrary strategy will determine the reproductive value of early successional fugitive species of tree, where early reproduction brings the chance of opportunistic escape to new successions starting elsewhere. It may be in the context of reproductive value

that the periodicities of seed production by trees take additional meaning. Periodicities are found most strongly in late successional species
of tree and there may be real advantage in a strategy of reproduction
that is not homoeostatic — that does not reliably balance reproductive
effort and the effort put into growth but allows quite violent oscillations
between the two. Quite apart from the advantages of alternately starving and oversaturating predators (Chapter 13) the reproductive value
of reliable year to year reproduction is probably only very slight in
organisms with a very long reproductive life.

The length of life of plants — death

There is a great literature about the lengths of life of plants, particularly
of trees. The existence of annual growth rings tempts botanists to establish records for old age in different species. There is a sensation-seek-
ing element in this and much of the literature on plant age establishes
the age of the oldest individual. This is not the sort of information that
usually interests demographers. The old animal in a population is usually almost irrelevant to population growth and has lost its reproductive
value. This is not necessarily true for plants and the oldest individuals
are often those that dominate a canopy or have formed extensive clones.
The oldest individuals in a plant population may have the greatest reproductive output, control the largest fraction of the resources, and
control the recruitment of new seedlings. There is probably no easy
parallel in higher animals to a situation in which the old control the
recruitment of the young, as is the case in the regeneration of mature
forests.

The data collected in Fig. 22/5 shows that it is the conifers that attain the greatest ages among trees and dominant members of a canopy
commonly reach an age of 200 and sometimes 400 years. The dicotyledonous trees are shorter lived though clones reach very good ages.
Populus tremuloides var. *aurea*, for example, forms clones that are
thought to be about 8000 years old (Cottam, 1954). A very few trees
are monocarpic: these are palms such as *Corypha elata* which lives for
about 40 years before flowering. A measured specimen attained a height
of 24 m before flowering and dying at age 44 (Corner, 1966; Zimmermann, 1973). A very odd form of terminal reproduction occurs in
Caryota urens, a tropical tree which builds a large trunk 18—30 m tall
and then starts to flower at about 15 years old. The flowering period

lasts 5—7 years and starts at the top of the tree; subsequent inflorescences develop from lateral buds lower and lower down the trunk until
the last flowers are borne at the base of the trunk and the tree then
dies (Corner, 1966; Purseglove, 1972).

The reproductive value of the seeds of a perennial is not an easy
matter to assess. The calculations by Fisher and others show that the
theoretical rate of population growth is dependent on the fertility and
death schedules of the parents, but do not allow for differing chances
of survival to reproduction of progeny born at different stages in the
life of the parent. There are such differences in some animals (e.g. the
increasing risk that offspring will bear chromosomal abnormalities as
the mother ages) but to this, for plants, has to be added a change in
the nature of the habitat during the growth of a population. The example
was quoted in Chapter 20 of a forest of *Abies balsamifera* in which 67%
of the seedlings that regenerated after a catastrophe came from the seed
shed in the first 28% of the total reproductive output of the parents.
This is a clear case in which precocious reproduction contributed to fitness — the later and larger reproductive period was relatively ineffective
in contributing descendants. In marked contrast, after the death of a
tree in a mixed-age population, the most likely colonists of the bare
space beneath it may often be the seeds of that tree if it had continued
bearing until old age. Under such a condition it may be the last dying
burst of reproduction that is critical for the multiplication of a particular genotype and the precocious seeder will be at no advantage. The
reproductive value of a forest tree depends also on whether the size
of populations is dependent on repeated episodes of invasion into new
territory; if so, precocity brings all the classic reproductive value to the
individual. If, however, the population is stabilized and seed is successful in establishing new plants mainly where they replace dying parents,
there is no premium on precocity and there may be a quite different
reproductive value, that of reproducing right to the year of death. The
reproductive value of the last seeds produced by a tree will presumably
be higher if (a) there is no bank of buried viable seeds from previous
generations and (b) there is no reserve of young saplings waiting to
occupy positions in the canopy.

Senescence and death in old age set puzzling problems to any biologist. The characteristic feature of senescence involve a breakdown in the
efficiency of functioning of an animal with age and this has been interpreted by Medawar (1952, 1955) as an evolutionary consequence of

the decline in reproductive value of an organism with age. The action of deleterious genes is thought to be postponed by modifiers and their expression drifts later into the life cycle, thus giving the attributes that we associate with senescence. It is not simply the case that deleterious factors are moved into periods of the life history with low reproductive value (Medawar, 1952, 1955); Hamilton (1966) has shown that there is no reason to suppose that such deleterious factors are moved forward into the juvenile phase where the reproductive value is also low. Rather, the expression of deleterious attributes occurs later and later in life — "senescence is an inevitable outcome of evolution" (Hamilton, 1966). These views are not easy to reconcile with the life of some perennial herbs. Plants with clonal growth show no apparent senescence and in species like *Pteridium aquilinum* there is no reason to suppose that the growth rate or fertility of clones has declined since genets became established in the Iron age! In the case of a clonal plant or a tree, being older commonly means being bigger and the old plant is usually the successful clone that has spread itself widely or the tree that has gained dominance in the canopy. The tree form brings with it necessary death (unless it is clonal) and in this case the causes of death seem to be closely tied to the side consequences of being large and having accumulated dead tissue. It is easier to interpret the senescence of trees as a consequence of the accumulating burden of respiratory tissue and disease-prone dead material than to the accumulating expression of sub-lethal genes. The very long lives that are known for some trees are probably short in comparison with the clones of some herbaceous plants — for the clonal herb old age has little meaning. The reproductive potential may theoretically increase indefinitely as it grows older and bigger. Somatic mutations affect meristems individually, not the whole functioning of the genet, and even the accumulation of viruses affects only parts of a genet if the connections between the parts decay. Although Hamilton (1966) has argued that in a population of organisms that increase their reproductive capacity exponentially with age there is still a tendency to accumulate senescent properties, there is remarkably little evidence that this happens in clonal plants. It may be that most plants in nature spend most of their lives in declining populations and that this gives a twist to natural selection so that it favours "eternal life" in herbaceous perennials.

It is of course an error to regard age as a function only of plants in the active phase of growth. Age takes on quite a new significance for organisms that have long-lived dormant seeds and it is easy in the case of some of the annual and biennial weeds of disturbed habitats to see that very long life involving an enormously elongated dormant juvenile period as buried seed and/or delayed reproduction may represent just the strategy that augments fitness in specific hazardous environments.

23

Community Structure and Diversity

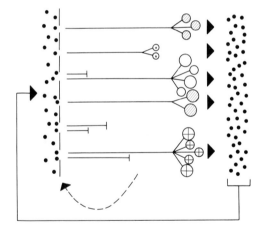

The plant population that is found growing at a point in space and time is the consequence of a catena of past events. The climate and the substrate provide the scenery and the stage for a cast of plant and animal players that come and go. The cast is large and many members play no part, remaining dormant. The remainder act out a tragedy dominated by hazard, struggle and death in which there are few survivors. The

appearance of the stage at any moment can only be understood in relation to previous scenes and acts, though it can be described and, like a
photograph of a point in the performance of a play, can be compared
with points in other plays. Such comparisons are dominated by the
scenery, the relatively unchanging backcloth of the action. It is not
possible to make much sense of the plot or the action as it is seen at
such a point in time. Most of our knowledge of the structure and diversity of plant communities comes from describing areas of vegetation at
points in time and by imposing for the purpose a human value of scale
on a system to which this may be irrelevant. The life and reproduction
of an individual plant is determined by events that impinge directly
and indirectly upon it; the scale of these events differs from species to
species. The scale at which an oak tree in a forest senses the presence of
its neighbours and the heterogeneity of its environment is different
from the scale relevant to a bluebell on the forest floor or a seedling
oak.

The analysis of pattern in plant communities shows that the appropriate scale for describing the pattern of one species in a community
may be quite different from that appropriate for another (e.g. Greig-
Smith, 1961). Both the morphology of genets and the dispersal pattern
of seeds may produce characteristic elements of pattern in the community. Each organism defines the scale of its environment; the "plant's-
eye" view is what is relevant to explain the distribution, adaptation
and the processes of change within species and within communities.
It is here that the analogy with players on a stage breaks down, as each
player in a population of organisms defines the size of its own stage.
A plant exerts an influence as far as its root system spreads, its canopy
shades and its products are dispersed. The animal in a plant community
exerts an influence determined by its range of dispersal and search. A
community that appears diverse to the observer of a quadrat of vegetation may be very monotonous to an organism within it. Individuals
of different species sense the diversity of the community in different
ways and on different scales. Just as an antelope "samples" the diversity of a savannah differently from a lizard, so a clonal plant with long
rhizome internodes samples the surrounding vegetation in a very different way from a tufted plant. A biological, as opposed to a geographical
description of community diversity must measure the ways in which
the plant senses its surroundings. For the population biologist, diversity is most interesting when it describes some element of the proba-

bility of encounters — is the community diverse for its inhabitants
or only for the observer? (see discussion in Hurlbert, 1971).

There is a variety of levels of diversity that a plant may meet amongst
its neighbours:

(i) *The somatic polymorphism of the parts of a genet.* There is a real
diversity in a stand of *Eucalyptus* spp. that derives from the different
form and arrangement of leaves on the juvenile branches compared with
the diversity of the mature form. Such a phenotypic diversity is as real
a component of the variety within the community, from the point of
view of an invading plant, as if the forest contained two different species
of tree. Similarly the presence of entire leaves on the rosettes of
Valeriana dioica but pinnate leaves on the flowering stems contributes
two distinct leaf forms to the variety of a herbaceous canopy. Examples
of this scale of diversity are illustrated in the extreme leaf polymorphism of desert shrubs (see Fig. 3/1), where a single species plays different roles with a different form in different seasons.

(ii) *The diversity of age-states within the community.* Even the different ages of plants of the same species within a community, e.g. an
even-aged stand and a mixed-aged stand of the same species, represent
different scales of diversity. The vegetative rosettes of *Digitalis purpurea*
on a woodland floor have a very different size, height and relationship
to neighbours from the flowering individuals; the morphological difference
between the age-states is as great as that between many species. During
year 1 + the foxglove behaves rather like a primrose, and later like a
golden-rod! The distribution of ages within a population may be one of
the elements of diversity that contributes to the stability of the community — at least in the sense that it permits or denies the chance of
rapid recovery after a disaster (Demetrius, 1975).

(iii) *The genetic variants within a species.* The diversity of a plant
community is inadequately described by the number and abundance
of the species within it. A major part of community diversity exists
at the intraspecific level. If generalizations are to be made about the
relation between community diversity and, for example, its stability
or productivity there is no special reason to pick the species as the
relevant element in the diversity. It has indeed been suggested that
where the number of species is low (loose species packing) the intraspecific diversity may be higher — a form of compensation (Van Valen,
1965, 1970).

The species is not the unit of precise local adaptation and the ecotype that represents a species in an area and its genetic diversity may not be the same even in closely neighbouring areas (Bradshaw and Jain, 1966; Snaydon, 1970): even a 1 m² quadrat of pasture containing *Trifolium repens* will usually include: (a) genetic polymorphism with respect to cyanogenic glycosides affecting acceptability to slugs and probably other predators; (b) genetic polymorphism with respect to leaf marks which have adaptive value because they appear to be used as feeding images by sheep; (c) genetic polymorphism with respect to susceptibility to infection by *Rhizobium* and further genetic variation that determines whether successful infections are also successful nitrogen-fixing associations; (d) genetic variation in leaf size (sheep select clover with respect to leaf size) and in persistence under grazing pressure; (e) genetic variation between clones in aggressiveness to different associated species of grass in the sward (Turkington, 1975); (f) polymorphism of incompatibility systems determining which combinations of parents can leave descendants. There are almost certainly also genetic variations affecting the length of stolons (and so clonal colonizing ability), the rate of stolon death, disease resistance and flowering frequency.

(iv) *The diversity of microsites within the habitat* may permit different species to occupy specialized microenvironments within the community. Often this underlying heterogeneity is difficult to define, but Fig. 23/1 describes a transect across permanent grassland in which the soil surface has characteristic ridges and furrows in a regular repetition. The floristic diversity of the community — under the same overall grazing regime — changes repeatedly from a very low value in the furrows to a high value in the ridges. It may be relevant in this case that the two species found in the furrows are *Agrostis stolonifera* and *Ranunculus repens,* both of which are stoloniferous species in which the genets explore and "sense" the habitat over a considerable distance, whereas the species on the ridges are mainly forms with no clonal growth (*Ranunculus bulbosus*) or with very localized intravaginal tillering (*Lolium perenne, Cynosurus cristatus*). A statement about the floristic richness of such a pasture as a whole has no real meaning in relation to the chances of interspecific encounter.

(v) Any measure of the *diversity* of a community ought perhaps sometimes to take into account groupings *at a higher level than the species,* e.g. generic and family diversity, or else escape completely from the taxonomic and systematic limits on description by considering the "diversity" of growth forms in an area or sample.

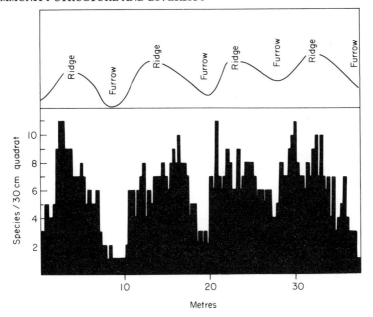

Fig. 23/1. Changes in the floristic richness of a grassland community near Oxford across a pattern of ridges and furrows. (From data of G. R. Sagar)

Elements contributing to the diversity of plant populations

It is helpful to consider plant populations as developmental sequences in time — in which disasters occur with different frequencies. The nature of a plant population then represents the stage in the working out of the consequences of a past episode. Figure 23/2 illustrates a theoretical community maturation curve in regions of different disaster frequency. No time scale is given — it is presumably some function of the length of life of the organisms concerned. No vertical scale is given — it is presumably something called degree of maturity, an approach to a theoretical asymptotic state, a K-condition in which resources are thoroughly exploited.

Populations that are always recovering from disasters are dominated by the "Gleason" forces (Gleason, 1926) of chance dispersal, proximity of parents, differences in dispersal distances of species and the raw physical features of the environment — soil and climate. The composition of such communities is controlled during the early phases of their development by the invasive and multiplication rates of the

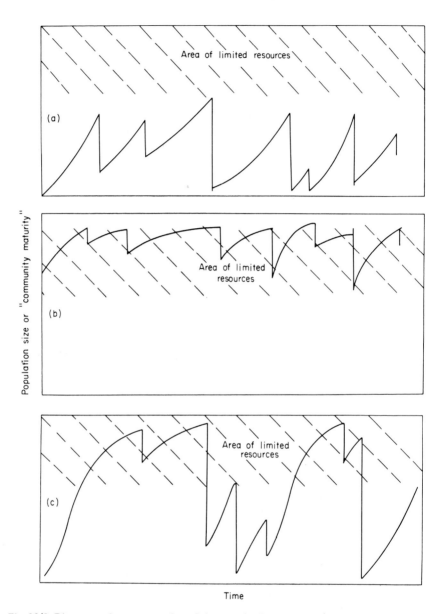

Fig. 23/2. Diagrammatic representation of the growth of populations (or the maturation of communities) in environments with disasters and resource limits: (a) always in r phase (b) always in K phase, (c) mixed.

colonists. There is a very close analogy between this r phase in the development of a plant community and the "founder effect" among the genotypes within a species that colonize a new "island". The initial composition of the population has little to do with adaptation or with interaction between the individuals. It is chance dominated in its beginnings and dominated by the rates of natural increase afterwards. There are few biological problems in the diversity of communities in such colonization stages. Limitless diversity of plant species can be contained provided that the species tolerate the climate and soil conditions and that a drawn-out struggle for existence between the individuals is prevented. Every botanic garden is witness to the incredible diversity of plant species that can survive side by side provided they are not too close together! In practice the stages of recovery of communities from disasters (like the early stages of colonization of a bare island) are overwhelmingly controlled by the dispersibility of propagules (in time as well as in space) and the intrinsic rates of natural increase of various populations. In a patchy mosaic of habitats, "a harlequin environment", the continual process of migrations and local extinctions is capable in itself of maintaining a high level of diversity (Skellam, 1951; Hutchinson, 1951; Hutchinson, 1965; Horn and MacArthur, 1972). A disaster is thought of as an unusual occurrence — a hazard in the life of the inhabitants of a community. Occasional fires, hurricanes and very extreme episodes of cold are catastrophes — they set back the development of a community in a drastic fashion. The winter of 1962/3 was the coldest in Britain for probably 700 years — this was a catastrophe for *Juncus acutus* — up to 88% of this very locally distributed plant were killed in one area of Devon (Hewett, 1971). Such catastrophes must not be confused with a situation in which an environmental disaster occurs with sufficient frequency and reliability that organisms have become adapted to it as normal expected experience in natural selection. Thus the weed flora of arable land is not really a catastrophe flora because the action of ploughing usually follows remorselessly year after year. The species composition of a long-established arable land is of forms that tolerate or are favoured by this type of disturbance. Even fire, if it is a reliable repeated hazard does not come into the category of a catastrophe; the flora that develops does not depend on new colonization phases after each episode but depends on buried viable seed (Sweeney, 1956).

In general, species that spend their time in habitats that are recovering from disasters are r-species: colonists, forms with high

dispersibility, high reproductive effort, with little of their soma devoted
to aggression and the preemption of resources. In such habitats there
are no special problems in the cohabitation of closely related species,
the wide variation in abundance of single species or wide variation in
the number of species. By definition any problems of resource shortage
in this situation are not created by the proximity of neighbours. When
a colonizing phase has continued long enough for individuals to interfere with each other, or if initial colonists were dense enough for this
to happen directly, a different phase in community development has
started and a series of quite new biological problems arise.

When individual plants are in such proximity that they make mutual
demands on limiting resources or influence each other's development
in other ways, the intrinsic rate of natural increase and the rate of dispersal have a declining influence on the composition of a community.
Instead, the consequences of interactions between organisms come to
dominate events. There is abundant experimental evidence (see e.g.
Chapter 11) that pairs of species grown together usually differ in their
ability to capture resources and to leave descendants. There is then a
biological problem of extreme interest in natural diversity: how does
it come about that one form or one species is not the unique winner in
the struggle for existence that develops?

Whittaker (1969) has likened a group of potentially competing species
to dancers:

> as additional dancers enter the floor, manoeuvres make space for them, with
> reductions of the dance areas of the remaining couples. There may come a
> time, however, when a part of the floor becomes so crowded that the rate at
> which new dancers enter is equalled by the rate of departure of couples dis-
> couraged or crowded off the floor.

The Lotka-Volterra equation for the relationships between two species
living in an environment with limited resources specifies the conditions
under which different sorts of dancer may remain together on the dance
floor (Chapter 1). Two types of dancer can persist together on the
crowded floor if each interferes more with its own sort than with the
other. Such dancers effectively occupy different or partly different
niches and can form a stable mixture.

Much of the thinking about species packing and the differentiation
of niches has been concerned with birds (e.g. MacArthur, 1971) and it
is now easy to envisage specialization between groups of birds in the
way in which they take foods. The range of foods available represents

a dimension within the n-dimensional niche hypervolume and a species may be adapted with some precision to a particular food size or type so that several species persist together in a habitat, each taking food within a circumscribed limit of choice (see e.g. Lack's study of Darwin's finches: Lack, 1947).

May (1973) has presented models of simple biological communities in which several species compete on a one-dimensional continuum of resources, e.g. food size:

> In a strictly unvarying (deterministic) environment, there is in general no limit to the degree of overlap, no limit to the number of dancers. However in a fluctuating (stochastic) environment, it is found that the average food sizes for species adjacent *on the same resource spectrum* [my italics] must differ by an amount roughly equal to the standard deviation in the food size taken by either individual species. This limit to species packing has a very weak (logarithmic) dependence on the degree of environmental variance.

This conclusion, of course, does not tell us how many species may be present in a community. For that, we would need to know whether there is a limit to the number of resource spectra.

The emphasis on food size as the relevant dimension of the niche is historical because of the strong role of ornithologists in the development of the theory. It is less easy to see how different species of plants may divide up a resource spectrum between them or rely on different spectra of resources. All green plants have essentially the same requirements for light, water and nutrients (Chapter 10), and with the exception of those that have access to microbially fixed atmospheric nitrogen, there is little opportunity for different species of plant to specialize on "food" resources. Moreover, a plant's resources are not packaged nor unitized as articles of food like that of animals. There may, however, be more subtlety in the exploitation of nutrient resources than appears at first sight and it is presumably perfectly possible for a variety of species depending on the same complex of nutrients to be limited each by its special sensitivity to the need for one of these nutrients. However, the specializations that we most commonly see within communities of plants are with respect to the dimensions of time and space in which resources are used. The resources needed by green plants for growth are distributed in an environment that is heterogeneous in space and time and most of the niche differentiation that has occurred has been interpreted in relation to this heterogeneity.

Diversity dependent on the use of different resources

The outstanding example of a group of plants that has escaped from
dependence on the common pool of resources is that group of species
that has access to atmospheric nitrogen by virtue of a symbiotic
association with nitrogen-fixing organisms. The legumes and species
such as *Hippophaë*, *Alnus* and *Azolla* have escaped from the otherwise
universal dependence of plants on fixed nitrogen. This makes possible
a floristic diversity in which the non-legumes in a community suffer
from each other's presence because they make demands on fixed nitrogen but the legumes are not deeply involved in this competition. It is
not surprising that in the various experiments made to determine the
mutual relationships of pairs of species it is mixtures of legumes and
non-legumes that reliably form self-stabilizing diversity (see Chapter 9).
Some plant species that form mycorrhizal associations (e.g. involving
Endogone) may also escape from the struggle for existence with other
species by obtaining, via the fungus, resources that are not available to
non-mycorrhizal associates. Insectivorous plants also escape from the
limitations of resources that affect more normal plants. They have
access to an unusual and specialized source of nitrogen and other elements and by-pass the general struggle for limited resources. It is rather
surprising that they are not more common. It is imaginable that higher
plants might have evolved to partition the spectrum of light, different
species specializing on different wavelengths, but this has not happened.

Diversity dependent on the lateral heterogeneity of environments

Perhaps the commonest differentiation between plant species is in relation to the lateral variation in conditions in a habitat. The resources
needed by plants are distributed in an environment that is heterogeneous both in space and in time. The specializations that permit
plants to form diverse communities are rarely in a particular type of
resource but usually involve seeking for the same resources at different
times or in different places. Instead of a spectrum of resources we find
a spectrum of conditions in which species are packed in a mosaic of
environmental gradients of pH, temperature, drought risk, etc. In any
one position on the gradient one form tends to be the winner in a
struggle for existence. There is a close analogy with the spatial (as opposed to resource) separation of species of warblers feeding at different
heights in a coniferous forest (MacArthur, 1958) and species of tit

(*Parus* spp.) feeding at different heights in British woodland (Hartley, 1953). An example of species packing among higher plants is that of oaks and other broad-leaved trees along an altitudinal gradient (Whittaker, 1969; see Fig. 23/3). An even more striking example is found

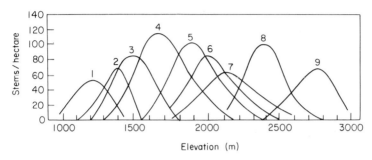

Fig. 23/3. The distribution of broad leaved tree species along the altitudinal gradient of the Santa Catalina Mountains, Arizona. Species 3 and 5 are evergreen oaks (subgenus *Erythrobalanus*) 2, 4 and 6 are evergreen oaks (subgenus *Lepidobalanus*) and 1, 7, 8 and 9 are other broad-leaved tree species. (From Whittaker, 1969)

in the genus *Solidago* where six species are strung out along a gradient of soil moisture. In old field successions the niches are broader and the range of overlap is greater than in mature prairies (Werner and Platt, 1976; Fig. 23/4). This suggests that in more mature communities it is interference between the species that narrows the range of habitats occupied and defines the niches more closely.

When species are separated out along a gradient of some environmental variable the question is immediately posed whether this separation represents "proximate" or "ultimate" factors (Baker, 1938). If the control of distribution is by proximate factors (i.e. present interaction), each of the species would be expected to occupy a much wider range in the absence of the others; if the distribution is controlled by ultimate factors (evolutionary consequences of past interaction narrowing the tolerance of the individuals) the distribution of each species would remain restricted even in the absence of the others.

A variety of species may persist in stable equilibrium within a habitat if each is preferentially favoured in some specific phase of the habitat mosaic. A grassland that has alternate well drained and badly drained patches may support *Ranunculus bulbosus* on the well drained and *R. repens* on the poorly drained areas (Fig. 5/10a, b); the relative frequency of the two species is then a function of the relative frequency of well and poorly drained patches. In such a situation the number of

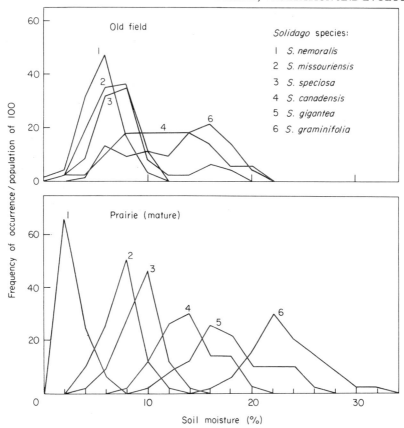

Fig 23/4. The distribution of six species of *Solidago* (golden-rod) in a 6 ha old-field in Michigan and a 16 ha portion of a prairie in Iowa. Soil moisture determinations were made between July and September. (From Werner and Platt, 1976)

species present is a function of the size of the area sampled and the scale of the underlying heterogeneity: both species appear to grow in the same habitat only because the choice of sample size for examining the habitat was not appropriate from the "plant's eye" view. From the point of view of *Ranunculus repens*, *Ranunculus bulbosus* does not grow in the same habitat, and vice versa: plants of the two species rarely make contact. This situation has been called "spurious cohabitation" (Harper *et al.*, 1961). If two species establish from seed in subtly different microsites (e.g. Fig. 5/7) the relative proportions will be determined by the relative frequencies of the types of microsites. Even if, during subsequent growth, the members of the two species interfere

with each other and make demands on limiting resources, every time
that a new colonization by seed occurs the proportions of the two
are corrected by the relative frequency of the two types of microsite.
Only if genets are long lived will this continued correction of the consequences of a struggle for existence be avoided. Only then will two
species which may differ profoundly in the establishment phase be
committed to a remorseless struggle for existence from which only
one form survives. A diversity of species that is controlled by the frequency of species-specific microsites for establishment is really another
example of "spurious cohabitation" in that, at a critical stage in the life
cycle, the two species are controlled by the number of different mosaic
patches in a heterogeneous environment. Any size of quadrat that
could realistically be used to measure diversity in vegetation must be
certain to include a heterogeneity at the scale relevant to seedling
establishment.

Diversity dependent on the vertical heterogeneity of environments

There is also vertical distribution of both conditions and resources in
plant habitats: one dimension of the n-dimensional niche hypervolume
is the depths of the soil at which nutrients and water are available for
extraction by different species. The two species of *Avena, A. fatua*
and *A. strigosa,* illustrate such a separation in the vertical exploitation
of the soil system. Some of the great floristic diversity of chalk grassland floras may be due in part to the great diversification of rooting
depths (see Salisbury, 1952).
    Above ground, the presence of vegetation creates a vertical gradient
of light intensity and this may be exploited by precise adaptation of
different species to different levels of radiation. Whittaker (1972) en-
visages a niche packing in relation to this gradient (Fig. 23/5); the
species lower in the canopy depend on the light that has not been trapped by the over canopy. The plants of the lower storey are dependent
on "the crumbs from the rich man's table". It is probably more common for understorey species to make use of the light when the canopy
species has not yet exposed its leaves.

Diversity dependent on the temporal division of the environment

Time is a further dimension of the niche hypervolume. Resources
needed for plant growth are unevenly distributed seasonally although

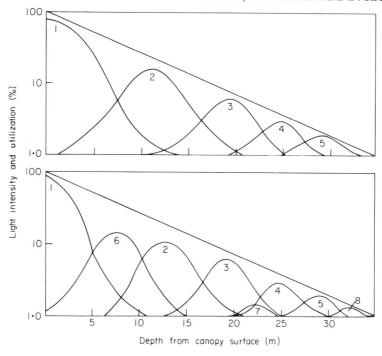

Fig. 23/5. Relations of species populations to a resource gradient (hypothetical). Light intensity
as a resource is plotted on a logarithmic scale on the ordinate; available light decreases with increasing depth from the top of the forest canopy as indicated by the
oblique lines. Species populations are adapted to different light intensities and have
their centres scattered along the gradient. The bell-shaped curves represent relative
utilization of light and relative population densities for the individual species. Five
species occupy the gradient in the upper part of the figure; through evolutionary time
three others enter and take positions along the gradient: Number 6, a strong competitor adapted to high light intensities fits in between 1 and 2, narrowing their
distributions and increasing the "packing" of the resource gradient. Number 7, a weak
competitor, while using a minor fraction of the resource, fits in with its adaptive
centre between 3 and 4. Number 8, a specialist adapted to very low light intensities,
uses the low extreme of the gradient not effectively occupied before. (From Whittaker, 1972)

there is in almost all climates one period that is, in a general sense,
"more favourable to plant growth". The division of the time dimension
of resources and conditions between a group of species is strikingly
illustrated in the early study by Salisbury (1916, 1918) of the exposure
of leaf area by different species within a woodland (Fig. 23/6). More
recently Falinska (1972) has recorded the seasonal distribution of leaf,

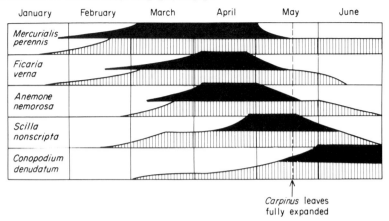

Fig. 23/6. The periods of vegetative growth (cross hatched) and of flowering in the field layer of oak–hornbeam woodlands (*Quercus–Corylus*) in Hertfordshire, England. (Redrawn from Salisbury, 1916)

flower and fruit in a Polish lime–hornbeam forest (*Tilia–Carpinus*). There is very great overlap in the period of leaf exposure (Fig. 23/7) and the sharpest distinctions are between the flowering (and to a lesser extent) the fruiting periods (Fig. 23/8). It is not easy to see the relevance of differentiation between flowering periods unless the plants are competing for pollinators or for seed dispersal agents — both these interpretations seem improbable. A similar seasonality that is more marked for flowering than for foliage is found in chalk grassland (Wells, 1972).

In sand-dune communities in North Wales, groups of autumn-germinating annuals complete their life cycle before the more vigorous perennial grasses and herbs have started active spring growth, so that there is a seasonal succession of activity. There are markedly different seasonal cycles of leaf production between grasses and between grasses and clovers in grasslands. One pair of species that shows a degree of complimentarity in the cycles of foliage production is *Lolium perenne* and *Trifolium repens*. Within an old field of permanent grassland with more than 20 species of grasses and dicotyledonous herbs present there was strong association between the distribution of *Trifolium repens* and *Lolium perenne*: if a leaf of one touched another species it was likely to be the other of this pair. *Lolium perenne* has a major peak of leaf production in early June and again in August–September, whereas *T. repens* has its peak period of leaf production in July, i.e. during the "midsummer

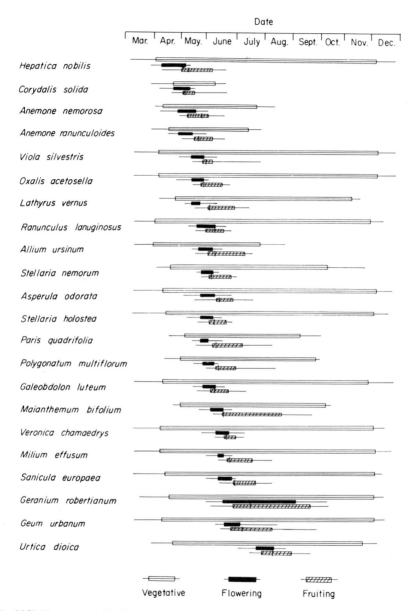

Fig. 23/7. The seasonal distribution of foliar exposure, flowering and fruiting in lime—hornbeam forest at Bielowieża, Poland. (From Falińska, 1972)

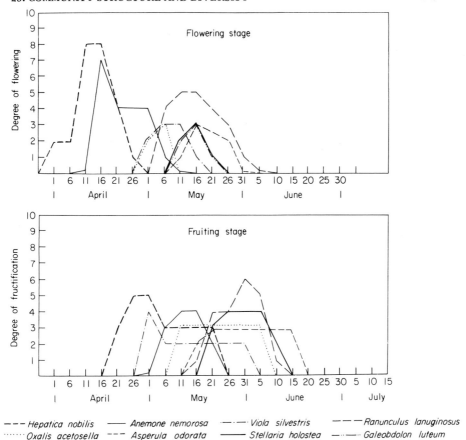

Fig. 23/8. The seasonal distribution of flowering and fruiting of herbs in lime—hornbeam forest at Bialowieża, Poland. (From Falińska, 1972)

gap" of *Lolium*. The close spatial relations that exist between these two species suggest that they fail to enter into an exclusive struggle for existence with each other because of such temporal niche separation. The fact that *T. repens* is a nitrogen fixer and *L. perenne* is among the most responsive of the grasses to nitrogen separates these two species in a second dimension (Turkington, 1975) and permits them to cohabit closely.

Interference and niche overlap in the field

The greater part of the diversity within plant populations that can be ascribed specifically to plant behaviour (i.e. excluding the activities of

animals) depends on differences in the behaviour of species that minimize inter- and maximize intraspecific interference. The differences
represent escape mechanisms from the demands made by neighbours.
It is rather easy to show that neighbouring plants interfere with each
other's activities in the field. Competition between plants is much
easier to demonstrate than competition between animals. The simplest
way to test whether plants in the field are under stress from the presence
of their neighbours is to remove the neighbours. All of forestry thinning practice is deliberate control of the influence from neighbours.
Two experiments made in grassland communities illustrate the intensity of mutual interference that occurs within a sward. A population
of *Plantago lanceolata* growing naturally within a dense grass sward
was treated with the selective herbicide, dalapon (2,2-dichloropropionic
acid), which removes grasses but does not harm *Plantago*. The result
of removing grasses was two-fold. There was a greatly increased flush
of seedlings of *Plantago* and a marked increase in reproductive vigour
of the already established plants, both in the numbers of seeds and
of ramets produced (Fig. 23/9). In this community of plants the grasses
may be said to occupy part of the fundamental niche of the plantains
and the development of the plantains is constrained by the presence
of the grass.

A comparable experiment was made with *Rumex acetosa* and *R.
acetosella* which both live in grassland communities. The sites chosen
for the experiments were hill grasslands in North Wales, on acid soils
of low nutrient status. The community containing *R. acetosa* was
composed predominantly of grasses (*Holcus lanatus, Festuca rubra,
F. ovina*) and had been protected from grazing for 2 years so that the
sward was tall (15—20 cm). The community containing *R. acetosella*
contained abundant *Festuca ovina* together with *Galium saxatile*. Both
communities contained a variety of other herbs. Plots within the
communities were treated to remove specific components: (i) dalapon
treatment to remove all grasses — this does some temporary harm to
Rumex spp. so the subsequent flush of growth represents recovery from
this damage plus the effects of removing the grasses; (ii) individual
plants of all dicots except *Rumex* were killed with 2,4-D applied
to plants individually; (iii) plots were sprayed with paraquat
(1,1-dimethyl-4,4'-bipyridylium-2A) which destroys all vegetation on
the plots except *Rumex* which recovers rapidly after suffering only a
leaf scorch; (iv) plants of *Rumex acetosa* and *R. acetosella* were re-

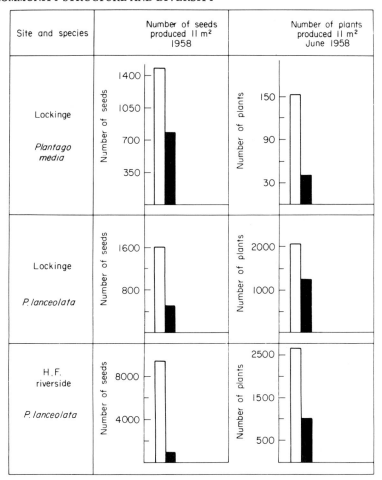

Fig. 23/9. The effects on *Plantago media* and *P. lanceolata* of the controlled herbicidal removal of grasses from mixed communities on permanent grassland. Plots were sprayed with 2:2-dichloropropionic acid on 3 October 1957. Open columns indicate sprayed plots and black columns unsprayed plots. The sites treated in this experiment were Lockinge, nr. Wantage, Berks, which is a seldom-grazed calcareous grassland community, and H. F. riverside, which is an alluvial meadow flooded by the Thames and subject to intense summer grazing. (After Sagar, 1959)

moved by spot treatment with Tordon 22K. In addition to the deliberate removal of components of the swards, seeds of *Rumex acetosa* or *R. acetosella* were deliberately sown in a series of the plots after the various treatments. Treatments were applied in the early summer of 1965 and harvested a year later and the effects are shown in

Fig. 23/10a, b.

Removal of the grasses had a spectacular effect on *Rumex acetosa.*
The number of shoot units increased 4-fold compared with the untreated

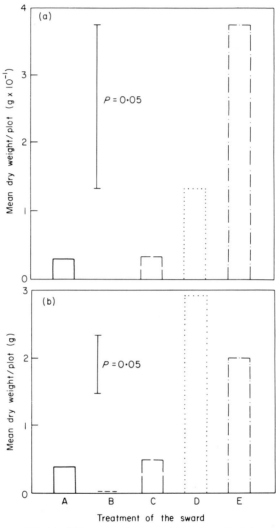

Fig. 23/10. The dry weight (g) of *Rumex acetosa* per plot at the end of the period of observa-
tion. (Putwain and Harper, 1970)
A = control
B = existing population of *Rumex* removed
C = all non-gramineous spp except *Rumex* removed
D = grasses removed
E = all species except *Rumex* removed

controls — though removal of the dicotyledonous population had no effect. Seedlings of *R. acetosa* established in abundance when the grasses were removed and the effect was increased if the dicots were removed also. These results are interpreted as a Venn diagram in Fig. 23/11a to show presumed niche relationships between the groups of species. The

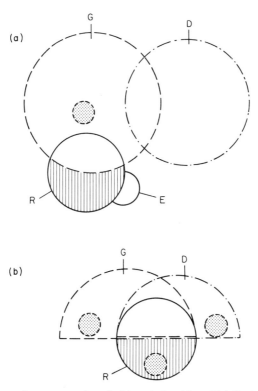

Fig. 23/11. Diagrammatic representation of niche relationships of (a) *Rumex acetosa* and (b) *R. acetosella* in mixed grassland swards. In each diagram the fundamental niches of gramineous species (G, − − −) and non-gramineous species excluding *Rumex* spp. (D, − · − ·) are indicated as overlapping areas. The fundamental niche of *Rumex* spp. (R) is shown as a continuous outline and the realized niche is hatched. (E) is that part of the fundamental niche of *R. acetosa* which is expressed in the presence of the non-gramineous species alone and which does not overlap the fundamental niches of (G) and (D). The fundamental niche of seedlings is shown by the stippled areas. (Modified from Putwain and Harper, 1970)

behaviour of *R. acetosella* was subtly different — growth was increased by removing the grasses and much more by removing both grasses and dicots (removal of the dicots alone had no significant effect). The establishment of seedlings was increased by the removal of grasses or of

dicots or of established plants of *R. acetosella*. Again an attempt is
made to illustrate these relationships in a niche diagram (Fig. 23/11b).

Experiments such as these can demonstrate with some precision
that plants are interfering with each other in a community, though
they do nothing to suggest in what dimensions of niche hypervolume
the interference is occurring. An alternative way of detecting interference in the field is to look for evidence of density-dependent processes affecting survival, growth and reproduction. Such evidence is
available from Sarukhán's study of *Ranunculus* spp. (Fig. 19/17)
and also from an analysis of the chance of seedling survival in natural
populations of *Plantago lanceolata* (Sagar 1959; see Table 23/I).

Table 23/I

The chance of survival of seedlings of *Plantago lanceolata* in relation to the
distance from the nearest "non-seedling plant" of *P. lanceolata*.

Fate of seedling present in June 1957	Number of individuals	Mean distance to nearest non-seedling	
Absent in August 1957	113	7.09	
Present in August 1957	81	8.93	sig. at $P < 0.01$

Indirect evidence that plants within a population interfere with each
other comes from a study of changing pattern of individuals within a
succession. Brook (1969) followed the scale of pattern of clumping of
individuals of a number of species within the first three years of a
succession on arable land. Populations started with strongly patterned
distributions of seedlings which over a single growing season thinned
to a more random distribution and the same effect was continued over
years. This is illustrated for *Ranunculus repens* in Table 23/II.

The direct method of removing neighbours from a plant and comparing its growth with a control is by far the most satisfactory means of
testing for interference or density stress in plant populations. Until such
an experiment has been done it is impossible to be sure that plants are
or are not influencing their neighbours. Even in populations of annuals
in desert communities, where a germinating seedling apparently almost
inevitably produces a plant, there is no adequate evidence that competition does not occur between individuals. It would not be difficult deliberately to thin some populations leaving isolated survivors to deter-

Table 23/II

Changes in the scale of pattern (the degree of clumping) of plants of *Ranunculus repens* establishing from seed in the first 3 years of an arable succession.
(From Brook, 1969)

Date	Scale of pattern (cm)	Intensity of pattern	Mean density
May 1966	160	12	3.87
	640	39	
June 1966	160	9	4.37
	640	33	
July 1966	160	5	4.18
	640	37	
August 1966	160	3	3.42
	640	34	
September 1966	160	4	2.51
	640	12	
March 1967	160	3.5	3.71
	640	10	
May 1967	160	2.5	2.48
	640	6.5	
August 1967	80	2.4	1.46
	320	3.4	
June 1968	80	1.5	1.69
	320	2.5	

mine if their size and reproductive activity is changed. Went (1973) has argued that the suppression of young trees within a dense forest is not due to competition but a situation in which the large trees provide an unfavourable environment for the small — this is a semantic problem in the use of the word competition which is almost always better replaced by other terms — density stress, interference, proximity of neighbours none of which carries the curious overburden of human associations that confuse the use of the word "competition" (Harper, 1961; Milne, 1961).

Plant form and the diversity of a community

The growth pattern of a genet can itself impose an order of diversity on a plant community. From the point of view of a bird flying in a forest canopy the environment is composed of clumps of the units derived from branched individual trees. From the point of view of a leaf in a forest the environment is composed almost exclusively of the leaves on other shoots of the same tree. For a frond of bracken almost all the other leaves that interfere with its activities are on the same rhizome system. Occasionally a frond will find itself under a tree or sapling and interspecific "confrontation" may then occur, but in general the nearest neighbour of a functional unit of a plant in nature is another such unit of the same plant. A study of the frequency of "contacts" made between leaves in a pasture (Turkington, 1975) showed that the overwhelming proportion of contact was intraspecific and presumably intra-clonal. The exceptions to this rule are the clonal species such as *Trifolium repens* which spend their life wandering in and making contacts with other species and the very growth form inhibits intra-clonal contact. In general, however, the plant's eye view of such a community is curiously narcissistic!

The analysis of pattern in plant communities emphasizes how strongly the scale is determined by the morphology of branching in clone-forming plants, the equivalent underground to the influence of branching pattern on the form of a tree canopy. A particularly striking example is the Indian cucumber (*Medeola virginiana*), a clonal perennial of woodland floors in the Eastern States of the U.S.A. and Canada. The angles at which buds grow out from the rhizome are rather strictly determined and tissues older than one year die so that independent ramets move out as an expanding "canopy" within the soil (Bell, 1974) (Fig. 23/12). The pattern of branching ensures that a developing colony has a characteristic form and Bell (1976) has shown that it is possible to predict the form of a growing genet by programming a computer with a graphic display screen: some common branching systems lead over simulated years to patterns of clumping, fairy rings of invasion and other patterns of shoot distribution that are well known in the field. An example of the complex interlocked rhizome system that may develop in nature is shown for *Calamagrostis neglecta* in Fig. 23/13.

Probably the most detailed study of the structure, and the population dynamics, of the parts of a clone-forming plant is that of *Pteridium aquilinum* by Watt (1945, 1970). This fern has a massive underground rhizome and is an aggressive invader of grasslands. Typically the plant forms an invading front or vanguard as it penetrates previously unoccupied grassland (Fig. 23/14). The vanguard consists of scattered short fronds attached deep in the ground to the rhizome system. The above-ground evidence of the spread of the plant is only the "tip of the iceberg": underground, the rhizomes may penetrate at least 1 m in ad-

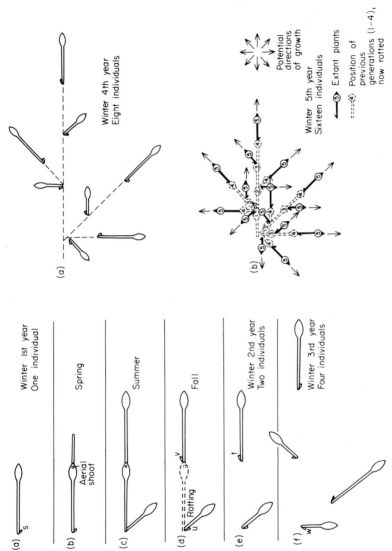

Fig. 23/12. The system of rhizome branching in the Indian Cucumber (*Medeola virginiana*) and its consequences in the pattern of shoots within a clone. (From Bell, 1974)

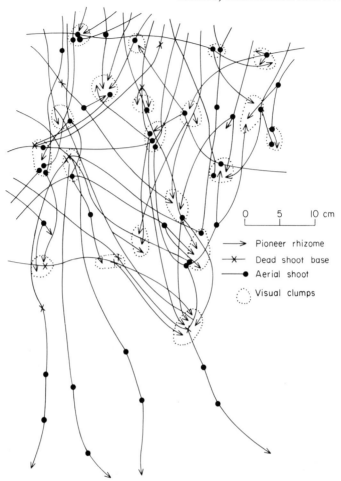

0   5   10 cm
⊢————⊢————⊢

⟶ Pioneer rhizome
—✕— Dead shoot base
●—— Aerial shoot
⟨⋯⟩ Visual clumps

Fig. 23/13. The arrangement of aerial shoots and rhizomes of *Calamagrostis neglecta* invading
        bare mud. (From Kershaw, 1962)

vance of the first emerged frond. Behind the vanguard is an "advancing
margin" of taller fronds borne on longer petioles and behind that a
continuous canopy of still more numerous and taller fronds representing
a maximum height — the crest of a wave of the advancing bracken population. Beneath this continuous canopy the grass dies, a mat of dead
fronds accumulates and the plant is now the exclusive occupant of the
area of land. Behind the crest of the wave is a zone of bracken that is
slightly shorter than at the crest and this zone is often about 30 m wide.

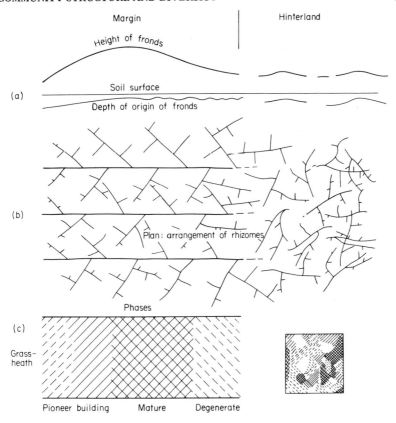

Fig. 23/14. The structure of a population of bracken (*Pteridium aquilinum*) fronds showing the changing phases behind an invading front, (a) the changes in height of fronds and continuity of cover (b) the plan arrangement of the rhizomes and (c) the linear sequence of phases at the advancing margin contrasted with the mosaic in the hinterland. (From Watt, 1947a)

Behind this the canopy suddenly gives way to a hinterland with patchily distributed fronds of all the types — vanguard, wave crest and patches of uniform fronds. Often concentric circles are formed in this hinterland region. The hinterland may be composed of the parts of a single plant but is also a population of multiple age or multiple growth stages. The rhizome system in bracken represents an astonishingly high proportion of the plant, up to 3 m of rhizome to each frond.

The spatial structure of the rhizome system derives from a sympodial growth pattern involving long shoots, 30—40 cm long, and short shoots (0.5—2 cm long) which bear the fronds. In the advancing front the long rhizomes tend to be aligned at right angles to the wave of advance but the arrangement is less orderly behind. At its greatest density there may be 9 m of rhizome underlying each m^2 of ground. Watt points out that such a sequence of zones seems unusual only because we are accustomed to thinking of perennial systems that grow vertically, usually trees. If we imagine a tree laid on its side and flattened, the pattern of growth is essentially like that of a bracken plant, complete with vanguard, advancing margin, degenerating zone and hinterland of variously aged shoots under the canopy. There is a sense in which Watt's study of bracken interprets the structure of forest canopies as well as the reverse!

Watt estimated that some of the oldest undecayed rhizomes in a bracken stand were 35 years old (in a brown earth soil where decay is rather rapid) and 72 years (in a podsol). Oinonen (1967) correlated clone size in bracken with local historical records of fire which is apparently needed for initial establishment from spores: the evidence suggested that individual clones in Finland reached 1400 years old and extended in one case over an area 474 x 292 m (Fig. 23/15). There is probably no angiosperm in which a single genet so completely dominates large areas of land except perhaps forest stands of *Populus tremuloides* (Alder, 1970), though aquatic angiosperms such as *Eichhornia* may possibly achieve comparable biomass and area dominance from a single seedling. *Pteridium aquilinum* is one of Coquillat's (1951) list of the five most common plants on the face of the earth, though there may not be a great many real individuals!

Other clone-forming species may also produce large areas of specific and genetic monotony. This is known in *Erica* spp. (Webb, 1954, 1958), *Festuca rubra* (Harberd, 1961) and *Holcus mollis* (Harberd, 1967). It might be expected that where a number of genets enter a community at some early stage in colonization these might, over a long period of a struggle for existence, steadily eliminate the less vigorous and the genetic composition of the populations become progressively more monotonous. This prediction is based on an analogy with the elimination of genets in a grass population (Fig. 6/29) or the genetic deprivation of ageing populations of *Primula* (Fig. 19/8). In fact some old communities contain a surprisingly great diversity of genotypes. A population of *Festuca rubra* in a water meadow at Barton Bridge, Yorkshire, contained an average of 5.2 clones per 15 x 15 cm quadrat. In this population the densest stands of *F. rubra* contained the lowest numbers of clones (Harberd and Owen, 1969) an observation compatible with

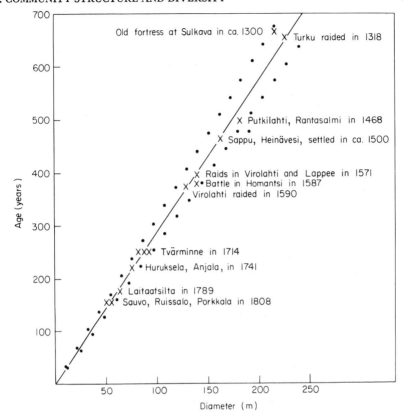

Fig. 23/15. The diameter of clones of bracken (*Pteridium aquilinum*) and their age calculated from the records of the last fire that occurred at the site and which permitted new establishment from spores. The notes at the right refer to historically dated fires. (From Oinonen, 1967b)

a steady process of elimination of clones where population pressures develop.

In *Trifolium repens* an exceptionally high density of clones exist in old grasslands and these are intermingled on such a fine scale that it is almost inconceivable that there is any progressive domination of the many by the more successful few. Rather, such high genetic diversity in a clonally growing species, is suggestive of a balanced polymorphism in which variety is itself favoured. Figure 23/16 shows some characteristic distributions of genotypes within a pasture, recognized in this case by the possession of leaf mark polymorphisms controlled by a multiple allellic series. So fine is the scale of intermingling of clones

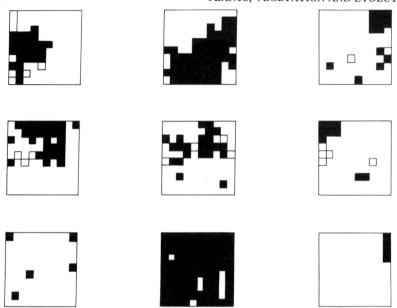

Fig. 23/16. The distribution of nine leaf mark morphs of *Trifolium repens* within a 1 m² quadrat of permanent grassland. (From Cahn and Harper, 1976a)

that a square decimeter of pasture most commonly contains three, four or five different clones (Fig. 23/17). The real diversity may be much greater than this because only the leaf mark genes were used to distinguish clones. There is evidence that sheep (perhaps other grazing animals such as pigeons) selectively predate between the different leaf marks — there is certainly selection because rumen samples contain a quite different proportion of morphs from that on offer in the field. It may be that there is apostatic selection that favours minority morphs and so maintains diversity in such a system. There is also evidence that, within a grassland, *Trifolium repens* may be differentiated into a variety of forms as a consequence of very local selection by the other species of plant in the pasture (Fig. 24/5). The existence of such a variety of apparently adaptive polymorphisms puts the study of diversity and the significance of intraspecific variation in local populations in a new perspective.

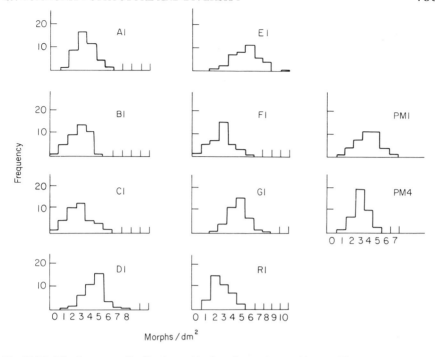

Fig. 23/17. The frequency distributions of leaf mark morphs per 10 cm x 10 cm square within
1 m² quadrats in various British grasslands. (From Cahn and Harper, 1976a)

A–G. Samples within a 60 year old permanent grassland at Henfaes, Aber, North
Wales.

R. Permanent grassland, Walland Marsh, Rye, Sussex.

P.M. Permanent grassland at least 800–900 years old. Port Meadow, Oxford.

The role of predators in the diversity of plant populations

Animals contribute to the diversity of vegetation as agents of dispersal,
selective feeders and creators of heterogeneity, e.g. footprints. They
may de-monotonize a community by causing or spreading epidemic
disease (the woodpecker for the spread of chestnut blight; scolytid
beetles for the spread of Dutch elm disease) and they may act as ploughs
creating local new foci for successions as in the burrowing activities of
rabbits, wild boar, gophers and, on a micro-scale, earthworms with their
casts (e.g. see Chapter 5). Clearly any animal activity that creates heterogeneity in the environment of plants is a source of heterogeneity in the
vegetation. For this reason alone a study restricted to the plants present
in a community is unlikely to account for its diversity. In the chalk

grassland communities of southern England, assemblages of species are intimately associated with rabbit burrows: *Atropa belladonna, Solanum dulcamara, Urtica dioica, Sambucus nigra.* These are unpalatable species, establishing in disturbed ground and maintaining populations where rabbits form warrens on chalk, graze selectively and disturb the soil. The diversity of these grasslands is a direct function of whether or not rabbits are present. A variety of other ways in which the activity of an animal determines the presence or absence or the abundance of a plant have been considered in Chapter 14. It is, however, as parts of a food chain that animals exert some of their most profound effects on diversity.

The animals and plants in a community are organized along food webs that originate with the plant population as the primary producer. Food webs are often highly branched structures and there are essentially three forms of relationship that make up the links in a food web. These may be called Iota, Gamma and Lambda links (Fig. 23/18).

The Iota linkage

This is the relationship between a prey species and its unique predator or parasite. This is the stuff of which models of predator–prey interaction are made. It is easy to visualize the sorts of interaction that may occur between a predator–prey system isolated from other organisms. When the prey density is low predator numbers are limited by a shortage of food. The prey population grows in the absence of severe predation, increasing the food supply of the predators; the predator population tracks the changing supply of prey by increasing. The increasing predation puts a check on the growth of the prey and a period of over-predation follows – the numbers of prey decline and food shortage again imposes a limit on the growth rate of the predators. If such an oscillation is violent there is a risk of extinction of the predator leaving the prey free, or extinction of the prey, in which case the predator, if he is a specialist, becomes extinct also. A classical equation describing predator–prey interaction was given by Lotka (1925) and Volterra (1926) which predicts that such a predator–prey system will oscillate. Lotka's equations can be expressed as:

$$dN_1/dt = r_1 N_1 - k N_1 N_2$$
$$dN_2/dt = r N_1 N_2 - d_2 N_2$$

where for a host—parasite system N_1 is the number of host population, N_2 is the number of the parasite population, d_2 is the death rate per head of parasites and r_1 is the birth rate minus the death rate per head of host from causes other than parasitism.

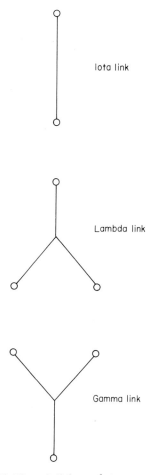

Iota link

Lambda link

Gamma link

Fig. 23/18. The unit linkages that compose food webs.

There are solutions to these equations that lead to extinction, to a unique situation of balance between the species and to constant oscillation with constant period and amplitude. This model has had enormous influence on ecological thinking and experimental design but May (1973) has pointed out that it is in many respects a very odd and unique mathematical model with the peculiarity that the oscillations "hunt" along a route that is determined by the last disturbance. Many of the characteristics of the predator–prey oscillations that are predicted by the Lotka–Volterra model are very strongly dependent on the length of time involved in the feedback between predators and prey, how rapidly a shortage of prey affects predator numbers and vice versa. Hutchinson (1948), and more recently May (1973), have examined the consequences of changing the time lag in this feedback process and have introduced other elements into the modelling of predator–prey interactions. What emerges from recent work of the modellers is the view that predator–prey systems may form stable limit cycles. The essential feature of such an interaction is that the stable system is itself a particular cycle, and if this is disturbed the populations interact to restore the same cyclic behaviour (Fig. 23/19d) (this is in strict contrast to the Lotka–Volterra model, in which disturbance of a cycle leads to a new cycle, (Fig. 23/19a).

The contribution of the modellers to understanding the diversity of communities has been to show that two species that exist in an iota link can persist indefinitely and that the stability of their cohabitation depends on their forming stable limit-cycles. May (1973) has examined the consequences of stringing two iota links together in a plant– herbivore–carnivore chain. The oscillations of plant and herbivore in a stable limit cycle are damped by adding the carnivore. There seem, there- fore, to be no theoretical problems in accounting in nature for diversity that is made up of long unbranched food chains; they can indeed be very stable.

The gamma linkage

One of the two forms of branched food chains is that involving two predators feeding on one species of prey. This is the condition represen- ted by theoretical models of competition between two species making demands on a common limiting resource. This is the model studied theoretically by Volterra and Lotka (Chapter 1) to which there are four classes of solution. Of species A and B feeding on C, (i) A may win at the expense of B, (ii) B win at the expense of A irrespective of their starting proportions, (iii) A may win at the expense of B (or vice versa) depending on the starting proportions or (iv) A and B may per- sist together in equilibrium. The condition for continued cohabitation is described in Chapter 1 — in essence it is that each of the species A

Fig. 23/19. Predator cycles. (a), (b) and (c) based on solutions of the Lotka-Volterra equations:
(a) a cycle that on disturbance adopts a new cycle dependent on the magnitude of
the disturbance, (b) oscillation moving towards a constant condition (progressively
damped), (c) oscillation leading to extinction of one or both components, (d) a stable
limit cycle which returns after disturbance to the original cycle.

and B should suffer more from the presence of members of its own kind
than from members of the other. This describes the sort of differences
in behaviour of two predators that might permit them both to contribute over long periods to the diversity of the fauna in the community.
The relationship between two such cohabiting species must be
frequency-dependent: each must suffer most from its own density

and the minority component is then always favoured. An accident that throws the numbers of the two species out of balance is redressed because the one more damaged suffers from fewer of its own sort of neighbour and therefore recovers faster.

The conditions in which two predators persist on one prey (the conditions for a gamma link to be stable) are that the numbers of the two predators are frequency dependent; otherwise the gamma link is unstable and breaks down to a simple iota link with the extinction of one predator. There is nothing intrinsically stable or stabilizing about a diversity of predators or about this sort of link in a food web. In fact, if one were to put a random collection of predator species together on a random selection of prey the most likely immediate reaction would not be the formation of an elegantly branched stable food web but its rapid degeneration to a restricted number of iota links. The stable branching of a food web (as a gamma link) depends on special relationships between the predators that are likely to come about only as a consequence of co-evolution. The sort of evolutionary process involved is the differentiation of the predators so that they eat different parts of the plant (e.g. the diversification of gall-forming insects on oak where each species has a specific gall site and there can be no vigorous competition between the species for such sites), or eat but do not destroy the plant at different times of year. May (1973) summarized this argument:

> Natural ecosystems, whether structurally complex or simple, are the products of a long history of coevolution of their constituent plants and animals. It is at least plausible that such intricate evolutionary processes have, in effect, sought out those relatively tiny and mathematically atypical regions of parameter space which endow the system with long-term stability.

It is interesting that when two species or races of *Drosophila* are forced into repeated encounter on limited food resources there is a rapid genetic change in both species such that neither interferes so strongly with the density regulation of the other (Seaton and Antonovics 1967; Barker, 1971, 1973). This is the kind of coevolutionary process that might stabilize a branch in a food chain.

Over long periods of time the gamma links in a food chain become more complex. This presumably depends on a continuous process of invasions, reaction and genetic change in both predator and prey. The historical component in the development of gamma linkages is illustrated in Fig. 23/20 in which Southwood has related the number of

insect species feeding on forest trees in Britain to the abundance of
the species in the Quaternary record.

It is possible that two predators feeding on one highly specialized
resource may form a stable system within a single stable limit cycle.
If one of the predators is more efficient at capturing prey at high density and the other is more efficient at low density the triplet of two
predators plus their prey might presumably persist together irrespective of any other form of niche differentiation (Utida, 1957).

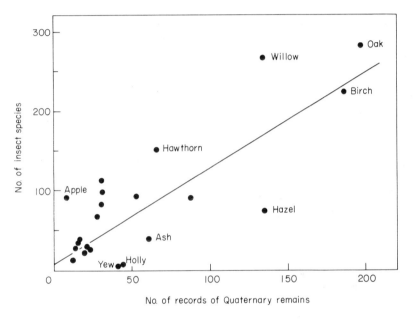

Fig. 23/20. The relationship between the number of species associated with a tree and its
abundance in recent geological times. (From Southwood, 1973)

The lambda linkage

One predator feeding on two species of plant prey is likely to predate
them differentially. It is virtually inconceivable that two plants are so
similar, so equally accessible and acceptable to a predator that the choice
of which to eat is entirely at random, and it is as difficult to imagine
two different species of plant that suffer equally from the same intensity
of predation. It is easy to show experimentally that, faced with a choice
between food plants, most animals have very decided preferences (Chap-

ter 16). Lambda linkages are therefore intrinsically unstable; the predator is a selective force dooming one component to extinction. Again
the obvious consequence of mixing a random collection of predators
and prey is that lambda branches in the network rapidly disappear, together with some of the species, and the system is likely to settle down
as a series of iota linkages. Lambda links in a food chain can be stabilized by specialized behaviour on the part of the animal. Frequency-
dependent choice of prey by a predator will stabilize a lambda link,
e.g. an animal that forms feeding images and takes preferentially the
form of prey that is most abundant will favour the rarer prey (whichever it is) and tend to form a stable lambda link. The wood pigeon grazing on a variety of crop plants behaves in the necessary way, switching
from one food when it is in short supply to another that is more
abundant and not hunting a preferred food to extinction.

Of the three links that form food webs only the iota link has intrinsic stabilizing properties. The gamma and lambda links are unstable
and require specialized (probably coevolved) relationships that make
them stable. There is nothing about branched food chains that makes
them stable: where we find branched links in nature the chances
are that some specialized stabilizing mechanism has evolved.

> There is no comfortable theorem assuring that increased diversity and com-
> plexity beget enhanced community stability; rather, as a mathematical generality,
> the opposite is true. The task, then, is to elucidate the devious strategies which
> make for stability in enduring natural systems. There will be no one simple
> answer to these questions.
>
> (May, 1973).

There is no question that the introduction of a predator into or its
removal from vegetation can have extremely profound consequences.
This is clear from the effects of biological control, the effects of acci-
dental introduction of pests and diseases and the consequences of catastrophes to a predator population; examples are given in Chapters
15–20. Paine (1966) suggested "that local animal diversity is related
to the number of predators in the system and their efficiency in preventing a single species from monopolizing some important limiting requisite". Intuitively, it seems obvious that a group of species making
similar demands on a limiting resource are doomed to enter a struggle
for existence in which, unless there are special balancing mechanisms
that have evolved, most are doomed to extinction. However, if each of

the species is limited in numbers by some other factor than the resource, for example each plant species is regulated in an iota link with its own specialist predator, the combined numbers of all the species may fall below that at which the food supply becomes limiting. Under such circumstances the fact that they all use the same resource is almost irrelevant to their continued cohabitation and each species remains a viable permanent member of the community. This is simply an example of Williamson's general law (Williamson, 1957) that the number of cohabiting species is determined by the number of independent controlling factors.

The most convincing demonstrations of the role of predation in the diversity of a population have been made in rocky intertidal zones (Paine, 1966). The predatory starfish *Pisaster ochraceus* was removed from a piece of shoreline 2 m high and 8 m long that had a rich fauna of chitons, limpets, acorn barnacles and *Mitella*. The immediate reaction was rapid colonization of available space by *Balanus* which in turn gave way to *Mytilus* and *Mitella*, leading to eventual domination by exclusive colonies of *Mytilus,* its epifauna and scattered clumps of adult *Mitella*. The benthic algae disappeared or were fast disappearing, the chitons and limpets had emigrated. The removal of *Pisaster* greatly reduced the diversity of the attached rock fauna and flora. The coral predating starfish *Acanthaster planci* apparently has a similar effect: where *Acanthaster* is present on reefs at high density the faunal diversity is high and where it is in low density a few monopolistic members of the fauna form a more monotonous community (Porter, 1972).

The sea urchins *Strongylocentrotus* spp. are herbivores grazing mainly on algae in the inter- and sub-tidal zones. The urchins *S. purpuratus* were removed from a series of rock pools, and *S. fransiscanus* was exclosed by cages from subtidal rocks or removed at monthly intervals. The result of removing urchins was a rapid increase in all plots of the number of algal species, forms normally excluded by preferential grazing by the urchins. Over the period of 2–3 years after exclusion of urchins a few species of brown algae progressively gained monopolistic occupation of the plots: *Hedophyllum sessile* in the intertidal areas and *Laminaria* spp. in the sub-tidal areas. In this case the two brown algae that monopolized the pools and the rocks in the absence of *Pisaster* were absent when *Pisaster* was present (Paine and Vadas, 1969).

These examples from marine environments may represent common roles for predators in determining vegetational diversity. There are three clear phases in the reaction:

(i) an initial phase in which the vegetation is maintained sparse by grazing and certain species are excluded by virtue of their high palatibility or intolerance of grazing.

(ii) a first reaction phase to removal of the herbivore in which the community is colonized by new species that could not tolerate grazing.

(iii) a struggle for existence in which the more aggressive of the plants assumes monopolistic dominance of the community.

The number of species present in the system is then in the order

$$(ii) > (iii) <, > \text{ or } = (i)$$

Phase (ii) is illustrated in the changing flora of a small island off the coast of North Wales (which had been studied in the seventeenth and eighteenth centuries and was well known to local naturalists) when in 1955 myxomatosis drastically reduced the rabbit population. Up to 1955 a total of 99 species of higher plant had been recorded on the island and in the period 1947–1954, 61 species had been found (the remainder of the 99 had presumably become extinct or been overlooked). In the year following death of the rabbits, 102 species of plant were recorded on the island and of these 33 species had never been found on the island before (see account in Harper, 1969). Sadly there has been no opportunity to see phase (iii) because after the rabbits had gone a plague of rats appeared on the island with new and quite different feeding habits.

In grassland habitats, grazing may increase or decrease the number of species present (Harper, 1969; see Chapter 17). In the hill grasslands of central Wales, the species richness of a pasture was greatly increased by the addition of fertilizers, but only when the activity of grazing animals was controlled. If sheep were allowed free access to a pasture the new species whose growth was made possible by the addition of fertilizers were generally very palatable and were eaten by the sheep as fast as they appeared. The herbivore in this case maintained monotony; it was not a diversifier.

Predation is only an effective diversifier if the animal predates unequally on the variety of available prey. Equivalent predation cannot increase the numbers of prey species (Van Valen, 1974). The effect of predation can be envisaged as acting upon a group of prey species that are potentially competitors for the same space (= resources).

Figure 23/21 shows a Venn diagram for the niches of six plant species. In community (a) which lacks a predator, species A is a monopolist and excludes species D, E, and F completely, partially excludes C and permits the presence of B. The introduction of a predator that eats A has the effect of reducing the space occupied (resources consumed) by A and in the now opened community, D, E and F can grow and species C can increase its numbers or mass (Fig. 23/21b). The third diagram (Fig. 23/21c) shows the effect of increasing the vigour of A, perhaps by the addition of a fertilizer to which it is especially responsive or the removal of a predator. The effect is to increase its monopoly of

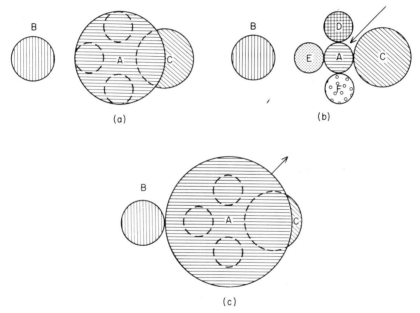

(a)

(b)

(c)

Fig. 23/21. Venn diagram of species relationships in a hypothetical mixed community. Each circle describes in 2-dimensional simplification, the fundamental niche of a species (expressed as its mass, productivity or population size). Continuous lines outline realized niches.

Species B, by virtue of differences in the manner in which its population is regulated, cohabits with A and neither excludes the other. Species C has a partial niche overlap with A, so that in the presence of A it is unable to develop its full population size (or range). Removal or partial suppression of A permits C to increase. D, E and F represent species whose fundamental niche lies wholly within that of the more aggressive A. Their presence depends on the removal or suppression of A. Thus the selective suppression of A (e.g. by grazing, fertilizers, herbicides), leads to a change from (a) to (b) and selective encouragement of A leads to a change from (a) to (c) — so determining the floristic composition of the community. (From Harper, 1970) resources and C now nearly disappears. In these models the predator is a diversifier of the plant environment because it chooses to predate the monopolist. However a further model is easily envisaged in which species A is highly unpalatable. In this case the introduction of a predator could swing model (b) into model (a). The effect of predation on the composition of a population of potential prey must then depend on the relative palatability of the prey or their different sensitivity to predation. A predator will increase the monopoly of an unpalatable species and will decrease that of a palatable one. This then poses the general question, is palatability generally related to vigour? It seems very likely that an organism that has developed spines or repellent compounds under continuous selection for unpalatability has done so by sacrificing some other attributes that might contribute to competitive ability or reproductive output (see Chapter 21). If this is the case removing a predator will usually let in a more vigorous monopolist so that the activities of a predator will most often, in the long run, monotonize a community.

The plant as a source of community diversity

The time scale of birth, growth and death of a plant creates a patch in the mosaic of the community that it occupies. The plant may shade, deplete nutrients, add organic matter, change the activities and distribution of predators, alter the flow paths of wind and so alter the patterns of seed dispersal and in countless ways make the environment different from what it was and different from the surroundings. Quite apart from mutual struggles for existence for limiting resources, the very presence, past or present, of a plant destroys homogeneity in a community and creates the potential for differently adapted species to be present. This element in the structure and diversity of communities was emphasized by Watt in a famous essay "Pattern and Process in the Plant Community"(Watt, 1947b). On brown earth soils under beech (*Fagus sylvatica*), ash (*Fraxinus excelsior*) regenerates more readily than beech, but under ash beech regenerates more successfully. This gives a "regeneration cycle" in which a whole beech—ash woodland may maintain an equilibrium condition, neither tree species succeeding at the expense of the other in the long term, though each succeeds the other in the short term and in a local patch of the mosaic. *Rubus fruticosus* agg. plays a critical part in this regeneration complex as it forms a dense ground cover under beech: ash seedlings, but not beech seedlings, can establish under its cover. Under ash *Rubus* is less conspicuous and beech seedlings then regenerate.

Similar cycles within vegetation have been described in *Calluna* heathland where the life cycle of the *Calluna* plant itself creates a cycle of environmental conditions into which other short-lived species fit. In raised bog communities a succession of *Sphagnum* species occupy higher and higher positions as they accumulate peat above the water table. Each species of *Sphagnum* is characteristic of a zone in this sequence which is both a succession and a zonation. Ultimately the tussocks are invaded by *Calluna vulgaris* and when the shrub dies the tussock breaks down to start the cycle again. The building and degeneration cycles are an intrinsic source of diversity within such a raised bog community (see Watt, 1947b).

Ecological diversity and Polymorphism

The explanations for the presence of polymorphisms within species in a population are effectively the same as those used to explain the diversity of species. Both are determined by selective forces working on the life and death of individuals; both depend on forces that work in a frequency-dependent fashion. The ecological geneticist (see Ford, 1971) recognizes three levels of intraspecific diversity within an interbreeding population. (i) Genetic diversity that is maintained at a level determined by the repeated mutation and the intensity of selection against the mutant genes within the population: the level of such diversity is maintained simply by the rate of mutation and the rate of elimination of the mutants. This is formally equivalent to the diversity of species in a plant population that is maintained by constant immigration and extinction from different communities in the neighbourhood. (ii) Transient polymorphism is a diversity that represents a point in the selective replacement of one form by another. This has its obvious equivalent in the diversity of species in a succession when some are on the way out and others on the way in. (iii) "Balanced" polymorphism — a diversity of genotypes present in a population that cannot be explained by the mutation rate nor as a phase in a process of replacement: it is this sort of polymorphism that in genetic terms calls for explanations based on frequency-dependent forces of selection. The equivalent in terms of the diversity of species is a heterogeneity in the habitat, a heterogeneity in the temporal availability of resources, a heterogeneity in the activity of predators, each or any of which may regulate species diversity in a frequency-dependent manner. In all discussions of the diversity of plant populations it is important that these three levels be clearly distinguished.

In conclusion it must be emphasized that present understanding of the process of evolution is based on arguments about the fate of individuals. Group selection, if it occurs, is the rare exception. There is nothing in the theory of evolution in natural selection that supposes any way in which "the species" reacts to or responds to events other than as the collective result of the behaviour of its individuals. Moreover, there is nothing in the process of evolution that should lead us to imagine some community goal, nothing to suggest that the collective evolution of the populations in a community is towards some ideal — community structure, stability, diversity, productivity, efficiency, information content, entropic level. There may be interesting correlations with species diversity at the level of the community or the unit of land but causal relationships will not be found from this type of correlation unless there is something very wrong with the theory of evolution through natural selection!

24

Natural Selection and the Population Biology of Plants

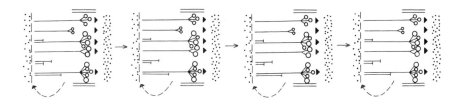

Nothing in biology makes sense except in the light of evolution.

(Dobzhansky, 1973)

Some organisms leave more descendants than others, and some of the characteristics that mark these "successful" ancestors are heritable. Hence the qualities of organisms tend to change over generations — this is the process of evolution. The fact that organisms leave more progeny than can possibly survive means both that there is usually an excess of individuals on which the process of natural selection can act *and* that pressures from overpopulation (inter- and intraspecific) are likely to be of prime importance in the process. In "The Origin of Species" Darwin places strong though not exclusive emphasis on the competitive inter-actions that arise from population pressures. This emphasis contrasts with that of Wallace (see Darwin, 1859) who saw natural selection more as a "struggle" against the elements, the physical forces of the environ-ment. It is therefore rather odd that, though it is Darwin's influence that

7

50 PLANTS, VEGETATION AND EVOLUTION

is usually acknowledged in the development of evolutionary theory, it is
a Wallacian attitude that has dominated the study of adaptation in plants.
Students of plant adaptation have concentrated on local differentiation
between populations in response to physical factors of the environment
(soil and climatic ecotypes) rather than on the consequences of Malthusian population growth that so impressed Darwin.

There is an important difference between adaptation to the physical
conditions of the environment and the evolutionary response to biotic
pressure. It is in the nature of adaptation to the physical environment
that a degree of ultimate adaptation is possible, provided that the environment does not change. A plant may evolve a cold-adapted strain,
a photoperiodically appropriate race, a copper-tolerant form, a calci-
colous ecotype and, apart from progressively perfecting adaptation, the
evolutionary game is then finished. In an unchanging physical environment adaptation is an asymptotic process of adjustment that slows
down as perfection is attained — there is a forseeable end to the evolutionary process. In contrast, in an environment dominated by the
biotic forces of competition, parasitism and predation, the evolutionary
game is potentially unending*. Organisms form part of each other's environment; a change in the nature of any one component changes the
environment and so changes the selective forces acting on others and
there is further feedback in turn as they change. This is the process in
which evolution is an existential game in which success is measured by
continuing to play the game rather than by winning. It is this sort of process, involving biotic interactions, that provides the evolutionary forces
that can account for the continuing speed of evolution and the evolution
of diversity. The fact that organisms living in different places are different is easy to explain by Wallacian forces. The question of how so
many sorts of organisms are able to persist together in the same "place"
is much more difficult to answer, is much more interesting; it demands
biotic interpretation and a Darwinian solution.

It is a task of the evolutionist to explain not only the differences
between species but also the differences that occur within populations.
The problems of biological diversity are not restricted to the conventional taxonomic level (see Chapter 23) but must be interpreted ultimately in the light of forces that affect the life and death or the fecun-

* The Red Queen has to keep running to stay in the same place (Van Valen, 1973; Maynard
Smith, 1976).

dity of individuals. One of the consequences of a population approach to the biology of plants is that it focuses attention at this level of individual behaviour; the population has no meaning except as the summed activities of its individuals and their interactions.

The study of adaptation has passed through a period in which it was barely respectable science. The Victorian excesses that followed Darwin involved fanciful adaptive interpretations for biological phenomena that had little base in hard science — the reaction to these excesses was a distrust of evolutionary speculation. "There is perhaps no more hopeful augury of the future of Biology than the increased and ever-increasing sobriety of biological speculation. Bold hypotheses are no doubt framed as profusely as ever but the speculator is made to feel that he must not set them forth in print until he can support them adequately" (Miall, 1912). Much scepticism about evolutionary speculation was appropriate so long as evidence suggested that the process was imperceptibly slow and not subject to the normal pace of scientific test. As late as 1932 Haldane (1932) was considering the consequences of a selective advantage of 1 in 10^6. The study of the ecological genetics of natural populations has since shown that forces of selection five orders of magnitude higher are readily detected (see Table 24/I; also Creed, 1971). These newer estimates of the force of natural selection, taken together with the estimates of high heritability of ecologically important attributes that are obtained by agronomists and others for plants brought from the wild, force the ecologist to treat natural populations as evolving systems in which it is wholly realistic to hypothesize and test adaptation. The process of evolution is clearly part of immediate ecological happenings. Ecology is concerned with evolution in action, not just with the interplay of its invariant results.

One very important consequence of recognizing that ecology is evolution in action is that it becomes extremely dangerous to generalize about the ecological properties of a species. There is no meaningful way in which a study made of one sample from a species can be said to represent that species until the range of variation within and between its populations has been established. It is quite unrealistic to assume that the biological properties of a species can be gauged from a study of a single or a few samples (cf. comparisons of the relative growth rates of a group of species made with samples from single populations: Grime and Hunt, 1975). In many of the attributes that contribute to success within·a population of neighbours, differences within species

Table 24/I

Coefficients of selection acting within various populations of grasses.
(From Jain and Bradshaw, 1966)

Example	Species	Habitats and coefficients of selection against unadapted types	Source of evidence
1	*Agrostis tenuis*	Mine, 0.95: pasture, 0.4	Growth of spaced plants on two soils
2	*A. tenuis*	Mine, 0.95: pasture, 0.05	Growth of plants under spaced and competitive conditions on two soils
3	*Anthoxanthum odoratum*	Mine, 0.99: pasture, 0.3	Growth of plants under spaced and competitive conditions on two soils
4	*Agrostis stolonifera*	Cliff, 0.8: pasture, 0.5	Growth of spaced plants under exposed cliff and garden conditions
5	*Anthoxanthum odoratum*	Unlimed, 0.09: limed, 0.27	Estimated from three competition experiments carried out under greenhouse conditions: under field conditions higher values seem likely

(or even between the parts of a plant) may indeed be greater than between species. Thus there is a greater difference in palatability between strains of *Lolium perenne* than between many pasture species (Fig. 13/5). The difference in germination requirement of seeds from the same plant of *Rumex crispus* is greater than between *R. crispus* and *R. obtusifolius*. The difference in mean seed number produced by autumn- and spring-germinating forms of *Papaver dubium* is greater (Chapter 3) than between many comparable species in the same and different genera.

The variation that is found within and between species living in the same area can be interpreted in relation to the sources and expression of genetic variation and the forces of selection that contain or maintain it.

Genetic variation within species with different breeding systems

The display of genetic variation within a population of a species depends on the breeding system which controls the relative roles of repetition and experimentation that occur in the production of progeny. The breeding system is itself under genetic control and it is reasonable to expect that the variety of breeding systems found in nature represents a variety of strategies of successful ancestors: that they are themselves evolved adapted systems.

Variation within populations of apomicts

Taraxacum officinale is an aggregate species containing many apomictic microspecies with strictly maternal inheritance. This is the situation *par excellence* in which populations might be expected to be genetically uniform, in which on a local scale a single apomictic race might be expected to have shown its superiority over others and the populations be genetically monotonous. In fact, natural populations that have been studied are found to contain an assortment of apomictic races, the mix varying from site to site (Kappert, 1954; Nilsson, 1947; Solbrig and Simpson, 1974). Apomictic races of *Taraxacum* can be distinguished by enzyme electrophoresis but the races that form mixtures differ also in features of obvious ecological relevance such as precocity of flowering, seed output and longevity. Three populations studied in Michigan (Solbrig and Simpson, 1974) contained at least four biotypes. Three populations in Germany contained 12 biotypes. There is comparable variation within quite small hedgerow communities of *Rubus fruticosus* another aggregate of apomictic races in which it appears to be the rule that quite small populations contain a mixture of several distinct races.

Genetic variation within populations of species with close inbreeding

Many annual plants are almost exclusively inbreeding, with anthers that rarely emerge and pollen that is shed directly onto the stigma. *Festuca microstachys* is such a grass and its coefficient of outbreeding has been estimated as 0%; tests made on 20 000 individuals showed no hybrids. Nevertheless families "from the same site showed a wide range of means for various characters, and data for combinations of characters showed that very large numbers of different genotypes coexists within each population" ". . . each such potentially uniform family appears to

be represented by only a single individual or at most very few individuals" (Kannenberg and Allard, 1967). The phenotypic and genetic
diversity of populations of *Festuca microstachys* can be compared with
two other grasses: *Avena fatua* which has 1—10% outcrossing and
Lolium multiflorum which is 100% outcrossing. The populations of
the highly inbred *Festuca microstachys* were as genetically variable as
the other species. In *Festuca microstachys* the data suggest that a
high level of genetic variation is maintained within a single species
and a single population and it is difficult to explain this without supposing that there are nearly as many adaptive optima within a habitat as
there are individuals.

Two experiments are particularly helpful in explaining the high level
of genetic variation that occurs in inbreeding populations. In the first
experiment, the F2 hybrids of 20 different forms of rice were blended
and the population was grown without conscious selection for eight
successive generations in three sites, one in California, one in Arkansas
and one in Texas (Adair and Jones, 1946). The populations were found
to have diverged and "one of the major patterns of change in response
to natural selection is rather rapid directional adjustment of measurement characters towards optimal values and the optimum is not the
same for different environments", but "plants with a wide range of
heading date and with a wide range of height survived at all locations".
"The second major effect of natural selection is therefore the preser-
vation of many different genotypes and not reduction to uniformity
as has been commonly supposed" (Allard *et al.,* 1968). Secondly, a
series of studies was made of the grass *Avena barbata* which has *ca* 2%
outbreeding. In the populations there was a consistent net reproductive
advantage to heterozygotes — often double that of the homozygotes.
"Selection acts to structure the genetic resources into highly interactive
allelic complexes." "A consequence of this is that an individual heterozygous at one locus is much more likely to be heterozygous for other
loci." Seedlings were collected from wild populations, grown first as
pot plants and later transplanted as spaced plants to optimal conditions
in a garden plot. The heterozygosity of the population was measured by
gel electrophoresis of esterases of the progeny of 66 randomly chosen
plants. These were compared with seedlings from individual plants
that survived and matured in the natural populations from the field. The
superiority of the heterozygotes in the natural crowded populations
was dramatically greater than in the carefully nursed spaced plants.

The heterozygotic superiority was due to their much higher survivor-ship rather than to differences in fecundity. It would appear that biotic interactions are again involved in maintaining a high degree of genetic heterogeneity within the populations, though in this case the diversity was at the level of coadapted gene complexes rather than inbred lineages: "viability interactions in which individuals heterozygous for these coadapted complexes are favoured are a major component of natural selection in *Avena barbata*" (Clegg and Allard, 1973). A remarkable example of the way in which density stress may favour heterozygotes comes from a study of the survivorship of rye in a density experiment (Fig. 24/1).

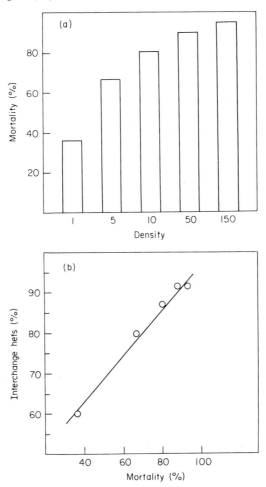

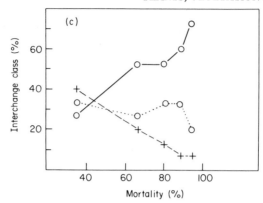

Fig. 24/1. The differential survivorship of interchange heterozygotes in a density experiment
with rye (*Secale cereale*). Populations were grown in 5 cm pots;
(a) the influence of density on mortality;
(b) the relationship between mortality and the proportion of interchange heterozygotes surviving. The heterozygotes are summed to include A and B, two
independent interchanges and also include AB;
(c) the relationship between mortality and the number of A and B heterozygotes
AB heterozygotes and homozygotes among the survivors. 0 − 0, A and B
heterozygotes; 0 . . .0, AB heterozygotes; + - - +, homozygotes. (From Bailey *et al.*,
1976)

Genetic variation within outbreeding populations

Variation within outbreeding populations is not surprising, but its
form is still unexpected. Discrete classes of phenotypes can often be
recognized in floral characters (heterostyly and colour variants) and in
vegetative characters. For example, in *Trifolium repens* there are polymorphisms for leaf marks, cyanogenic glycosides, *Rhizobium* host
strains and incompatibility alleles. These variants appear to be of selective importance because a parallel polymorphism occurs in other
species (e.g. *Trifolium pratense*), clones exist in the frequency of the
characters and in some cases there is direct proof of the selective value
of the morph characters (see Chapter 23). The coexistence of many
forms within a local cross-fertilizing population strongly suggests disruptive selection in which there are perhaps very many adaptive optima. In an important sense, the existence of polymorphic differentiation in outbreeding species argues, even more strongly than for inbreeding species, for a subtle differentiation of niches within the species
and within single populations.

Long-lived plants tend to be outbreeders; tight inbreeding systems are common in annuals though there are a number of exceptions. The length of life may be important in understanding the diversity of genotypes that may be present in a single outbreeding population. If the establishment of new genets occurs very rarely (and the evidence marshalled in Chapters 19 and 20 strongly supports this view), the few survivors from each generation may represent the variety of adaptive optima present within the habitat. Thus within a perennial grassland community, clones (e.g. *Trifolium repens*) are potentially immortal and spread widely. The outbreeding system of such species is successful only if it adds new genets to the community that are superior to those that are already there.

Variation within dioecious species

Separation of the sexes is one extreme form of outbreeding and presumably a constant hindrance to the development of local micro- and co-adaptation. Any process that maximizes genetic interchange such as sex or an incompatibility system tends to blur local differentiation within the species or, looking at it another way, demands very high selective pressures to maintain local adaptation. Where there is sex differentiation there is, however, the possibility that the two sexes may take different ecological roles. There are at least four cases in which this seems to have happened.

(a) *Rumex acetosella*. Male plants are shorter, more precocious in vegetative growth and senesce earlier than females. The tall females therefore occupy a different seasonal niche in the grassland environment (Putwain and Harper, 1972; and see Fig. 9/13). *Rumex acetosa* seems to behave in a similar manner.

(b) *Asparagus officinale*. Male and female plants are sufficiently different (at least in cultivation) for one sex (the male) to be preferred as a crop. The male produces a greater yield of shoots slightly earlier than the female. The mortality risk is greater for females than males, perhaps because the role of seed filling puts a greater strain on the plants and so increases the risk of death from other causes. It has been shown that the high death rate of the females cannot be accounted for by competition from the males (Bouwkamp and McCully, 1972). As in *Rumex acetosella* the males senesce precociously and so presumably avoid competing with the females at the end of the growing season — the seasonal niche is partitioned between the sexes.

(c) *Cannabis sativa.* Male plants senesce earlier than females.

(d) *Mercurialis perennis.* There have been two reports that the sexes of this perennial species occupy different phases, the shaded and the more open habitats in the ground flora of woodlands (Muckerji, 1936; Abeywickrama, 1949).

In the previous chapter it was pointed out that diversity at the species or intra-specific level may arise because there is constant immigration and emigration *or* because the community is in a transient state *or* because there are balancing, frequency-dependent forces acting within the community. It is important to remember that intra-specific diversity may be affected in outbreeding species by immigration of pollen (haploid immigration); it does not necessarily require seed immigrants to a plant population to maintain a degree of genetic diversity within it. Nevertheless, the existence of a high degree of diversity within inbreeding and apomictic communities makes it clear that gene flow is not a necessary condition for extremely local genetic diversity to arise. It looks as though, almost irrespective of the breeding system, diversity is maintained within populations and maintained either as coadapted inbred lines, apomictic races, coadapted gene complexes in partly outbreeding species or ecological differentiation between the sexes in outbreeding species. To understand this level of diversity it is important to know something of the nature of the selective forces that act within natural populations.

Forces of selection acting within populations of plants

A very large part of what we know about adaptation concerns speciation, the evolution of barriers that restrict gene flow between populations, the evolution of breeding systems, subtleties in pollination mechanisms, floral attraction, chiasma frequency etc. It is rather strange that the adaptations about which we know most are those that control the rate of release of variation as if this was an end in itself! The rate of release of variation is important in evolution only if the variations affect the number of descendants that an individual will leave. The study of population biology ought to display those forces that are important at the level of the life of the individual and what sort of variation is important in determining survivorship and reproduction.

It is not the role of this chapter to discuss the evolution of local differentiation to climates and soils, nor to discount its importance, but

rather to pick from knowledge of the population biology of plants
those forces that, on a local scale, seem likely to dominate the chance
that an individual will leave descendants. The scale is local because
natural selection acts on individuals in the context of their immediate
experience. The other evolutionary processes of divergence in isolation
and speciation have received full treatment in the splendid texts of
Stebbins (1950, 1974), Grant (1963), Mayr (1963) and Dobzhansky
(1970).

The forces of selection have been described as (i) *directional selection*
in which an adaptive optimum for a property does not coincide with
the modal frequency of that property in the phenotypes of the population — selection then acts most strongly against one tail of the frequency distribution of the phenotypes; (ii) *stabilizing selection* in which
the adaptive optimum for the population coincides with the mode of
the distribution of phenotypes and both tails of the frequency distribution are selected against; and (iii) *disruptive selection* in which two (or
more) adaptive optima exist and selection acts against the norm (Fig.
24/3) (Mather, 1953). These broad categories of selective force are in
a sense statistical rather than biological categories; the biological categories need to take into account the nature as well as the direction of
selection. A number of generalized biological categories can be recognized.

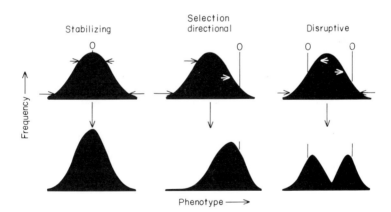

Fig. 24/2. The three basic types of selection. The optimum expressions of the phenotype are
indicated by O, and the directions and relative magnitudes of the selective forces by
the horizontal arrows. Selection is heavier against phenotypes farther away from the
optimum. (After Mather, 1953)

r-Selection

During phases in which a population increases after a crash or after a new invasion, the individuals are by definition free from interference from neighbours of their own species. During the phase of population increase the pre-eminent force selecting the individuals that contribute most descendants to subsequent generations is their fecundity and the effect of fecundity is enhanced by precocity (provided that generation follows hard on the heels of generation). Progeny produced late in an individual's life are relatively unimportant (have low reproductive value; see Chapter 22) and there is nothing in r-selection to favour length of life unless the survivorship of juvenile stages is especially at risk. We may therefore expect to find r-selection favouring fecundity and pre-cocity and, associated with this, selection for the allocation of a large fraction of resources to seed production: during this phase there is no premium on attributes that confer competitive ability. We might also expect that during colonizing phases the action of predators, particu-larly specialist predators, will be slight and so we would expect only weak, if any, selection for mechanisms of predator defence. It seems indeed to be a general rule that populations that have evolved in condi-tions of repeated r-phases do not have specialized (although there may be generalized) defence mechanisms such as repellent or toxic chemicals, and generally lack specialized host relations with pests or pathogens. (Where specialized pathogens attack colonizing species there is usually an alternate perennial host, as in the heteroecious rusts, which gives continuity to the pathogen population and so keeps up the selective pressure on the colonizing host through crashes and recovery phases.)

There are other forces of selection that may favour some of the same features as r-selection. In very short growing seasons an ephemeral growth habit may be favoured (desert annuals) or a population may evade interference from perennial neighbours by making growth in a short autumn and spring burst. Such life cycles, although short, are in many other respects quite ill-fitted for rapid colonizing episodes and have seed longevity, major periods of seed dormancy, low fecundity and poor dispersal mechanisms, none of which are r-qualities.

K-selection

An increasing population may reach densities at which individuals begin to interfere with each other. The "struggle for existence" that ensues contributes a special element to the process of natural selection, for the chance of an individual leaving descendants now depends on the share of needed resources that can be captured from competitors. Precocity and high fecundity are now relatively unimportant compared to factors that confer aggressiveness, e.g. height, a perennial or quickly renewable canopy, larger but fewer seeds. In dense communities both inter- and intraspecific encounters favour individuals with longer life, longer juvenile periods spent in gaining competitive ascendancy, a woody habit or clonal growth and closed internal cycling of nutrients within the individual rather than their wasteful release to successors. The struggle ceases to favour the fecund, short-lived and instead favours selfish and conservative growth strategies. Such K selection can be seen interspecifically as vegetational succession and intraspecifically as natural selection.

Most of the evidence that density stress has selective consequences comes from experimental studies with insects. In mixed populations *Drosophila simulans* eliminated *D. melanogaster*: a population of *D. simulans* that had achieved this ascendancy once was subsequently able to eliminate *D. melanogaster* more quickly. The competitive ability of *D. simulans* had been improved as a result of the encounter (Moore, 1952). A population of *Drosophila melanogaster* that had eliminated two mutant genes was subsequently able to eliminate the same mutants more rapidly (Buzatti-Traverso, 1955). Blowflies (*Phaenicia sericata*) that had been reared in dense populations together with houseflies (*Musca domestica*) suffered drastically in the mixture but the survivors were better able to persist in such a mixture or even to gain ascendancy (Pimentel, 1964; Pimentel *et al.*, 1965). This sort of interaction produces what Pimentel calls the genetic feedback mechanism. In a mixture of two species, A and B, when A is in the majority it meets mainly intraspecific selection which does not improve (and may reduce) its superiority with respect to B. However, the minority component of the mixture, B, meets mainly interspecific selection which favours the forms of B most resistant or tolerant to the presence of A. Under these circumstances selection improves the interspecific competitive ability of B so that eventually it gains ascendancy over A. As soon as that happens, B becomes the predominant member of the mixture and so meets mainly intraspecific interference, and it is now primarily the force of interspecific competition that acts upon A because it is the minority component. This can, in theory, produce an oscillation between two species, each in turn gaining dominance in the mixture as it changes genetically.

There is abundant evidence that "competitive ability" is heritable in
Drosophila (Mather and Cooke, 1962; Gale, 1964) and also in *Tribolium*
(Lerner and Ho, 1961). Competitive ability has been shown to be heritable in rice (Sakai and Gotoh, 1955), indeed Sakai (1961) considered
that it was a property that could be considered as more than the sum
of specific analysable growth attributes such as height and leafiness.
Most of these experimental studies have been made to show the
inheritance of properties that make one form superior to another;
that enable one form to win in a struggle for existence under crowded
conditions. There is, however, a quite different evolutionary solution
to competitive stress — to avoid it.

Selection for ecological combining ability

If two non-interbreeding populations are thrown together under
density stress, selection may favour a divergence in behaviour so that
each population ceases to make such heavy demands on the resources
needed by the other. This is the process of niche diversification (or
annidation — Ludwig, 1959). Clear demonstrations of this phenomenon
come again from studies with *Drosophila melanogaster*. Wild type and
the mutant "dumpy" were reared together in equiproportioned mixtures
and mating between the two forms was prevented. At the end of each
generation, virgin wild type was mated with wild type and virgin dumpy
mated with dumpy and the populations were started again in a second
generation of equiproportioned mixture; at the same time pure populations of dumpy and of wild type were maintained, with each generation
started at the same overall density as the mixed populations. After a
period of five generations of selection in mixed and in pure cultures a
series of tests was made in which *Drosophila*, selected in a mixed population, was grown with *Drosophila* that had not had this experience.
The results are shown in Fig. 24/3 and it is clear that both selected
dumpy and selected wild type had improved in their performance against
the unselected forms. Moreover, in a mixture of selected dumpy with
selected wild type neither suffered so much from the presence of the
other as was the case with unselected forms. The two strains of *Drosophila* had evolved "ecological combining ability" (Seaton and Antonovics, 1967). This experiment has since been repeated using quite different strains of *Drosophila* but with essentially the same results (Barker,

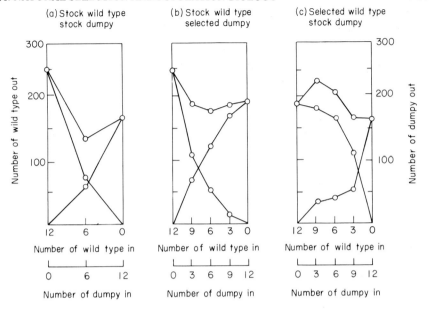

(a) Stock wild type
stock dumpy

(b) Stock wild type
selected dumpy

(c) Selected wild type
stock dumpy

Number of wild type out

Number of dumpy out

Number of wild type in

Number of dumpy in

Fig. 24/3. The relationship between the numbers of fertilized female *Drosophila* (wild type and "dumpy") introduced into a culture (=IN) and the number of descendants after one generation (=OUT):

(a) populations with no previous experience of mixed culture;

(b) wild type with selected "dumpy", i.e. after five generations of mixed culture with wild type;

(c) "dumpy" with selected wild type, i.e. after five generations of mixed culture with "dumpy".

(From Seaton and Antonovics, 1967)

1973). These experiments show the evolution of differences between parts of a population that permit them to evade the full force of a struggle for existence and are perhaps of the same nature as the differences within populations of *Festuca microstachys*. Evidence for this view comes from an experiment with barley. Eight families were selected from a single random plant taken from the F18 generation of a mass-propagated, highly heterogeneous population that had been developed by intercrossing in all combinations 31 barley varieties of wide geographic origin (Composite Cross V). Representatives from the eight families were grown in pure populations and in mixtures with differently arranged neighbours. Of the various combinations of genotypes, 40% showed large and statistically significant increases in yield in mixtures compared with their performance in pure stands.

> The genotypes from Composite Cross V . . . have a history of as many as 18
> generations of natural selection in competition with a large and diverse array of
> genotypes. Apparently, selection in such mixed populations favours the survival
> of genotypes which have superior 'ecological' combining ability, in the sense that
> they are good competitors and at the same time, good neighbours.
>
> (Allard and Adams, 1969).

Simulations of the long-term consequences of the interaction between
genotypes in Composite Cross V (assuming no out-breeding) predict that
they form stable mixtures. At the same time, Allard and Adams had
grown pure stands and paired mixtures of four varieties of barley (and in
a separate experiment four varieties of wheat) but "only two of the differences were significant and the average increase in yield in mixture was
less than 2% in 1959—60 and less than 1% in 1960—61)". The interaction
between the varieties which had no history of growing together was
very minor compared with that between the apparently coadapted
races of Composite Cross V. "The data presented here indicate that
intergenotypic interactions are an important force in maintaining
genetic diversity in populations (see also Allee *et al.,* 1949). Further,
the frequency with which such interactions occur in plant and animal
associations suggests that they may well be among the major forces
involved" (Allard and Adams, 1969).

There are two examples that illustrate the ways in which such
niche selection may occur in nature. Populations of plants and of seed
of *Potentilla erecta* were selected from areas in which *Agrostis—Festuca*
pasture was immediately adjacent to an area of *Molinia caerulea* grassland. The plants and seedlings were grown in a field trial and the populations from the two habitats were clearly different, even when sampled
only 2.7 m on either side of a habitat boundary (Fig. 24/4). This suggests
that the environment of *Agrostis—Festuca* or *Molinia* grassland selected
different growth habits from the population of *Potentilla erecta* even
within close populations of this outbreeding species (Watson, 1969).
Populations of *Trifolium repens* were sampled from a 1 ha field of permanent grassland at Aber, North Wales. The grassland was a floristically diverse community and contained four abundant grass species as
well as others. The main grasses were *Lolium perenne, Holcus lanatus,
Cynosurus cristatus* and *Agrostis tenuis.* The samples of *Trifolium
repens* were deliberately chosen from within clumps dominated by
these four grasses, and were grouped to represent the genotypes that
grow in association with each grass. The four clover populations were
multiplied clonally to give sufficient material for two tests. In the

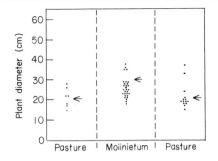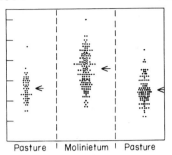

Fig. 24/4. The diameter of plants of *Potentilla erecta* sampled from *Agrostis—Festuca* pasture and from *Molinia caerulea* grassland in close proximity. (a) Transplants from the field to an experimental garden; (b) plants grown in an experimental garden from seed collected in the field. → = mean. (From Watson, 1969).

first test the clones were planted out as phytometers into the pasture from which they had been sampled: each clover population was planted in association with each of the four grass species growing naturally in the sward. Both in respect of dry weight and plant survivorship it was clear that the clover populations behaved differently. Table 24/II shows the results for dry weight of the clover phytometers. The four clover populations differed ($P < 0.01$) in their performance with the different grasses. The four grasses differed in their effect on the clovers ($P < 0.01$). Moreover, the interaction between the four grasses and the four clovers was significant at $P < 0.05$ and this was conspicuously due to the principle diagonals of the interaction table; each clover variety tended to perform better when grown with the species of grass from which it was originally sampled (Turkington, 1975). Even more convincing evidence of local coadaptation between the clovers and the grasses came from a second test in which the four clover populations were grown in all possible combinations with the four species of grass in a glasshouse experiment. In this case the variations in substrate that undoubtedly occurred (though they were very minor) in the field were eliminated by using a standard potting compost throughout the experiment. The performance of the four clover populations is shown in Fig. 24/5. Each clover type grew best in association with the grass from which it had originally been sampled, with the exception of the clover from *Holcus lanatus* which performed equally well with *Lolium perenne* or *Holcus lanatus* as companion. No attempt was made in these experiments to test for variation within the species of grass: it may be that an old permanent pasture contains variation

Table 24/II

The growth made by phytometers of *Trifolium repens* sampled from the neighbourhood of four pasture grasses in an old permanent grassland and transplanted into sites dominated by the same grasses, in all combinations of clover sample with grass species. (From Turkington, 1975).

Yield of clover (g)

Dominant grass in the site of the phytometer (Y)

Dominant grass in the site of origin of the clover samples (X)	Lolium perenne	Holcus lanatus	Cynosurus cristatus	Agrostis tenius	Mean
Lolium	**0.51**	0.05	0.53	0.24	0.33
perenne	(0.25)	(0.22)	(0.44)	(0.41)	
Holcus lanatus	0.30	0.35	0.34	0.94	0.48
	(0.37)	(0.32)	(0.60)	(0.63)	
Cynosurus	0.50	0.08	**0.87**	0.20	0.41
cristatus	(0.32)	(0.27)	(0.51)	(0.54)	
Agrostis	0.57	1.15	1.29	**1.83**	1.21
tenuis	(0.93)	(0.81)	(1.50)	(1.59)	
Mean	0.47	0.41	0.76	0.81	

Values in parentheses are expected values calculated for no interaction; the values in heavy type are examples in which the interaction gives performance higher than expectation. The interaction is significant at $P = 0.05$ and the phytometer origins and transplant positions are significant at $P = 0.01$.

within the grass species in their reaction to clover as well as vice versa and that the relationship is a close evolutionary coadaptation involving all the species. The experiment does, however, point to a level of biological differentiation within a single species in a small area in which a variety of forms is adapted to live with different associated species. When it is remembered that the individual genets of *Trifolium repens* may live to great ages, that the species is an outbreeder, that only a very few new genets are recruited into the growing population each year and that the growth form of clover allows individual genets to migrate continuously through the sward tracking local favourable areas, such a fine scale of coadaptation might indeed be expected. This would appear to be a case of a truly "Darwinian" evolution in which it is the force of interference from population pressures that drives natural selection. It suggests, moreover, that the variety of

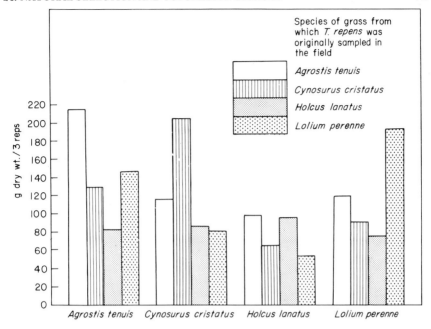

Fig. 24/5. The growth made by populations of *Trifolium repens* sampled in the field from the
neighbourhood of four different grass species and grown in glasshouse conditions
in mixed populations with each of the species of grass. (From Turkington, 1975)

associated species present in a community is itself a disruptive force
of selection. The varied habits of the grasses (different tussock structures, different annual cycles of growth and leaf production are perhaps
the most important) select clones of clover with compatible habits and
where there are several species of grass present there will be a corresponding selection for at least an equal number of forms of clover.

> . . . The more diversified the descendants from any one species become in struc-
> ture, constitution and habits, by so much will they be better enabled to seize on
> many widely diversified places in the polity of nature, and so enabled to increase
> in numbers.   (Darwin, 1859)

Selection by the activity of predators and pathogens

An individual plant that is more prone than its neighbours to attack
by pathogens and/or predators will almost inevitably have lowered fit-

ness. The variety of ways in which a host may gain resistance, immunity or a degree of tolerance to such attack immediately defines equivalent ways in which the predator or pathogen might counter the defence. Just as in the coevolution of competing species at the same trophic level, any change in one component immediately changes the selective forces acting on the other. Two categories of defence can be recognized: "horizontal" resistance in which there is a generalized defence against a range of pests (or races of a pathogen) and "vertical" resistance in which each host/pest reaction depends on a specific interaction (van der Plank, 1963).

Any system that involves continuous genetic feedback between competitors or between pests, pathogens and their hosts must be dependent on the nature of the breeding systems of both the participants in the process of coadaptation. Coevolution requires continuous genetic tracking by each component of the genetic changes in the other. Levin (1975) has argued that, for this reason, outbreeding systems are favoured by selection pressure from pests. This argument goes some way to explain the occurrence of outbreeding, high chiasma frequency and generally high recombination values in many species. This had appeared to be a puzzle because there have been so many convincing arguments suggesting that the effects of natural selection will usually favour tight linkage between interacting groups of coadapted genes (e.g. Lewontin, 1971; Feldman, 1972). The question was posed by Turner (1967), "Why does the genotype not congeal?"; for clearly in many species it remains a largely open system of relatively free genetic combination. Levin's argument invokes pest pressures: "I propose that the persistent tracking of plant hosts by multiple pathogens and herbivores is a prime factor which prohibits the congealing of the genomes of species especially those in closed communities." As pointed out earlier the problems of pest pressures are generally less severe in colonizing or fugitive species which continually escape to new areas — it is within crowded K-selected communities that epidemic pest attack develops most easily and in which coevolution and consequently a high recombination frequency are favoured.

> The simple genetic bases for host resistance and pathogen or pest virulence, the vulnerability of species with restricted recombinations to disease epidemics and pest outbreaks, and the ability of pathogens and pests to rapidly evolve resistance to new natural or synthetic biocides provide the empirical superstructure for postulating why the genomes of K-selected species do not congeal.
>
> (Levin, 1975)

The evolutionary consequences of crashes, cycles and catastrophes

The biologist who gains his knowledge of the forces of selection by reading about selection experiments might be forgiven for thinking that the process is usually constant and relentless (e.g. adaptation in laboratory conditions to a high temperature, the presence of a toxin or a new food supply). Organisms in nature live in environments that contain rhythms and unpredictabilities, patterns and noise. The environment varies both in space and in time and although it is apparently rather easy for many outbreeding organisms to adapt to a stable or reliably rhythmic environment, nature sets an infinitely more difficult problem of adaptation to environments that are unstable: the physically changing unpredictabilities of the climate, the micro- and megapatterns of the soil, the changing interactions between neighbours at the same trophic level within a community and the interactions between predators and prey. It is of course wrong to assume that all differences between individuals within a population are necessarily the result of disruptive or coevolutionary selection — each case requires formal experimental proof as in the experiments with populations of barley or with clover (Chapter 24). There are other forces that may provide such variation, e.g. gene flow, seed immigration and the effect of catastrophes may be to produce genetic changes between and within populations that are not themselves adaptive. An evolutionary "catastrophe" is taken to be "an event producing a subversion of the order or system of things" (*Shorter Oxford English Dictionary*, 1944). A major fluctuation in the environment that caused a drastic reduction in population size and local extinctions might be thought of as an evolutionary catastrophe, an occurrence so abnormal that it can be regarded as outside the normal range of varied experiences that determine fitness. In this sense a hurricane is not an evolutionary catastrophe for a New England forest: it is part of "normal" selectional experience because it occurs on average at 60—70 year intervals which is less than a single lifetime for a forest tree. In my usage it is a "disaster" to the individuals, not a "catastrophe" to the population. After an evolutionary catastrophe the genetic composition of the population may have been seriously depleted and in the population explosion that follows the genotypes of the few survivors are multiplied giving the equivalent of a "founder effect". Such an effect is particularly likely when the plants have no seed longevity and poor seed dispersal so that a catas-

trophe has a persistent local effect. Such conditions are thought to hold
for several annuals of semi-arid regions, e.g. *Clarkia rubicunda* and other
species of the genus, also for the annual sand dune grass *Vulpia membranacea* (Watkinson, 1975). In these cases localized differentiation
occurs between populations and in *Clarkia* the differences are even
associated with chromosomal barriers to hybridization (Lewis, 1962;
1966; Bartholomew *et al.*, 1973).

Clearly "catastrophe" is a relative term and the scale on which an
evolutionary catastrophe is measured must presumably involve an
episode so severe that a population is brought very near to extinction.
It must presumably also be so rare that its nature has little or no lasting
selective effect so that if and when it recurred the disaster to the population would be as great. Somewhere along a scale of frequency and
intensity of hazards must come a stage at which the hazard is sufficiently
part of the "order or system of things" for the responses to it to be
thought of as contributions to fitness and for the hazards to become
part of natural selection. The fitness of an organism is measured by the
descendants that it leaves, measured over a number of generations. The
short-term fitness experienced by one or two generations during an
unusual period or during one phase in a cycle is unimportant; what
matters is the long term. This is why Thoday (1953; see also Thoday,
1975) emphasized that it is really necessary to measure fitness over
about 10^8 years. This represents a time span different for organisms
with different life cycles but within which major populational events
like predator—prey cycles, successions and coevolutionary phases involving races of pathogens and host may repeat themselves many times. The
effect of such repeated episodes in the selectional history of a population may be regarded as disruptive selection (see discussion in Thoday,
1972) or alternatively as selection with more or less frequently changing direction so that the heterogeneity in the populations that results
might be thought of as a maintained transient polymorphism. Certainly
both spatial and temporal variation in the environment can maintain a
higher average heterozygosity and a greater average number of alleles
per locus than occur in more constant environments (Powell, 1971).

Among the most obvious changes which succeeding generations of
plants may experience is the alternation between colonizing and
density-stressed conditions, and correspondingly an alternation between r- and K-selection. Among the ancestors of the plants that we
now know have been those that have passed through phases of popula-

tion growth, and most will have passed through phases of density stress. The ancestors of some populations will have lived mainly during periods of recovery from hazards and periods of explosive local colonizations. The ancestors of others will have spent most of their time in density-stressed, resource-limited communities. The properties that favour leaving descendants during an r-phase are quite different from those in a K-phase. Presumably the genetic qualities of each population integrate the selective experience of its ancestors, weighted or biased by that of the most recent ancestors.

Among all the evolutionary dilemmas that face higher plants one of the most acute arises because most plants occupy stages in successions. There are two alternatives that might increase the fitness of an organism in a succession: (i) to climb its own successional tree by reacting to the selective pressures put on it from later stages in the succession and so to persist longer in the succession (i.e. to respond to K-selection); (ii) to develop more efficient mechanisms to escape from the succession and to discover and colonize suitable early stages of successions elsewhere (Harper *et al.*, 1961). Where such disruptive force has acted on a population and the force is very strong, or a breeding barrier can be seized upon as a focus of differentiation, the population may divide and the two halves follow the two horns of the dilemma. Salisbury (1942) has listed closely related pairs of species that occupy different stages in successions and it appears that it is often a polyploid that moves to occupy a more advanced stage and the diploid specializes as the more fugitive species.

There is no doubt that the individuals that escape from crowded colonies are often unrepresentative of the bulk of the population. The *Drosophila* that escape from crowded culture vessels bear genetic properties for "escape" and the quality can be selected (Sakai, 1958). The voles that escape from overcrowded populations are genetically different from the main colonies (Krebs *et al.*, 1973). The forms of *Agropyron spicatum* that colonize newly opened habitats are different from those in closed communities and, as a colonizing open population changes to a closed stage in the succession, the representation of forms of the species changes: in the colonizing phases forms are favoured which have a caespitose habit and this is the form that predominates on newly bared areas. After a dense plant cover develops selection eliminates caespitose habit in favour of rhizomatous forms (Daubenmire, 1960b).

A dilemma of natural selection also occurs for species that occupy patches in mosaic communities when the phases of the mosaic are continually replacing each other. This is essentially succession on a much smaller scale but provides much the same dilemma. In the raised bog community a succession of species of *Sphagnum* occupy zones on developing peat hummocks (Osvald, 1923; Tansley, 1949). The hummocks eventually break down to form hollows and a new hummock then begins to form and repeat the characteristic succession and zonation of species. A population of each species is sandwiched between two others and the population of each is therefore under competitive pressure from the one above it and the one below. The selective forces exerted by these two neighbours are most unlikely to be the same and there is, therefore, presumably continuous disruptive selection acting on the population in the middle. This phenomenon of the selectional sandwich is presumably repeated wherever a zonation of species involves competitive interfaces.

Cycles of predator and prey provide forces of natural selection that alternate in direction over time. Whether cycles in a herbivore are caused by its hyperpredators or parasites or by direct interaction between the herbivore and its plant prey, the plant population will experience alternate phases in which selection is primarily by the predator and phases when other forces are dominant. The direction of evolutionary change that maximizes fitness in the presence of the abundant predator will be quite different from the direction during predator crashes. Presumably, if the life cycle of the plant prey is long relative to the population cycles of the predator, the opposing forces of selection become diffused into an optimal compromise that maximizes fitness. When the life cycle of the plant is short relative to the predator's population cycle, the changing directions of selection create a dilemma: the condition for fitness keeps changing, no one adaptive solution is "right" all the time.

The nature of the breeding system affects the role of natural selection in populations that are unstable, that suffer disasters and that alternate periods of population growth with periods of competitive restraint or decline. During phases in which a population is increasing rapidly without competitive restraints an outbreeding population tends to display much variation. This is particularly the case when a temporary colonizing event occurs and there may be temporary selection for fugitive properties. At the end of the colonizing episode the array of genetic forms departs from its original value and the fitness of the individuals will have been lowered when the population returns to density-stressed conditions. For a population that spends its time in repeated explosions, followed by density stress, conditions that allow genetic variants to accumulate alternate with conditions in which fitness means conforming to a strict adaptive line. After each explosion the population returns to the "normal" condition less fit, unless in some way the display of variation has been controlled during the colonizing phase. We may perhaps interpret the predominantly inbreeding habit of colonizers and fugitives as a protection against genetic display during colonizing episodes and against the too rapid "tracking" of adaptive optima that are only transient.

A slightly different emphasis was put on essentially the same argument by Stebbins (1958):

> Plant populations characterised by self-fertilization or apomixis are usually found in temporary habitats, and the plants themselves tend to have particularly efficient methods of seed dispersal. Typical examples among self-fertilizers are the prickly lettuce (*Lactuca scariola*) and sow thistle (*Sonchus oleraceus*) and among apomicts are the dandelion (*Taraxacum* species) and hawkweed (*Hieracium* species). Such species have populations which fluctuate greatly in size; often a single vigorous immigrant into a new locality may in one or two generations give rise to a large population. Under such conditions the constancy provided by self-fertilization or apomixis may have a positive selective value in a rapidly expanding population, those genotypes will succeed best which produce the largest possible number of highly adapted descendants. If, on the other hand, a population remains constant in size, and its individuals are so long lived that they produce progeny in far greater numbers than are able to survive, then the advantage will go to those individuals which are able to produce a relatively small proportion of highly fit descendants, even at the expense of a large number of poorly adapted progeny. On this basis, we should expect that our dominant forest trees should be cross-fertilized and highly heterozygous; which is, in fact, the case.

Close inbreeding or apomixis are ways in which temporary forces of selection are prevented from throwing the long-term adaptation of a population off course, just as parthenogenesis prevents new recombinants from becoming established during a population explosion of aphids or rotifers. Two other important properties can have the same effect in higher plants. The presence of a buried seed bank and a bank of dormant buds on rhizome systems or on tree trunks allows individual genets to persist through temporary hazards that may shift the adaptation of new seedling recruits off course. The speed with which a newly recruited population of seedlings can be dramatically changed by, for example, a temporary grazing regime has been illustrated (Chapter 14). The longevity of genets, particularly in dormant stages, is a strong safeguard against the too sensitive adaptive tracking of transient changes in the environment.

Selection in a patchy environment

Just as the life span of a plant or a seed sets a scale to temporal variations in the environment so the dispersal mechanisms of a plant sets a scale to its spatial heterogeneity. There are three ways in which plants may sense the diversity of their environment and these determine the selective forces that are met.

(i) The form of the individual genet determines the way in which it meets neighbours. The clonal habit may be such as to form a sparse, wandering vegetative body, decaying in the older regions so that the genet continually explores through other vegetation in a "guerilla" strategy as a series of more or less separate modules. Such a growth form maximizes interclonal and interspecific encounters. In contrast, a clone based on short rhizome internodes or intra-vaginal tillering behaves as a cohort of modules with a tight advancing front (a "phalanx" strategy) and in which most of the "neighbours" are intraclonal and much of the competition is within the phenotype. In a woody community there is a similar contrast between the tight branching system of a spruce tree in which each needle and each branch is most likely to contact others from the same genet and the liane which wanders as a genet through much of the interspecific variety of a canopy and sometimes the forest floor as well. The form of the plant determines either that much of the variation in the environment is encompassed in the experience of a single individual or that the same environmental variation is experienced as different forces acting on different individuals. It is interesting to consider growth forms as a way in which the spatial variation in a habitat is sampled in the same way as longevity samples its temporal variation.

(ii) The mechanism of seed dispersal determines the range of environmental heterogeneity that is sampled by potential descendants. Many desert and dune annuals have extremely poor seed dispersal — seeds land close to the parent and selection will then tend to narrow the range of variation to a very local scale — small, discrete, very locally adapted populations (e.g. *Arabidopsis thaliana*; Jones 1971a, b). In contrast, a vigorously wind-dispersed species (e.g. *Tussilago farfara*) exploits and "senses" the environment over distances of several kilometres that includes a quite different mosaic of communities and potential neighbours than those met by *Arabidopsis*.

(iii) The mechanism of pollen dispersal determines the range of environment that is sampled by the genetic system and the speed and distance that genetic material may travel (though the haploid phase of environmental exploration cannot of course act further than has already been colonized by the diploid phase of the species). It is an important difference between higher plants and higher animals that the haploid phase of plants can travel much further than the diploid so that plants of the same species can "interfere" genetically with each other's progeny at distances much greater than the distance between individuals. This is an important reason why for plants it is necessary to distinguish between populations that coexist (are sympatric) and have potential for exchange of genetic information and populations that *cohabit* and have the potential for making demands on the same pool of limiting resources (Harper *et al.*, 1961).

Levins (1968) has shown that the evolutionary decision whether to adopt specialist, generalist or polymorphic adaptation depends on the grain of the environment, the way in which the population samples the diversity of its environment. There are at least seven categories of solution to the dilemmas of maximizing fitness in an environment that varies in space or time:

(i) The formation of local specialized races. Within a community this demands an inbreeding system, or (for an outbreeding system) a very high reproductive output and intense local selection. (The cases of *Festuca microstachys* and *Trifolium repens* described in this chapter illustrate this response which is appropriate for a spatial scale of heterogeneity.)

(ii) The development of genetic polymorphism with large blocks of linked coadapted genes or super-genes.

(iii) The maintenance of a high degree of genetic variance within the population, coupled either with longevity of growing plants or with a large seed bank which provides a memory of past adapted genotypes that can be called upon when conditions that have been previously experienced recur. (It is appropriate for an environment of temporal heterogeneity.)

(iv) The act of speciation and the formation of distinct breeding groups that can evolve independently to occupy different parts of the community mosaic (or the formation of apomictic races) — appropriate to spatial heterogeneity.

(v) The development of somatic polymorphism in which a single genotype develops a variety of phenotypes each adapted to a different spatial or temporal phase in the community (e.g. leaf polymorphism between seasons or ages of a plant, seed polymorphism in which different sorts of seeds are borne on the same plant).

(vi) The evolution of a phenotypic plasticity that enables individuals to react during their development in an adaptive fashion to the specific conditions that they experience. (This is appropriate for temporal and spatial heterogeneity.)

(vii) The evolution of heterozygotic superiority. This is likely to arise when succeeding generations meet alternating conditions and the heterozygote is most fit because it alone produces some progeny that are successful in either environment.

Which direction is followed by a particular evolving population will depend in part on the genetic archetype of the species — some changes are easier for one species than another because of the different potential that each has for genetic change; in part it depends on the nature of the breeding system and the accessibility of new genetic recombinations from populations elsewhere, for gene flow can both provide genetic material for local adaptation and hinder too rapid a local deviation from the species norm.

Fitness, adaptation and productivity

Adaptation is a word too loosely used in ecological writing. Often to say that a feature of an organism's life or form is adaptive is to say no more than that the feature appears to be a good thing, judged on the basis of an anthropomorphic attitude to the problems that the organism is seen to face. More accurately, adaptations are those features of an organism that in the past improved the fitness of its ancestors and so were transmitted to descendants. Adaptation is always retrospective. Fitness itself is relative — it is defined by the numbers of descendants left by an individual relative to its fellows. An organism will be more fit if its activities reduce the number of descendants left by neighbours, *even if the activities do nothing to the number of descendants that it itself leaves.* The point is easily made by considering the evolution of height in plants. Within a population of plants growing densely and absorbing the larger part of incident light, success depends on placing leaves high in the canopy and shading and suppressing neighbours. There is no intrinsic advantage to the individual from being high (there are some real disadvantages in the amount of non-reproductive tissue to be supported), only an advantage from being higher than neighbours. It is being higher, not just high, that pays. Similarly a genetic change that gave a plant a larger and earlier root system might bring no advantages to the possessor other than the relative advantage over the neighbours that it is able to deprive. If an activity of an organism brings no direct benefit but hinders the chance that neighbours will leave descendants, the activity will increase fitness — it will be "adaptive".

This argument may be important in understanding evolutionary processes. Often the process is seen as in some way optimizing the behaviour of descendants — in some way making them "better" or "adjusted to the environment". There is in fact nothing innate in a process that maximizes evolutionary fitness, that necessarily "optimizes" physiological function. Indeed a genetic change that resulted in an organism immobilizing mineral nutrients in old tissue until it died instead of returning them to the cycle within the ecosystem would almost certainly confer fitness provided that potentially competing neighbours were deprived of needed nutrients by this activity.

A theory of natural selection that is based on the fitness of individuals leaves little room for the evolution of populations or species towards some optimum, such as better use of environmental resources, higher productivity per area of land, more stable ecosystems, or even for the view that plants in some way become more efficient than their ancestors. Instead, both the study of evolutionary processes and of the natural behaviour of populations suggest that the principles of "beggar my neighbour" and "I'm all right Jack" dominate all and every aspect of evolution. Nowhere does this conclusion have more force than when man takes populations that have evolved in nature under criteria of individual fitness, grows them in culture as populations and then applies quite different criteria of performance — productivity per unit area of land. Natural selection is about individuals and it would be surprising if the behaviour that favoured one individual against another was also the behaviour that maximized the performance of the population as a whole. For this to happen, selection would have to act on groups. It is an interesting thought that group selection which is believed to be extremely rare or absent in nature (Maynard Smith, 1964) may be the most proper type of selection for improving the productivity of crop and forest plants. Plant breeding would then be concerned to undo the results of selection for the selfish qualities of individual fitness and focus on the performance of populations.

References

Abbott, H. G. and Quink, T. F. (1970). Ecology of eastern white pine seed caches made by small forest mammals. *Ecology* **51**, 271–278.

Åberg, E., Johnson, I. J. and Wilsie, C. P. (1943). Associations between species of grasses and legumes. *J. Am. Soc. Agron.* **35**, 357–369.

Abeywickrama, B. S. (1949). "A study of the variation in the field layer vegetation of two Cambridgeshire woods". Ph.D. thesis, University of Cambridge.

Abrahamson, W. G. and Gadgil, M. D. (1973). Growth form and reproductive effort in golden rods (*Solidago*, Compositae). *Am. Nat.* **107**, 651–661.

Adair, C. R. and Jones, J. W. (1946). Effect of environment on the characteristics of plants surviving in a bulk hybrid population of rice. *J. Am. Soc. Agron.* **38**, 708–716.

Adams, M. W. (1967). Basis of yield component compensation in crop plants with special reference to the field bean, *Phaseolus vulgaris. Crop Sci.* **7**, 505–510.

Adams, S., Strain, B. R. and Adams, M. S. (1970). Water-repellent soils, fire and annual plant cover in a desert scrub community of southeastern California. *Ecology* **51**, 696–700.

Alder, G. M. (1970). "Age-profiles of aspen forests in Utah and northern Arizona". M.S. thesis, University of Utah.

Allard, R. W. and Adams, J. (1969). Population studies in predominantly self-pollinating species. XIII. Intergenotypic competition and population structure in barley and wheat. *Am. Nat.* **103**, 621–645.

Allard, R. W., Babbel, G. R., Clegg, M. T. and Kahler, A. L. (1972). Evidence for coadaptation in *Avena barbata. Proc. Natn. Acad. Sci. U.S.A.* **69**, 3043–3048.

Allard, R. W., Jain, S. K. and Workman, P. L. (1968). The genetics of inbreeding populations. *Adv. Genet.* **14**, 55–131.

Allee, W. C., Emerson, A. E., Park, T. and Schmidt, K. P. (1949). "Principles of Animal Ecology". Saunders, Philadelphia.

779

Allen, J. A. and Clarke, B. (1968). Evidence for apostatic selection by wild passerines. *Nature, Lond.* **220**, 501—502.

Amen, R. D. (1966). The extent and role of seed dormancy in alpine plants. *Q. Rev. Biol.* **41**, 271—281.

Amen, R. D. (1968). A model of seed dormancy. *Bot. Rev.* **34**, 1—31.

Anderson, M. C. (1964). Light relations of terrestrial plant communities and their measurement. *Biol. Rev.* **39**, 425—486.

Anderson, M. C. (1966). Some problems of the simple characterization of the light climate in plant communities. *In* "Light as an Ecological Factor" (R. Bainbridge, G. C. Evans and O. Rackham, eds) *Symp. Br. Ecol. Soc.* 77—90.

Ando, T., Sakaguchim, K., Narita, T. and Satoo, S. (1962). Growth analysis on the natural stands of Japanese Red Pine (*Pinus densiflora* Sieb. et Zucc.) 1. *Bull. Gov. Forest exp. Stn.* **144**, 30 pp.

Andres, L. A. and Goeden, R. D. (1975). The biological control of weeds by introduced natural enemies. *In* "Biological Control" (C. B. Huffaker, ed.), pp. 143—164. Plenum Press, New York.

Andrewartha, H. G. (1957). The use of conceptual models in population ecology. *Cold Spring Harbour Symp. Quant. Biol.* **22**, 219—36.

Andrewartha, H. G. and Birch, L. C. (1954). "The Distribution and Abundance of Animals". University of Chicago Press, Chicago.

Antonovics, J. (1972). Population dynamics of the grass *Anthoxanthum odoratum* on a zinc mine. *J. Ecol.* **60**, 351—366.

Arber, A. (1950). "The Natural Philosophy of Plant Form". Cambridge University Press, London.

Archer, M. (1973). The species preferences of grazing horses. *J. Br. Grassl. Soc.* **28**, 123—138.

Arnold, G. W. (1962). Effects of pasture maturity on the diet of sheep. *Aust. J. agric. Res.* **13**, 701—706.

Arnold, G. W. (1964). Some principles in the investigation of selective grazing. *Proc. Aust. Soc. Anim. Prod.* **5**, 258—271.

Arnold, G. W., Ball, J., McManus, W. R. and Bush, I. G. (1966). Studies on the diet of the grazing animal. 1. Seasonal changes in the diet of sheep grazing on pastures of different availability and composition. *Aust. J. agric. Res.* **17**, 543—556.

Arthur, A. E. (1969). "Variation in natural populations of *Papaver dubium*". Ph.D. thesis, University of Birmingham.

Arthur, A. E., Gale, J. S. and Lawrence, K. J. (1973). Variation in wild populations of *Papaver dubium*. VII. Germination time. *Heredity* **30**, 189—197.

Ashby, E. and Wangermann, E. (1951). Studies in the morphogenesis of leaves. VII Part 1. Effect of light intensity and temperature on the cycle of ageing and rejuvenation in the vegetative life history of *Lemna minor*, and Part II. Correlative effects of fronds in *Lemna minor*. *New Phytol.* **50**, 186—209.

Ashby, E., Wangermann, E. and Winter, E. J. (1949). Studies in the morphogenesis of leaves. III. Preliminary observations on vegetative growth in *Lemna minor*. *New Phytol.* **48**, 374—381.

Ashton, D. H. (1956). "Studies on the autoecology of *E. regnans*". Ph.D. thesis, University of Melbourne, Australia.

Ashton, P. H. (1956). Speciation among tropical forest trees: some deductions in the light of recent evidence. *Biol. J. Linn. Soc.* 1, 155–196.

Askew, R. R. (1962). The distribution of galls of *Neuroterus* (Hym. Cynipidae) on oak. *J. Anim. Ecol.* 31, 439–445.

Aspinall, D. (1960). An analysis of competition between barley and white persicaria. II. Factors determining the course of competition. *Ann. appl. Biol.* 48, 637–654.

Auclair, A. N. and Cottam, G. (1971). Dynamics of black cherry (*Prunus serotina* Erhr.) in Southern Wisconsin oak forests. *Ecol. Monogr.* 41, 153–177.

DeBach, P. (1964). "Biological Control of Insect Pests and Weeds". (P. DeBach, ed., assisted by E. I. Schlinger). Chapman and Hall, London.

Baeumer, K. and De Wit, C. T. (1968). Competitive interference of plant species in monocultures and mixed stands. *Neth. J. agric. Sci.* 16, 103–122.

Bailey, R. J., Rees, H. and Jones, L. M. (1976). Interchange heterozygotes versus homozygotes. *Heredity* 37, 109–112.

Baker, H. (1937). Alluvial meadows: a comparative study of grazed and mown meadows. *J. Ecol.* 25, 408–420.

Baker, J. R. (1938). The evolution of breeding seasons. "Evolution", Essays presented to E. S. Goodrich (Oxford) 161–177.

Bakker, D. (1960). A comparative life-history study of *Cirsium arvense* (L) Scop. and *Tussilago farfara* (L.), the most troublesome weeds in the newly reclaimed polders of the former Zuiderzee. *In* "The Biology of Weeds" (J. L. Harper, ed.) *Symp. Br. ecol. Soc.* 1, 205–222.

Bandeen, J. D. and Buchholtz, K. P. (1967). Competitive effects of quack grass upon corn as modified by fertilisation. *Weeds* 15, 220–223.

Barallis, G. (1965). Aspects ecologiques des mauvaises herbes dans les cultures annuelles. C. r. 3ème Conf. COLUMA, Paris.

Barallis, G. (1968). Ecology of blackgrass. *Proc. 9th Br. Weed Control Conf.* 6–8.

Barber, S. A., Walker, J. M. and Vasey, E. H. (1962). Principles of ion movement through the soil to the plant root. *Trans. Int. Soil Sci. Comm.* II and IV, New Zealand, 121–124.

Barclay-Estrup, P. (1970). The description and interpretation of cyclical processes in a heath community. II. Changes in biomass and shoot production during the *Calluna* cycle. *J. Ecol.* 58, 243–249.

Barger, G. and Blackie, J. J. (1937). Alkaloids of *Senecio*. III. Jacobine, Jacodine and Jaconine. *J. chem. Soc.* 584–586.

Barker, J. S. F. (1971). Ecological differences and competitive interaction between *Drosophila melanogaster* and *Drosophila simulans* in small laboratory populations. *Oecologia* (Berl.) 8, 139–156.

Barker, J. S. F. (1973). Natural selection for coexistence or competitive ability in laboratory populations of *Drosophila*. *Egypt. J. Genet. Cytol.* 2, 288–315.

Barrett, L. I. (1931). Influence of forest litter on the germination and early survival of chestnut oak, *Quercus montana* Willd. *Ecology* 12, 476–484.

Barrow, M. D., Costin, A. B. and Lake, P. (1968). Cyclical changes in an Australian fjaeldmark community. *J. Ecol.* **56**, 89—96.

Bartholomew, B. (1970). Bare zone between California shrub and grassland communities: the role of animals. *Science, N.Y.* **170**, 1210—1212.

Bartholomew, B., Eaton, L. C. and Raven, P. H. (1973). *Clarkia rubicunda*: a model of plant evolution in semi-arid regions. *Evolution* **27**, 505—517.

Bazzaz, F. A. and Harper, J. L. (1976). Relationship between plant weight and numbers in mixed populations of *Sinapis arvensis* (L.) Rabenh. and *Lepidium sativum* L. *J. appl. Ecol.* **13**, 211—216.

Beard, J. S. (1946). The Mora forests of Trinidad, British West Indies. *J. Ecol.* **33**, 173—192.

Beatley, J. (1969). Dependence of desert rodents on winter annuals and precipitation. *Ecology* **50**, 722—724.

Bell, A. D. (1974). Rhizome organisation in relation to vegetative spread in *Medeola virginiana*. *J. Arnold Arboretum* **55**, 458—468.

Bell, A. D. (1976). Computerized vegetative mobility in rhizomatous plants. *In* "Automata, Languages and Development" (A. Lindemayer and G. Rosenberg, eds). North-Holland, Amsterdam.

Belyea, R. M. (1952). Death and deterioration of balsam fir weakened by spruce budworm defoliation in Ontario. *J. Forestry* **50**, 729—738.

Benecke, P. and Mayer, R. (1971). Aspects of soil water behaviour as related to beech and spruce stands. — Some results of water balance investigations. *In* "Integrated Experimental Ecology" (H. Ellenberg, ed.), pp. 153—163. Chapman and Hall, London.

Bent, A. C. (1939). Life histories of north American woodpeckers. *U.S. National Museum, Bulletin.* **174**, Washington, D.C.

Berg, R. Y. (1975). Myrmecochorous plants in Australia and their dispersal by ants. *Aust. J. Bot.* **23**, 475—508.

Bergh, J. P. van den (1968). An analysis of yields of grasses in mixed and pure stands. *Versl. Landbouwk. Onderz.* **714**, 1—71.

Bergh, J. P. van den and de Wit, C. T. (1960). Concurrentie tussen timothee en reukras. *Inst. voor Biol. en Scheikundig Onderz. Landbouw. Wageningen. Mededeling.* **121**, 155—165.

Beukema, J. J. (1968). Predation by the three-spined stickleback (*Gasterosteus aculeatus* L.): the influence of hunger and experience. *Behaviour* **31**, 1—126.

Beveridge, A. E. (1964). Dispersal and destruction of seed in central North Island Podocarp forests. *Proc. N.Z. ecol. Soc.* **11**, 48—55.

Bingham, J. (1967). Investigations on the physiology of yield in winter wheat, by comparisons of varieties and by artificial variations in grain number per ear. *J. agric. Sci.* **68**, 411—422.

Birch, L. C. (1948). The intrinsic rate of natural increase of an insect population. *J. Anim. Ecol.* **17**, 15—26.

Bishop, J. A. and Korn, M. E. (1969). Natural selection and cyanogenesis in white clover. *Heredity* **24**, 423—430.

Blaak, G. (1972). Prediction of precocity in the oil palm (*Elaeis guineensis* Jacq.) *Euphytica* **21**, 22—26.

Black, C. C. (1971). Ecological implications of dividing plants into groups with distinct photosynthetic production capacities. *In*: "Advances in Ecological Research" (J. B. Gragg, ed.), Vol. 7, pp. 87–114. Academic Press, London and New York.

Black, C. C., Chen, T. M. and Brown, R. H. (1969). Biochemical basis for plant competition. *Weed Sci.* 17, 338–344.

Black, J. N. (1956a). The influence of seed size and depth of sowing on pre-emergence and early vegetative growth of subterranean clover (*Trifolium subterraneum* L.). *Aust. J. agric. Res.* 7, 98–109.

Black, J. N. (1956b). The distribution of solar radiation over the earth's surface. *Arch. Met. Geophys. Bioklim.* B 7, 165–189.

Black, J. N. (1957). The early vegetative growth of three strains of subterranean clover (*Trifolium subterraneum* L.) in relation to size of seed. *Aust. J. agric. Res.* 8, 1–14.

Black, J. N. (1958). Competition between plants of different initial seed sizes in swards of subterranean clover (*Trifolium subterraneum* L.) with particular reference to leaf area and the light microclimate. *Aust. J. agric. Res.* 9, 299–318.

Black, J. N. (1959). Seed size in herbage legumes. *Herb. Abstr.* 29, 235–241.

Black, J. N. (1960a). The relationship between illumination and global radiation. *Trans. R. Soc. S. Aust.* 83, 83–87.

Black, J. N. (1960b). The significance of petiole length, leaf area, and light interception in competition between strains of subterranean clover (*Trifolium subterraneum* L.) grown in swards. *Aust. J. agric. Res.* 11, 277–291.

Black, J. N. (1960c). An assessment of the role of planting density in competition between red clover (*Trifoloum pratense* L.) and lucerne (*Medicago sativa* L.) in the early vegetative stage. *Oikos* 11, 26–42.

Black, J. N. (1963). The interrelationship of solar radiation and leaf area index in determining the rate of dry matter production of swards of subterranean clover (*Trifolium subterraneum* L.) *Aust. J. agric. Res.* 14, 20–38.

Black, J. N. and Wilkinson, G. N. (1963). The role of time of emergence in determining the growth of individual plants in swards of subterranean clover (*Trifolium subterraneum* L.) *Aust. J. agric. Res.* 14, 628–638.

Black, M. (1969). Light-controlled germination of seeds. *In* "Dormancy and Survival" (H. W. Woolhouse, ed.) *Symp. Soc. exp. Biol.* 23, 193–217.

Black, M. and Wareing, P. F. (1955). Growth studies in woody species. VII. Photoperiodic control of germination in *Betula pubescens* Ehrh. *Physiologia Pl.* 8, 300–316.

Blackburn, W. H. and Tueller, P. T. (1970). Pinyon and Juniper invasion in black sagebrush communities in East-Central Nevada. *Ecology* 51, 841–848.

Blackman, G. E. and Wilson, G. L. (1951a). Physiological and ecological studies in the analysis of plant environment. VI. The constancy for different species of a logarithmic relationship between net assimilation rate and light intensity and its ecological significance. *Ann. Bot.* 15, 64–94.

Blackman, G. E. and Wilson, G. L. (1951b). Physiological and ecological studies in the analysis of plant environment. VII. An analysis of the differential effects of light intensity on the net assimilation rate, leaf-area ratio, and relative growth rate of different species. *Ann. Bot.* 15, 374–408.

784   REFERENCES

Blackman, G. E. and Wilson, G. L. (1954). Physiological and ecological studies in the analysis of plant environment. IX. Adaptive changes in the vegetative growth and development of *Helianthus annuus* induced by an alteration in light level. *Ann. Bot.* 18, 72–94.

Bleasdale, J. K. A. (1966). Plant growth and crop yield. *Ann. appl. Biol.* 57, 173–182.

Bleasdale, J. K. A. and Nelder, J. A. (1960). Plant population and crop yield. *Nature, Lond.* 188, 342.

Böcher, T. W. (1961). Experimental and cytological studies on plant species. VI. *Dactylis glomerata* and *Anthoxanthum odoratum*. *Bot. Tidsskr.* 56, 314–355.

Boddington, M. J. and Metterick, D. F. (1974). The distribution, abundance, feeding habit and population biology of the immigrant triclad *Dugesia polychroa* (Platyhelminthes: Turbellaria) in Toronto harbour, Canada. *J. Anim. Ecol.* 43, 681–699.

Bode, H. R. (1958). Beiträge zur Kenntnis allelopathischer Erscheinungen bei einigen Juglandaceen. *Planta* 51, 440–480.

Boller, E. F. and Bush, G. L. (1974). Evidence for genetic variation in populations of the European cherry fruit fly, *Rhagoletis cerasi* (Diptera: Tephritidae) based on physiological parameters and hybridization experiments. *Entomologia exp. appl.* 17, 279–293.

Bormann, F. H. (1960). Individuality in forest trees. *Dart. Alumni Magazine* February issue 1960, 43–45.

Bormann, F. H. (1961). Intraspecific root grafting and the survival of eastern white pine stumps. *For. Sci.* 7, 248–255.

Bormann, F. H. (1966). The structure, function and ecological significance of root grafts in *Pinus strobus* L. *Ecol. Monogr.* 36, 1–26.

Bormann, F. H. and Graham, B. F. (1959). The occurrence of natural root grafting in eastern white pine, *Pinus strobus* L., and its ecological implications. *Ecology* 40, 678–691.

Borthwick, H. A., Hendricks, S. B., Toole, E. H. and Toole, V. K. (1954). Action of light on lettuce-seed germination. *Bot. Gaz.* 115, 205–225.

Bosch, C. A. (1971). Redwoods: a population model. *Science, N.Y.* 172, 345–349.

Bosch, R. van den and Telford, A. D. (1964). Environmental modifications and biological control. *In* "Biological Control of Insect Pests and Weeds" (P. de Bach ed.) Chapter 16. Chapman and Hall, London.

Botkin, D. B., Jordan, P. A., Dominski, A. S., Lowendorf, H. S. and Hutchinson, G. E. (1973). Sodium dynamics in a northern ecosystem (moose, wolves, plants) *Proc. Natn. Acad. Sci. U.S.A.* 70, 2745–2748.

Bouwkamp, J. C. and McCully, J. E. (1972). Competition and survival in female plants of *Asparagus officinalis* L. *J. Am. Soc. hort. Sci.* 97, 74–76.

Bradley, R. T., Christie, J. M. and Johnston, D. R. (1966). Forest Management Tables. *Bookl. For. Commn.* 16, 1–218.

Bradshaw, A. D. (1959). Population differentiation in *Agrostis tenuis*. Sibth. II. The incidence and significance of infection by *Epichloe typhina*. *New Phytol.* 58, 310–315.

Bradshaw, A. D. (1965). Evolutionary significance of phenotypic plasticity in plants. *Adv. Genet.* **13**, 115—155.

Bradshaw, A. D. and Jain, S. K. (1966). Evolutionary divergence among adjacent plant populations. I. The evidence and its theoretical analysis. *Heredity* **21**, 407—441.

Brenchley, W. E. (1918). Buried weed seeds. *J. agric. Sci.* **9**, 1—31.

Brenchley, W. E. and Warington, K. (1930). The weed seed population of arable soil. I. Numerical estimation of viable seeds and observations on their natural dormancy. *J. Ecol.* **18**, 235—272.

Brenchley, W. E. and Warington, K. (1930). The weed seed population of arable soil. II. Influence of crop, soil, and methods of cultivation upon the relative abundance of viable seeds. *J. Ecol.* **21**, 103—127.

Brenchley, W. E. and Warington, K. (1936). The weed seed population of arable soil. III. The re-establishment of weed species after reduction by fallowing. *J. Ecol.* **24**, 479—501.

Brenchley, W. E. and Warington, K. (1945). The influence of periodic fallowing on the prevalence of viable weed seeds in arable soil. *Ann. appl. Biol.* **32**, 285—296.

Brierley, J. K. (1955). Seasonal fluctuations in the oxygen and carbon dioxide concentrations in beech litter with reference to the salt uptake of beech mycorrhizas. *J. Ecol.* **43**, 404—408.

Briese, D. T. (1974). "Ecological studies on an ant community in a semi-arid habitat (with emphasis on seed-harvesting species)". Ph.D. thesis, Australian National University.

Broadbent, L. (1957). "Investigation of Virus Diseases of Brassica Crops". Cambridge University Press, London and New York.

Brook, J. M. (1969). "Studies of pattern and succession in seed populations". Ph.D. thesis, University of Wales.

Brougham, R. W. (1962). The leaf growth of *Trifolium repens* as influenced by seasonal changes in the light environment. *J. Ecol.* **44**, 448—459.

Brougham, R. W. and Harris, W. (1967). Rapidity and extent of changes in genotypic structure induced by grazing in a rye grass population. *N.Z. Jl. agric. Res.* **10**, 56—65.

Brown, R. H., Cooper, R. B. and Blaser, R. E. (1966). Effects of leaf age on efficiency. *Crop Science* **6**, 206—209.

Brussard, P. F., Levin, S. A., Miller, C. N. and Whittaker, R. H. (1971). Redwoods: a population model debunked. *Science, N.Y.* **174**, 435—436.

Buchholtz, K. P. (1971). The influence of allelopathy on mineral nutrition. *In* "Biochemical Interactions among Plants". (U.S. Nat. Comm. for I.B.P., eds.) 86—89. Natn. Acad. Sci. Washington, D.C.

Budd, A. C., Chepil, W. S. and Doughty, J. L. (1954). Germination of weed seeds. III. The influence of crops and fallow on the weed seed population of the soil. *Can. J. agric. Sci.* **34**, 18—27.

Buechner, H. K. and Dawkins, C. H. (1961). Vegetation changes induced by elephants and fire in Murchison Falls National Park, Uganda. *Ecology* **42**, 752—766.

Buell, J. H. (1945). The prediction of growth in uneven-aged timber stands on the basis of diameter distribution. *Duke Univ. Sch. Forest. Bull.* 11, 70 pp.

Bullock, J. A. (1967). The insect factor in plant ecology. *J. Indian bot. Soc.* 46, 323–330.

Bunting, A. H. (1959). Some reflections on the ecology of weeds. *In* "The Biology of Weeds" (J. L. Harper, ed.) *Symp. Br. ecol. Soc.* 1, 11–26. Blackwell, Oxford.

Burdon, J. J. and Chilvers, G. A. (1974). Fungal and insect parasites contributing to niche differentiation in mixed species stands of Eucalypt saplings. *Aust. J. Bot.* 22, 103–114.

Burdon, J. J. and Chilvers, G. A. (1975). Epidemiology of damping-off disease (*Pythium irregulare*) in relation to density of *Lepidium sativum* seedlings. *Ann. appl. Biol.* 81, 135–143.

Burrows, F. M. (1973). Calculation of the primary trajectories of plumed seeds in steady winds with variable convection. *New. Phytol.* 72, 647–664.

Burrows, F. M. (1975a). Wind-borne seed and fruit movement. *New Phytol.* 75, 405–418.

Burrows, F. M. (1975b). Calculation of the primary trajectories of dust seeds, spores and pollen in unsteady winds. *New Phytol.* 75, 389–403.

Bush, G. L. (1969a). Mating behaviour, host specificity, and the ecological significance of sibling species in frugivorous flies of the genus *Rhagoletis* (Diptera, Tephritidae). *Am. Nat.* 103, 669–672.

Bush, G. L. (1969b). Sympatric host race formation and speciation in frugivorous flies of the genus *Rhagoletis* (Diptera, Tephritidae). *Evolution* 23, 237–251.

Buxton, J. (1950). "The Redstart". Collins, London.

Buzatti-Traverso, A. A. (1955). Evolutionary changes in components of fitness and other polygenic traits in *Drosophila melanogaster*. *Heredity* 9, 153–86.

Cahn, M. A. and Harper, J. L. (1976a). The biology of the leaf mark polymorphism in *Trifolium repens* L. I. Distribution of phenotypes at a local scale. *Heredity* 37, 309–325.

Cahn, M. A. and Harper, J. L. (1976b). The biology of the leaf mark polymorphism in *Trifolium repens* L. II. Evidence for the selection of leaf marks by rumen fistulated sheep. *Heredity* 37, 327–333.

Cain, A. J. (1969). *In* "Speciation in tropical environments". *Biol. J. Linn. Soc.* 1, 234.

Cameron, E. (1935). A study of the natural control of ragwort (*Senecio jacobaea* L.) *J. Ecol.* 23, 265–322.

Campbell, M. H. and Swain, F. G. (1973). Effect of strength, tilth and heterogeneity of the soil surface on radicle entry of surface-sown seeds. *J. Br. Grassl. Soc.* 28, 41–50.

Canfield, R. H. (1957). Reproduction and life span of some perennial grasses of Southern Arizona. *J. Range Management* 10, 199–203.

Cannon, J. R., Corbett, N. H., Haydock, K. P., Tracey, J. G. and Webb, L. J. (1962). An investigation of the effect of the dehydroangustione present in the leaf litter of *Backhousia angustifola* on the germination of *Araucaria cunninghamii* etc. *Aust. J. Bot.* 10, 119–128.

Cantlon, J. E. (1969). The stability of natural populations and their sensitivity to technology. *In* "Diversity and Stability in Ecological Systems". *Brookhaven Symp. Biol.* 22, 197–203.

Carron, L. T. (1968). "An outline of forest mensuration, with special reference to Australia". Australian National University Press, Canberra.

Caswell, H., Reed, F., Stephenson, S. N. and Werner, P. A. (1973). Photosynthetic pathways and selective herbivory: a hypothesis. *Am. Nat.* **107**, 465–480.

Cavers, P. B. (1963). "Comparative biology of *Rumex obtusifolius* L. and *R. crispus* L. including the variety *trigranulatus*". Ph.D. thesis, University of Wales.

Cavers, P. B. and Harper, J. L. (1966). Germination polymorphism in *Rumex crispus* and *Rumex obtusifolius*. *J. Ecol.* **54**, 367–382.

Chalbi, N. (1967). Biometrie et analyse quantitative de la competition entre genotypes chez la Luzerne. *Ann. Amélior. Plantes* **17**, 119–159.

Champness, S. S. (1949). Notes on the buried seed populations beneath different types of ley in their seedling year. *J. Ecol.* **37**, 51–56.

Champness, S. S. and Morris, K. (1948). The population of buried viable weed seeds in relation to contrasting pasture and soil types. *J. Ecol.* **36**, 149–173.

Chancellor, R. J. (1965). Weed seeds in the soil. *Rep. A.R.C. Weed Res. Org.* (1960–4), 15–19.

Chapman, H. H. (1945). Effect of overhead shade on loblolly pine seedlings. *Ecology* **26**, 274–282.

Charles, A. H. (1968). Some selective effects operating upon white and red clover. *J. Br. Grassl. Soc.* **23**, 20–25.

Charlesworth, B. and Giesel, J. T. (1972). Selection in populations with overlapping generations. II Relations between gene frequency and demographic variables. *Am. Nat.* **106**, 388–401.

Charnov, E. L. and Schaffer, W. M. (1973). The life historical consequences of natural selection: Coles' result revisited. *Am. Nat.* **107**, 791–793.

Chattaway, M. M. (1958). The regenerative powers of certain eucalypts. *Victoria Nat.* **75**, 45–46.

Chettleburgh, M. R. (1952). Observations on the collection and burial of acorns by Jays in Hainault Forest. *British Birds* **45**, 359–364.

Chew, R. M. and Chew, A. E. (1965). The primary productivity of a desert shrub (*Larrea tridentata*). *Ecol. Monogr.* **35**, 355–375.

Chew, R. M. and Chew, A. E. (1970). Energy relationships of the mammals of a desert shrub (*Larrea tridentata*) community. *Ecol. Monogr.* **40**, 1–21.

Chippindale, H. G. and Milton, W. E. J. (1934). On the viable seeds present in the soil beneath pastures. *J. Ecol.* **22**, 508–531.

Chitty, D., Pimentel, D. and Krebs, C. J. (1968). Food supply of overwintered voles. *J. anim. Ecol.* **37**, 113–120.

Christie, B. R. and Kalton, R. R. (1960). Recurrent selection for seed weight in Bromegrass, *Bromus inermis*, Leyss. *Agron. J.* **52**, 575–578.

Churchill, G. B., John, H. H., Duncan, D. P. and Hodson, A. C. (1964). Long-term effects of defoliation of Aspen by the forest tent caterpillar. *Ecology* **45**, 630–633.

Clapham, A. R., Tutin, T. and Warburg, E. F. (1952). "Flora of the British Isles". Cambridge University Press, London.

Clarke, B. (1962). Balanced polymorphism and the diversity of sympatric species. *In* "Taxonomy and Geography" (D. Nichols, ed.), pp. 47–70. Systematics Association, London.

Clatworthy, J. N. (1960). "Studies on the nature of competition between closely related species". D.Phil. thesis, University of Oxford.

Clatworthy, J. N. and Harper, J. L. (1962). The comparative biology of closely related species living in the same area. V. Inter- and intraspecific interference within cultures of *Lemna* spp. and *Salvinia natans*. *J. exp. Bot.* 13, 307—324.

Clegg, M. T. and Allard, R. W. (1973). Viability versus fecundity selection in the slender Wild Oat, *Avena barbata* L. *Science, N.Y.* 181, 667—668.

Clements, F. E. (1916). Plant succession: an analysis of the development of vegetation. *Carnegie Inst. Wash. Publ.* 242, 1—512.

Clements, F. E. and Goldsmith, G. W. (1924). The phytometer method in ecology. *Carnegie Inst. Wash. Publ.* 356, 1—106.

Clements, F. E. and Weaver, J. E. (1924). Experimental vegetation. *Carnegie Inst. Wash. Publ.* 355, 1—172.

Clements, F. E., Weaver, J. E. and Hanson, H. C. (1929). Competition in cultivated crops. *Carnegie Inst. Wash. Publ.* 398, 202—233.

Clements, J. R. (1964). The potential influence of soil splashing by raindrops on seedling survival. *For. Chron.* 40, 512—513.

Cody, M. L. (1966). A general theory of clutch size. *Evolution* 20, 174—184.

Cohen, D. (1966). Optimising reproduction in a randomly varying environment. *J. theor. Biol.* 12, 119—129.

Cole, L. C. (1954). The population consequences of life history phenomena. *Q. Rev. Biol.* 29, 103—137.

Collins, W. J. and Aitken, Y. (1970). The effect of leaf removal on flowering time in subterranean clover. *Aust. J. agric. Res.* 21, 893—903.

Colwell, R. (1951). The use of radioactive isotopes in determining spore distribution patterns. *Am. J. Bot.* 38, 511—523.

Coquillat, M. (1951). Sur les plantes les plus communes a la surface du globe. *Bull. Men. Soc. Lyon* 20, 165—170.

Corner, E. J. H. (1966). "The Natural History of Palms". Weidenfeld and Nicholson, London.

Coulter, J. M. and Chamberlain, C. J. (1909). "Morphology of Angiosperms". D. Appleton & Co., New York.

Cowlishaw, S. J. and Alder, F. E. (1960). The grazing preferences of cattle and sheep. *J. agric. Sci., Camb.* 54, 157—165.

Crawford-Sidebotham, T. J. (1972). The role of slugs and snails in the maintenance of the cyanogenesis polymorphisms of *Lotus corniculatus* L. and *Trifolium repens* L. *Heredity* 28, 405—411.

Creed, R., ed. (1971). "Ecological Genetics and Evolution: Essays in Honour of E. B. Ford". Blackwell, Oxford.

Cremer, K. W. (1965). Dissemination of seed from *Eucalyptus regnans*. *Aust. For.* 33—37.

Cumming, B. G. (1959). Extreme sensitivity of germination and photoperiodic reaction in the genus *Chenopodium* (Tourn.) L. *Nature, Lond.* 184, 1044—1045.

Cumming, B. G. (1963). Germination, as influenced by light and temperature, particularly in *Chenopodium* spp. *Int. Symp. Physiol., Ecol. and Biochem. of Germination.* 1—6.

Cunningham, T. M. (1960). "The natural regeneration of *E. regnans*". Bull. 1, School of Forestry, University of Melbourne, Australia.

Curtin, R. A. (1964). Stand density and the relationship of crown width to diameter and height in *Eucalyptus obliqua*. *Aust. For.* 91—105.

Curtin, R. A. (1970). Dynamics of tree and crown structure in *Eucalyptus obliqua*. *For. Sci.* 16, 321—328.

Curtis, J. T. and Greene, H. C. (1953). Population changes in some native orchids of southern Wisconsin, especially in the University of Wisconsin Arboretum. *The Orchid Journal* 2, 152—155.

Curtis, R. O. and Reukema, D. L. (1970). Crown development and site estimates in a Douglas fir plantation spacing test. *For. Sci.* 16, 287—301.

Dahlgren, R. (1971). Multiple similarity of leaf between two genera of Cape plants, *Cliffortia* L. (Rosaceae) and *Aspalanthus* L. (Fabaceae). *Bot. Notiser* 124, 292—304.

Danilow, D. (1953). Einfluss der Samenerzeugung auf die Struktur der Jahrringe. *Allg. Forstz.* 8, 454—455.

Darlington, C. D. and Janaki Ammal, E. K. (1945). Adaptive iso-chromosomes in *Nicandra*. *Ann. Bot.* 9, 267—281.

Darlington, H. T. (1931). The 50-year period for Dr. Beal's seed viability experiment. *Am. J. Bot.* 18, 262—265.

Darlington, H. T. (1951). The seventy-year period of Dr. W. J. Beal's seed viability experiment. *Am. J. Bot.* 38, 379—381.

Darlington, H. T. and Steinbauer, G. P. (1961). The eighty-year period for Dr. Beal's seed viability experiment. *Am. J. Bot.* 48, 321—325.

Darwin, C. (1859). "The Origin of Species". Harvard Facsimile 1st edn. 1964.

Darwin, C. (1881). "The Formation of Vegetable Mould Through the Action of Worms". John Murray, New York.

Darwin, C. and Wallace, A. R. (1858). On the tendency of species to form varieties: and on the perpetuation of varieties and species by natural selection. *J. Linn. Soc. (Zool.)* 3, 45.

Daubenmire, R. F. (1952). Forest vegetation of northern Idaho and adjacent Washington, and its bearing on concepts of vegetation classification. *Ecol. Monogr.* 22, 301—330.

Daubenmire, R. F. (1960a). A seven year study of cone production as related to xylem layers in *Pinus ponderosa*. *Am. Midl. Nat.* 64, 187—193.

Daubenmire, R. F. (1960b). An experimental study of variation in the *Agropyron spicatum—A. inerme* complex. *Bot. Gaz.* 122, 104—108.

Davidson, J. L. and Donald, C. M. (1958). The growth of swards of subterranean clover with particular reference to leaf area. *Aust. J. agric. Res.* 9, 53—72.

Davidson, J. L. and Milthorpe, F. L. (1966a). Leaf growth in *Dactylis glomerata* following defoliation. *Ann. Bot.* 30, 174—184.

Davidson, J. L. and Milthorpe, F. L. (1966b). The effect of defoliation on the carbon balance in *Dactylis glomerata*. *Ann. Bot.* 30, 186—198.

Davidson, J. L. and Quick, J. P. (1961). The influence of dissolved gypsum on pasture establishment on irrigated sodic clays. *Aust. J. agric. Sci.* 12, 100—110.

Davies, H. (1966). Molehills and pasture reversion. *J. Br. Grassl. Soc.* 21, 148—149.

Davis, R. B. (1966). Spruce—fir forests of the coast of Maine. *Ecol. Monogr.* 36, 79—94.

Dawkins, H. C. (1963). Crown diameters: their relation to bole diameter in tropical forest trees. *Commonw. For. Rev.* **42**, 318–333.

Deevey, E. S. (1947). Life tables for natural populations of animals. *Q. Rev. Biol.* **22**, 283–314.

Demetrius, Ll. (1975). Reproductive strategies and natural selection. *Am. Nat.* **109**, 243–249.

Dethier, V. G. (1959). Food plant distribution and density and larval dispersal as factors affecting insect populations. *Can. Ent.* **91**, 581–596.

Dewers, R. S. and Moehring, D. (1970). Effect of soil water stress on initiation of ovulate primordia in loblolly pine. *For. Sci.* **16**, 219–221.

Dickinson, A. G. and Jinks, J. L. (1956). A generalized analysis of diallel crosses. *Genetics* **41**, 65–78.

Dickson, J. G. (1923). Influence of soil temperature and moisture on the development of the seedling blight of wheat and corn caused by *Gibberella saubinetii*. *J. agric. Res.* **23**, 837–870.

Dickson, J. G. (1928). The relation of temperature to the development of disease in plants. *Am. Nat.* **62**, 311–313.

Diem, J. E. and McGregor, J. L. (1971). Redwoods: a population model debunked. *Science, N.Y.* **174**, 436.

Dixon, A. F. G. (1971a). The role of aphids in wood formation. I. The effect of the sycamore aphid, *Drepanosiphum platanoides* (Schr.) (Aphididae), on the growth of sycamore, *Acer pseudoplatanus* (L.). *J. appl. Ecol.* **8**, 165–179.

Dixon, A. F. G. (1971b). The role of aphids in wood formation. II. The effect of the lime aphid, *Eucallipterus tiliae* L. (Aphididae), on the growth of the lime, *Tilia* x *vulgaris* Hayne. *J. appl. Ecol.* **8**, 393–409.

Dobben, W. H. van (1951). Enkele beschouwingen naar aanleiding van mengcultuur. *Maanbl. Landb. Voorl.* **8**, 89–100.

Dobben, W. H. van (1952). Proefnemingen met mengcultuur van haver en gerst in 1951. *C.I.L.O. Gest Versl. Interprov. Proeren* **39**, 1–19.

Dobben, W. H. van (1953). Proefnemingen met mengcultuur van haver en gerst in 1952. *C.I.L.O. Gest. Versl. Interprov. Proeren* **42**, 1–24.

Dobzhansky, Th. (1970). "Genetics of the Evolutionary Process". Columbia University Press, New York.

Dobzhansky, Th. (1973). Nothing in biology makes sense except in the light of evolution. *Am. Biol. Teacher* March 1973, 125–129.

Dodd, A. P. (1940). "The Biological Campaign Against Prickly Pear". Commonwealth Prickly Pear Board, Brisbane, Australia.

Dollinger, P. M., Ehrlich, P. R., Fitch, W. L. and Breedlove, D. E. (1973). Alkaloid and predation patterns in colorado lupine populations. *Oecologia* (Berl.) **13**, 191–204.

Donald, C. M. (1946). Competition between pasture species, with reference to the hypothesis of harmful root interactions. *J. Coun. scient. ind. Res. Aust.* **19**, 32–37.

Donald, C. M. (1951). Competition among pasture plants. I. Intra-specific competition among annual pasture plants. *Aust. J. agric. Res.* **2**, 355–376.

Donald, C. M. (1958). The interaction of competition for light and for nutrients. *Aust. J. agric. Res.* **9**, 421–435.

Donald, C. M. (1959). The production and life span of seed of subterranean clover (*Trifolium subterraneum* L.). *Aust. J. agric. Res.* 10, 771—787.

Donald, C. M. (1961). Competition for light in crops and pastures. *In* "Mechanisms in Biological Competition". (F. L. Milthorpe, ed.) *Symp. Soc. exp. Biol.* 15, 283—313.

Donald, C. M. (1963). Competition among crop and pasture plants. *Adv. Agron.* 15, 1—118.

Draper, A. D. and Wilsie, C. P. (1965). Recurrent selection for seed size in Birdsfoot Trefoil, *Lotus corniculatus* L. *Crop. Sci.* 5, 313—315.

Duffey, E., Morris, M. G., Sheail, J., Ward, L. K., Wells, D. A. and Wells, T. C. E. (1974). "Grassland Ecology and Wildlife Management". Chapman and Hall, London.

Duncan, D. P. and Hodson, A. C. (1958). Influence of the forest tent caterpillar upon the aspen forests of Minnesota. *For. Sci.* 4, 71—93.

Durrant, A. (1965). Analysis of reciprocal differences in diallell crosses. *Heredity* 20, 573—607.

Duvel, J. W. T. (1902). Seeds buried in soil. *Science, N.Y.* 17, 872—873.

Eber, W. (1971). The characterisation of the woodland light climate. *In* "Integrated Experimental Ecology" (H. Ellenberg, ed.). Chapman and Hall, London. 143—152.

Egli, D. B., Pendleton, J. W. and Peters, D. B. (1970). Photosynthetic rate of three soybean communities as related to carbon dioxide levels and solar radiation. *Agron. J.* 62, 411—414.

Ehrlich, P. R. and Birch, L. C. (1967). The "Balance of Nature" and "Population Control". *Am. Nat.* 101, 97—108.

Ehrlich, P. R. and Gilbert, L. E. (1973). Population structure and dynamics of the tropical butterfly *Heliconius ethilla*. *Biotropica* 5, 69—82.

Ehrlich, P. R. and Raven, P. H. (1965). Butterflies and plants: a study in coevolution. *Evolution* 18, 586—608.

Ekberg, L. (1970). Studies in the genus *Allium*. III. Wind dispersal of *Allium* bulbs. *Bot. Notiser* 123, 115—118.

Ellern, S. J., Harper, J. L. and Sagar, G. R. (1970). A comparative study of the distribution of the roots of *Avena fatua* and *A. strigosa* in mixed stands using a ^{14}C labelling technique. *J. Ecol.* 58, 865—868.

Ellison, L. (1960). Influence of grazing on plant succession of rangelands. *Bot. Rev.* 26, 1—78.

Elton, C. (1958). "The Ecology of Invasions by Plants and Animals". Methuen, London.

Emlen, J. M. (1966). The role of time and energy in food preference. *Am. Nat.* 100, 611—617.

Emlen, J. M. (1968). Optimal choice in animals. *Am. Nat.* 102, 385—389.

England, F. J. W. (1965). Interaction in mixtures of herbage grasses. *Scottish Plant Breeding Station Rec.* 1965, 125—149.

England, F. J. W. (1968). Competition in mixtures of herbage grasses. *J. appl. Ecol.* 5, 227—242.

Ennik, G. C. (1960). De concurrentie tussen witte klaver en Engels raaigras bij verschillen in lichtintensiteit en vochtvoorziening. *Inst. voor Biol. en Scheikundig. Onderz. Landbouw. Wageningen.* Mededeling 109, 37–50.

Epling, C. and Lewis, H. (1952). Increase of the adaptive range of the genus *Delphinium. Evolution* 6, 253–267.

Evans, D. W. and Canode, C. L. (1971). Influence of nitrogen fertilization, gapping and burning on seed production of Newport Kentucky Bluegrass. *Agron. J.* 63, 575–580.

Evans, G. C. (1956). An area survey method of investigating the distribution of light intensity in woodlands, with particular reference to sunflecks, including an analysis of data from rainforest in Southern Nigeria. *J. Ecol.* 44, 391–428.

Evans, J. M. (1974). "Population variability and its ecological significance in *Arabidopsis thaliana* (L.) Heynh.". Ph.D. thesis, University of Leicester.

Evans, L. T. (1971). Evolutionary, adaptive and environmental aspects of the photosynthetic pathway: assessment. *In* "Photosynthesis and Photorespiration" (M. D. Hatch, C. B. Osmond and R. O. Slatyer, eds.). Wiley, New York.

Evenari, M. (1949). Germination inhibitors. *Bot. Rev.* 15, 153–194.

Evenari, M. (1965). Light and seed dormancy. *Encyclopedia of Plant Physiology* 15 (2), 804–847.

Exell, A. W. (1931). The longevity of seeds. *Gdnrs' Chron.* 89, 283.

Fail, H. (1956). The effect of rotary cultivation on the rhizomatous weeds. *J. agric. Engng Res.* 1, 68–80.

Falinska, K. (1972). Fenologiczna reakcja gatunkow na zroznicowanie gradow. *Phytocoenosis* 1, 5–35.

Feeny, P. P. (1968). Effect of oak leaf tannins on larval growth of the winter moth *Operophtera brumata. J. Insect Physiol.* 14, 805–817.

Feeny, P. P. (1969). Inhibitory effect of oak leaf tannins on the hydrolysis of proteins by trypsin. *Phytochemistry* 8, 2119–2126.

Feeny, P. P. (1970). Seasonal changes in oak leaf tannins and nutrients as a cause of spring feeding by winter moth caterpillars. *Ecology* 51, 565–581.

Feeny, P. P. and Bostock, H. (1968). Seasonal changes in the tannin content of oak leaves. *Phytochemistry* 7, 871–880.

Fehr, W. R. and Weber, C. R. (1968). Mass selection by seed size and specific gravity in soybean populations. *Crop Sci.* 8, 551–554.

Feldman, M. W. (1972). Selection for linkage modification: I. Random mating populations. *Theor. Pop. Biol.* 3, 324–346.

Feller, W. (1940). On the logistic law of growth and its empirical verifications in biology. *Acta biotheor.* A5, 51–66.

Fery, R. L. and Janick, J. (1971). Response of corn (*Zea mays* L.) to population pressure. *Crop Sci.* 11, 220–224.

Fielding, J. M. (1964). Seed dissemination in forests of *Pinus radiata. Aust. For. Res.* 1, 48–50.

Fielding, J. M. (1965). The viability of the seed in cones on trees of *Pinus radiata. Aust. For. Res.* 1, 22–23.

Finlay, K. W. and Wilkinson, G. N. (1963). The analysis of adaptation in a plant-breeding programme. *Aust. J. agric. Res.* 14, 742–754.

Fisher, R. A. (1929, revised 1958). "The Genetical Theory of Natural Selection"

(Revised edition). Dover Press, New York.

Flint, L. H. and MacAlister, E. D. (1935). Wavelengths of radiation in the visible spectrum inhibiting the germination of light sensitive lettuce seed. *Smithson. misc. Collns* 94, 1–11.

Flower-Ellis, J. G. K. (1971). Age structure and dynamics in stands of bilberry (*Vaccinium myrtillus* L.) Avdel f. Skogsekol., Rapp. Upps. 9, 108 pp.

Ford, E. B. (1971). "Ecological Genetics" (3rd edition). Chapman and Hall, London.

Fortanier, E. J. (1973). Reviewing the length of the generation period and its shortening particularly in tulips. *Scientia Hort.* 1, 107–116.

Foster, J. (1964). "Studies on the population dynamics of the daisy, *Bellis perennis*". Ph.D. thesis, University of Wales.

Frankton, C. and Bassett, I. J. (1968). The genus *Atriplex* (Chenopodiaceae) in Canada, Part I. Three introduced species: *A. heterosperma*, *A. oblongifolia*, and *A. hortensis*. *Can. J. Bot.* 46, 1309–1313.

Fresco, L. F. M. (1973). A model for plant growth. Estimation of the parameters of the logistic formation. *Acta. bot. neerl.* 22, 486–489.

Frick, K. E. and Holloway, J. K. (1964). Establishment of the cinnabar moth, *Tyria jacobaeae*, on tansy ragwort in the western United States. *J. econ. Entomol.* 57, 152–154.

Friedman, J. (1971). The effect of competition by adult *Zygophyllum dumosum* Boiss. on seedlings of *Artemisia herba-alba* Asso. in the Negev desert of Israel. *J. Ecol.* 59, 775–782.

Fritz, E. (1929). Some popular fallacies concerning California redwood. *Madroño* 1, 221–223.

Gadgil, M. (1971). Dispersal: population consequences and evolution. *Ecology* 52, 253–261.

Gadgil, M. and Bossert, W. H. (1970). The life historical consequences of natural selection. *Am. Nat.* 104, 1–24.

Gadgil, M. D. and Solbrig, O. T. (1972). The concept of *r*- and *K*-selection: evidence from wild flowers and some theoretical considerations. *Am. Nat.* 106, 14–31.

Gale, J. S. (1964). Competition between three lines of *Drosophila melanogaster*. *Heredity* 19, 681–699.

Galil, J. (1967). On the dispersal of the bulbs of *Oxalis cernua* Thunb. by mole rats (*Spalax ehrenbergi* Nehring). *J. Ecol.* 55, 787–792.

Gallais, A. (1970). Modèle pour l'analyse des relations binaires. *In* "Biometrie-Praximetrie", XI, 2–3, 51–80.

Garrett, S. D. (1956). "Biology of Root-infecting Fungi". Cambridge University Press, London and New York.

Garrett, S. D. (1960). Inoculum potential. *In* "Plant Pathology" (J. G. Horsfall and A. E. Dimond, eds), Vol. 3, pp. 23–56. Academic Press, New York and London.

Garrett, S. D. (1970). "Pathogenic Root-infecting Fungi". Cambridge University Press, London.

Gashwiler, J. S. (1967). Conifer seed survival in a western Oregon clearcut. *Ecology*

48, 431–433.

Gashwiler, J. S. (1970). Further study of conifer seed survival in a western Oregon clearcut. *Ecology* 51, 849–854.

Gäumann, E. (1946). "Pflanzliche Infektionslehre". Verlag Birkhauser, Basel.

Gause, G. F. (1934). "The Struggle for Existence". Waverly Press, Baltimore, U.S.A.

Gerhold, H. D. (1966). Selection for precocious flowering in *Pinus sylvestris. U.S. Forest Serv. Res. Pap.* NC-6, 4–7.

Ghent, A. W. (1958). Studies of regeneration in forest stands devastated by the spruce budworm. II. Age, height, growth and related studies of balsam fir seedlings. *For. Sci.* 4, 135–146.

Giesel, J. T. (1972). Sex ratio, rate of evolution and environmental heterogeneity. *Am. Nat.* 106, 380–387.

Gibson, I. A. S. (1956). Sowing density and damping-off in pine seedlings. *E. Afr. agric. J.* 21, 183–188.

Gilbert, J. M. (1958). "Eucalypt-rainforest relationships and the regeneration of eucalypts". Ph.D. thesis, University of Tasmania.

Gilbert, J. M. (1959). Forest succession in the Florentine Valley, Tasmania. *Pap. Proc. R. Soc. Tas.* 93, 129–151.

Gilbert, L. E. (1971). Butterfly–plant coevolution: has *Passiflora adenopoda* won the selectional race with Heliconine butterflies? *Science, N.Y.* 172, 585–586.

Gilbert, L. E. (1975). Ecological consequences of a coevolved mutualism between butterflies and plants. *1st Int. Cong. Syst. & Evoln. Biol.* Boulder, Colorado (1973) Symp. 5, 210–240.

Gillett, J. B. (1962). Pest pressure, an underestimated factor in evolution. *Systematics Association Publications* No. 4, 37–46.

Gilpin, M. E. and Ayala, F. J. (1973). Global models of growth and competition. *Proc. Nat. Acad. Sci.* 70, 3590–3593.

Gimingham, C. H. (1971). *Calluna* heathlands: use and conservation in the light of some ecological effects of management. *In* "The Scientific Management of Animal and Plant Communities for Conservation" (E. Duffey and A. S. Watt, eds). *Symp. Br. Ecol. Soc.* 11, 91–103.

Ginzo, H. D. and Lovell, P. H. (1973a). Aspects of the comparative physiology of *Ranunculus bulbosus* L. and *Ranunculus repens* L. I. Response to nitrogen. *Ann. Bot.* 37, 753–764.

Ginzo, H. D. and Lovell, P. H. (1973b). Aspects of the comparative physiology of *Ranunculus bulbosus* L. and *Ranunculus repens* L. II. Carbon dioxide assimilation and distribution of photosynthates. *Ann. Bot.* 37, 765–776.

Gleason, H. A. (1926). The individualistic concept of the plant association. *Bull. Torrey bot. Club.* 53, 7–26.

Godwin, H. (1936). Studies in the ecology of Wicken Fen. III. The establishment and development of fen scrub. *J. Ecol.* 24, 82–116.

Good, R. (1974). "Geography of Flowering Plants". Longman, London.

Goodall, D. W. (1967). Computer simulation of changes in vegetation subject to grazing. *J. Indian bot. Soc.* 46, 356–362.

Goodall, D. W. (1969). Simulating the grazing situation. *In* "Biomathematics. Vol. 1, Concepts and Models of Biomathematics" (F. Heinmets, ed.), pp. 211–236. Dekker, New York.

Goode, J. E. (1956). *In Rep. East Malling Res. Stn* for 1955, 69.

Goodman, G. T. and Perkins, D. F. (1959). Mineral uptake and retention in Cotton-grass (*Eriophorum vaginatum* L.). *Nature, Lond.* 184, 467–468.

Goodrum, P. D., Reid, V. H. and Boyd, C. E. (1971). Acorn yields, characteristics, and management criteria of oaks for wildlife. *J. Wildl. Mgmt* 35, 520–532.

Goss, W. L. (1924). The vitality of buried weed seeds. *J. agric. Res.* 29, 349–362.

Govier, R. N. (1966). "The interrelationships of the hemiparasites and their hosts, with special reference to *Odontites verna* (Bell.) Dum.". Ph. D. thesis, University of Wales.

Graber, R. E. (1970). Natural seed fall in white pine (*Pinus strobus* L.) stands of varying density. *U.S.D.A. Forest Service Res. Note.* NE-119, 1–6.

Graham, B. F. (1959). "Root grafts in eastern white pine, *Pinus strobus* L.: Their occurrence and ecological implications". Ph.D. thesis, Duke University.

Graham, S. A. (1929). "Principles of Forest Entomology". McGraw-Hill, New York and London.

Gram, E. (1960). Quarantines. *In* "Plant Pathology" (J. G. Horsfall and A. E. Dimond, eds), Vol. 3, pp. 313–356. Academic Press, New York and London.

Grant, S. A. and Hunter, R. F. (1966). The effects of frequency and season of clipping on the morphology, productivity and chemical composition of *Calluna vulgaris* (L.) Hall. *New Phytol.* 65, 125–133.

Grant, V. (1963). "The Origin of Adaptations". Columbia University Press, New York.

Gray, R. and Bonner, J. (1948). An inhibitor of plant growth from the leaves of *Encelia farinosa*. *Am. J. Bot.* 34, 52–57.

Gray, R. and Bonner, J. (1948). Structure determination and synthesis of a plant growth inhibitor, 3-acetyl-6-methoxybenzaldehyde found in the leaves of *Encelia farinosa*. *J. Am. chem. Soc.* 70, 1249–1253.

Greenwood, D. J. (1969). Effect of oxygen distribution in the soil on plant growth. *In* "Root Growth" (W. J. Whittingham, ed.), pp. 202–223. Proc. 15th Easter School Ag. Sci. Nottingham, 1968.

Greig-Smith, P. (1957). "Quantitative Plant Ecology". Butterworths, London.

Greig-Smith, P. (1961). The use of pattern analysis in ecological investigations. *In* "Mathematical Analysis of Plant Communities". *Recent Advances in Botany* 2, 1354–1358.

Griffin, J. R. (1971). Oak regeneration in the upper Carmel valley, California. *Ecology* 52, 862–868.

Griffiths, M. and Barker, R. (1966). The plants eaten by sheep and by kangaroos grazing together in a paddock in south-western Queensland. *C.S.I.R.O. Wildlife Res.* 11, 145–167.

Grime, J. P. and Hunt, R. (1975). Relative growth-rate: its range and adaptive significance in a local flora. *J. Ecol.* 63, 393–422.

Grime, J. P. and Jeffrey, D. W. (1965). Seedling establishment in vertical gradients of sunlight. *J. Ecol.* 53, 621–642.

Grümmer, G. (1955). "Die gegenseitige Beinflussung höherer Pflanzen-Allelopathie". Fischer, Jena.

Grümmer, G. (1961). The role of toxic substances in the interrelationships between higher plants. *In* "Mechanisms in biological competition" (F. L. Milthorpe, ed.) *Symp. Soc. Exp. Biol.* 15, 219–228.

Grümmer, G. and Beyer, H. (1960). The influence exerted by species of *Camelina* on flax by means of toxic substances. *In* "The Biology of Weeds" (J. L. Harper, ed.) *Br. ecol. Soc. Symp.* 1, 153—157.

Gunary, D. (1968). Discussion on mineral nutrient supply from soils. *In* "Ecological Aspects of the Mineral Nutrition of Plants" (I. H. Rorison, ed.) *Br. ecol. Soc. Symp.*, pp. 149—152. Blackwell, Oxford.

Gutterman, Y., Wiztum, A. and Evenari, M. (1967). Seed dispersal and germination in *Blepharis persica. Israel J. Bot.* 16, 213—234.

Gysel, L. W. (1971). A 10-year analysis of beechnut production and use in Michigan. *J. Wildl. Mgmt* 35, 516—519.

Haber, E. S. (1950). Longevity of the seed of sweet corn inbreds and hybrids. *Proc. Am. Soc. hort. Sci.* 55, 410—412.

Haddad, S. Y. (1968). "Studies on clones of *Agropyron repens* (L.) Beauv. with particular reference to herbicide response". M.Sc. thesis, University of Wales.

Hafez, E. S. E. and Schein, M. W. (1962). "The Behaviour of Domestic Animals" (E. S. E. Hafez, ed.), Bailliere, Tindall and Cox, London.

Hagner, S. (1965). Cone crop fluctuations in Scots Pine and Norway Spruce. *Stud. For. suec. Skogshogski.* Stockholm. 33, 1—21.

Hairston, N. G., Smith, F. E. and Slobodkin, L. B. (1960). Community structure, population control, and competition. *Am. Nat.* 94, 421—425.

Haizel, K. A. and Harper, J. L. (1973). The effects of density and the timing of removal on interference between barley, white mustard and wild oats. *J. appl. Ecol.* 10, 23—31.

Halbach, K. (1971). Redwoods: a population model debunked. *Science, N.Y.* 174, 436.

Haldane, J. B. S. (1932). "The Causes of Evolution". Longmans, Green, London.

Haldane, J. B. S. (1953). Animal populations and their regulation. *New Biol.* (Penguin Books) 15, 9—24.

Hall, R. L. (1974a). Analysis of the nature of interference between plants of different species. I. Concepts and extension of the De Wit analysis to examine effects. *Aust. J. agric. Res.* 25, 739—747.

Hall, R. L. (1974b). Analysis of the nature of interference between plants of different species. II. Nutrient relations in a Nandi *Setaria* and Greenleaf *Desmodium* association with particular reference to potassium. *Aust. J. agric. Res.* 25, 749—756.

Hallé, F. and Oldeman, R. A. A. (1970). "Essai sur l'Architecture et la Dynamique de Croissance des Arbres Tropicaux." Masson, Paris.

Hamilton, B. A., Hutchinson, K. J., Annis, P. C. and Donnelly, J. B. (1973). Relationships between the diet selected by grazing sheep and the herbage on offer. *Aust. J. agric. Res.* 24, 271—277.

Hamilton, R. A. (1956). Utilisation of grassland. *Outlook on Agriculture* 1, 5—11.

Hamilton, W. D. (1966). The moulding of senescence by natural selection. *J. theor. Biol.* 12, 12—45.

Hansen, K. (1911). Weeds and their vitality. *Ugeskr. Landm.* 56, 149.

Harberd, D. J. (1961). Observations on population structure and longevity of *Festuca rubra* L. *New Phytol.* 60, 184—206.

Harberd, D. J. (1963). Observations on natural clones of *Trifolium repens* L. *New Phytol.* **62**, 198–204.

Harberd, D. J. (1967). Observations on natural clones of *Holcus mollis*. *New Phytol.* **66**, 401–408.

Harberd, D. J. and Owen, M. (1969). Some experimental observations on the clone structure of a natural population of *Festuca rubra* L. *New Phytol.* **68**, 93–104.

Harper, J. L. (1955). The influence of the environment on seed and seedling mortality VI. Effects of the interaction of soil moisture content and temperature on the mortality of maize grains. *Ann. appl. Biol.* **43**, 696–708.

Harper, J. L. (1956). The evolution of weeds in relation to the resistance to herbicides. *Proc. 3rd Br. Weed Control Conf.* (Blackpool) **1**, 179–188.

Harper, J. L. (1957). Biological flora of the British Isles, *Ranunculus acris* L. *Ranunculus repens* L. *Ranunculus bulbosus* L. *J. Ecol.* **45**, 289–342.

Harper, J. L. (1958a). Famous plants – 8. The Buttercup. *New Biol.* **26**, 30–46.

Harper, J. L. (1958b). The ecology of ragwort (*Senecio jacobaea*) with especial reference to control. *Herbage Abstr.* **28**, 152–157.

Harper, J. L. (1959). The ecological significance of dormancy. *Proc. IV Int. Congr. Crop. Prot.* (Hamburg, 1957), pp. 415–420.

Harper, J. L. (1961). Approaches to the study of plant competition. *In* "Mechanisms in Biological Competition" (F. L. Milthorpe, ed.) *Symp. Soc. exp. Biol.* **15**, 1–39.

Harper, J. L. (1964a). The nature and consequence of interference amongst plants. *In* "Genetics Today". *Proc. XI. Internat. Cong. Genet.* **2**, 465–482.

Harper, J. L. (1964b). The individual in the population. *J. Ecol.* **52**, (Suppl.) 149–158.

Harper, J. L. (1966). The reproductive biology of the British poppies. *In* "Reproductive Biology and Taxonomy of Vascular Plants" (J. G. Hawkes, ed.) pp. 26–39. Botanical Society of the British Isles, London.

Harper, J. L. (1967). A Darwinian approach to plant ecology. *J. Ecol.* **55**, 247–270.

Harper, J. L. (1969). The role of predation in vegetational diversity. *In* "Diversity and Stability in Ecological Systems", *Brookhaven Symp. in Biology* **22**, 48–62.

Harper, J. L. (1970). Grazing, fertilizers and pesticides in the management of grasslands. *In* "The Scientific Management of Animal and Plant Communities for Conservation" (E. Duffey, and A. S. Watt, eds), pp. 15–31. 11th Symp. Br. ecol. Soc.

Harper, J. L. (1974). A centenary in population biology. *Nature, Lond.* **252**, 526–527.

Harper, J. L. and Benton, R. A. (1966). The behaviour of seeds in soil, part 2. The germination of seeds on the surface of a water supplying substrate. *J. Ecol.* **54**, 151–166.

Harper, J. L. and Chancellor, A. P. (1959). The comparative biology of closely related species living in the same area. IV. *Rumex*: interference between individuals in populations of one and two species. *J. Ecol.* **47**, 679–695.

Harper, J. L. and Clatworthy, J. N. (1963). The comparative biology of closely related species. VI. Analysis of the growth of *Trifolium repens* and *T. fragiferum* in pure and mixed populations. *J. exp. Bot.* **14**, 172–190.

Harper, J. L. and Gajic, D. (1961). Experimental studies of the mortality and plasticity of a weed. *Weed Res.* **1**, 91–104.

Harper, J. L. and McNaughton, I. H. (1960). The inheritance of dormancy in inter- and intraspecific hybrids of *Papaver*. *Heredity* **15**, 315–320.

Harper, J. L. and McNaughton, I. H. (1962). The comparative biology of closely related species living in the same area. VII. Interference between individuals in pure and mixed populations of *Papaver* species. *New Phytol.* **61**, 175–188.

Harper, J. L. and Obeid, M. (1967). Influence of seed size and depth of sowing on the establishment and growth of varieties of fiber and oil seed flax. *Crop. Sci.* **7**, 527–532.

Harper, J. L. and Ogden, J. (1970). The reproductive strategy of higher plants. 1. The concept of strategy with special reference to *Senecio vulgaris* L. *J. Ecol.* **58**, 681–698.

Harper, J. L. and Sagar, G. R. (1953). Some aspects of the ecology of buttercups in permanent grassland. *Proc. Br. Weed Control Conf.* **1**, 256–265.

Harper, J. L. and White, J. (1971). The dynamics of plant populations. *Proc. adv. Study Inst. Dynamics Numbers Popul.* (*Oosterbeek, 1970*) 41–63.

Harper, J. L. and White, J. (1974). The demography of plants. *A. Rev. Ecol. Syst.* **5**, 419–463.

Harper, J. L. and Wood, W. A. (1957). *Senecio jacobaea* L. *In* "Biological Flora of the British Isles". *J. Ecol.* **45**, 617–637.

Harper, J. L., Landragin, P. A. and Ludwig, J. W. (1955). The influence of environment on seed and seedling mortality. II. The pathogenic potential of the soil. *New Phytol.* **54**, 119–131.

Harper, J. L., Clatworthy, J. N., McNaughton, I. H. and Sagar, G. R. (1961). The evolution and ecology of closely related species living in the same area. *Evolution* **15**, 209–227.

Harper, J. L., Williams, J. T. and Sagar, G. R. (1965). The behaviour of seeds in soil. Part 1. The heterogeneity of soil surfaces and its role in determining the establishment of plants from seed. *J. Ecol.* **53**, 273–286.

Harper, J. L., Lovell, P. H. and Moore, K. G. (1970). The shapes and sizes of seeds. *Ann. Rev. ecol. Syst.* **1**, 327–356.

Harris, P. (1974). A possible explanation of plant yield increases following insect damage. *Agroecosystems* **1**, 219–225.

Hartig, R. (1889). Über den Einfluss der Samenproducktion auf Zuwachsgrösse und Reservestoffvorrat der Baüme. *Allgem. Forst. und Jagd-ztg.* **65**, 13–17.

Hartley, P. H. T. (1953). An ecological study of the feeding habits of the English titmice. *J. Anim. Ecol.* **22**, 261–288.

Hartsema, A. M. (1961). Influence of temperatures on flower formation and flowering of bulbous and tuberous plants. *Encyclopedia of Plant Physiology* **16**, 123–167.

Hasan, S. and Wapshere, A. J. (1973). The biology of *Puccinia chondrillina* a potential biological control agent of skeleton weed. *Ann. appl. Biol.* **74**, 325–332.

Hatto, J. and Harper, J. L. (1969). The control of slugs and snails in British cropping systems, specially grassland. *Int. Copper Res. Assn. Project* **115A**, 1–25.

Hawkes, R. B. (1968). The cinnabar moth, *Tyria jacobaeae*, for control of tansy ragwort. *J. econ. Entomol.* 61, 499–501.

Hawthorn, W. R. (1973). "Population dynamics of two weedy perennials, *Plantago major* L. and *P. rugelii* Decne." Ph.D. thesis, University of Western Ontario. London, Ontario, Canada.

Heimburger, C. C. (1945). Comment on the budworm outbreak in Ontario and Quebec. *For. Chron.* 21, 114–126.

Heinselman, M. L. (1971). The natural role of fire in northern conifer forest. *In* "Fire in the Northern Environment — a Symposium". U.S.D.A. Forest Service, Pacific Northwest Forest Range. Exp. Stn, 61–72.

Henry, J. D. and Swan, J. M. A. (1974). Reconstructing forest history from live and dead plant material — an approach to the study of forest succession in southwestern New Hampshire. *Ecology* 55, 772–783.

Hermann, R. K. and Chilcote, W. W. (1965). Effect of seedbeds on germination and survival of Douglas-Fir. *Res. Pap. (For. Mgmt. Res.) Ore. For. Res. Lab.* 4, 1–28.

Hett, J. M. (1971). A dynamic analysis of age in sugar maple seedlings. *Ecology* 52, 1071–1074.

Hett, J. M. and Loucks, O. L. (1968). Application of life-table analyses to tree seedlings in Quetico Provincial Park, Ontario. *For. Chron.* 44, 29–32.

Hewett, D. G. (1971). The effects of the winter of 1962/3 on *Juncus acutus* at Braunton Burrows, Devon. *Rep. & Trans. Devon. Assoc. Advan. Sci. Lit. & Ant.* 102, 193–201.

Hewston, L. J. (1964). "Effect of seed size on germination, emergence and yield of some vegetable crops". M.Sc. thesis, University of Birmingham.

Heydecker, W. (1956). Establishment of seedlings in the field. I. Influence of sowing depth on seedling emergence. *J. hort. Sci.* 31, 76–88.

Higazy, M. K. M. T. (1962). Shortening the juvenile phase for flowering. *Meded. LandHoogesch. Wageningen* 62, 1–53.

Hirano, S. and Kira, T. (1965). Influence of autotoxic root exudation on the growth of higher plants grown at different densities (Intraspecific Competition Among Higher Plants XII). *J. Biol. Osaka City Univ.* 16, 27–44.

Hirano, S. and Morioka, S. (1964). On the interrelation of the growth retarding activity of the root-excretion among various kinds of fruit tree. *J. Jap. Soc. hort. Sci.* 33, 287–290.

Hiroi, T. and Monsi, M. (1966). Dry-matter economy of *Helianthus annuus* communities grown at varying densities and light intensities. *J. Fac. Sci. Univ. of Tokyo* 9, 241–285.

Hodgkinson, K. C. (1974). Influence of partial defoliation on photosynthesis, photorespiration and transpiration by lucerne leaves of different ages. *Aust. J. Pl. Physiol.* 1, 561–578.

Hodgson, G. L. and Blackman, G. E. (1957a). An analysis of the influence of plant density on the growth of *Vicia faba*. Part I. The influence of density on the pattern of development. *J. exp. Bot.* 7, 147–165.

Hodgson, G. L. and Blackman, G. E. (1957b). An analysis of the influence of plant density on the growth of *Vicia faba*. Part 2. The significance of competition for light in relation to plant development at different densities. *J. exp. Bot.* **8**, 195–219.

Hodgson, J. (1966). The frequency of defoliation of individual tillers in a set stocked sward. *J. Br. Grassl. Soc.* **21**, 258–263.

Holland, A. A. and Moore, C. W. E. (1962). The vegetation and soils of the Bollon District in south-western Queensland. *C.S.I.R.O. Aust. Div. Plant Ind. Tech. Pap. 17*.

Holland, P. G. (1969). Weight dynamics of *Eucalyptus* in the mallee vegetation of southeast Australia. *Ecology* **50**, 212–219.

Holliday, R. J. (1960). Plant population and crop yield. *Fld Crop Abstr.* **13**, 159–167.

Holliday, R. J. and Putwain, P. D. (1974). Variation in the susceptibility to simazine in three species of annual weeds. *Proc. 12th Brit. Weed Cont. Conf.* **2**, 649–654.

Holloway, J. K. (1964). Projects in biological control of weeds. *In* "Biological Control of Pests and Weeds" (P. DeBach, ed.), pp. 650–670. Chapman and Hall, London.

Holmsgaard, E. (1955). Tree ring analysis of Danish Forest Trees. *Det. Forstl. Forvögsvaesen Danmark.* **22**, 1–246.

Holt, B. R. (1972). Effect of arrival time on recruitment, mortality and reproduction in successional plant populations. *Ecology* **53**, 668–673.

Honing, J. A. (1930). Nucleus and plasma in the heredity of the need of light for germination in *Nicotiana* seeds. *Genetica* **12**, 441–468.

Hope-Simpson, J. (1940). Studies of the vegetation of the English chalk. VI. Late stages in succession leading to chalk grassland. *J. Ecol.* **28**, 386–402.

Hopkinson, J. M. (1964). Studies on the expansion of the leaf surface. IV. The carbon and phosphorus economy of a leaf. *J. exp. Bot.* **15**, 125–137.

Horn, H. S. (1976). Succession, in "Theoretical Ecology" (R. M. May, ed.), pp. 187–204. Blackwell Scientific Publications, Oxford.

Horn, H. S. and MacArthur, R. H. (1972). Competition among fugitive species in a harlequin environment. *Ecology* **53**, 749–752.

Horne, F. R. (1953). The significance of weed seeds in relation to crop production. *Proc. 1st Brit. Weed Cont. Conf.* 372–398.

Horton, K. W. (1964). Deer prefer Jack Pine. *J. For.* **62**, 497–499.

Hough, J. S. (1953). Studies on the common spangle gall of oak. III. The importance of the stage in laminar extension of the leaf. *New Phytol.* **52**, 229–237.

Howarth, S. E. and Williams, J. T., *Chrysanthemum segetum, In* Biological Flora of the British Isles". *J. Ecol.* **60**, 573–584.

Hozumi, K. and Shinozaki, K. (1970). Studies on the frequency distribution of the weight of individual trees in a forest stand. II. Exponential distribution. *Jap. J. Ecol.* **20**, 1–9.

Hozumi, K., Shinozaki, K. and Tadaki, Y. (1968). Studies on the frequency distribution of the weight of individual trees in a forest stand. I. A new approach toward the analysis of the distribution function and the − 3/2 th. power distribution. *Jap. J. Ecol.* **18**, 10–20.

Hozumi, K. and Ueno, Y. (1954). Spacing experiments on the growth of root vegetables. *Mem. Col. Agric. Kyoto Univ.* **70** (Hort. Ser. No. 3), 1–20.

Huffaker, C. B. (1958). Experimental studies on predation: dispersion factors and predator-prey oscillations. *Hilgardia* **27**, 343–383.

Huffaker, C. B. (1964). Fundamentals of Biological Weed Control. *In* "Biological Control of Insect Pests and Weeds" (P. DeBach and E. I. Schlinger, eds) Ch. 20, pp. 74—117. Chapman and Hall, London.

Huffaker, C. B. (1971). "Biological Control". Plenum Press, New York.

Huffaker, C. B. (1975). Biological control in the management of pests. *Agroecosystems* 2, 15—31.

Hughes, E. G. (1974). Coolibah trees return to life after 68 years 'dead'. *Arid Zone Newsletter*, C.S.I.R.O., Perth 166.

Hughes, R. D. (1975). Introduced dung beetles and Australian pasture ecosystems. *J. appl. Ecol.* 12, 819—837.

Humphries, E. C. (1963). Dependence of net assimilation rate on root growth of isolated leaves. *Ann. Bot.* 27, 175—183.

Hunter, R. F. (1964). Home range behaviour in hill sheep. "Grazing in Marine and Terrestrial Environments" (D. J. Crisp, ed.), pp. 155—171. *4th Symp. Brit. Ecol. Soc.* Blackwell, Oxford. 155—171.

Hurlbert, S. H. (1971). The nonconcept of species diversity: a critique and alternative parameters. *Ecology* 52, 577—586.

Hutchinson, G. E. (1948). Circular causal systems in ecology. *Ann. N. Y. Acad. Sci.* 50, 221—246.

Hutchinson, G. E. (1951). Copepodology for the ornithologist. *Ecology* 32, 571—577.

Hutchinson, G. E. (1965). "The Ecological Theater and the Evolutionary Play". Yale University Press, Newhaven, Conn.

Hutchinson, G. E. and Deevey, E. S. (1949). Ecological studies on populations. *Surv. Biol. Progr.* 1, 325—359.

Hyde, E. O. C. (1954). The function of the hilum in some Papilionaceae in relation to the ripening of the seed and the permeability of the testa. *Ann. Bot.* N.S. 18, 241—256.

Isaak, L. A. (1943). "Reproduction habits of Douglas Fir". Charles Larthroy Park Forestry Foundation, Washington, D.C.

Iseley, D. (1952). Employment of tetrazolium chloride for determining viability of small grain seed. *Proc. Assoc. Off. Seed Analysts* 42, 143—153.

Isikawa, S. (1954). Light sensitivity against germination. 1. Photoperiodism of seeds. *Bot. Mag. Tokyo* 67, 51—54.

Ivins, J. D. (1952). The relative palatability of herbage plants. *J. Br. Grassl. Soc.* 7, 43—54.

Ivins, J. D. (1955). The palatability of herbage. *Herbage Abstr.* 25, 75—79.

Jack, W. H. (1971). The influence of tree spacing on Sitka spruce growth. *Irish Forestry* 28, 13—33.

Jacquard, P. (1968). Manifestation et nature des relations sociales chez les végétaux superieurs. *Oecol. Plant.* 3, 137—168.

Jacquard, P. and Caputa, J. (1970). Comparaison de trois modèles d'analyse des relations sociales entre espèces vegetales. *Ann. Amélior. Plantes* 20, 115—158.

Jain, S. K. and Bradshaw, A. D. (1966). Evolutionary divergence among adjacent plant populations. 1. The evidence and its theoretical analysis. *Heredity* 21, 407—441.

Jameson, D. J. (1963). Response of individual plants to harvesting. *Bot. Rev.* 29, 532—594.

Janzen, D. H. (1967). Synchronization of sexual reproduction of trees within the dry season in central America. *Evolution* 21, 620–637.

Janzen, D. H. (1968). Host plants as islands in evolutionary and contemporary time. *Am. Nat.* 102, 592–594.

Janzen, D. H. (1969). Seed eaters versus seed size, number, toxicity and dispersal. *Evolution* 23, 1–27.

Janzen, D. H. (1970). Herbivores and the number of tree species in tropical forests. *Am. Nat.* 104, 501–528.

Janzen, D. H. (1971). *Cassia grandis* L. beans and their escape from predators: a study in tropical predator satiation. *Ecology* 52, 964–979.

Janzen, D. H. (1972). Escape in space of *Sterculia apetala* seeds from the bug *Dysdercus fasciatus* in a Costa Rican deciduous forest. *Ecology* 53, 350–361.

Janzen, D. H. (1976). Why bamboos wait so long to flower. *A Rev. Ecol. Syst.* 7, 347–91.

Jayakar, S. D. (1970). A mathematical model for interaction of gene frequencies in a parasite and its host. *Theor. Pop. Biol.* 1, 140–164.

Jelinowska, A. (1967). Investigations on the influence of barley-cover crop on undersown lucerne (in Polish) *Pamietnik Pulawski-Prace Iung. Zeszyt* 26, 119–179.

Johnston, T. J., Pendleton, J. W., Peters, D. B. and Hicks, D. R. (1969). Influence of supplemental light on apparent photosynthesis, yield and yield components of soybeans (*Glycine max* L.) *Crop Sci.* 9, 577–581.

Jones, D. A. (1962). Selection of the acyanogenic form of the plant *Lotus corniculatus* L. by various animals. *Nature* 193, 1109–1110.

Jones, D. A. (1966). On the polymorphism of cyanogenesis in *Trifolium repens* L. 1. Selection by animals. *Can. J. Genet. Cytol.* 8, 556–567.

Jones, D. A. (1972). On the polymorphism of cyanogenesis in *Lotus corniculatus* L. IV. The Netherlands. *Genetica* 43, 394–406.

Jones, D. A. (1973). On the polymorphism of cyanogenesis in *Lotus corniculatus*. V. Denmark. *Heredity* 30, 381–386.

Jones, Ll. I. (1967). Studies on hill land in Wales. *Tech. Bull. Welsh Pl. Breed. Stn* 2, 1–179.

Jones, M. E. (1971a). The population genetics of *Arabidopsis thaliana*. I. The breeding system. *Heredity* 27, 39–50.

Jones, M. E. (1971b). The population genetics of *Arabidopsis thaliana*. II. Population structure. *Heredity* 27, 51–58.

Jones, M. G. (1933a). Grassland management and its influence on the sward. *Emp. J. exp. Agric.* 1, 43–57.

Jones, M. G. (1933b). Grassland management and its influence on the sward. II. The management of a clovery sward and its effects. *Emp. J. exp. Agric.* 1, 122–128.

Jones, M. G. (1933c). Grassland management and its influence on the sward. III. The management of a "grassy" sward and its effects. *Emp. J. exp. Agric.* 1, 224–234.

Jones, M. G. (1933d). Grassland management and its influence on the sward. IV. The management of poor pastures. V. Edaphic and biotic influences on pastures. *Emp. J. exp. Agric.* 1, 362–367.

Jones, M. G. (1933e). Grassland management and its influence on the sward. *Jl R. agric. Soc.* 94, 21–41.

803

Jones, M. G. and Jones, Ll. I. (1930). The effect of varying the periods of
  rest in rotational grazing. *Bull. Welsh Pl. Breed. Stn* 11, 38–59.

Kannenberg, L. W. and Allard, R. W. (1967). Population studies in predominantly
  self-pollinated species. VIII. Genetic variability in the *Festuca microstachys*
  complex. *Evolution* 21, 227–240.

Kappert, H. (1954). Experimentelle Untersuchungen über die Variabilität eines
  Totalapomikten. *Ber. dt. bot. Ges.* 67, 325–334.

Kasanaga, H. and Monsi, M. (1954). On the light-transmission of leaves and its
  meaning for production of dry matter in plant communities. *Jap. J. Bot.* 14,
  304–324.

Kaufman, J. H. (1962). Ecological and social behaviour of the coati *Nasau nasau*
  on Barro Colorado Island, Panama. *Univ. Calif. Publ. Zool.* 60, 95–222.

Kays, S. and Harper, J. L. (1974). The regulation of plant and tiller density in a
  grass sward. *J. Ecol.* 62, 97–105.

Kershaw, K. A. (1962). Quantitative ecological studies from Landmannahellir,
  Iceland. II. The rhizome behaviour of *Carea bigelowii* and *Calamagrostis
  neglecta. J. Ecol.* 50, 171–179.

Kershaw, K. A. (1964). "Quantitative and Dynamic Ecology". Arnold, London.

Kerster, H. W. (1968). Population age structure in the prairie forb *Liatris
  aspera. BioScience* 18, 430–432.

Khan, A. A. and Sagar, G. R. (1969). Alteration of the pattern of distribution of
  photosynthetic products in the tomato by manipulation of the plant. *Ann.
  Bot.* 33, 753–762.

Khan, M. I. (1967). "The genetic control of canalisation of seed size in plants".
  Ph.D. thesis, University of Wales.

Kidd, F. (1914). The controlling influence of carbon dioxide in the maturation,
  dormancy, and germination of seeds. *Proc. R. Soc.* B 87, 408–421, 609–625.

Kidd, F. and West, C. (1917). The controlling influence of carbon dioxide. IV.
  On the production of secondary dormancy in seeds of *Brassica alba* following
  treatment with carbon dioxide, and the relation of this phenomenon to the
  question of stimuli in growth processes. *Ann. Bot.* 31, 457–487.

Kira, T., Ogawa, H. and Shinozaki, K. (1953). Intraspecific competition among
  higher plants. 1. Competition-density-yield inter-relationships in regularly
  dispersed populations. *J. Inst. Polytech. Osaka Cy. Univ.* D. 4, 1–16.

Kittredge, J. (1944). Estimation of the amount of foliage of trees and shrubs.
  J. For. 42, 905–912.

Klopfer, P. H. (1962). "Behavioural Aspects of Ecology". Prentice-Hall, London.

Knapp, R. (1954). "Experimentelle Soziologie der höheren Pflanzen." Eugen
  Ulmer, Stuttgart.

Knight, W. E. and Hollowell, E. A. (1962). Response of crimson clover to
  different defoliation intensities. *Crop Sci.* 2, 124–127.

Knipe, D. and Herbel, C. H. (1966). Germination and growth of some semi-
  desert grassland species treated with aqueous extract from creosotebush.
  Ecology 47, 775–781.

Knuchel, H. (1953). "Planning and Control in the Managed Forest". Oliver and
  Boyd, Edinburgh.

Koch, R. Complete Works. George Thieme, Leipzig, 1912, 1, 650–660. (*10th
  International Medical Congress*, Berlin, 1890).

Koller, D. and Roth, N. (1963). Germination regulating mechanisms in some desert seeds. Part 7. *Panicum turgidum* (Gramineae). *Israel J. Bot.* 12, 64–73.

Koller, D. and Roth, N. (1964). Studies on the ecological and physiological significance of amphicarpy in *Gymnarhena micrantha* (Compositae). *Am. J. Bot.* 51, 26–35.

Koyama, H. and Kira, T. (1956). Intraspecific competition among higher plants. VIII. Frequency distribution of individual plant weight as affected by the interaction between plants. *J. Inst. Polytech. Osaka Cy. Univ.* 7, 73–94.

Krajicek, J. E., Brinkman, K. A. and Gingrich, S. F. (1961). Crown competition — a measure of density. *For. Sci.* 7, 35–42.

Krebs, C. J., Gaines, M. S., Keller, B. L., Myers, J. H. and Tamarin, R. H. (1973). Population cycles in small rodents. *Science, N.Y.* 179, 35–41.

Kropač, Z. (1966). Estimation of weed seeds in arable soil. *Pedobiologia* 6, 105–128.

Kulman, H. M. (1971). Effect of insect defoliation on growth and mortality of trees. *A. Rev. Entom.* 16, 289–324.

Kuroiwa, S. (1960). Ecological and physiological studies on the vegetation of Mt. Shimagare. V. Intraspecific competition and productivity difference among tree classes in the *Abies* stand. *Bot. Mag., Tokyo* 73, 165–174.

Lacey, W. S. (1957). A comparison of the spread of *Galinsoga parviflora* and *G. ciliata* in Britain. *In* "Progress in the Study of the British Flora" (J. E. Lousley, ed.), pp. 109–115. (B.S.B.I. Conference Reports 5) Arbroath.

Lack, D. (1947). "Darwin's Finches". Cambridge University Press.

Lack, D. (1954). "The Natural Regulation of Animal Numbers". Clarendon Press, Oxford.

Lack, D. (1964). "The Life of the Robin". Witherby, London.

Lack, D. (1966). "Population Studies of Birds". Oxford University Press, London.

Lang, A. (1965). Effect of some internal and external conditions on seed germination. *Encyclopedia of Plant Physiol.* 15, 848–893.

Lang, A. L., Pendleton, J. W. and Dungan, G. H. (1956). Influence of population and nitrogen level on yield and protein contents of nine corn hybrids. *Agron. J.* 48, 284–289.

Langer, R. H. M. (1972). "How Grasses Grow". Arnold, London.

Large, E. C. (1940). "The Advance of the Fungi". Cape, London.

Larson, M. M. and Schubert, G. H. (1970). Cone crops of ponderosa pine in central Arizona, including the influence of Albert squirrels. *U.S.D.A. Forest Service Res. Pap. RM-58*, 15 pp.

Leach, L. D. (1947). Growth rates of host and pathogen as factors determining the severity of pre-emergence damping-off. *J. agric. Res.* 75, 161–179.

Leak, W. B. (1964). An expression of diameter distribution for unbalanced, uneven aged stands and forests. *Forest Sci.* 10, 39–50.

Lemon, E. R. and Wiegand, C. L. (1962). Soil aeration and plant root relations. II. Root respiration. *Agron. J.* 54, 171–175.

Lerner, I. M. and Dempster, E. R. (1962). Indeterminism in interspecific competition. *Proc. Natn. Acad. Sci. U.S.A.* 48, 821–826.

Lerner, I. M. and Ho, F. K. (1961). Genotype and competitive ability in *Tribolium* species. *Am. Nat.* 95, 329–343.

Leslie, P. H. (1945). On the use of matrices in certain population mathematics. *Biometrika* **33**, 183–212.

Lester, D. T. (1967). Variation in cone production of red pine in relation to weather. *Can. J. Bot.* **45**, 1683–1691.

Levin, D. A. (1973). The age structure of a hybrid swarm in *Liatris* (Compositae). *Evolution* **27**, 532–535.

Levin, D. A. (1975). Pest pressure and recombination systems in plants. *Am. Nat.* **109**, 437–451.

Levins, R. (1968). "Evolution in Changing Environments". Princeton University Press, Princeton, N.J.

Levins, R. (1969). Dormancy as an adaptive strategy. *In* "Dormancy and Survival" (H. W. Woolhouse, ed.) *Symp. Soc. exp. Biol.* **23**, 1–10.

Lewis, H. (1962). Catastrophic selection as a factor in speciation. *Evolution* **16**, 257–271.

Lewis, H. (1966). Speciation in flowering plants. *Science, N.Y.* **152**, 167–172.

Lewontin, R. C. (1965). Selection for colonizing ability. *In* "The Genetics of Colonizing Species" (H. G. Baker and G. L. Stebbins, eds), pp. 77–91. Academic Press, New York and London.

Lewontin, R. C. (1971). The effect of genetic linkage on the mean fitness of a population. *Proc. Natn. Acad. Sci. U.S.A.* **68**, 984–986.

Libby, W. F. (1951). Radiocarbon dates, 11. *Science, N.Y.* **114**, 291–296.

Lindstrom, E. W. (1942). Inheritance of seed longevity in maize inbreds and hybrids. *Genetics* **27**, 154.

Linkola, K. (1935). Über die Dauer und Jahresklassenverhältnisse des Jungenstadiums bei einigen Wiesenstauden. *Acta forest. fenn.* **42**, 1–56.

Linnaeus, C. (1762). *Amoenitates Academica* 2, 203–240 and in *Pan Suecicus* 30 submittit N.L. Hesselgren (Upsaliae, 1749).

Lippert, R. D. and Hopkins, H. H. (1950). Study of viable seeds in various habitats in mixed prairies. *Trans. Kansas Acad. Sci.* **53**, 355–364.

Litav, M. and Harper, J. L. (1967). A method for studying the spatial relationship between the root systems of two neighbouring plants. *Plant and Soil* **26**, 389–392.

Livingstone, R. B. (1972). Influence of birds, stones and soil on the establishment of pasture juniper, *Juniperus communis*, and red cedar *J. virginiana* in New England pastures. *Ecology* **53**, 1141–1147.

Livingstone, R. B. and Allessio, M. L. (1968). Buried viable seed in successional field and forest stands, Harvard Forest, Mass. *Bull. Torrey bot. Club* **95**, 58–69.

Lock, J. M. (1972). The effects of hippopotamus grazing on grasslands. *J. Ecol.* **60**, 445–467.

Longman, K. A. (1969). The dormancy and survival of plants in the humid tropics. *In* "Dormancy and Survival" (H. W. Woolhouse, ed.) *Symp. Soc. exp. Biol.* **23**, 471–488.

Lotka, A. J. (1925). "Principles of Physical Biology". Waverly Press, Baltimore.

Ludwig, J. W., Bunting, E. S. and Harper, J. L. (1957). The influence of environment on seed and seedling mortality. III. The influence of aspect on maize germination. *J. Ecol.* **45**, 205–224.

Ludwig, J. W. and Harper, J. L. (1958). The influence of the environment on seed and seedling mortality. VIII. The influence of soil colour. *J. Ecol.* **46**, 381–389.

Ludwig, W. (1959). Die Selectionstheorie. *In* "Die Evolutionen der Organismen". (G. Herberer, ed.), pp. 662–712. Fischer, Stuttgart.

Luxmoore, R. J., Stolzy, L. H. and Letey, J. (1970a). Oxygen diffusion in the soil-plant system. I. A model, *Agron. J.* **62**, 317–322.

Luxmoore, R. J., Stolzy, L. H. and Letey, J. (1970b). Oxygen diffusion in the soil-plant system. II. Respiration rate, permeability, and porosity of consecutive excised segments of maize and rice roots. *Agron. J.* **62**, 322–324.

Luxmoore, R. J., Stolzy, L. H. and Letey, J. (1970c). Oxygen diffusion in the soil-plant system. III. Oxygen concentration profiles, respiration rates, and significance of plant aeration predicted for maize roots. *Agron. J.* **62**, 325–329.

Luxmoore, R. J., Stolzy, L. H. and Letey, J. (1970d). Oxygen diffusion in the soil-plant system. IV. Oxygen concentration profiles, respiration rates, and radial oxygen losses predicted for rice roots. *Agron. J.* **62**, 329–332.

Lykke, J. (1965). Elk and forest. *Norsk Skogbr.* **11**, 151–153.

MacArthur, R. H. (1958). Population ecology of some warblers of northeastern coniferous forests. *Ecology* **39**, 599–619.

MacArthur, R. H. (1971). Patterns of terrestrial bird communities. *In* "*Avian Biology*" (D. S. Farner and J. R. King, eds), Vol. 1, pp. 189–221. Academic Press, New York and London.

MacArthur, R. H. (1972). "Geographical Ecology: Patterns in the Distribution of Species". Harper and Row, New York.

MacArthur, R. H. and Connell, J. H. (1956). "The Biology of Populations". Wiley, New York.

MacArthur, R. H. and Levins, R. (1964). Competition, habitat selection, and character displacement in a patchy environment. *Proc. Natn. Acad. Sci.* **51**, 1207–1210.

MacArthur, R. H. and Pianka, E. R. (1966). On optimal use of a patchy environment. *Am. Nat.* **100**, 603–609.

MacArthur, R. H. and Wilson, E. O. (1967). "The Theory of Island Biogeography". Princeton University Press, Princeton, N.J.

McClure, F. A. (1967). "The Bamboos – a Fresh Perspective". Harvard University Press, Cambridge, Mass.

McClure, H. E. (1966). Flowering, fruiting and animals in the canopy of a tropical rain forest. *Malay. Forester* **29**, 192–203.

Macfadyen, A. (1957). "Animal Ecology – Aims and Methods". Pitman, London.

McGee, C. E. and Della-Bianca, L. (1967). Diameter distributions in natural yellow-poplar stands. *U.S. Forest Service Res. Pap. SE-25*, 7 pp.

McGilchrist, C. A. (1965). Analysis of competition experiments. *Biometrics* **21**, 975–985.

McGilchrist, C. A. and Trenbath, B. R. (1971). A revised analysis of competition experiments. *Biometrics* **27**, 659–671.

MacLusky, D. S. (1960). Some estimates of the areas of pasture fouled by the excreta of dairy cows. *J. Br. Grassl. Soc.* **15**, 181−188.

McNaughton, I. H. (1960). "The comparative biology of closely related species living in the same area, with special reference to the genus *Papaver*". D.Phil. thesis, University of Oxford.

McQuilkin, W. E. (1940). The natural establishment of pine in abandoned fields in the Piedmont Plateau region. *Ecology* **21**, 135−147.

McQuillan, M. J. (1974). Influence of plant husbandry on rice plant hoppers (Hemiptera; Delphacidae) in the Solomon Islands. *Agroecosystems* **1**, 339−358.

McRill, M. and Sagar, G. R. (1973). Earthworms and seeds. *Nature, Lond.* **243**, 482.

McRill, M. (1974). "Some botanical aspects of earthworm activity". Ph.D. thesis, University of Wales.

Mack, R. (1976). Survivorship of *Cerastium atrovirens* at Aberffraw, Anglesey. *J. Ecol.*, **64**, 309−312.

Mack, R. and Harper, J. L. (1977). Interference in dune annuals: spatial pattern and neighbourhood effects. *J. Ecol.*, **65**, 345−363.

Maggs, D. H. (1964). Growth rates in relation to assimilate supply and demand. I. Leaves and roots as limiting regions. *J. exp. Bot.* **15**, 574−583.

Maguire, W. P. (1965). Are ponderosa pine cone crops predictable? *J. Forest.* **54**, 778−779.

Major, J. and Pyott, W. T. (1966). Buried viable seeds in California bunchgrass sites and their bearing on the definition of a flora. *Vegetatio Acta Geobotanica* **13**, 253−282.

Malone, C. R. (1967). A rapid method for the enumeration of viable seeds in the soil. *Weeds* **15**, 381−382.

Malthus, T. R. (1798). "An Essay on the Principle of Population". London.

Mansfield, T. A. (1968). Carbon dioxide compensation points in maize and *Pelargonium*. *Physiol. Pl.* **21**, 1159−1162.

Marshall, C. and Sagar, G. R. (1965). The influence of defoliation on the distribution of assimilates in *Lolium multiflorum* Lam. *Ann. Bot.* **29**, 365−372.

Marshall, C. and Sagar, G. R. (1968). The distribution of assimilates in *Lolium multiflorum* Lam. following differential defoliation. *Ann. Bot.* **32**, 715−719.

Marshall, D. R. and Jain, S. K. (1967). Cohabitation and relative abundance of two species of wild oats. *Ecology* **48**, 656−659.

Martin, D. J. (1964). Analysis of sheep diet utilizing plant epidermal fragments in faeces samples. *In* "Grazing in Marine and Terrestrial Environments". (D. J. Crisp, ed.) *Symp. Br. Ecol. Soc.* **4**, pp. 173−188 Blackwell, Oxford.

Martin, P. and Rademacher, B. (1960). Untersuchungen zur Frage der Wurzelallel-opathie von Kulturpflanzen und Unkräutern. *Beiträge Biol. Pflanz.* **35**, 213−237.

Massey, A. B. (1925). Antagonism of the walnuts (*Juglans nigra* and *J. cinerea*) in certain plant associations. *Phytopathology* **15**, 773−784.

Mather, K. (1953). The genetical structure of populations. *In* "Evolution" (R. Brown and J. F. Danielli, eds) *Symp. Soc. exp. Biol.* **7**, 66−95.

Mather, K. (1973). "Genetical Structure of Populations". Chapman and Hall, London.

Mather, K. and Cooke, P. (1962). Differences in competitive ability between genotypes of *Drosophila*. *Heredity* 17, 381–407.

Matthews, J. D. (1963). Factors affecting the production of seed by forest trees. *Forest. Abstr.* 24, 1–13.

Maun, M. A. and Cavers, P. B. (1971). Seed production and dormancy in *Rumex crispus*. I. The effects of removal of cauline leaves at anthesis. *Can. J. Bot.* 49, 1123–1130.

May, R. M. (1973). "Stability and Complexity in Model Ecosystems". Princeton University Press, Princeton, N.J.

Mayer, A. M. and Poljakoff-Meyber, A. (1963). "The Germination of Seeds". Pergamon, Oxford.

Maynard Smith, J. (1964). Group selection and kin selection. *Nature, Lond.* 201, 1145–1147.

Maynard Smith, J. (1976). A comment on the Red Queen. *Am. Nat.* 110, 325–330.

Mayr, E. (1963). "Animal Species and Evolution". Harvard University Press, Cambridge, Mass.

Mead, R. (1966). A relationship between individual plant spacing and yield. *Ann. Bot.* 30, 301–309.

Medawar, P. B. (1952). "An Unsolved Problem in Biology". H. K. Lewis, London.

Medawar, P. B. (1955). *CIBA Fdn. Colloq. Ageing* 1, 4.

Medway, L. (1972). Phenology of a tropical rain forest in Malaya. *Biol. J. Linn. Soc.* 4, 117–146.

Mellanby, K. (1968). The effects of some mammals and birds on regeneration of oak. *J. appl. Ecol.* 5, 359–366.

Merz, E. (1959). Pflanzen und Raupen. Uber einige Prinzipien der Futterwahl bei Gross-schmetterlingsraupen. *Biol. Zbl.* 78, 152–188.

Meyer, H. A. (1952). Structure, growth and grain in balanced uneven aged forests. *J. Forest.* 50, 85–92.

Meyer, J. A. and Stevenson, D. D. (1949). The structure and growth of virgin beech-birch-maple-hemlock forests in northern Pennsylvania. *J. agric. Res.* 67, 465–484.

Meyer, J. H., Lofgreen, G. P. and Hull, J. L. (1957). Selective grazing by sheep and cattle. *J. Anim. Sci.* 16, 766–772.

Miall, L. C. (1912). "The Early Naturalists, Their Lives and Work (1530–1789)". Macmillan, London.

Miller, G. R., Nicholson, I. A., McCowan, D., Peterson, I. S., Parish, T., Badenoch, C. O., Cummins, R. P., Miles, A. M. and Moyer, S. M. (1970). Grazing and the regeneration of shrubs and trees. *Nat. Conservancy Research in Scotland, Rept.* 1968–70, pp. 33–35.

Milne, A. (1961). Definition of competition among animals. *In* "Mechanisms in Biological Competition" (F. L. Milthorpe, ed.) *Symp. Soc. exp. Biol.* 15, 40–61.

Milthorpe, F. L. (1961). The nature and analysis of competition between plants of different species. *In* "Mechanisms in Biological Competition" (F. L. Milthorpe, ed.) *Symp. Soc. exp. Biol.* 15, 330–355.

Milthorpe, F. L. and Moorby, J. (1974). "An Introduction to Crop Physiology". Cambridge University Press, London.

Milton, W. E. J. (1936). The buried viable seeds of enclosed and unenclosed hill land. *W.P.B.S. Bull. Series H* 14, 58—84.

Milton, W. E. J. (1939). The occurrence of buried viable seeds in soils at different elevations and on a salt marsh. *J. Ecol.* 27, 149—159.

Milton, W. E. J. (1940). The effect of manuring, grazing and liming on the yield, botanical and chemical composition of natural hill pastures. *J. Ecol.* 28, 326—356.

Milton, W. E. J. (1947). The composition of natural hill pasture, under controlled and free grazing cutting and manuring. *Welsh J. Agric.* 14, 182—195.

Mirov, N. T. (1936). Germination behaviour of some California plants. *Ecology* 17, 667—672.

Mohamed, B. F. and Gimingham, C. H. (1970). The morphology of vegetative regeneration in *Calluna vulgaris. New Phytol.* 69, 743—750.

Molisch, H. (1938). "The Longevity of Plants". Transl. E. H. Fulling, New York.

Monsi, M. and Saeki, T. (1953). Über den Lichtfaktor in den Pflanzengesell-schaften und seine Bedeutung für die Stoffproduktion. *Jap. J. Bot.* 14, 22—52.

Moore, J. A. (1952). Competition between *Drosophila melanogaster* and *Drosophila simulans.* II. The improvement of competitive ability through selection. *Proc. Natn. Acad. Sci. U.S.A.* 38, 381—407.

Morris, R. F. (1959). Single-factor analysis in population dynamics. *Ecology* 40, 580—588.

Morris, R. F. (1963a). The dynamics of epidemic spruce budworm populations. *Mem. ent. Soc. Can.* 31, 1—332.

Morris, R. F. (1963b). Predictive population equations based on key factors. *Mem. ent. Soc. Can.* 32, 16—21.

Morris, R. M. (1969). The pattern of grazing in "continuously" grazed swards. *J. Br. Grassl. Soc.* 24, 65—70.

Mortimer, A. M. (1974). "Studies of germination and establishment of selected species with special reference to the fates of seeds". Ph.D. thesis, University of Wales.

Morton, E. S. (1973). On the evolutionary advantages and disadvantages of fruit eating in tropical birds. *Am. Nat.* 107, 3—22.

Moss, R., Miller, G. R. and Allen, S. E. (1972). Selection of heather by captive red grouse in relation to the age of the plant. *J. appl. Ecol.* 9, 771—782.

Mount, A. B. (1964). The interdependence of the Eucalypts and forest fires in Southern Australia. *Aust. Forestry* 28, 166—172.

Mountfort, G. (1957). "The Hawfinch". Collins, London.

Muckerji, S. K. (1936). Contributions to the autecology of *Mercurialis perennis* L. *J. Ecol.* 24, 38—81.

Muller, C. H. (1953). The association of desert annuals with shrubs. *Am. J. Bot.* 40, 53—60.

Muller, C. H. (1965). Inhibitory terpenes volatilised from *Salvia* shrubs. *Bull. Torrey Bot. Club.* 93, 332—351.

Muller, C. H. and del Moral, R. (1966). Soil toxicity induced by terpenes from *Salvia leucophylla. Bull. Torrey bot. Club.* 93, 130—137.

Muller, C. H., Muller, W. H. and Haines, B. L. (1964). Volatile growth inhibitors produced by aromatic shrubs. *Science, N.Y.* 143, 471—473.

Muller, W. H. (1965). Volatile materials produced by *Salvia leucophylla*: effects on seedling growth and soil bacteria. *Bot. Gaz.* **126**, 195–200.

Muller, W. H. and Muller, C. H. (1956). Association patterns involving desert plants that contain toxic products. *Am. J. Bot.* **43**, 354–361.

Müllverstedt, R. (1963). Untersuchungen über die Keimung von Unkrautsamen in Abhängigkeit vom Sauerstoffpartialdruck. *Weed Res.* **3**, 154–163.

Murphy, G. I. (1968). Pattern in life history and the environment. *Am. Nat.* **102**, 391–403.

Murton, R. K. (1965). "The Woodpigeon". Collins, London.

Murton, R. K., Isaacson, A. J. and Westwood, N. J. (1963). The use of baits treated with α-chloralose to catch wood pigeons. *Ann. appl. Biol.* **52**, 271–293.

Murton, R. K., Isaacson, A. J. and Westwood, N. J. (1966). The relationships between wood pigeons and their clover food supply and the mechanism of population control. *J. appl. Ecol.* **3**, 55–93.

Murton, R. K., Westwood, N. J. and Isaacson, A. J. (1964). The feeding habits of the wood pigeon *Columba palumbus*, stock dove. *C. oenas* and turtle dove *Streptopelia turtur. Ibis* **106**, 174–188.

Nägeli, C. (1874). Verdrängung der Pflanzenformen durch ihre Mitbewerber. *Sitzb. Akad. Wiss. München* **11**, 109–164.

Naylor, R. E. L. (1970a). Weed predictive indices. *Proc. Br. Weed Control Conf.* **10**, 26–29.

Naylor, R. E. L. (1970b). The prediction of blackgrass infestations. *Weed Res.* **10**, 296–299.

Naylor, R. E. L. (1972a). Aspects of the population dynamics of the weed *Alopecurus myosuroides* Huds. in winter cereal crops. *J. appl. Ecol.* **9**, 127–139.

Naylor, R. E. L. (1972b). Biological Flora of the British Isles. 129. *Alopecurus myosuroides* Huds (*A. agrestis* L.) *J. Ecol.* **60**, 611–622.

Neales, T. F. and Incoll, L. D. (1968). The control of leaf photosynthesis rate by the level of assimilate concentration in the leaf: a review of the hypothesis. *Bot. Rev.* **34**, 107–125.

Neilson, J. A. (1964). Autoradiography for studying individual root systems in mixed herbaceous stands. *Ecology* **45**, 644–646.

New, J. K. (1958). A population study of *Spergula arvensis*. Part 1. Two clines and their significance. *Ann. Bot.* **22**, 457–477.

New, J. K. (1959). A population study of *Spergula arvensis*. Part 2. Genetics and breeding behaviour. *Ann. Bot.* **23**, 23–33.

New, J. K. (1961). Biological flora of the British Isles. No. 76. *Spergula arvensis*. *J. Ecol.* **49**, 205–215.

Newton, I. (1967). The feeding ecology of the bullfinch (*Pyrrhula pyrrhula* L.) in southern England. *J. Anim. Ecol.* **36**, 721–744.

Nicholson, A. J. (1933). The balance of animal populations. *J. Anim. Ecol.* **2** (Suppl.) 132–178.

Nicholson, A. J. (1958). Dynamics of insect populations. *A. Rev. Ent.* **3**, 107–136.

Nicholson, A. J. and Bailey, V. A. (1935). The balance of animal populations. *Proc. zool. Soc. Lond.* Part *1*, 551–598.

Nilsson, H. (1947). Totale Inventierung der Mikrotypen eines Minimal areals von *Taraxacum officinale. Hereditas* **33**, 119–142.

Noble, J. C. (1975). The effects of Emus (*Dromaius novaehollandiea* Latham) on the distribution of the Nitre bush (*Nitraria billardieri* DC.) *J. Ecol.* 63, 979–984.

Norrington-Davies, J. (1967). Application of diallel analysis to experiments in plant competition. *Euphytica* 16, 391–406.

Norrington-Davies, J. (1968). Diallel analysis of competition between grass species. *J. agric. Sci., Camb.* 71, 223–231.

Norrington-Davies, J. and Hutto, J. M. (1972). Diallel analysis of competition between diploid and tetraploid genotypes of *Secale cereale* grown at two densities. *J. agric. Sci., Camb.* 78, 251–256.

Numata, M., Hayashi, K., Komura, T. and Oki, K. (1964). Ecological studies on the buried-seed population in the soil as related to plant succession. 1. *Jap. J. Ecol.* 14, 207–215.

Nye, P. H. (1968). The soil model and its application to plant nutrition. *In* "Ecological Aspects of the Mineral Nutrition of Plants" (I. H. Rorison, ed.), *9th Symp. Br. Ecol. Soc.*, pp. 105–114. Blackwell, Oxford.

Obeid, M. (1965). "Experimental models in the study of interference in plant populations". Ph.D. thesis, University of Wales.

Obeid, M., Machin, D. and Harper, J. L. (1967). Influence of density on plant to plant variations in Fiber Flax, *Linum usitatissimum. Crop Sci.* 7, 471–473.

Ødum, S. (1965). Germination of ancient seeds; floristical observation and experiments with archaeologically dated soil samples. *Dansk Botanisk Arkiv.* 24, 1–70.

Ogden, J. (1968). "Studies on reproductive strategy with particular reference to selected composites". Ph.D. thesis, University of Wales.

Ogden, J. (1970). Plant population structure and productivity. *Proc. N.Z. ecol. Soc.* 17, 1–9.

Oinonen, E. (1967a). Sporal regeneration of bracken in Finland in the light of the dimensions and age of its clones. *Acta forest. fenn.* 83, 3–96.

Oinonen, E. (1967b). The correlation between the size of Finnish bracken (*Pteridium aquilinum* (L.) Kuhn) clones and certain periods of site history. *Acta forest. fenn.* 83, 1–51.

Oldeman, R. A. A. (1972). "L'Architecture de la Forêt Guyanaise". Thèse, Université des Sciences et Techniques du Languedoc.

Olmsted, N. W. and Curtis, J. D. (1947). Seeds of the forest floor. *Ecology* 28, 49–52.

Olson, J. S. (1958). Rates of succession and soil changes on southern Lake Michigan sand dunes. *Bot. Gaz.* 119, 125–170.

Oosting, H. J. (1956). "The Study of Plant Communities". W. H. Freeman, San Francisco.

Oosting, H. J. and Billings, W. D. (1951). A comparison of virgin spruce-fir forest in the Northern and Southern Appalachian systems. *Ecology* 32, 84–103.

Oosting, H. J. and Humphreys, M. E. (1940). Buried viable seed in a successional series of old fields and forest soils. *Bull. Torrey bot. Club.* 67, 253–273.

Opie, J. (1968). Predictability of individual tree growth using various definitions of competing basal area. *For. Sci.* 14, 314–323.

Orians, G. H. and Janzen, D. H. (1974). Why are embryos so tasty? *Am. Nat.* 108, 581–592.

Orshan, G. (1954). Surface reduction and its significance as a hydroecological factor. *J. Ecol.* **42**, 443–444.

Orshan, G. (1963). Seasonal dimorphism of desert and mediterranean chamaephytes and its significance as a factor in their water economy. *In* "The Water Relations of Plants" (A. J. Rutter and F. H. Whitehead, eds.), pp. 207–222. *3rd Symp. Br. ecol. Soc.* Blackwell, Oxford.

Osvald, H. (1923). "Die Vegetation des Hochmoores Komosse". Akad. Abhandlungen, Uppsala.

Osvald, H. (1947). Equipment of plants in the struggle for space. (In Swedish) *Växtodling* **2**, 288–303.

Ovington, J. D. (1957). Dry matter production by *Pinus sylvestris* L. *Ann. Bot.* **21**, 287–314.

Owen, M. (1971). The selection of feeding site by white-fronted geese in Winter. *J. appl. Ecol.* **8**, 905–919.

Oxley, E. R. H. (1977). "The population dynamics of the foxglove, *Digitalis purpurea* L." (in prep.)

Paine, R. T. (1966). Food web complexity and species diversity. *Am. Nat.* **100**, 65–75.

Paine, R. T. and Vadas, R. L. (1969). The effects of grazing by sea urchins *Strongylocentrus* spp., on benthic algal populations. *Limnol. Oceanogr.* **14**, 710–719.

Palmblad, I. G. (1968). Competition in experimental studies on populations of weeds with emphasis on the regulation of population size. *Ecology* **49**, 26–34.

Park, T. (1955). Experimental competition in beetles, with some general implications. *In* "The Numbers of Man and Animals", pp. 69–84. Institute of Biology, Oliver and Boyd, Edinburgh.

Parnell, J. R. (1966). Observations on the population fluctuations and life histories of the beetles *Bruchidius ater* (Bruchidae) and *Apion fuscirostre* (Curculionidae) on broom (*Sarothamnus scoparius*). *J. Anim. Ecol.* **35**, 157–188.

Passet, J. (1971). "*Thymus vulgaris* L. Chemotaxonomie et biogenèse monoterpenique". Thèse, Faculté de Pharmacie, Université de Montpellier.

Passioura, J. B. (1972). The effect of root geometry on the yield of wheat growing on stored water. *Aust. J. agric. Res.* **23**, 745–752.

Pawsey, C. K. (1960). Cone production reduced, apparently by drought in the south-east of south Australia. *Aust. Forest.* **24**, 74–75.

Pearl, R. and Reed, L. J. (1920). On the rate of growth of the population of the United States since 1790 and its mathematical representation. *Proc. Natn. Acad. Sci., U.S.A.* **6**, 275–288.

Pearsall, W. H. and Bengry, R. P. (1940). The growth of *Chlorella* in darkness and in nutrient solution. *Ann. Bot.* **4**, 365–377.

Pelton, J. (1953). Studies on the life history of *Symphoricarpos occidentalis* Hook. in Minnesota. *Ecol. Monogr.* **23**, 17–39.

Pelton, J. (1956). A study of seed dormancy in eighteen species of high altitude Colorado plants. *Butler Univ. bot. Studies* **13**, 74–84.

Pemadasa, M. A. and Lovell, P. H. (1974). Factors affecting the distribution of annuals in the dune system at Aberffraw, Anglesey. *J. Ecol.* 62, 379–402.

Penman, H. L. (1956). Evaporation: an introductory survey. *Neth. J. agric. Sci.* 4, 9–29.

Peterken, G. F. (1966). Mortality of holly (*Ilex aquifolium*) seedlings in relation to natural regeneration in the New Forest. *J. Ecol.* 54, 259–269.

Peterken, G. F. and Tubbs, C. R. (1965). Woodland regeneration in the New Forest, Hampshire, since 1650. *J. appl. Ecol.* 2, 159–170.

Pharis, R. P. and Morf, W. (1969). Precocious flowering of Coastal and Giant Redwood with gibberellins A_3, $A_{4/7}$ and A_{13}. *BioScience* 19, 719–720.

Pianka, E. R. (1974). "Evolutionary Ecology". Harper and Row, New York.

Pickering, S. U. and the Duke of Bedford (1919). "Science and the Fruit Grower". Macmillan, London.

Pijl, L. van der (1969). "Principles of Dispersal in Higher Plants". Berlin, Springer Verlag, Berlin.

Pimentel, D. (1964). Population ecology and the genetic feedback mechanism. *In* "Genetics Today" *Proc. XIth Int. Cong. Genet.* The Hague, Netherlands (1963). Pergamon Press, Oxford, pp. 483–488.

Pimentel, D. and Al-Hajidh, R. (1965). Ecological control of a parasite population by genetic evolution in the parasite–host system. *Ann. Ent. Soc. Am.* 58, 1–6.

Pimentel, D., Feinburg, E. H., Wood, P. W. and Hayes, J. T. (1965). Selection, spatial distribution and the coexistence of competing fly species. *Am. Nat.* 99, 97–109.

Pimentel, D., Nagel, W. P. and Madden, J. L. (1963). Space–time structure of the environment and the survival of parasite–host systems. *Am. Nat.* 97, 141–167.

Plank, J. E. van der (1960). Analysis of epidemics. *In* "Plant Pathology" (J. G. Horsfall and A. E. Dimond, eds), Vol. 3, pp. 229–289. Academic Press, New York and London.

Plank, J. E. van der (1963). "Epidemics and Control". Academic Press, New York and London.

Poole, A. L. and Cairns, D. (1940). Botanical aspects of ragwort (*Senecio jacobaea* L.) control. *Bull. N.Z. Dep. sci. industr. Res.* 82, 1–66.

Porter, J. W. (1972). Predation by *Acanthaster* and its effect on coral diversity. *Am. Nat.* 106, 487–492.

Powell, J. R. (1971). Genetic polymorphism in varied environments. *Science, N.Y.* 174, 1035–1036.

Pozzera, G. (1959). Relations between cone production by *P. pinea* and certain meteorological factors. *Ital. Forest. Mont.* 14, 196–206. (Cited from *Forest. Abstr.* 21, 1963.)

Priestley, J. H. and Pearsall, W. H. (1922). An interpretation of some growth curves. *Ann. Bot.* 36, 239–249.

Proebsting, F. L. and Gilmour, A. E. (1941). The relation of peach root toxicity to the re-establishment of peach orchards. *Proc. Am. Soc. hort. Sci.* 38, 21–26.

Prokopy, R. J., Moericke, V. and Bush, G. L. (1973). Attraction of apple maggot flies to odor of apples. *Env. Ent.* 2, 743–749.

Pryor, L. D. (1959). Species distribution and association in Eucalyptus. *In*

"Biogeography and Ecology in Australia" (A. Keast, R. L. Crocker and C. S. Christian, eds). W. Junk, The Hague.

Puckridge, D. W. and Donald, C. M. (1967). Competition among wheat plants sown at a wide range of densities. *Aust. J. agric. Res.* **18**, 193–211.

Purseglove, J. W. (1972). "Tropical Crops: Monocotyledons". Longmans, London.

Putwain, P. D. and Harper, J. L. (1970). Studies in the dynamics of plant populations. III. The influence of associated species on populations of *Rumex acetosa* L. and *R. acetosella* L. in grassland. *J. Ecol.* **58**, 251–264.

Putwain, P. D. and Harper, J. L. (1972). Studies in the dynamics of plant populations. V. Mechanisms governing the sex ratio in *Rumex acetosa* and *R. acetosella. J. Ecol.* **60**, 113–129.

Putwain, P. D., Machin, D. and Harper, J. L. (1968). Studies in the dynamics of plant populations. II. Components and regulation of a natural population of *Rumex acetosella* L. *J. Ecol.* **56**, 421–431.

Rabotnov, T. A. (1956). Some data on the content of living seeds in soils of meadow communities. *In* "Akademika V.N. Sukacheva k 75-letiyo so duia rozhdeniia". Akad Nauk SSSR, Moskva-Leningrad, 481–499. (In Russian.)

Rabotnov, T. A. (1956). The life cycle of *Heracleum sibiricum* L. *Bull. Moscow Soc. Natur.* **61**, 73–81. (In Russian.)

Rabotnov, T. A. (1958). The life cycle of *Ranunculus acer* L. and *R. auricomus* L. *Bull. Moscow Soc. Natur.* **63**, 77–86. (In Russian.)

Rabotnov, T. A. (1960). Some problems in increasing the proportion of leguminous species in permanent meadows. *Proc. 8th Int. Grassl. Congr.* 260–264.

Rabotnov, T. A. (1969). On coenopopulations of perennial herbaceous plants in natural coenoses. *Vegetatio* **19**, 87–95.

Rabotnov, T. A. and Saurina, N. I. (1971). The density and age composition of certain populations of *Ranunculus acris* L. and *R. auricomus* L. *Bot. Zh.* **56**, 476–484. (In Russian.)

Radford, P. J. (1967). Growth analysis formulae — their use and abuse. *Crop Sci.* **7**, 171–175.

Raup, H. M. (1957). Vegetational adjustment to the instability of site. *Proc. Pap. Union Conserv. Nature Nat. Resour.* 36–48.

Rehfeldt, G. E., Stage, A. R. and Bingham, R. T. (1971). Strobili development in Western White pine; periodicity, prediction and association with weather. *Forest Sci.* **17**, 454–461.

Reineke, L. H. (1933). Perfecting a stand-density index for even-aged forests. *J. agric. Res.* **46**, 627–638.

Reynoldson, T. B. and Davies, R. W. (1970). Food niche and coexistence in lake-dwelling triclads. *J. Anim. Ecol.* **39**, 599–617.

Reynoldson, T. B. and Bellamy, L. S. (1973). Interspecific competition in lake-dwelling triclads; a laboratory study. *Oikos* **24**, 301–313.

Reynoldson, T. B. and Young, J. O. (1965). The food of four species of lake-dwelling triclads. *J. Anim. Ecol.* **32**, 175–191.

Rhodes, I. (1968). Yield of contrasting ryegrass varieties in monoculture and mixed culture. *J. Br. Grassl. Soc.* **23**, 156–158.

Rhodes, I. (1970). The production of contrasting genotypes of perennial ryegrass (*Lolium perenne* L.) in monocultures and mixed cultures of varying complexity. *J. Br. Grassl. Soc.* **25**, 285–288.

Rice, E. L. (1974). "Allelopathy". Academic Press, New York and London.

Richards, P. W. (1952). "The Tropical Rainforest". Cambridge University Press, London.

Ridley, H. N. (1930). "The Dispersal of Plants Throughout the World". L. Reeve & Co., Ashford, Kent.

Righter, F. I. (1939). Early flower production among the pines. *J. Forest.* **37**, 935–938.

Roach, B. M. B. (1928). On the influence of light and of glucose on the growth of a soil alga. *Ann. Bot.* **42**. 317–345.

Roberts, E. H. (1960). The viability of cereal seed in relation to temperature and moisture. *Ann. Bot.* **24**, 12–30.

Roberts, E. H. (1961). The viability of rice seed in relation to temperature, moisture content, and gaseous environment. *Ann. Bot.* **25**, 381–390.

Roberts, H. A. (1958). Studies on the weeds of vegetable crops. 1. Initial effects of cropping on the weed seeds in the soil. *J. Ecol.* **46**, 759–768.

Roberts, H. A. (1964). Emergence and longevity in cultivated soil of seeds of some annual weeds. *Weed Res.* **4**, 296–307.

Roberts, H. A. (1968). The changing population of viable weed seeds in an arable soil. *Weed. Res.* **8**, 253–256.

Roberts, H. A. (1970). Viable weed seeds in cultivated soils. *Rep. natn. Veg. Res. Stn* (1969) 25–38.

Roberts, H. A. and Feast, P. M. (1972). Fate of seeds of some annual weeds in different depths of cultivated and undisturbed soil. *Weed Res.* **12**, 316–324.

Roberts, H. A. and Feast, P. M. (1973). Emergence and longevity of seeds of annual weeds in cultivated and undisturbed soil. *J. appl. Ecol.* **10**, 133–143.

Roberts, H. A. and Stokes, F. G. (1966). Studies on the weeds of vegetable crops. VI. Seed populations of soil under commercial cropping. *J. appl. Ecol.* **3**, 181–190.

Robinson, E. L. and Kust, C. A. (1962). Distribution of Witchweed seeds in soil. *Weeds* **10**, 335.

Rockwood, L. L. (1973). The effect of defoliation on seed production of six Costa Rican tree species. *Ecology* **54**, 1363–1369.

Roe, A. L. (1967). Seed dispersal in a bumper spruce seed year. *U.S. Forest Service Res. Pap. (Int.)* **39**, 1–10.

Rohmeder, E. (1967). Beziehungen zwischen Frucht-bzw. Samenerzeugung und Holzerzeugung der Waldbäume. *Allg. Forstzeitschr.* **22**, 33–39.

Rollin, S. F. and Johnston, F. A. (1961). Our laws that pertain to seeds. *In* "Seeds. Yearbook of Agriculture, 1961". U.S.D.A. Washington, D.C. 482–492.

Rorison, I. H. (1967). A seedling bioassay on some soils in the Sheffield area. *J. Ecol.* **55**, 725–741.

Rosensweig, M. L. and Sterner, P. W. (1970). Population ecology of desert rodent communities: body size and seed husking as bases for heteromyid coexistence. *Ecology* **51**, 217–224.

Ross, M. A. (1968). "The establishment of seedlings and the development of patterns in grassland". Ph.D. thesis, University of Wales.

Ross, M. A. and Harper, J. L. (1972). Occupation of biological space during seedling establishment. *J. Ecol.* **60**, 77—88.

Roughgarden, J. (1972). Evolution of niche width. *Am. Nat.* **106**, 683—718.

Rousvoal, D. and Gallais, A. (1973). Comportement en association biniaire de cinq espèces d'une prairie permanente. *Oecol. Plant.* **8**, 279—300.

Roy, D. F. (1966). Silvical characteristics of Redwood. *U.S. Forest Service Res. Pap. PSW-28*, 20 pp.

Rudolph, T. D. (1966). Stimulation of earlier flowering and seed production in jack pine seedlings through greenhouse and nursery culture. *U.S. Forest Service Res. Pap. NC-6*, 80—83.

Ruiter, L. de (1952). Some experiments on the camouflage in stick caterpillars. *Behaviour* **4**, 222—232.

Runge, M. (1971). Investigations of the content and the production of mineral nitrogen in soils. *In* "Integrated Experimental Ecology" (H. Ellenberg, ed.) pp. 191—202. Chapman and Hall, London.

Russell, E. W. (1966). "Soil Conditions and Plant Growth". Longmans, London.

Sacher, G. A. (1967). The complementarity between development and aging — experimental and theoretical considerations. *In* "Interdisciplinary Perspectives of Time" (E. M. Weyer, ed.). *Ann. N.Y. Acad. Sci.* **138**, 680—712.

Saeki, T. (1960). Interrelationships between leaf amount, light distribution and total photosynthesis in a plant community. *Bot. Mag. Tokyo* **73**, 55—63.

Saeki, T. and Kuroiwa, S. (1959). On the establishment of a vertical distribution of photosynthetic systems in a plant community. *Bot. Mag. Tokyo* **72**, 27—35.

Sagar, G. R. (1959). "The biology of some sympatric species of grassland". D.Phil. thesis, University of Oxford.

Sagar, G. R. and Harper, J. L. (1960). Factors affecting the germination and early establishment of Plantains (*Plantago lanceolata, P. media* and *P. major.*) *In* "The Biology of Weeds" (J. L. Harper, ed.). *Br. Ecol. Soc.Symp.* **1**, 236—245.

Sagar, G. R. and Harper, J. L. (1961). Controlled interference with natural populations of *Plantago lanceolata, P. major* and *P. media. Weed Res.* **1**, 163—176.

Sakai, K-I (1953). Studies on competition in plants. 1. Analysis of the competitional variance in mixed plant populations. *Jap. J. Bot.* **14**, 161—168.

Sakai, K-I (1955). Competition in plants and its relation to selection. *Cold Spring Harbor Symp. Quant. Biol.* **20**, 137—157.

Sakai, K-I. (1957). Studies on competition in plants. VII. Effect on competition of a varying number of competing and non-competing individuals. *J. Genet.* **55**, 227—234.

Sakai, K-I. (1958). Studies on competitions in plants and animals. IX. Experimental studies on migration in *Drosophila melanogaster. Evolution* **12**, 98—103.

Sakai, K-I. (1961). Competitive ability in plants: its inheritance and some related problems. *Symp. Soc. exp. Biol.* 15, 245–263.

Sakai, K-I. and Gotoh, K. (1955). Studies on competition in plants. IV. Competitive ability of F_1 hybrids in barley. *J. Hered.* 46, 139–143.

Salisbury, E. J. (1916). The Oak-Hornbeam woods of Hertfordshire I & II. *J. Ecol.* 4, 83–120.

Salisbury, E. J. (1918). The Oak-Hornbeam woods of Hertfordshire III & IV. *J. Ecol.* 6, 14–52.

Salisbury, E. J. (1942). "The Reproductive Capacity of Plants". Bell, London.

Salisbury, E. J. (1952). "Downs and Dunes". Bell, London.

Salisbury, E. J. (1958). *Spergularia salina* and *Spergularia marginata* and their heteromorphic seeds. *Kew Bull.* 1, 41–51.

Salisbury, E. J. (1961). "Weeds and Aliens". Collins, London.

Sandfaer, J. (1968). Induced sterility as a factor in the competition between barley varieties. *Nature, Lond.* 218, 241–243.

Sandfaer, J. (1970a). Barley stripe mosaic virus as the cause of the sterility interaction between barley varieties. *Hereditas* 64, 150–152.

Sandfaer, J. (1970b). An analysis of the competition between some barley varieties. *Danish Atomic Energy Commission, Risö Report*, 230, Rosskilde.

Sang, J. H. (1950). Population growth in *Drosophila* cultures. *Biol. Rev.* 25, 188–219.

Sarukhán, J. (1971). "Studies on plant demography". Ph.D. thesis, University of Wales.

Sarukhán, J. (1974). Studies on plant demography: *Ranunculus repens* L., *R. bulbosus* L. and *R. acris* L. II. Reproductive strategies and seed population dynamics. *J. Ecol.* 62, 151–177.

Sarukhán, J. and Gadgil, M. (1974). Studies on plant demography: *Ranunculus repens* L., *R. bulbosus* L. and *R. acris* L. III. A mathematical model incorporating multiple modes of reproduction. *J. Ecol.* 62, 921–936.

Sarukhán, J. and Harper, J. L. (1973). Studies on plant demography: *Ranunculus repens* L., *R. bulbosus* L. and *R. acris*. I. Population flux and survivorship. *J. Ecol.* 61, 675–716.

Saurina, N. I. (1972). Dynamics of numbers of individuals and age spectrum in some populations of *Ranunculus acris* and *Ranunculus auricomus*. *Bull. Moscow Soc. Natur.* 77, 115–120. (In Russian.)

Savage, Z. (1966). Citrus yield per tree by age. *Fla. Agr. Exp. Sta. Agr. Exten. Serv. Econ. Ser.* 66, 9 pp.

Schaffer, W. M. (1974). Optimal reproductive effort in fluctuating environments. *Am. Nat.* 108, 783–790.

Schmidtling, R. C. (1969). Reproductive maturity related to height of loblolly pine. *U.S.D.A. Forest Service Res. Note* SO-94.

Schoener, T. W. (1969a). Models of optimal size for solitary predators. *Am. Nat.* 103, 277–312.

Schoener, T. W. (1969b). Optimal size and specialisation in constant and fluctuating environments: an energy–time approach. *In* "Diversity and Stability in Ecological Systems". *Brookhaven Symp. Biol.* 22, 103–114.

Schorger, A. W. (1966). "The Wild Turkey, its History and Domestication".

University of Oklahoma Press.

Schwanitz, F. (1957). "Die Entstehung der Kulturpflanzen". Berlin. (Published by Harvard University Press, Cambridge, Massachusetts.)

Seaton, A. J. P. and Antonovics, J. (1967). Population interrelationships. I. Evolution in mixtures of *Drosophila* mutants. *Heredity* (Lond.) 22, 19–33.

Selman, M. (1970). The population dynamics of *Avena fatua* (Wild Oats) in continuous spring barley. *Proc. Br. Weed Control Conf.* 10, 1176–1188.

Sen, D. N., Chawan, D. D. and Chatterji, U. N. (1968). Diversity in germination of seeds in *Calotropis procera* R. Br. population. *Österr. Bot. Z.* 115, 6–17.

Sharitz, R. R. and McCormick, J. F. (1972). Population dynamics of two competing annual plant species. *Ecology* 54, 723–740.

Sharp, W. M. (1958). Evaluating mast yields in the oaks. *Pennsylvania State Univ. agric. Exp. Stn Bull.* 635, 1–22.

Sharp, W. M. and Sprague, V. G. (1967). Flowering and fruiting in the white oaks. Pistillate flowering, acorn development, weather and yields. *Ecology* 48, 243–251.

Shaw, M. W. (1968a). Factors affecting the natural regeneration of sessile oak (*Quercus petraea*) in North Wales. I. A preliminary study of acorn production, viability and losses. *J. Ecol.* 56, 565–583.

Shaw, M. W. (1968b). Factors affecting the natural regeneration of sessile Oak (*Quercus petraea*) in North Wales. II. Acorn losses and germination under field conditions. *J. Ecol.* 56, 647–660.

Sheail, J. (1971). "Rabbits and Their History". David and Charles, Newton Abbot.

Sheldon, J. C. (1974). The behaviour of seeds in soil. III. The influence of seed morphology and the behaviour of seedlings on the establishment of plants from surface lying seeds. *J. Ecol.* 62, 47–66.

Sheldon, J. C. and Burrows, F. M. (1973). The dispersal effectiveness of the achenepappus units of selected Compositae in steady winds with convection. *New Phytol.* 72, 665–675.

Shibles, R. M. and Weber, C. R. (1965). Leaf area, solar radiation interception and dry matter production by soybeans. *Crop Sci.* 5, 575–577.

Shinozaki, K. and Kira, T. (1956). Intraspecific competition among higher plants. VII. Logistic theory of the C-D. Effect. *J. Inst. Polytech, Osaka Cy. Univ.* 7, 35–72.

Shinozaki, K. and Kira, T. (1961). The C-D Rule, its theory and practical uses (Intraspecific competition among higher plants. X.) *J. Biol., Osaka Cy. Univ.* 12, 69–82.

Shinozaki, K., Yoda, K., Hozumi, K. and Kira, T. (1964). A quantitative analysis of plant form – the pipe model theory. I. Basis analyses. II. Further evidence of the theory and its application in forest ecology. *Jap. J. Ecol.* 14, 97–105, 133–139.

Sibma, L., Kort, J. and de Wit, C. T. (1964). Experiments on competition as a means of detecting possible damage by nematodes. *Jaarb. I.B.S.* 1964, 119–124.

Singh, L. B. (1960). "The Mango". Leonard Hill, London.

Skellam, J. G. (1951). Random dispersal in theoretical populations. *Biometrika* 38, 196—218.

Slatyer, R. O. (1970). Comparative photosynthesis, growth and transpiration of two species of *Atriplex*. *Planta* 93, 175—189.

Slobodkin, L. B. (1964a). "Growth and Regulation of Animal Populations". Holt, Rinehart and Winston, New York.

Slobodkin, L. B. (1964b). Experimental populations of hydrida. *J. Ecol.* (Suppl.) 52, 131—148.

Slobodkin, L. B. (1964c). The strategy of evolution. *Am. Sci.* 52, 342—357.

Slobodkin, L. B. (1968). Toward a predictive theory of evolution. *In* "Population Biology and Evolution" (R. C. Lewontin, ed.), pp. 187—205. Syracuse University Press, N.Y.

Slobodkin, L. B. and Rapoport, A. (1974). An optimal strategy of evolution. *Q. Rev. Biol.* 49, 181—199.

Slobodkin, L. B., Smith, F. E. and Hairston, N. G. (1967). Regulation in terrestrial ecosystems, and the implied balance of nature. *Am. Nat.* 101, 109—124.

Small, J. (1918). The origin and development of the Compositae. IX. Fruit dispersal. *New Phytol.* 17, 200—230.

Smirnova, O. V. and Toropova, N. H. (1972). The great life cycle of *Galeobdolon luteum* Huds. *Bull. Moscow Soc. Natur.* 77, 76—87. (In Russian.)

Smith, C. C. (1968). The adaptive nature of social organisation in the genus of tree squirrels, *Tamiasciurus*. *Ecol. Monogr.* 38, 31—63.

Smith, C. C. (1970). The coevolution of pine squirrels (*Tamiasciurus*) and conifers. *Ecol. Monogr.* 40, 349—371.

Smith, H. S. (1935). The role of biotic factors in the determination of population densities. *J. econ. Ent.* 28, 873—898.

Snaydon, R. W. (1970). Rapid population differentiation in a mosaic environment. I. The response of *Anthoxanthum odoratum* populations to soils. *Evolution* 24, 257—269.

Snaydon, R. W. (1971). An analysis of competition between plants of *Trifolium repens* L. populations collected from contrasting soils. *J. appl. Ecol.* 8, 687—697.

Snow, D. W. (1964). A possible selective factor in the evolution of fruit seasons in tropical forest. *Oikos* 15, 274—281.

Soane, I. D. and Clarke, B. (1973). Evidence for apostatic selection by predators using olfactory clues. *Nature, Lond.* 241, 62—64.

Solbrig, O. T. and Simpson, B. B. (1974). Components of regulation of a population of dandelions in Michigan. *J. Ecol.* 62, 473—486.

Southwood, T. R. E. (1973). The insect/plant relationship — an evolutionary perspective. *In* "Insect/Plant Relationships" (H. F. van Emden, ed.). *Symp. R. ent. Soc.* 6, 3—30.

Spedding, C. R. W. (1971). "Grassland Ecology". Clarendon Press, Oxford.

Spencer, D. A. (1964). Porcupine population fluctuations in past centuries revealed by dendrochronology. *J. appl. Ecol.* 1, 127—149.

Spring, P. E., Brewer, M. L., Brown, J. R. and Fanning, M. E. (1974).
Population ecology of loblolly pine *Pinus taeda* in an old field community.
Oikos 25, 1—6.

Spurr, S. H. (1964). "Forest Ecology". The Ronald Press, New York.

Stalter, R. (1971). Age of a mature pine (*Pinus taeda*) stand in South
Carolina. *Ecology* 52, 532—533.

Stapp, C. (1961). "Bacterial Plant Pathogens". (Translated by A. Schoenfeld).
Oxford University Press.

Stark, R. H., Hafenrichter, A. L. and Klages, K. H. (1949). The production of
seed and forage by mountain brome as influenced by nitrogen and age of
stand. *Agron. J.* 63, 575—580.

Stebbins, G. L. (1950). "Variation and Evolution in Plants". Columbia
University Press, New York.

Stebbins, G. L. (1958). Longevity, habitat and release of genetic variability in
the higher plants. *Cold Spring Harb. Symp. quant. Biol.* 23, 365—378.

Stebbins, G. L. (1974). "Flowering Plants — Evolution above the Species
Level". Harvard University Press, Cambridge, Mass.

Sterk, A. A. (1969). Biosystematic studies of *Spergularia media* and
Spergularia marina in the Netherlands. Part 1. The morphological
variability of *S. media*. *Acta bot. neerl.* 18, 325—338.

Sterk, A. A. and Duijkhuizen, L. (1972). The relation between the genetic
determination and the ecological significance of the seed wing in *Spergularia
media* and *S. marina*. *Acta bot. neerl.* 21, 481—490.

Stern, W. R. (1965). The effect of density on the performance of individual
plants in subterranean clover swards. *Aust. J. agric. Res.* 16, 541—555.

Stern, W. R. and Donald, C. M. (1962). Light relationships in grass-clover
swards. *Aust. J. agric. Res.* 13, 599—614.

Steveninck, R. F. M. van (1957). Factors affecting the abscission of reproductive
organs in yellow lupins (*Lupinus luteus* L.) Part 1. The effect of different
patterns of flower removal. *J. exp. Bot.* 8, 373—381.

Stevens, O. A. (1932). The number and weight of seeds produced by weeds.
Am. J. Bot. 19, 784—794.

Stevens, O. A. (1957). Weights of seeds and numbers per plant. *Weeds* 5, 46—55.

Stolzy, L. H. and Barley, K. P. (1968). Mechanical resistance encountered by
roots entering compact soil. *Soil Sci.* 105, 297—301.

Streeter, D. T. (1970). The effects of public pressure on the vegetation of
chalk downland at Box Hill, Surrey. *In* "The Scientific Management of
Animal and Plant Communities for Conservation" (E. Duffey and A. S. Watt,
eds). *11th Symp. Br. ecol. Soc.* Blackwell, Oxford.

Studhalter, R. A. (1955). Tree growth: some historical chapters. *Bot. Rev.* 21, 1—72.

Sukatschew, W. (1928). Einige experimentelle Untersuchungen über den Kampf
ums Dasein zwischen Biotypen derselben Art. *Z. indukt. Abstamm.-u.
VererbLehre* 45, 54—74.

Sukatschew, W. N. (1928). "Plant Communities". Nauk, Moscow. (In Russian.)

Summerfield, R. J. (1972). Biological inertia — an example. *J. Ecol.* 60,
793—798.

Summerfield, R. J. (1973). Factors affecting the germination and establishment of seedlings of *Narthecium ossifragum* in mire ecosystems. *J. Ecol.* **61**, 387–398.

Summerhayes, V. S. (1951). "Wild Orchids of Britain". Collins, London.

Sunderland, N. (1960). Germination of the seeds of angiospermous root parasites. *In* "The Biology of Weeds" (J. L. Harper, ed.) *Symp. Br. ecol. Soc.* **1**, 83–93.

Swanberg, P. O. (1951). Food storage, territory and song in the thick billed nutcracker. *Proc. Xth Internatl. Ornith. Congr. Uppsala*, June 1950. 545–554.

Sweeney, J. R. (1956). Responses of vegetation to fire: a study of the herbaceous vegetation following chaparral fires. *Univ. Calif. Publ. Bot.* **28**, 143–250.

Sweet, G. B. and Wareing, P. F. (1966). Role of plant growth in regulating photosynthesis. *Nature, Lond.* **210**, 77–79.

Sylvén, N. (1906). On de svenska dikotyledonernas första förstärkningsstadium eller utveekling fran frö till blomning. 1. *K. Sv. Vet.-Akad. Handl.* **40**

Symonides, E. (1974). Populations of *Spergula vernalis* Willd. on dunes in the Torun basin. *Ekol. pol.* **22**, 379–416.

Tadros, T. M. (1957). Evidence of the presence of an edaphobiotic factor in the problem of serpentine tolerance. *Ecology* **38**, 14–23.

Takeda, T. (1961). Studies on the photosynthesis and production of dry matter in the community of rice plants. *Jap. J. Bot.* **17**, 402–437.

Tamm, C. O. (1948). Observations on reproduction and survival of some perennial herbs. *Bot. Notiser* **3**, 305–321.

Tamm, C. O. (1956a). Composition of vegetation in grazed and mown sections of a former hay meadow. *Oikos* **7**, 144–157.

Tamm, C. O. (1956b). Further observations on the survival and flowering of some perennial herbs. 1. *Oikos* **7**, 274–292.

Tamm, C. O. (1972a). Survival and flowering of some perennial herbs. II. The behaviour of some orchids on permanent plots. *Oikos* **23**, 23–28.

Tamm, C. O. (1972b). Survival and flowering of perennial herbs. III. The behaviour of *Primula veris* on permanent plots. *Oikos* **23**, 159–166.

Tansley, A. G. (1917). On competition between *Galium saxatile* L. (*G. hercynicum* Weig.) and *Galium sylvestre* Poll. (*G. asperum* Schreb.) on different types of soil. *J. Ecol.* **5**, 173–179.

Tansley, A. G. (1949). "The British Isles and their Vegetation". Cambridge University Press, Cambridge.

Tansley, A. G. and Adamson, R. S. (1925). Studies of the vegetation of the English chalk. III. The chalk grasslands of the Hampshire-Sussex border. *J. Ecol.* **13**, 177–223.

Taylorson, R. B. and Borthwick, H. A. (1969). Light filtration by foliar canopies: significance for light-controlled weed seed germination. *Weed Sci.* **17**, 48–51.

Thoday, J. M. (1953). Components of fitness. *In* "Evolution" (R. Brown and J. F. Danielli, eds) *Symp. Soc. exp. Biol.* **7**, 96–113.

Thoday, J. M. (1972). Disruptive selection. *Proc. R. Soc.* B. **182**, 109—143.

Thoday, J. M. (1975). Non-Darwinian evolution and biological programs. *Nature, Lond.* **255**, 675—677.

Thomas, A. S. (1960). The trampling animal. *J. Br. Grassl. Soc.* **15**, 89—93.

Thomson, A. J. (1969). Yields and tiller numbers of four perennial ryegrass varieties grown as monocultures and certain mixtures in microplots. *J. agric. Sci., Camb.* **73**, 321—328.

Thorsteinson, A. J. (1960). Host selection in phytophagous insects. *A. Rev. Ent.* **5**, 193—218.

Thurston, J. M. (1957). Morphological and physiological variation in wild oats (*Avena fatua* L. and *A. ludoviciana* Dur.) and in hybrids between wild and cultivated oats. *J. agric. Sci., Camb.* **49**, 260—274.

Tinbergen, L. (1960). The natural control of insects in pine woods. 1. Factors influencing the intensity of predation by songbirds. *Archs néerl. Zool.* **13**, 265—343.

Tinker, P. B. (1968). The transport of ions in the soil around plant roots. *In* "Ecological Aspects of the Mineral Nutrition of Plants" (I. H. Rorison, ed.). *9th Symp. Br. ecol. Soc.*, 135—147. Blackwell, Oxford.

Toole, E. H. and Brown, E. (1946). Final results of the Duvel buried seed experiment. *J. agric. Res.* **72**, 201—210.

Tothill, J. D. (1958). Some reflections on the cause of insect outbreaks. *Proc. 10th Internat. Congr. Ent.* (1956) 4, 525—531.

Trenbath, B. R. (1974a). Neighbour effects in the genus *Avena* II. Comparison of weed species. *J. appl. Ecol.* **11**, 111—125.

Trenbath, B. R. (1974b). Biomass productivity of mixtures. *Adv. Agron.* **26**, 177—210

Trenbath, B. R. (1975). Neighbour effects in the genus *Avena* III. A diallel approach. *J. appl. Ecol.* **12**, 189—200.

Trenbath, B. R. (1976). Application of a growth model to problems of the productivity and stability of mixed stands. *Proc. XII int. Grassld. Cong., Moscow* (1974), Vol. 1, Part 2, pp. 546—558.

Trenbath, B. R. and Harper, J. L. (1973). Neighbour effects in the genus *Avena*. I. Comparison of crop species. *J. appl. Ecol.* **10**, 379—400.

Trevis, L. (1958). Interrelations between the harvester ant *Veromessor pergandei* (Mayr) and some desert ephemerals. *Ecology* **39**, 695—704.

Tripathi, R. S. and Harper, J. L. (1973). The comparative biology of *Agropyron repens* L. (Beauv.) and *A. caninum* L. (Beauv.) 1. The growth of mixed populations established from tillers and from seeds. *J. Ecol.* **61**, 353—368.

Triplett, G. B. and Tesar Jr. M. B. (1960). Effects of compaction, depth of planting, and soil moisture tension on seedling emergence of alfalfa. *Agron. J.* **52**, 681—684.

Troughton, A. (1960). Growth correlation between the root and shoots of grass plants. *Proc. 8th Int. Grassl. Cong.* 280—283.

Tukey, H. B. Jr. (1971). Leaching of substances from plants. *In* "Biochemical

Interactions Among Plants" (U.S. Nat. Comm. for IBP, eds, pp. 25–32). Natn. Acad. Sci. Washington, D.C.

Turcek, F. J. (1966). Rediscovery of seeds cached by *Garrulus glandarius* and *Nucifraga caryocatactes*. *Waldhygiene* 6, 215–217.

Turkington, R. A. (1975). "Relationships between neighbours among species of permanent grassland". Ph.D. thesis, University of Wales.

Turner, J. R. (1967). Why does the genotype not congeal? *Evolution* 21, 645–656.

Turner, R. G. (1969). Heavy metal tolerance in plants. *In* "Ecological Aspects of Mineral Nutrition in Plants". (I. H. Rorison, ed.) *9th Symp. Br. ecol. Soc.*, 399–410.

Twamley, B. E. (1967). Seed size and seedling vigour in birdsfoot trefoil. *Can. J. Plant Sci.* 47, 603–609.

Uranov, A. A., Grigorieva, V. M., Egorova, V. N., Ermakova, I. M. and Matveev, A. R. (1970). The variability and dynamics of age spectra in some meadow plants. *In* "Theoretical Problems of Phytocenology and Biogenocenology" (to the 90th birthday of V. N. Sukachev), pp. 194–214. Nauk, Moscow. (In Russian.)

Uranov, A. A. and Smirnova, O. V. (1969). Classification and basic features of the development of perennial plant populations. *Bull. Moscow Soc. Natur.* 74, 119–134. (In Russian.)

Usher, M. B. (1966). A matrix approach to the management of renewable resources, with special reference to selection forests. *J. appl. Ecol.* 3, 355–367.

Usher, M. B. (1973). "Biological Management and Conservation". Chapman and Hall, London.

Utida, S. (1957). Population fluctuation, an experimental and theoretical approach. *Cold Spring Harb. Symp. quant. Biol.* 22, 139–151.

Van Valen, L. (1965). Morphological variation and width of ecological niche. *Am. Nat.* 99, 377–390.

Van Valen, L. (1970). Variation and niche width re-examined. *Am. Nat.* 104, 587–590.

Van Valen, L. (1973). A new evolutionary law. *Evolut. Theory* 1, 1–30.

Van Valen, L. (1974). Predation and species diversity. *J. theor. Biol.* 44, 19–21.

Vallis, I., Haydock, K. P., Ross, P. J. and Henzel, E. F. (1967). Isotopic studies on the uptake of nitrogen by pasture plants. III. The uptake of small additions of ^{15}N-labelled fertiliser by Rhodes grass and Townsville Lucerne. *Aust. J. agric. Res.* 18, 865–877.

Vandermeer, J. H. (1975). A graphical model of insect seed predation. *Am. Nat.* 109, 147–160.

Varley, G. C. (1967). Estimation of secondary production in species with an annual life cycle. *In* "Secondary Production of Terrestrial Ecosystems" (K. Petrusewicz, ed.), pp. 447–457. Warsaw.

Varley, G. C. and Gradwell, G. R. (1962). The effect of partial defoliation by caterpillars on the timber production of oak trees in England. *XI Int. Cong. Ent.* Wien, 1960, 2, 211–214.

Vegis, A. (1961). Samenkeimung und vegetative Entwicklung der Knospen.

Encyclopedia of Plant Physiology **16**, 168−298.

Vegis, A. (1964). Dormancy in higher plants. *A. Rev. Pl. Physiol.* **15**, 185−224.

Verhagen, A. M. W., Wilson, J. H. and Britten, E. J. (1963). Plant production in relation to foliage illumination. *Ann. Bot.* **27**, 627−640.

Verheij, E. W. M. (1968). Yield-density relationships in apple: Results of a planting system experiment in Hungary. *Instituut Voor Tuinbouwtechniek − Wageningen* **37**, 1−31.

Verhulst, P. F. (1844). Recherches mathématiques sur la loi d'accroissement de la population. *Mem. Acad. r. Bruxelles* **18**, 1−58.

Verhulst, P. F. (1846). Deuxième memoire sur la loi d'accroissement de la population. *Mem. Acad. Roy. Bruxelles* **20**, 1−52.

Verschaeffelt, E. (1910). The cause determining the selection of food in some herbivorous insects. *Proc. Acad. Sci., Amsterdam* **13**, 536−542.

Vezina, P. E. and Boulter, D. W. K. (1966). The spectral composition of near ultraviolet and visible radiation beneath forest canopies. *Can. J. Bot.* **44**, 1267−1284.

Visser, T. (1964). Juvenile phase and growth of apple and pear seedlings. *Euphytica* **13**, 119−129.

Visser, T. (1965). On the inheritance of the juvenile period in apple. *Euphytica* **14**, 125−134.

Visser, T. (1967). Juvenile period and precocity of apple and pear seedlings. *Euphytica* **16**, 319−320.

Visser, T. (1970). The relation between growth, juvenile period and fruiting of apple seedlings, and its use to improve breeding efficiency. *Euphytica* **19**, 293−302.

Volterra, V. (1926). Variazioni e fluttuazioni del numero d'individui in specie animali conviventi. *Mem. Acad. Lincei.* **2**, 31−113.

Volterra, V. (1931). "Leçons sur la Théorie mathématique de la Lutte puor la Vie". Cahiers Scientifiques No. 7 Gauthiers-Villars, Paris.

Wager, H. G. (1938). Growth and survival of plants in the Arctic. *J. Ecol.* **26**, 390−410.

Wapshere, A. J. (1970). The effect of human intervention on the distribution and abundance of *Chondrilla juncea* L. *Proc. Adv. Study Inst. Dynamics Popul. (Oosterbeek)* 469−477.

Wapshere, A. J. (1971). Selection and biological control organisms of weeds. *Commonwealth Inst. of Biological Control, Mis. Pub. No. 6*, 56−61.

Wapshere, A. J., Hasan, S., Wahba, W. K. and Caresche, L. (1974). The ecology of *Chondrilla juncea* in the Western Mediterranean. *J. appl. Ecol.* **11**, 783−800.

Wareing, P. F., Khalifa, M. M. and Treharre, K. J. (1968). Rate-limiting processes in photosynthesis at saturating light intensities. *Nature, Lond.* **220**, 453−457.

Watkinson, A. R. (1975). "The population biology of a dune annual, *Vulpia membranacea*". Ph.D. thesis, University of Wales.

Watson, D. J. (1958). The dependence of net assimilation rate on leaf area index. *Ann. Bot.* **22**, 37−54.

Watson, P. (1969). Evolution in closely adjacent populations. VI. An entomophilous species, *Potentilla erecta* in two contrasting habitats. *Heredity* **24**, 407–422.

Watt, A. S. (1919). On the causes of failure of natural regeneration in British oakwoods. *J. Ecol.* **7**, 173–203.

Watt, A. S. (1940). Contributions to the ecology of bracken (*Pteridium aquilinum*). 1. The rhizome. *New Phytol.* **39**, 401–422.

Watt, A. S. (1943). Contributions to the ecology of bracken (*Pteridium aquilinum*). II. The frond and the plant. *New Phytol.* **42**, 103–126.

Watt, A. S. (1945). Contributions to the ecology of bracken (*Pteridium aquilinum*). III. Frond types and the make-up of the populations. *New Phytol.* **44**, 156–178.

Watt, A. S. (1947a). Contributions to the ecology of bracken (*Pteridium aquilinum*). IV. The structure of the community. *New Phytol.* **46**, 97–121.

Watt, A. S. (1947b). Pattern and process in the plant community. *J. Ecol.* **35**, 1–22.

Watt, A. S. (1970). Contributions to the ecology of bracken (*Pteridium aquilinum*). VII. Bracken and Litter. The cycle of change. *New Phytol.* **69**, 431–449.

Watt, K. E. F. (1968). "Ecology and Resource Managment". McGraw-Hill, New York.

Watts, C. H. S. (1968). The foods eaten by wood mice (*Apodemus sylvaticus*) and bank voles (*Clethrionomys glareolus*) in Wytham Woods, Berks. *J. Anim. Ecol.* **37**, 25–41.

Way, M. J. and Cammell, M. (1970). Aggregation behaviour in relation to food utilization by aphids. *In* "Animal Populations in Relation to their Food Resources" (A. Watson, ed.), pp. 229–247. *10th Symp. Br. ecol. Soc.* Blackwell, Oxford.

Weaver, J. E. (1919). The ecological relations of roots. *Carneg. Inst. Wash. Publn.* **286**, 1–128.

Webb, D. A. (1954). Notes on four Irish heaths. Part 1. *Irish Nat.* **11**, 187.

Webb, D. A. (1958). Notes on four Irish heaths. Part 2. *Proc. bot. Soc. Br. Isl.* **3**, 105.

Webb, L. J., Tracey, J. G. and Haydock, K. P. (1961). The toxicity of *Eremophila mitchellii* Benth. leaves in relation to the establishment of adjacent herbs. *Aust. J. Sci.* **24**, 244–245.

Webb, L. J., Tracey, J. G. and Haydock, K. P. (1967). A factor toxic to seedlings of the same species associated with living roots of the non-gregarious subtropical rain forest tree *Grevillea robusta*. *J. appl. Ecol.* **4**, 13–25.

Welbank, P. J. (1960). Toxin production from *Agropyron repens*. In "The Biology of Weeds" (J. L. Harper, ed.), pp. 158–164. *10th Symp. Br. ecol. Soc.* Blackwell, Oxford.

Welbank, P. J. (1961). A study of the nitrogen and water factors in competition with *Agropyron repens* (L.) Beauv. *Ann. Bot.* **25**, 116–137.

Welbank, P. J. (1962). The effects of competition with *Agropyron repens* and of nitrogen and water-supply on the nitrogen content of *Impatiens parviflora*. *Ann. Bot.* 26, 361–373.

Welbank, P. J. (1963). Toxin production during decay of *Agropyron repens* (couch grass) and other species. *Weed Res.* 3, 205–214.

Welbank, P. J. (1963). A comparison of competitive effects of some common weed species. *Ann. appl. Biol.* 51, 107–125.

Welbank, P. J. (1964). Competition for nitrogen and potassium in *Agropyron repens*. *Ann. Bot.* 28, 1–15.

Wells, T. C. E. (1967). Changes in a population of *Spiranthes spiralis* (L.) Chevall. at Knocking Hoe National Nature Reserve, Bedfordshire, 1962–5. *J. Ecol.* 55, 83–99.

Wells, T. C. E. (1972). Ecological studies on calcareous grasslands. *Monks Wood exp. Stn. Rep.* (1969–71), 44–46.

Went, F. W. (1942). The dependence of certain annual plants on shrubs in Southern Californian deserts. *Bull. Torrey bot. Club* 69, 100–114.

Went, F. W. (1955). The ecology of desert plants. *Scient. Am.* 192, 68–75.

Went, F. W. (1957). "The Experimental Control of Plant Growth". Chronica Botanica, Waltham, Mass.

Went, F. W. (1973). Competition among plants. *Proc. Natn. Acad. Sci., U.S.A.* 70, 585–590.

Werner, P. A. (1975a). Predictions of fate from rosette size in Teasel (*Dipsacus fullonum* L.). *Oecologia* 20, 197–201.

Werner, P. A. (1975b). A seed trap for determining patterns of seed deposition in terrestrial plants. *Can. J. Bot.* 53, 810–813.

Werner, P. A. and Platt, W. J. (1976) Ecological relationships of co-occurring golden rods (*Solidago*: Compositae) *Am. Nat.* 110, 959–971.

Wesson, G. and Wareing, P. F. (1967). Light requirements of buried seeds. *Nature, Lond.* 213, 600–601.

Wesson, G. and Wareing, P. F. (1969). The induction of light sensitivity in weed seeds by burial. *J. exp. Bot.* 20, 413–425.

West, N. E. (1968). Rodent-influenced establishment of ponderosa pine and bitterbrush seedlings in central Oregon. *Ecology* 49, 1009–1011.

White, J. (1968). "Studies on the behaviour of plant populations in model systems". M.Sc. thesis, University of Wales.

White, J. and Harper, J. L. (1970). Correlated changes in plant size and number in plant populations. *J. Ecol.* 58, 467–485.

White, J. H. and French, N. (1968). Leather jacket damage to grassland. *J. Br. Grassl. Soc.* 23, 326–329.

Whittaker, R. H. (1969). Evolution of diversity in plant communities. *In* "Diversity and Stability in Ecological Systems". *Brookhaven Symp. Biol.* 22, 178–196. Springfield, Va. National Bureau of Standards, U.S. Dept. of Commerce.

Whittaker, R. H. (1972). Evolution and measurement of species diversity. *Taxon* 21, 213–251.

Whittaker, R. H. (1974). "The design and stability of plant communities". *1st Int. Cong. Ecol. The Hague, Sept. 1974.* PUDOC, Wageningen.

Whittington, W. J. and O'Brien, T. A. (1968). A comparison of yields from plots sown with a single species or a mixture of grass species. *J. appl. Ecol.* 5, 209–213.

Wilcott, J. C. (1973). "A seed demography model for finding optimal strategies for desert annuals". Ph.D. thesis, Utah State University, Logan, Utah.

Williams, C. B. (1967). Spruce budworm damage symptoms related to radial growth of Grand Fir, Douglas Fir, and Engelmann Spruce. *For. Sci.* 13, 274–285.

Williams, E. J. (1962). The analysis of competition experiments. *Aust. J. biol. Sci.* 15, 509–525.

Williams, G. C. (1966). "Adaptation and Natural Selection". Princeton University Press, Princeton, N.J.

Williams, J. T. and Harper, J. L. (1965). Seed polymorphism and germination. 1. The influence of nitrates and low temperatures on the germination of *Chenopodium album. Weed Res.* 5, 141–150.

Williams, O. B. (1970). Population dynamics of two perennial grasses in Australian semi-arid grassland. *J. Ecol.* 58, 869–875.

Williams, O. B. and Roe, R. (1975). Management of arid grasslands for sheep: plant demography of six grasses in relation to climate and grazing. *Proc. ecol. Soc. Aust.* 9, 142–156.

Williams, R. F. (1955). Redistribution of mineral elements during development. *A. Rev. Pl. Physiol.* 6, 25–42.

Williams, W. A. (1963). Competition for light between annual species of *Trifolium* during the vegetative phase. *Ecology* 44, 475–485.

Williams, W. A. (1967). Seedling growth of a hypogeal legume, *Vicia dasycarpa*, in relation to seed weight. *Crop Sci.* 7, 163–164.

Williamson, M. H. (1957). An elementary theory of interspecific competition. *Nature, Lond.* 180, 422–425 and 181, 1415.

Williamson, M. H. (1972). "The Analysis of Biological Populations". Arnold, London.

Willson, M. F. (1971). Seed selection in some north American finches. *Condor* 73, 415–429.

Wilson, B. F. (1968). Red Maple stump sprouts: development in the first year. *Harvard Forest Paper* 18, 1–10.

Wilson, D. E. and Janzen, D. H. (1972). Predation on *Scheelea* palm seeds by Bruchid beetles: seed density and distance from the parent palm. *Ecology* 53, 954–959.

Wilson, E. O. and Bossert, W. H. (1971). "A Primer of Population Biology". Sinauer Ass. Inc., Stamford, Conn.

Wilson, R. E. and Rice, E. L. (1968). Allellopathy as expressed by *Helianthus annuus* and its role in old-field succession. *Bull. Torrey bot. Club* 95, 432–448.

Winkworth, R. E. (1970). Soil water regime of an arid grassland community, Australia. *Agric. Meteorol.* 7, 387–399.

Wit, C. T. de (1960). On competition. *Versl. Landbouwk. Onderz.* 66, 1–82.

Wit, C. T. de (1965). Photosynthesis of leaf canopies. *Versl. Landbouwk. Onderz.* **663**, 1—57.

Wit, C. T. de, Tow, P. G. and Ennik, G. C. (1966). Competition between legumes and grasses. *Versl. Landbouwk. Onderz.* **687**, 3—30.

Witkamp, H. (1925). De ifzehaut als geologische indikator. *Trop. Natuur.* **14**, 97—103.

Womark, D. and Thurman, R. L. (1962). Effect of leaf removal on the grain yield of wheat and oats. *Crop Sci.* **2**, 423—426.

Woodell, S. R. J., Mooney, H. A. and Hill, A. J. (1969). The behaviour of *Larrea divaricata* (Creosote bush) in response to rainfall in California. *J. Ecol.* **57**, 37—44.

Woolhouse, H. W. ed. (1969). Dormancy and Survival. *Symp. Soc. exp. Biol.* **23**, 1—598.

Wright, J. L. and Lemon, E. R. (1966a). Photosynthesis under field conditions. VIII. Analysis of windspeed fluctuation data to evaluate turbulent exchange within a corn crop. *Agron. J.* **58**, 255—261.

Wright, J. L. and Lemon, E. R. (1966b). Photosynthesis under field conditions. IX. Vertical distribution of photosynthesis within a corn crop. *Agron. J.* **58**, 265—268.

Yelverton, C. S. and Quay, T. L. (1959). Food habits of the Canada goose. *Game Division, N. Carolina Wildlife Resources Commissn. Raleigh, N. Carolina.* 1—44.

Yocom, H. A. (1968). Short leaf pine seed dispersal. *J. Forest.* **66**, 422.

Yoda, K., Kira, T. and Hozimu, K. (1957). Intraspecific competition among higher plants. IX. Further analysis of the competitive interaction between adjacent individuals. *J. Inst. polytech., Osaka Cy Univ.* **8**, 161—178.

Yoda, K., Kira, T., Ogawa, H. and Hozumi, K. (1963). Self thinning in overcrowded pure stands under cultivated and natural conditons. *J. Biol. Osaka Cy Univ.* **14**, 107—129.

Zentmyer, G. A., Wallace, P. P. and Horsfall, J. G. (1944). Distance as a dosage factor in the spread of Dutch elm disease. *Phytopathology* **34**, 1025—1033.

Zimmermann, M. H. (1973). Transport problems in arborescent monocotyledons. *Q. Rev. Biol.* **48**, 314—321.

Zohary, M. (1962). "Plant Life of Palestine, Israel and Jordon". Ronald Press, New York.

Author Index

Numbers in italics refer to pages on which authors' names are quoted in a Figure or a Table

Species Index

SPECIES INDEX

R. bicolor, 472
redbacked vole, see *Clethrionomys gapperi*
redcurrant, 48
red fescue S.59, see *Festuca rubra*
red maple, 615, *617*
red pine, 96, *419*
redstarts, see *Phoenicurus phoenicurus*
redwing, see *Turdus musicus*
Resedaceae, 413
Reseda alba, 115
R. odorata, 653
Rhagoletis, 411
R. pomonella, 411, *412*
Rhamnus frangula, 48
Rhaphanus raphanistrum, 51
Rhinanthus, 672
R. crista-galli, 437
Rhizobium, 260, 261, 262, *263*, *284*, 285, 369, *484*, 708, 756
Rhizoctonia, 100, 101, 147, 174, 493
rhodes grass, see *Chloris gayana*
Rhus glabra, 668
R. typhina, 24, *25*, 544
Ribes, 493
ribgrass, see *Plantago lanceolata*
rice, 100, 241, 245, *266*, 409, 425, 754, 762
Ribes, 412
robin, American, see *Turdus migratorius*
robin, English, see *Erithacus rubecula*
Rosaceae, 67, 417, *665*
rotifer, 590
rough-stalked meadowgrass, see *Poa trivialis*
rubber, see *Hevea brasiliensis*
Rubus spp., 95
R. fructicosus, 415
R. fructicosus agg., 746, 747, 753
Rudbeckia hirta, 370
Rumex acetosa, 139, *140*, 143, 722, 723, *724*, *724*, *725*, *725*, 757
R. acetosella, 143, 240, 290, *290*, 291, *291*, 596, *597*, 722, 723, *725*, *725*, 726, 757
R. crispus, 51, 72, *73*, 99, 172, 295, 296, *297*, *298*, 402, 403, 549, 669, 752
R. obtusifolius, *73*, 76, 172, 295, 296, *297*, *298*, 752
rye, see *Secale cereale*

ryegrass, see *Lolium perenne*

S

Saccharum officinale, *326*
sagebush, 414, 604, 605
Salicornia spp., 636
Salix, 418
Salsola kali, *326*
Salvia spp., 376, 377
S. columbariae, 80
S. leucophylla, *378*, 376
Salvinia, 4, 7, 11, 15, 16, *18*, 19
S. natans, 5, 7, *9*, 10, 15, 16, *18*, 19
Sambucus nigra, 736
Sanguisorba tenuifolia, *309*
Sanicula europaea, 562, *563*, 720
Santiria laevigata, 471, *472*
Sasa nipponica, *309*
Saxifraga, 238
S. nivalis, 556
S. tridactylites, 241, 243
Scabiosa columbaria, 139, *140*
Scandix pecten-veneris, 85
Scenedesmus, 4, *5*
Scheelea, 479
Scilla nonscripta, *719*
Sclerotinia, 484
scolytid beetles, 735
Scolytus, *484*, 485
Scots pine, see *Pinus sylvestris*
Scrophulariaceae, *665*
sea urchin, see *Strongylocentrotus* spp.
Secale cereale, *266*, 756
Sedum smallii, 525, *526*, *527*, 528, *528*, *529*
seed fly, see *Euaresta aequalis*
Senecio spp., 122, 413, 414
S. jacobaea, 37, 38, *128*, 415, 437, *459*, 501, 505, 506
S. squalidus, 44
S. sylvaticus, 205, *206*, *207*, 208
S. viscosus, 44, 205, *206*, *207*, 413
S. vulgaris, 44, *45*, 57, *107*, *168*, 520, 657, *658*, *659*, 683, 684
Sequoia, 684
S. sempervirens, 604, 623, *624*, 625
Sequoiadendron, 684
Sesame vulgaris, *175*, 176, *178*
Setaria anceps, 365, *367*, *368*, 368, 369

W

X

Y

Z

Subject Index

A

Aberffraw, 531, 535, 536, 538, 539
abortion, of seeds, 402, 667
abrasion, of seeds, 68
abscission, of seeds, 122, 203
abundance and reproductive capacity, 648, 649
accumulated support tissue, 20
3-acetyl, 6-methoxybenzaldehyde, 375
achene, 43–48, 51, 70–72, 122, 220, 670–672
acid fuchsin, 234
acid grassland, 138, 596
acid soil, 139, 622, 722
acidic habitats, 209
acorns, germination, 116
adaptation, 749–751, 769
   fitness, productivity, 776–778
adaptive space, 673–675
additive experiments, 249–255
adventitious root systems, 20
aerenchyma, 19
aerobic respiration, 342
aerodynamic drag, 42, 43
aerodynamics of seeds, 42–47
aeroplanes, 46
aestivation, 504
Africa, 487

age
   classes of trees, 600, 601, 635
   distribution, 11, 12
   shoot volume in shrubs, 636–638
   & size of trees, 600, 601, 634, 635
   specific death rates, *Ranunculus*, 578
   specific fecundity schedule, 681
   states in communities, 707
   structure, 21, 22, 550–561, 575, 595, 600, 601, 678 (*see* life table)
      of forest, 602–636
      of shrubs, 636–643
agricultural crops, *see* crops
agricultural soil, 133
agriculture, 50, 92, 98, 237, 301, 385, 433, 435, 469, 512, 543, 657
aggressive species, 243, 245, 249–267, 301, 302, 338, 349, 351, 370, 374, 392, 486, 503, 551, 667, 708, 712, 744, 761
agronomy, 24, 152, 222, 270, 435, 438, 516, 693, 751
airborne seeds, *see* wind dispersal
air currents, *see* wind
alien thinning, 171, 183–187, 293, 295–298
alkaloids, 396, 414, 430, 652
alleles, 733, 756, 770
allelopathy, 372

857

868

floristic diversity, (contd.)
  and grazing, 435–437, 736
flour beetles, 15, 16, 413
flow diagrams of population dynamics,
  584, 585
flowering, 11, 20, 56, 110, 183, 201,
    204, 207–211, 213–218,
    682–684, 689, 708
  annuals, 520, 525–527, 532,
    537–542, 653, 656
  in bamboo, 695, 696
  biennials, 542–547
  in geophytes, 689
  grasses, 595
  and grazing, 397, 403, 405, 407
  herbaceous perennials, 551, 558,
    560, 688, 696
  in meadows, 438
  orchids, 567, 568
  period, 56–58, 696, 697
  vs. vegetative growth, 653–664
  in woodland, 718–721
flowers, 152, 195, 202, 203, 218, 219,
    250, 251, 416, 442, 452
  buds, 520
  sexuality, 654, 655
  shoots, 494
  stalk, 43
fluctuations, conifer cone production,
  690, 691
flux, 33, 34, 500, 517, 518, 521, 583,
  596
fodder, 196, 440
food
  chain, see food web
  collecting habit, 410
  and niche dimensions, 712, 713
  reserves in seeds, 42, 577–580, 651,
    652, 672, 673
  supply, 21–23
  web, 471, 736–743
footprints, 735
  and seedling establishment, 127
forage, 406, 407, 444, 447, 448
foraging units, 20
forces of selection, 751, 752, 758–
  768, 769
forest, 28, 36, 38–41, 48, 58, 92–94,
    105, 138, 144–147, 176, 183,
    185–187, 218–235, 245, 246,
    374, 393, 398, 459, 461, 465,

    485, 493, 496, 557, 560, 652,
    707, 718–720, 727, 774
forest age structure, 602–636
  clearings, 38–40
  fires, 144, 145
  flowering, 696–698
  integration of individuals, 234–235
  nurseries, 197
  phenology, 470–473
  population dynamics, 595, 599–643
    (Ch. 20), 649, 654, 655, 741
  yield, 152, 185–187
foresters, 152, 183, 185, 186, 221–225,
    398, 516, 601
forestry, 231–234, 237, 385, 512, 657
  management tables, 185–187
form
  and community diversity, 727–735
form density and reproduction,
  195–235 (Ch. 7)
  influence of density, 195–235 (Ch. 7)
fossil seeds, 665
founder effect, 711, 769
France, 494
free-floating aquatics, 4–11, 15, 16
frequency
  and density, 291–301
  dependence, 12
  dependent interactions, models,
    285–291
  distribution of plant weight, 162,
    163
  independent interactions, models,
    287–291
fronds, 4, 6, 7–11, 15–19, 516, 728,
  731
frost, 64, 68, 80, 520
fruit
  colour, 480
  growers, 654
  trees, 231
  yield in orchards, 231–233
fruiting, in woodland, 718–721
fruits, 42, 47, 67, 152, 329, 399, 403,
    405, 409–412, 416, 664, 695
  as food, 422, 430, 452–454
  plumed, 42, 43
  and predators, 457–482 (Ch. 15)
    695
  rotation, 46, 47
fugitive properties, 34, 60